Metaharmonic Lattice
Point Theory

PURE AND APPLIED MATHEMATICS

A Program of Monographs, Textbooks, and Lecture Notes

MONOGRAPHS AND TEXTBOOKS IN PURE AND APPLIED MATHEMATICS

Recent Titles

Santiago Alves Tavares, Generation of Multivariate Hermite Interpolating Polynomials (2005)

Sergio Macías, Topics on Continua (2005)

Mircea Sofonea, Weimin Han, and Meir Shillor, Analysis and Approximation of Contact Problems with Adhesion or Damage (2006)

Marwan Moubachir and Jean-Paul Zolésio, Moving Shape Analysis and Control: Applications to Fluid Structure Interactions (2006)

Alfred Geroldinger and Franz Halter-Koch, Non-Unique Factorizations: Algebraic, Combinatorial and Analytic Theory (2006)

Kevin J. Hastings, Introduction to the Mathematics of Operations Research with *Mathematica®*, Second Edition (2006)

Robert Carlson, A Concrete Introduction to Real Analysis (2006)

John Dauns and Yiqiang Zhou, Classes of Modules (2006)

N. K. Govil, H. N. Mhaskar, Ram N. Mohapatra, Zuhair Nashed, and J. Szabados, Frontiers in Interpolation and Approximation (2006)

Luca Lorenzi and Marcello Bertoldi, Analytical Methods for Markov Semigroups (2006)

M. A. Al-Gwaiz and S. A. Elsanousi, Elements of Real Analysis (2006)

Theodore G. Faticoni, Direct Sum Decompositions of Torsion-Free Finite Rank Groups (2007)

R. Sivaramakrishnan, Certain Number-Theoretic Episodes in Algebra (2006)

Aderemi Kuku, Representation Theory and Higher Algebraic K-Theory (2006)

Robert Piziak and P. L. Odell, Matrix Theory: From Generalized Inverses to Jordan Form (2007)

Norman L. Johnson, Vikram Jha, and Mauro Biliotti, Handbook of Finite Translation Planes (2007)

Lieven Le Bruyn, Noncommutative Geometry and Cayley-smooth Orders (2008)

Fritz Schwarz, Algorithmic Lie Theory for Solving Ordinary Differential Equations (2008)

Jane Cronin, Ordinary Differential Equations: Introduction and Qualitative Theory, Third Edition (2008)

Su Gao, Invariant Descriptive Set Theory (2009)

Christopher Apelian and Steve Surace, Real and Complex Analysis (2010)

Norman L. Johnson, Combinatorics of Spreads and Parallelisms (2010)

Lawrence Narici and Edward Beckenstein, Topological Vector Spaces, Second Edition (2010)

Moshe Sniedovich, Dynamic Programming: Foundations and Principles, Second Edition (2010)

Drumi D. Bainov and Snezhana G. Hristova, Differential Equations with Maxima (2011)

Willi Freeden, Metaharmonic Lattice Point Theory (2011)

Metaharmonic Lattice Point Theory

Willi Freeden

University of Kaiserslautern

Germany

CRC Press

Taylor & Francis Group

Boca Raton London New York

CRC Press is an imprint of the
Taylor & Francis Group, an **informa** business

A CHAPMAN & HALL BOOK

CRC Press
Taylor & Francis Group
6000 Broken Sound Parkway NW, Suite 300
Boca Raton, FL 33487-2742

First issued in paperback 2019

© 2011 by Taylor & Francis Group, LLC
CRC Press is an imprint of Taylor & Francis Group, an Informa business

No claim to original U.S. Government works

ISBN-13: 978-1-4398-6184-4 (hbk)
ISBN-13: 978-1-138-38210-7 (pbk)

Visit the Taylor & Francis Web site at
http://www.taylorandfrancis.com

and the CRC Press Web site at
http://www.crcpress.com

Contents

Preface   xi

About the Book   xv

About the Author   xvii

List of Symbols   xix

List of Figures   xxiii

1 Introduction   1
1.1 Historical Aspects 1
1.2 Preparatory Ideas and Concepts 3
1.3 Tasks and Perspectives 5

2 Basic Notation   9
2.1 Cartesian Nomenclature 9
2.2 Regular Regions 12
2.3 Spherical Nomenclature 13
2.4 Radial and Angular Functions 15

3 One-Dimensional Auxiliary Material   17
3.1 Gamma Function and Its Properties 17
3.2 Riemann–Lebesgue Limits 31
3.3 Fourier Boundary and Stationary Point Asymptotics 34
3.4 Abel–Poisson and Gauß–Weierstraß Limits 38

4 One-Dimensional Euler and Poisson Summation Formulas   45
4.1 Lattice Function 46
4.2 Euler Summation Formula for the Laplace Operator 54
4.3 Riemann Zeta Function and Lattice Function 63
4.4 Poisson Summation Formula for the Laplace Operator 68
4.5 Euler Summation Formula for Helmholtz Operators 76
4.6 Poisson Summation Formula for Helmholtz Operators 82

5 Preparatory Tools of Analytic Theory of Numbers   **87**
5.1 Lattices in Euclidean Spaces 88
5.2 Basic Results of the Geometry of Numbers 92
5.3 Lattice Points Inside Circles 98
5.4 Lattice Points on Circles 105
5.5 Lattice Points Inside Spheres 113
5.6 Lattice Points on Spheres 118

6 Preparatory Tools of Mathematical Physics   **121**
6.1 Integral Theorems for the Laplace Operator 122
6.2 Integral Theorems for the Laplace–Beltrami Operator 133
6.3 Tools Involving the Laplace Operator 139
6.4 Radial and Angular Decomposition of Harmonics 144
6.5 Integral Theorems for the Helmholtz–Beltrami Operator . . 180
6.6 Radial and Angular Decomposition of Metaharmonics 192
6.7 Tools Involving Helmholtz Operators 215

7 Preparatory Tools of Fourier Analysis   **223**
7.1 Periodical Polynomials and Fourier Expansions 224
7.2 Classical Fourier Transform 227
7.3 Poisson Summation and Periodization 229
7.4 Gauß–Weierstraß and Abel–Poisson Transforms 232
7.5 Hankel Transform and Discontinuous Integrals 243

8 Lattice Function for the Iterated Helmholtz Operator   **247**
8.1 Lattice Function for the Helmholtz Operator 248
8.2 Lattice Function for the Iterated Helmholtz Operator 255
8.3 Lattice Function in Terms of Circular Harmonics 256
8.4 Lattice Function in Terms of Spherical Harmonics 265

9 Euler Summation on Regular Regions   **269**
9.1 Euler Summation Formula for the Iterated Laplace Operator 270
9.2 Lattice Point Discrepancy Involving the Laplace Operator . 278
9.3 Zeta Function and Lattice Function 282
9.4 Euler Summation Formulas for Iterated Helmholtz Operators 294
9.5 Lattice Point Discrepancy Involving the Helmholtz Operator 299

10 Lattice Point Summation   **303**
10.1 Integral Asymptotics for (Iterated) Lattice Functions 304
10.2 Convergence Criteria and Theorems 308
10.3 Lattice Point-Generated Poisson Summation Formula 312
10.4 Classical Two-Dimensional Hardy–Landau Identity 314
10.5 Multi-Dimensional Hardy–Landau Identities 317

11 Lattice Ball Summation   **323**
11.1 Lattice Ball-Generated Euler Summation Formulas 324
11.2 Lattice Ball Discrepancy Involving the Laplacian 328
11.3 Convergence Criteria and Theorems 331
11.4 Lattice Ball-Generated Poisson Summation Formula 337
11.5 Multi-Dimensional Hardy–Landau Identities 338

12 Poisson Summation on Regular Regions   **343**
12.1 Theta Function and Gauß–Weierstraß Summability 344
12.2 Convergence Criteria for the Poisson Series 350
12.3 Generalized Parseval Identity 355
12.4 Minkowski's Lattice Point Theorem 359

13 Poisson Summation on Planar Regular Regions   **361**
13.1 Fourier Inversion Formula 362
13.2 Weighted Two-Dimensional Lattice Point Identities 365
13.3 Weighted Two-Dimensional Lattice Ball Identities 379

14 Planar Distribution of Lattice Points   **385**
14.1 Qualitative Hardy–Landau Induced Geometric Interpretation 386
14.2 Constant Weight Discrepancy 391
14.3 Almost Periodicity of the Constant Weight Discrepancy . . . 396
14.4 Angular Weight Discrepancy 406
14.5 Almost Periodicity of the Angular Weight Discrepancy . . . 408
14.6 Radial and Angular Weights 409
14.7 Non-Uniform Distribution of Lattice Points 415
14.8 Quantitative Step Function Oriented Geometric Interpretation 421

15 Conclusions   **429**
15.1 Summary 429
15.2 Outlook 430

Bibliography   **431**

Index   **443**

Preface

These lecture notes are the result of an interrelated "transfer" of methods, settings, and tools of (spherically oriented) geomathematics and of (periodically reflected) analytic theory of numbers. The essential ingredients of mathematical (geo-)physics in this work are special function systems of the Laplace equation and the Helmholtz equation, i.e., harmonic and metaharmonic functions, problem-adapted constructions of Green's functions, and eigenvalue-based solution theory in terms of "Green type" integral formulas. Surprisingly, these fundamental techniques relevant for geomathematical research in gravitation, magnetics, geothermal research, etc. enable us to recover significant topics of lattice point theory in Euclidean spaces (such as Hardy–Landau identities determining the total number of lattice points inside spheres, weighted (radial and angular) lattice point summation, non-uniform distribution of lattice points, etc.). Even more, multi-dimensional alternating series become attackable by convergence criteria relating the specific oscillation properties of a summand to an appropriate choice of a Helmholtz operator. In addition, new classes of lattice point identities can be developed by adapted procedures of periodization within "Green type" integral formulas, i.e., Euler and Poisson summation.

More specifically, the main objectives of this work are multi-dimensional generalizations of the Euler summation formula by suitably interpreting the classical "Bernoulli polynomials" as Green's functions and by appropriately establishing the link to Zeta and Theta functions. The multi-dimensional Euler summation formulas are generated on arbitrary lattices by the conversion of the Helmholtz wave equation into an associated integral equation based on the concept of Green's functions as a bridging tool. In doing so, we are able to compare weighted sums of functional values for a prescribed system of lattice points with the corresponding integral over the function, plus a remainder term that is adaptable to the (oscillating) function under consideration. The remainder term is particularly useful for two aspects of multi-variate lattice point theory, viz. to guarantee the convergence of multi-dimensional alternating series and to formulate appropriate criteria for the validity of the Poisson summation formula. Since the infinite lattice point sums occurring in our approach usually offer the pointwise, but refuse the absolute convergence, the specification of the multi-dimensional summation process is a decisive feature. Throughout this book, with respect to the rotational invariance of the Laplace operator, (pointwise) convergence is understood in the spherical sense. In other words, multi-dimensional summation is consistently extended over balls, if the series expansion under consideration turns out to refuse absolute convergence.

The title of our work can be reformulated in more detail as the Helmholtz equation induced verification of the Hardy–Landau type lattice point identities with particular interest in characterizing radial and angular distributions of (planar) lattice points. Altogether, the book can be characterized briefly as a lecture note in the analytic theory of numbers in Euclidean spaces based on methods and procedures of mathematical physics. Its essential purpose is to establish multi-dimensional Euler and Poisson summation formulas corresponding to (iterated) Helmholtz operators for the adaptive determination and calculation of formulas and identities involving weighted lattice point numbers.

The roots of the book are threefold: (i) the basic results due to L.J. Mordell on one-dimensional Euler and Poisson summation formulas as well as the one-dimensional Zeta and Theta function (ii) the work by C. Müller on two-dimensional periodical Euler (Green) functions and their representation in the framework of complex analysis, and (iii) my own work on multi-dimensional generalizations of the Euler summation formula to elliptic operators and some attempts to extend the multi-dimensional Poisson summation formula to regular ("potato"-like) regions. In consequence, the number theoretical understanding of the book requires that the reader has mastered some material usually covered in courses on elliptic partial differential equations and special functions of mathematical physics, especially related to the theory of iterated Laplace as well as Helmholtz equations. The book can be used as a graduate text or as a reference for researchers.

The idea of writing this book first occurred to me while teaching graduate courses given during the last years at the University of Kaiserslautern, when I presented various topics on Green's functions in different fields of geomathematical application. Indeed, the lecture notes represent the link between my former PhD activities at the RWTH Aachen in analytic theory of numbers and my present work in geomathematics at the University of Kaiserslautern.

The preparation of the final version of this work was supported by important remarks and suggestions of many colleagues. I am deeply obliged to Z. Nashed, Orlando, USA, and T. Sonar, Braunschweig, Germany, for friendly collaboration and continuous support over the last years. It is a great pleasure to express my particular appreciation to my colleague G. Malle, University of Kaiserslautern, Germany, who helped me to clarify some concepts. I am indebted to M. Schreiner, NTB Buchs, Switzerland and M. A. Slawinski, Memorial University of Newfoundland, St. John's NL, Canada, for helpful comments and remarks.

Thanks also go to my co-workers, especially to M. Augustin, C. Gerhards, M. Gutting, S. Möhringer, and I. Ostermann, for eliminating inconsistencies in an earlier version. I am obliged to L. Hämmerling, Aachen, for providing me with the phase-dependent numerical computation and graphical illustration (Figure 14.5) of the radial distribution of lattice points in the plane.

The cover illustration shows the geoid of the Earth (i.e., the equipotential surface at sea level as it will be seen by the satellite GOCE) imbedded in a three-dimensional lattice. The "geoidal potato" constitutes a typical (geophysically relevant) regular region as discussed in this work. I am obliged to R. Haagmans, Head, Earth Surfaces and Interior Section, Mission Science Division, ESA–European Space Agency, ESTEC, Noordwijk, the Netherlands, for providing me with the image (ESA ID number SEMLXEOA90E).

I wish to express my particular gratitude to Claudia Korb, Geomathematics Group, TU Kaiserslautern, for her support in handling the typing job.

Finally, it is a pleasure to acknowledge the courtesy and the ready cooperation of Taylor & Francis and all staff members there who were involved in the publication of the manuscript. My particular thanks go to Bob Stern, Amber Donley, and Karen Simon.

Willi Freeden

Kaiserslautern

About the Book

This book is dedicated to the work of Claus Müller, Rheinisch–Westfälische Technische Hochschule Aachen (RWTH Aachen). His fascinating lecture on the two-dimensional Euler Summation Formula and its applications to the analytic theory of numbers held in the winter semester of 1969/1970 (and derived from C. Müller [1954a]) and his excellent guidance leading to the (unpublished) "Diplom" and "Staatsexamen" theses, at the RWTH Aachen, motivated the author more than four decades later to publish this book on metaharmonic lattice point theory. It presents a mathematical collection of promising fruits for the cross-fertilization of two disciplines, namely classical analytic and geometric number theory and future oriented geomathematics involving geophysically relevant regions (such as ball, "geoid(al potato)", (real) Earth's body).

Claus Müller (born February, 1920 in Solingen, Germany, died February, 2008 in Aachen, Germany).

About the Author

Personal data sheet: Studies in mathematics, geography, and philosophy at the RWTH Aachen, 1971 "Diplom" in mathematics, 1972 "Staatsexamen" in mathematics and geography, 1975 PhD in mathematics (see W. Freeden [1975]), 1979 "Habilitation" in mathematics (see W. Freeden [1979]), 1981/1982 Visiting Research Professor at The Ohio State University, Columbus (Department of Geodetic Science and Surveying), 1984 Professor of Mathematics at the RWTH Aachen (Institute of Pure and Applied Mathematics), 1989 Professor of Technomathematics (Industrial Mathematics), 1994 Head of the Geomathematics Group, 2002-2006 Vice-President for Research and Technology at the University of Kaiserslautern, 2009 Editor in Chief of the *International Journal on Geomathematics (GEM)*, member of the editorial board of five journals, author of more than 135 papers, several book chapters, and six books.

Willi Freeden (born March, 1948 in Nettetal–Kaldenkirchen, Germany).

List of Symbols

$\mathbb{N}_0$ .... set of non-negative integers

$\mathbb{N}$ .... set of positive integers

$\mathbb{Z}$ .... set of integers

$\mathbb{R}$ .... set of real numbers

$\mathbb{C}$ .... set of complex numbers

$\Re(s)$ .... real part of $s \in \mathbb{C}$

$\Im(s)$ .... imaginary part of $s \in \mathbb{C}$

$\mathbb{R}^q$ .... q-dimensional Euclidean space

x, y, z .... elements of $\mathbb{R}^q$

$x \cdot y$ .... scalar product of vectors

$|x|$ .... Euclidean norm of $x \in \mathbb{R}^q$

$\epsilon^i,\ i = 1, \ldots, q$ .... canonical orthonormal basis in $\mathbb{R}^q$

δ_{ij} .... Kronecker symbol

$\mathrm{C}^{(k)}, \mathrm{L}^p$ .... classes of functions

F, G .... scalar-valued functions

f, g .... vector-valued functions

$\mathbf{i}$ .... identity tensor

$\mathbf{t}$ .... orthogonal matrix

$\mathbf{t}^T$ .... transpose of the matrix $\mathbf{t}$

$\det \mathbf{t}$ .... determinant of the matrix $\mathbf{t}$

Γ .... point set in $\mathbb{R}^q$

$\partial\Gamma$ .... boundary of Γ

$\overline{\Gamma}$ .... closure of Γ

$F|M$ .... restriction of F to M

∇ .... gradient

$\nabla \cdot$ .... divergence

Δ .... Laplace operator

$\Delta + \lambda$ .... Helmholtz operator

$\{\ldots\}$ .... set of the elements

$\emptyset$ .... empty set

$\mathbb{B}^q$ .... unit ball in $\mathbb{R}^q$ around 0

$\mathbb{B}^q_N(y)$ .... ball in $\mathbb{R}^q$ with radius N around y

$\mathbb{B}^q_{\rho, N}(y)$ .... ball ring with radii ρ and N around y

$\mathbb{S}^{q-1}$ .... unit sphere in $\mathbb{R}^q$ around 0

$\mathbb{S}^{q-1}_N(y)$ .... sphere in $\mathbb{R}^q$ with radius N around y

t, φ .... polar coordinates in $\mathbb{R}^2$

ξ, η, ζ .... elements of $\mathbb{S}^{q-1}$

$dV, dV_{(q)}$ .... volume element in $\mathbb{R}^q$

$dS, dS_{(q-1)}$ .... surface element in $\mathbb{R}^q$

∇^* .... surface gradient in $\mathbb{R}^q$

$\nabla^* \cdot$ .... surface divergence in $\mathbb{R}^q$

Δ^* .... Laplace–Beltrami operator in $\mathbb{R}^q$

$\Delta^* + \lambda$ .... Helmholtz–Beltrami operator in $\mathbb{R}^q$

$(\Delta^*)^\wedge(n)$ .... $\mathbb{S}^{q-1}$-symbol of the Laplace–Beltrami operator in $\mathbb{R}^q$

$(\Delta^* + \lambda)^\wedge(n)$ ..... $\mathbb{S}^{q-1}$-symbol of the Helmholtz–Beltrami operator in $\mathbb{R}^q$

$G(\Delta^* ; \cdot)$ .... $\mathbb{S}^{q-1}$-sphere function for Δ^*

$G(\Delta^* + \lambda ; \cdot)$ .... $\mathbb{S}^{q-1}$-sphere function for $\Delta^* + \lambda$

$\mathbb{Z}^q$ .... unit lattice in $\mathbb{R}^q$

$\tau\mathbb{Z}^q$ .... dilated unit lattice in $\mathbb{R}^q$

$\mathbb{Z}^q + \{x\}$ .... translated unit lattice based at x in $\mathbb{R}^q$

Λ .... (general) lattice in $\mathbb{R}^q$

Λ^{-1} .... inverse lattice of Λ in $\mathbb{R}^q$

$\mathcal{F}$ .... fundamental cell of Λ in $\mathbb{R}^q$

$\mathcal{F}^{-1}$ .... fundamental cell of Λ^{-1} in $\mathbb{R}^q$

$\|\mathcal{F}\|$ .... volume of the fundamental cell Λ

g .... lattice points of Λ

h .... lattice points of Λ^{-1}

$\text{Spect}_\Delta(\Lambda)$ .... eigenspectrum of the Laplace operator in $\mathbb{R}^q$

$\Phi_h, \ h \in \Lambda^{-1}$ .... (orthonormal) Λ-periodical polynomials in $\mathbb{R}^q$

$C_\Lambda^{(k)}, L_\Lambda^p$ .... classes of Λ-periodical functions in $\mathbb{R}^q$

$F_\Lambda^\wedge(h)$ .... Fourier transform of F at $h \in \Lambda^{-1}$ with respect to Λ

$F_{\mathbb{R}^q}^\wedge(x)$ .... Fourier transform of F at $x \in \mathbb{R}^q$ with respect to $\mathbb{R}^q$

$\mathcal{X}_\mathcal{M}$ .... characteristic function for a set $\mathcal{M}$

$\tau\Lambda$ .... dilated point lattice in $\mathbb{R}^q$

$\Lambda - \{x\}$ .... shifted point lattice in $\mathbb{R}^q$

$(\Delta)^\wedge(h)$ .... Λ-symbol of the Laplace operator in $\mathbb{R}^q$

$(\Delta + \lambda)^\wedge(h)$ .... Λ-symbol of the Helmholtz operator in $\mathbb{R}^q$

$G(\Delta ; \cdot)$ .... Λ-lattice function for Δ

$G(\Delta + \lambda ; \cdot)$ .... Λ-lattice function for $\Delta + \lambda$

$\mathcal{M} + \Lambda$ .... figure lattice associated to $\mathcal{M}$ by Λ

$\mathbb{B}_\tau^q + \Lambda$ .... ball lattice associated to $\mathbb{B}_\tau^q$ by Λ

$\mathcal{G}$ .... regular region in $\mathbb{R}^q$

$\overline{\mathcal{G}} = \mathcal{G} \cup \partial\mathcal{G}$ .... closure of a regular region in $\mathbb{R}^q$

$\partial\mathcal{G}$ .... boundary of the regular region $\mathcal{G}$ in $\mathbb{R}^q$

$\|\mathcal{G}\|$ .... volume of the regular region $\mathcal{G}$ in $\mathbb{R}^q$

$\alpha(x)$ .... solid angle at x subtended by $\partial\mathcal{G}$

o, O, Ω .... Landau symbols

$\sim$ .... asymptotically equal

$\simeq$ .... (B^2-)Besicovitch almost periodical

$\lfloor \delta \rfloor$ .... the largest integer $\leq \delta$ (floor-function)

$\lceil \delta \rceil$ .... the smallest integer $\geq \delta$ (ceil-function)

L .... L function

ζ .... Zeta function

$\vartheta, \Theta, \theta$ .... Theta functions

p, p_i .... primes

p^m, p_i^l .... prime powers

$d | n$ .... d divides n

$d \nmid n$ .... d does not divide n

$u \equiv (v \bmod m)$ .... m is divisor of $u - v$, i.e., $m | u - v$

$u \not\equiv (v \bmod m)$ .... m is not divisor of $u - v$, i.e., $m \nmid u - v$

(n, m) .... the greatest common divisor (gcd) of n and m

$\{n, m\}$ .... the least common multiple (lcm) of n and m

$n = \prod_p p^{l(p)}$ .... the prime factorization of an integer $n > 1$

.... with p running through all primes and exponents $l(p)$

$r(n) (= r_2(n))$ .... the solution number of the equation

.... $n_1^2 + n_2^2 = n$ with $n_1, n_2 \in \mathbb{Z}, n \in \mathbb{N}_0$,

$r_q(n)$ .... the solution number of the equation

.... $n_1^2 + \ldots + n_q^2 = n$ with $n_1, \ldots, n_q \in \mathbb{Z}, n \in \mathbb{N}_0$,

$\#_\Lambda (\overline{\mathcal{G}}) = \sum_{\substack{g \in \overline{\mathcal{G}} \\ g \in \Lambda}} 1$ .... total number of lattice points of Λ inside $\overline{\mathcal{G}}$

$P^\lambda (F; \overline{\mathcal{G}})$ .... Λ-lattice point discrepancy of F in $\overline{\mathcal{G}}$ w.r.t. $\Delta + \lambda$

$P (\overline{\mathcal{G}}) = P^0 (1; \overline{\mathcal{G}})$ .... Λ-lattice point discrepancy in $\overline{\mathcal{G}}$ w.r.t. Δ

$P^\lambda_\tau (F; \overline{\mathcal{G}})$ .... Λ-lattice τ-ball discrepancy of F in $\overline{\mathcal{G}}$ w.r.t. $\Delta + \lambda$

$P_\tau (\overline{\mathcal{G}}) = P^0_\tau (1; \overline{\mathcal{G}})$ .... Λ-lattice τ-ball discrepancy in $\overline{\mathcal{G}}$ w.r.t. Δ

List of Figures

2.1 Typical (geomathematically relevant) regular region ("geoidal potato"). 13

3.1 The illustration of the coordinate transformation relating the Beta and the Gamma functions. 21
3.2 The Gamma function on the real line $\mathbb{R}$. 28

4.1 The integer lattice $\mathbb{Z}$. 46
4.2 The fundamental cell $\mathcal{F}$ of the integer lattice $\mathbb{Z}$. 46
4.3 The illustration of the $\mathbb{Z}$-lattice function $G(\Delta; \cdot)$ for Δ. 53
4.4 The derivative $G(\nabla; x) = \nabla G(\Delta; x)$ of the $\mathbb{Z}$-lattice function $G(\Delta; x)$. 54
4.5 The illustration of the integration interval in Euler summation. 55

5.1 Two-dimensional lattice and its fundamental cell $\mathcal{F}$. 88
5.2 An illustration of a three-dimensional lattice. 89
5.3 A two-dimensional figure lattice associated to a rectangle. . . 92
5.4 An example of a two-dimensional ball lattice (i.e., circle lattice). 93
5.5 An example of a two-dimensional covering lattice. 94
5.6 An example of a two-dimensional filling lattice. 94
5.7 An example of a convex set in $\mathbb{R}^2$. 96
5.8 The geometric situation of Minkowski's theorem. 97
5.9 Lattice points inside a circle. 98
5.10 Northwest edges. 99
5.11 The polyhedral set $\mathbb{P}_N^2$. 99
5.12 Two-dimensional lattice point sum versus circle area. 100
5.13 The region D implying Lemma 5.2. 107

6.1 Three-dimensional cube $\mathcal{G} = (-1,1)^3$. 128
6.2 The geometric situation as discussed by Lemma 6.5. 139
6.3 The three cases under consideration in Lemma 6.8. 142
6.4 The two integration paths of the Hankel functions $H_n^{(1)}(q; \cdot)$ (left) and $H_n^{(2)}(q; \cdot)$ (right), respectively. 203

8.1 The geometric illustration for Equation (8.7) in Euclidean space $\mathbb{R}^2$. 250

9.1   The geometric situation of Euler summation in Equation (9.1).   270

11.1   Lattice points (left), lattice balls (right).   323

12.1   Lattice points of a 3D-lattice inside a regular region (such as a "potato").   343

13.1   Lattice points of a 2D-lattice inside a regular region (such as a planar "potato slice").   361

14.1   Circular rings around the origin of fixed width (left), a special sector within circular circles of fixed width (right).   416

14.2   Configuration generating circular rings with $\tau = \frac{1}{4}$, $w = 0$. . .   422

14.3   Configuration generating circular rings with $\tau = \frac{1}{4}$, $w = \frac{1}{4}$. . .   422

14.4   Configuration generating circular rings with $\tau = \frac{1}{4}$, $w = \frac{1}{2}$. . .   422

14.5 $C_\delta(w; \frac{1}{4})$ (continuous curve) and its approximation $C_\delta^{appr}(w; \frac{1}{4})$ (dashed curve).   424

1

Introduction

CONTENTS

1.1 Historical Aspects 1
1.2 Preparatory Ideas and Concepts 3
1.3 Tasks and Perspectives 5

1.1 Historical Aspects

Leonhard Euler (1707–1783) discovered his powerful "summation formula" in the early 1730s. He used it in 1736 to compute the first 20 decimal places for the alternating sum

$$\sum_{g=0}^{\infty} \frac{(-1)^g}{2g+1} = 1 - \frac{1}{3} + \frac{1}{5} - \frac{1}{7} + - \dots = \frac{\pi}{4}. \tag{1.1}$$

Since, aside from the geometric series, very few infinite series then had a known sum, Euler's remarkable sum enticed mathematicians like G. Leibniz (1646–1716) and the Bernoulli brothers Jakob (1654–1705) and Johann (1667–1748) to seek sums of other series, particularly the sum of the reciprocal squares. But it was L. Euler, within the next two decades up to 1750, who did a "broadening of the context" to formulate his "summation formula" for the general sum

$$\sum_{g=0}^{n} F(g) = \sum_{\substack{g \in [0,n] \\ g \in \mathbb{Z}}} F(g) = \sum_{\substack{0 \le g \le n \\ g \in \mathbb{Z}}} F(g) \tag{1.2}$$

(with n possibly infinite). More concretely, under the assumption of second order continuous derivatives of F on the interval $[0, n]$, $n \in \mathbb{N}$, Euler succeeded in finding the summation formula

$$\sum_{\substack{0 \le g \le n \\ g \in \mathbb{Z}}} F(g) - \frac{1}{2} \left(F(0) + F(n) \right) \tag{1.3}$$

$$= \int_0^n F(x) \, dx + \frac{1}{12} \left(F'(n) - F'(0) \right) + \underbrace{\int_0^n \left(-\frac{1}{2} B_2(x) \right) F''(x) \, dx}_{=G(\Delta;x)},$$

1

where B_2 given by

$$B_2(x) = (x - \lfloor x \rfloor)^2 - (x - \lfloor x \rfloor) + \frac{1}{6} \tag{1.4}$$

is the "Bernoulli function" of degree 2. In particular, Euler's new setting also encompassed the quest for closed formulas for sums of powers

$$\sum_{k=0}^{n} k^l \simeq \int_0^n x^l \, dx, \tag{1.5}$$

which had been sought since antiquity for area and volume investigations. In addition, this setting provided a canonical basis for the introduction of the Zeta function.

From the mathematical point of view, Euler's summation formula is a fine illustration of how a generalization can lead to the solution of seemingly independent problems. The particular structure of his summation formula also captures the delicate details of the connection between integration, i.e., "continuous summation", and its various discretizations, viz. summation. Obviously, it subsumes and resolves the appropriate bridge between continuous and discrete summation within a single exposition. But it should be pointed out that Leonhard Euler himself used this interrelation between continuous and discrete sums only for estimating sums and series by virtue of integrals. It was actually Colin Maclaurin (1698–1746), who discovered the summation formula (1.3) independently in 1742, to use it for the evaluation of integrals in terms of sums.

Altogether, the classical Euler summation formula provides a powerful tool of connecting integrals and sums. It can be used in diverse areas to approximate integrals by finite sums, or conversely to evaluate finite sums and infinite series based on the integral calculus.

More specifically, the Euler summation formula offers two important perspectives:

- to compute (slowly) converging infinite series as well as to specify convergence criteria for (alternating) infinite series and to verify limits and asymptotic relations of infinite lattice point sums,

- to evaluate integrals (numerically) as well as to estimate and to optimize the error and to provide multi-dimensional settings of constructive approximation.

1.2 Preparatory Ideas and Concepts

In this book we follow Euler's interest, i.e., the first of the aforementioned perspectives including its applications to relevant lattice point sums of analytic theory of numbers. The essential idea is based on the interpretation of the Bernoulli function (1.4) occurring in the classical (one-dimensional) Euler summation formula (1.3) by means of mathematical physics as the Green function $G(\Delta; \cdot)$ for the (one-dimensional) Laplace operator Δ corresponding to the "boundary condition" of $\mathbb{Z}$-periodicity (note that $\Delta = (\frac{d}{dx})^2$ is the operator of the second order derivative). More concretely, the periodical Green function $G(\Delta; \cdot)$ for the Laplace operator Δ is constructed so as to have the bilinear expansion

$$G(\Delta; x - y) = \sum_{\substack{\Delta^\wedge(h) \neq 0 \\ h \in \mathbb{Z}}} \frac{e^{2\pi ihx} e^{-2\pi ihy}}{-\Delta^\wedge(h)}, \quad x, y \in \mathbb{R}, \tag{1.6}$$

where the sequence $\{\Delta^\wedge(h)\}_{h \in \mathbb{Z}}$ forms the spectrum $\mathrm{Spect}_\Delta(\mathbb{Z})$ of the Laplace operator Δ, i.e.,

$$(\Delta + \Delta^\wedge(h)) \, e^{2\pi ihx} = 0, \quad x \in \mathbb{R}, \tag{1.7}$$

where

$$\Delta^\wedge(h) = 4\pi^2 h^2, \quad h \in \mathbb{Z}. \tag{1.8}$$

In doing so, the Bernoulli function - in the jargon of mathematical physics, the Green function - acts as a connecting tool to convert a differential equation involving the Laplace operator corresponding to periodical boundary conditions into an associated integral equation, i.e., the Euler summation formula (1.3). Observing the special values

$$G(\Delta; 0) = G(\Delta; n) = \sum_{\substack{\Delta^\wedge(h) \neq 0 \\ h \in \mathbb{Z}}} \frac{1}{-\Delta^\wedge(h)} = -\frac{1}{12} \tag{1.9}$$

and the explicit representation of the Fourier series expansion (1.6) we are able to reformulate the Euler summation formula (1.3). Partial integration yields (by letting $F'(x) = \nabla F(x)$, $F''(x) = \Delta F(x)$)

$$\sum_{\substack{0 \leq g \leq n \\ g \in \mathbb{Z}}} F(g) - \frac{1}{2}(F(0) + F(n)) \tag{1.10}$$

$$= \int_0^n F(x)\, dx + \sum_{\substack{h \neq 0 \\ h \in \mathbb{Z}}} \left(\frac{\nabla F(n) - \nabla F(0)}{4\pi^2 h^2} - \int_0^n \frac{e^{2\pi ihx}}{4\pi^2 h^2} \Delta F(x)\, dx \right)$$

$$= \int_0^n F(x)\, dx - \lim_{N \to \infty} \sum_{\substack{|h| \leq N \\ h \neq 0}} \frac{1}{2\pi ih} \int_0^n \nabla F(x)\, e^{2\pi ihx}\, dx,$$

such that the Poisson summation formula comes into play

$$\sum_{\substack{0 \leq g \leq n \\ g \in \mathbb{Z}}} F(g) - \frac{1}{2}\left(F(0) + F(n)\right) = \lim_{N \to \infty} \sum_{\substack{|h| \leq N \\ h \in \mathbb{Z}}} \int_0^n F(x)\, e^{2\pi i h x}\, dx. \qquad (1.11)$$

Surprisingly, in spite of their apparent dissimilarity, the Euler summation formula (1.3) and the Poisson summation formula (1.11) are equivalent for twice continuously differentiable functions on the interval $[0, n]$. Moreover, the Green function for the Laplace operator and the "boundary condition" of $\mathbb{Z}$-periodicity acts as the canonical bridge between both identities.

The "building blocks" of the bridge between the two equivalent formulas (1.3) and (1.11) are the *defining constituents of the Green function* $G(\Delta; \cdot)$, which can be uniquely characterized in the following way:

(*Periodicity*) $G(\Delta; \cdot)$ is continuous in $\mathbb{R}$ and $\mathbb{Z}$-periodical

$$G(\Delta; x) = G(\Delta; x + g), \quad x \in \mathbb{R},\ g \in \mathbb{Z}, \qquad (1.12)$$

(*Differential equation*) $\Delta G(\Delta; \cdot)$ "coincides" apart from an additive constant with the Dirac function(al)

$$\Delta_x G(\Delta; x) = -1, \quad x \in \mathbb{R} \backslash \mathbb{Z}, \qquad (1.13)$$

(*Characteristic singularity*) $G(\Delta; \cdot)$ possesses the singularity of the fundamental solution of the (one-dimensional) Laplace operator

$$G(\Delta; x) - \frac{1}{2}|x| = O(1), \quad x \to 0, \qquad (1.14)$$

(*Normalization*) $G(\Delta; \cdot)$ integrated over a whole period interval of length 1 is assumed to be zero

$$\int_{-\frac{1}{2}}^{\frac{1}{2}} G(\Delta; x)\, dx = 0. \qquad (1.15)$$

Even more generally, for *arbitrary intervals* $[a, b] \subset \mathbb{R}$, $a < b$, and *arbitrary* twice continuously differentiable (weight) functions F on $[a, b]$, the constituents (1.12)–(1.15) of the Green function $G(\Delta; \cdot)$ enable us to guarantee the equivalence of the Euler summation formula

$$\sum_{\substack{a \leq g \leq b \\ g \in \mathbb{Z}}} {}' F(g) = \int_b^a F(x)\, dx + \{F(x)\ (\nabla G(\Delta; x)) - G(\Delta; x)\ (\nabla F(x))\}\,|_a^b$$

$$+ \int_a^b G(\Delta; x)\, \Delta F(x)\, dx \qquad (1.16)$$

and the Poisson summation formula

$$\sum_{\substack{a \leq g \leq b \\ g \in \mathbb{Z}}} {}' F(g) = \lim_{N \to \infty} \sum_{\substack{|h| \leq N \\ h \in \mathbb{Z}}} \int_a^b F(x)\, e^{2\pi i h x}\, dx, \qquad (1.17)$$

where we have used the abbreviation

$$\sum_{\substack{a \le g \le b \\ g \in \mathbb{Z}}} {}' F(g) = \sum_{\substack{a < g < b \\ g \in \mathbb{Z}}} F(g) + \frac{1}{2} \sum_{\substack{g=a,b \\ g \in \mathbb{Z}}} F(g), \qquad (1.18)$$

and the second sum on the right side of (1.18) occurs only if a and/or b is an integer.

Of particular interest in lattice point theory is the special case of a constant weight function (i.e., $F = 1$), i.e., the *one-dimensional "Hardy–Landau identity"*

$$\sum_{\substack{a \le g \le b \\ g \in \mathbb{Z}}} {}' 1 = b - a + \sum_{\substack{|h| \ne 0 \\ h \in \mathbb{Z}}} \int_a^b e^{2\pi i h x} \, dx. \qquad (1.19)$$

The formula (1.19) compares the number of $\mathbb{Z}$-lattice points inside an interval $[a, b]$ with the length $b - a$ of the interval $[a, b]$ under the explicit knowledge of the remainder term (usually called, the $\mathbb{Z}$-*lattice point discrepancy*) as a one-dimensional alternating series.

In addition, the close relation between the Hardy–Landau summation and the metaharmonicity of the summands becomes obvious in the identity (1.19) since the function $e(h \cdot) = e^{2\pi i h \cdot}, h \in \mathbb{Z}$, satisfies *the one-dimensional Helmholtz equation* $\left(\Delta + 4\pi^2 h^2 \right) e(h x) = 0$, $x \in \mathbb{R}$, corresponding to the "wave number" $\Delta^{\wedge}(h) = 4\pi^2 h^2$, $h \in \mathbb{Z}$; i.e., the $\mathbb{Z}$-periodical polynomial $e(h \cdot)$, $h \in \mathbb{Z}$, is *metaharmonic* in $\mathbb{R}$.

1.3 Tasks and Perspectives

This book is devoted to the generalization of the univariate features, settings, and methods involving Euler and Poisson summation to higher dimensions. The key points are the defining properties (1.12)–(1.15) of the $\mathbb{Z}$-periodical Green function, which can be easily transferred to the multi-dimensional case (in contrast to the bilinear expansion (1.6)). Our tasks actually show plenty of essential aspects, namely the supply of multi-variate tools for the Laplace operator and Helmholtz operators, the feasible construction of the multi-dimensional Green function with respect to (iterated) Laplace and Helmholtz operators and the "boundary condition" of periodicity, the realization of associated Euler summation formulas, the introduction and explanation of special function systems of harmonic as well as metaharmonic nature such as spherical harmonics as well as Bessel and Kelvin functions, the formulation of adequate convergence criteria for multi-dimensional alternating sums, some pointwise inversion procedures of Fourier and other integral transforms such as Gauß–Weierstraß and Abel–Poisson transforms, suitable concepts to establish the validity of the multi-dimensional Poisson summation formula, and finally their applications to problems of the analytic and geometric theory of numbers, for example, in the field of radial and angular lattice point distribution.

Altogether, keeping in mind the physically motivated character of the key ingredient, i.e., the Green function with respect to the "boundary condition" of $\mathbb{Z}$-periodicity for Euler summation, we are able to guarantee various directions of extension to

- replace the one-dimensional lattice $\mathbb{Z}$ (consisting of the integers) by *multi-dimensional "lattices"* Λ (such as $\mathbb{Z}^q, \tau\mathbb{Z}^q$, etc).

- consider instead of the finite one-dimensional interval $[a, b]$ *other "geometries"* for summation/integration such as the fundamental cell $\mathcal{F}$ of a multi-dimensional lattice Λ, a (regular) region ("potato") $\mathcal{G}$ in $\mathbb{R}^q$, or the whole Euclidean space $\mathbb{R}^q$ (of course, under additional asymptotic relations at infinity).

- substitute the operator of the second order derivative, i.e., the one-dimensional Laplace operator by *special elliptic differential operators* such as iterated Laplace or Helmholtz operators.

- transfer the classical "boundary conditions" of periodicity into *other boundary conditions* (e.g., of Dirichlet's/Neumann's type).

- "blow up" the multi-dimensional lattice points to *"lattice balls"* for establishing lattice ball analogues of Euler and Poisson summation formulas.

The critical ingredients of our approach to Euler summation are twofold:

On the one hand, the multi-dimensional generalization of the Bernoulli function, i.e., the Green function for a Helmholtz operator and "periodical boundary conditions", is not (yet) available as an elementary function. In addition, the multi-dimensional counterparts of the Fourier series expansion (1.6) are divergent. This is the reason why we first condense the original Bernoulli functions to their constituting properties (see (1.12) – (1.15) for more details) in order to find a setup of a uniquely determined definition in terms of specific features. In turn, the constituting properties of a Green function $G(\Delta + \lambda; \cdot)$ (namely, boundary condition, differential equation, characteristic singularity, normalization) can be used as the keystones to characterize the role of Green's function in generalized variants of the Euler summation formula.

On the other hand, from a structural point of view (every generalization of) the Euler summation formula rests on the basic idea to relate a sum of values of a function at finitely or infinitely many successive nodes to certain sums involving (Helmholtz) derivatives of this function. Clearly, this makes things complicated. But it opens the perspective to stop sums with an expectation that the particular value in which one is interested lies between any partial sum and another one, all of them being explicitly calculable. In doing so, the summation formula provides appropriate approximation, and the (infinite) series or integrals can also be attacked asymptotically even if they diverge. Moreover, the "wave number" $\lambda \in \mathbb{R}$ of an Helmholtz operator $\Delta + \lambda$ may be used to adapt the operator to the specific properties of a summand showing an alternating character, e.g., in order to force the convergence of the associated infinite series.

Following our approach of establishing extensions of the Euler summation formula by specifying particular classes of Green's function within a certain framework (once more, boundary condition, Helmholtz differential operator, singularity and normalization (if necessary)), we are able to make essential scientific progress in both formulating convergence criteria for a multi-dimensional series in adaptation to the (oscillating) properties of the summand and representing the series in terms of certain volume and surface integrals to come up with relevant lattice point identities of number theoretical significance.

Indeed, the list of significant topics and innovative results based on the Euler summation with respect to Helmholtz operators is long. It enables us to

- interrelate Green functions to Zeta and Theta functions,

- develop convergence criteria for (alternating) multi-dimensional series (always understood here in spherical summation),

- formulate adapted conditions for the validity of the multi-dimensional Poisson summation formula in Euclidean spaces,

- outline Euler summation formulas for regular ("potato"-like) regions,

- deduce Poisson summation formulas for regular ("potato"-like) regions in the sense of Gauß–Weierstraß or Abel–Poisson summability,

- verify extended Hardy–Landau identities for "lattice point" sums as well as "lattice ball" sums in spheres,

- derive asymptotic relations for weighted lattice point sums,

- explain non-uniform radial and angular distributions of lattice points.

- develop comparisons of asymptotic laws between lattice ball and lattice point sums.

As essential ingredients to establish our results we need a number of auxiliary means and tools such as

- fundamental solutions for the iterated Laplace equation, asymptotic laws for spherical integrals involving Green functions with respect to (iterated) Helmholtz operators, the appropriate ball averaging of the Green functions,

- an (alternative) approach to the theory of spherical harmonics,

- fundamental solutions for the iterated (spherical) Laplace–Beltrami equation, spherical integral formulas for the Beltrami operator, discrepancy representations by means of the Green function on the sphere for the Beltrami operator,

- the (metaharmonic) theory of cylinder functions (Bessel, Hankel, Kelvin, Neumann functions, etc.), asymptotic rules for entire solutions of the Helmholtz equation (i.e., the reduced wave equation),

- integral transforms, e.g., Abel–Poisson transform, Gauß–Weierstraß transform,

- the Fourier inversion formula for discontinuous functions, however, in the pointwise sense,

- Hankel transform involving discontinuous integrals in terms of Bessel functions,

- functional equations of Zeta and Theta functions,

- lattice point and lattice ball discrepancies as specific expressions in terms of periodical Green functions,

- the theory of almost periodicity in the (B^2)-Besicovitch sense,

- asymptotic expansions for weighted "lattice point" and "lattice ball" discrepancies,

- "width" and "phase" dependent quantification of planar radial lattice point distributions.

2

Basic Notation

CONTENTS

2.1 Cartesian Nomenclature 9
            Differential Operators 10
            Multi-Indices 11
2.2 Regular Regions 12
2.3 Spherical Nomenclature 13
            Differential Operators 14
            Spheres and Balls 14
2.4 Radial and Angular Functions 15

2.1 Cartesian Nomenclature

Throughout this book we base our considerations on the following notational background.

The letters $\mathbb{N}, \mathbb{N}_0, \mathbb{Z}, \mathbb{R}$, and $\mathbb{C}$ denote the set of positive, non-negative integers, integers, real numbers, and complex numbers, respectively.

As usual, we write $x, y, \ldots$ to represent the elements of the q-dimensional (real) Euclidean space $\mathbb{R}^q$ ($q \geq 1$). In Cartesian coordinates we have the component representation (q-tuples of real numbers)

$$x = \begin{pmatrix} x_1 \\ \vdots \\ x_q \end{pmatrix}, \qquad y = \begin{pmatrix} y_1 \\ \vdots \\ y_q \end{pmatrix}. \tag{2.1}$$

If necessary we write $x_{(q)}$ instead of x to point out that x is an element of $\mathbb{R}^q$. The canonical orthonormal system in $\mathbb{R}^q$ is denoted by $\epsilon^1, \ldots, \epsilon^q$. More explicitly,

$$\epsilon^1 = \begin{pmatrix} 1 \\ 0 \\ \vdots \\ 0 \end{pmatrix}, \quad \ldots, \quad \epsilon^q = \begin{pmatrix} 0 \\ \vdots \\ 0 \\ 1 \end{pmatrix}. \tag{2.2}$$

Any $x \in \mathbb{R}^q$ may be represented in Cartesian coordinates $x_i, i = 1, \ldots, q$, by

$$x = \sum_{i=1}^{q} x_i \epsilon^i. \tag{2.3}$$

In Cartesian coordinates the *inner (scalar) product* of two elements $x, y \in \mathbb{R}^q$ is given by

$$x \cdot y = x^T y = \sum_{i=1}^{q} x_i y_i. \tag{2.4}$$

Clearly,

$$x^2 = |x|^2 = x \cdot x = x^T x, \quad x \in \mathbb{R}^q, \tag{2.5}$$

i.e., *the norm* in $\mathbb{R}^q$ is given

$$|x| = \sqrt{x \cdot x} = \sqrt{x^T x}, \quad x \in \mathbb{R}^q. \tag{2.6}$$

Given a vector $a \in \mathbb{R}^q$ and a set $\mathcal{M} \subset \mathbb{R}^q$. Let $\mathcal{M} + \{a\}$ denote the set of all points $y = x + a$, as x runs through the points of $\mathcal{M}$. $\mathcal{M} + \{a\}$ is the *translate* of the set $\mathcal{M}$ by a. More generally, if $\mathcal{N}$ denotes some set of vectors from $\mathbb{R}^q$ then by $\mathcal{M} + \mathcal{N}$ we understand the set of all points $y = x + a$ for an arbitrary x from $\mathcal{M}$ and an arbitrary a from $\mathcal{N}$.

If $\mathcal{G}$ is a set of points in $\mathbb{R}^q$, $\partial \mathcal{G}$ will denote its *boundary*. The set $\overline{\mathcal{G}} = \mathcal{G} \cup \partial \mathcal{G}$ is called the *closure* of $\mathcal{G}$. A set $\mathcal{G} \subset \mathbb{R}^q$ is called a *region* if and only if it is open and connected.

By a scalar or vector function (field) on a region $\mathcal{G} \subset \mathbb{R}^q$, we mean a function that assigns to each point of $\mathcal{G}$, a scalar or vectorial function value, respectively. Unless otherwise specified, all functions are assumed to be complex valued. It will be of advantage to use the following general scheme of notation:

capital letters F, G : scalar functions,
lower-case letters f, g : vector fields.

The restriction of a scalar-valued function F or a vector-valued function f to a subset M of its domain is denoted by $F|M$ or $f|M$, respectively. For a set S of functions, we set $S|M = \{F|M \mid F \in S\}$.

Differential Operators

Let $\mathcal{G} \subset \mathbb{R}^q$ be a region. Suppose that $F : \mathcal{G} \to \mathbb{C}$ is differentiable. $\nabla F : x \mapsto (\nabla F)(x)$, $x \in \mathcal{G}$, denotes the *gradient* of F on $\mathcal{G}$. The *partial derivatives* of F at $x \in \mathcal{G}$, sometimes briefly written as $F_{|i}$, $i \in \{1, \ldots, q\}$, are given by

$$F_{|i}(x) = \frac{\partial F}{\partial x_i}(x) = (\mathrm{grad}_x \, F)(x) \cdot \epsilon^i = (\nabla_x F)(x) \cdot \epsilon^i = ((\nabla_x F)(x))_i \; . \tag{2.7}$$

Let $u : \mathcal{G} \to \mathbb{C}^q$ be a vector field, and suppose, in addition, that u is differentiable at a point $x \in \mathcal{G}$. The partial derivatives of u at $x \in \mathcal{G}$ are given by

$$u_{i|j}(x) = \frac{\partial u_i}{\partial x_j}(x) = \epsilon^i \cdot (\nabla u)(x) \epsilon^j \; . \tag{2.8}$$

The *divergence* of u at $x \in \mathcal{G}$ is the scalar value

$$\nabla_x \cdot u(x) = \mathrm{div}_x u(x) = \mathrm{tr}\ (\nabla u)(x)\ . \tag{2.9}$$

Thus, we have the identity

$$\nabla_x \cdot u(x) = \mathrm{div}_x u(x) = \sum_{i=1}^{q} u_{i|i}(x)\ . \tag{2.10}$$

Let F be a differentiable scalar field on $\mathcal{G}$, and suppose, in addition, that ∇F is differentiable at $x \in \mathcal{G}$. Then we introduce the *Laplace operator (Laplacian)* of F at $x \in \mathcal{G}$ by

$$\Delta_x F(x) = \mathrm{div}_x\left((\mathrm{grad}_x\ F)(x)\right) = \nabla_x \cdot \left((\nabla_x F)\,(x)\right). \tag{2.11}$$

Analogously, we define the *Laplacian* of a vector field $f : \mathcal{G} \to \mathbb{C}^q$ (with ∇f being differentiable at $x \in \mathcal{G}$) by

$$\Delta_x f(x) = \mathrm{div}_x\left((\mathrm{grad}_x\ f)(x)\right) = \nabla_x \cdot \left((\nabla_x f)\,(x)\right). \tag{2.12}$$

Clearly, for sufficiently often differentiable fields F, f, we have

$$\Delta_x F(x) = \sum_{i=1}^{q} F_{|i|i}(x), \tag{2.13}$$

$$\Delta_x f(x) \cdot \epsilon^i = \sum_{j=1}^{q} f_{i|j|j}(x)\ . \tag{2.14}$$

Multi-Indices

Let $\alpha = (\alpha_1, \ldots, \alpha_q)^{\mathrm{T}}$ be a q-tuple of non-negative integers $\alpha_1, \ldots, \alpha_q$, i.e., $\alpha \in \mathbb{N}_0{}^q$. We set

$$\alpha! = \alpha_1! \cdot \ldots \cdot \alpha_q!, \tag{2.15}$$

$$[\alpha] = \alpha_1 + \ldots + \alpha_q, \tag{2.16}$$

$$|\alpha| = \sqrt{\alpha_1^2 + \ldots + \alpha_q^2}. \tag{2.17}$$

We say $\alpha = (\alpha_1, \ldots, \alpha_q)^{\mathrm{T}}$ is a q-dimensional *multi-index of degree n* if $[\alpha] = n$. As usual, we set

$$x^\alpha = x_1^{\alpha_1} \cdot \ldots \cdot x_q^{\alpha_q}, \quad x \in \mathbb{R}^q, \ \alpha \in \mathbb{N}_0^q, \tag{2.18}$$

$$(\nabla_x)^\alpha = \left(\frac{\partial}{\partial x_1}\right)^{\alpha_1} \cdots \left(\frac{\partial}{\partial x_q}\right)^{\alpha_q} = \frac{\partial^{[\alpha]}}{(\partial x_1)^{\alpha_1} \ldots (\partial x_q)^{\alpha_q}}. \tag{2.19}$$

Clearly, for $[\alpha] = [\beta]$, we have

$$(\nabla_x)^\alpha x^\beta = \begin{cases} 0 & , \quad \alpha \neq \beta, \\ \alpha! & , \quad \alpha = \beta. \end{cases} \tag{2.20}$$

In this notation of multi-indices we have

$$\left(\sum_{i=1}^{q} x_i \right)^n = \sum_{[\alpha]=n} \frac{n!}{\alpha!} x^\alpha. \tag{2.21}$$

As is well known, for $x, y \in \mathbb{R}^q$, the *binomial theorem* reads

$$(x \cdot y)^n = \left(\sum_{i=1}^{q} x_i y_i \right)^n = \sum_{[\alpha]=n} \frac{n!}{\alpha!} x^\alpha y^\alpha. \tag{2.22}$$

2.2   Regular Regions

A bounded region $\mathcal{G} \subset \mathbb{R}^q$ is called *regular*, if its boundary $\partial\mathcal{G}$ is an orientable piecewise smooth Lipschitzian manifold of dimension $q-1$ (for more details about regular regions the reader is referred to textbooks of vector analysis). Examples are ball, cube, other polyhedra, geoid(al potato), (real) Earth's body, etc.

$F \in \mathrm{C}^{(k)}(\overline{\mathcal{G}})$, $0 \leq k \leq \infty$, means that the function $F : \overline{\mathcal{G}} \to \mathbb{C}$ is k-times continuously differentiable in $\overline{\mathcal{G}} = \mathcal{G} \cup \partial\mathcal{G}$. By convention, $F \in \mathrm{C}^{(k-1)}(\overline{\mathcal{G}}) \cap \mathrm{C}^{(k)}(\mathcal{G})$ means that the function $F : \overline{\mathcal{G}} \to \mathbb{C}$ is $(k-1)$-times continuously differentiable in $\overline{\mathcal{G}}$ such that $F|\mathcal{G}$ is k-times continuously differentiable.

The *volume of a regular region* $\mathcal{G} \subset \mathbb{R}^q$ is given by

$$\|\mathcal{G}\| = \int_{\mathcal{G}} dV_{(q)}(x), \tag{2.23}$$

where

$$dV_{(q)}(x) = dx_1 \ldots dx_q \tag{2.24}$$

is the volume element.

The *area of the boundary* $\partial\mathcal{G}$ of a regular region $\mathcal{G} \subset \mathbb{R}^q$ is given by

$$\|\partial\mathcal{G}\| = \int_{\partial\mathcal{G}} dS_{(q-1)}(x), \tag{2.25}$$

where $dS_{(q-1)}(x)$ is the surface element

$$\begin{aligned} dS_{(q-1)}(x) &= x_1 \, dx_2 \ldots dx_q \\ &\quad -x_2 \, dx_1 dx_3 \ldots dx_q \\ &\quad + - \quad \cdots \\ &\quad +(-1)^{q-1} x_q \, dx_1 \ldots dx_{q-1}. \end{aligned} \tag{2.26}$$

Remark 2.1. *Throughout this work, for integration in the q-dimensional Euclidean space and on the boundary surface $\partial \mathcal{G}$ in $\mathbb{R}^q$, we use the traditional (non-oriented) notations dV and dS, respectively. If the dimension and the variable of integration must be specified, the notations $dV_{(q)}(x)$ and $dS_{(q-1)}(x)$ are used, respectively.*

FIGURE 2.1
Typical (geomathematically relevant) regular region ("geoidal potato").

2.3 Spherical Nomenclature

As usual, the unit sphere in $\mathbb{R}^q$ is denoted by $\mathbb{S}^{q-1}$:

$$\mathbb{S}^{q-1} = \left\{ x \in \mathbb{R}^q \middle| \; |x| = 1 \right\}. \tag{2.27}$$

Each $x \in \mathbb{R}^q$, $x = (x_1, \dots, x_q)^{\mathrm{T}}$, $|x| \neq 0$, admits a *representation in polar coordinates* of the form

$$x = r\xi, \; r = |x|, \; \xi = (\xi_1, \dots, \xi_q)^{\mathrm{T}}, \tag{2.28}$$

where $\xi \in \mathbb{S}^{q-1}$ is the uniquely determined (unit) vector of x.

Using the canonical orthonormal basis $\epsilon^1, \ldots, \epsilon^q$ in $\mathbb{R}^q$ (more accurately, $\epsilon^1_{(q)}, \ldots, \epsilon^q_{(q)}$ in $\mathbb{R}^q$) we are able to write $\xi_{(q)} \in \mathbb{S}^{q-1}$, $q \geq 3$, in the form

$$\xi_{(q)} = t\epsilon^q_{(q)} + \sqrt{1-t^2}\,\xi_{(q-1)}, \; t \in [-1,1], \; \xi_{(q-1)} \in \mathbb{S}^{q-2}, \quad (2.29)$$

$$\xi_{(2)} = (\cos\varphi, \sin\varphi)^T, \; \xi_{(2)} \in \mathbb{S}^1, \; \varphi \in [0, 2\pi). \quad\quad\quad (2.30)$$

Differential Operators

By means of polar coordinates $x_{(q)} = r\xi_{(q)}$, $r = |x_{(q)}|$, $\xi_{(q)} \in \mathbb{S}^{q-1}$, the *gradient* ∇ in $\mathbb{R}^q$ can be represented in the form

$$\nabla_{x_{(q)}} = \xi_{(q)}\frac{\partial}{\partial r} + \frac{1}{r}\nabla^*_{\xi_{(q)}}, \quad\quad\quad (2.31)$$

where ∇^* is the *surface gradient* on $\mathbb{S}^{q-1}$. Moreover, in terms of spherical coordinates the *Laplace operator (Laplacian)* $\Delta = \nabla \cdot \nabla$ in $\mathbb{R}^q$

$$\Delta_{x_{(q)}} = \left(\frac{\partial}{\partial x_1}\right)^2 + \ldots + \left(\frac{\partial}{\partial x_q}\right)^2 \quad\quad\quad (2.32)$$

has the representation

$$\Delta_{x_{(q)}} = r^{1-q}\frac{\partial}{\partial r}r^{q-1}\frac{\partial}{\partial r} + \frac{1}{r^2}\Delta^*_{\xi_{(q)}}, \quad\quad\quad (2.33)$$

where Δ^* describes the *Laplace–Beltrami operator* of the unit sphere $\mathbb{S}^{q-1}$ recursively given by

$$\Delta^*_{\xi_{(q)}} = (1-t^2)\left(\frac{\partial}{\partial t}\right)^2 - (q-1)t\frac{\partial}{\partial t} + \frac{1}{1-t^2}\Delta^*_{\xi_{(q-1)}}, \; q \geq 3, \; (2.34)$$

$$\Delta^*_{\xi_{(2)}} = \left(\frac{\partial}{\partial\varphi}\right)^2 \quad\quad\quad (2.35)$$

(if no confusion is likely to arise the Laplace–Beltrami operator is simply called the Beltrami operator).

Clearly,

$$\Delta^* = \nabla^* \cdot \nabla^*, \quad\quad\quad (2.36)$$

where $\nabla^*\cdot$ is the *surface divergence* on $\mathbb{S}^{q-1}$ (for more details concerning the differential operators in the three-dimensional case see, e.g., W. Freeden, M. Schreiner [2009]).

Spheres and Balls

The *sphere in $\mathbb{R}^q$ with radius R around $y \in \mathbb{R}^q$* is denoted by $\mathbb{S}^{q-1}_R(y)$

$$\mathbb{S}^{q-1}_R(y) = \{x \in \mathbb{R}^q \mid |x-y| = R\}, \quad\quad\quad (2.37)$$

and $\mathbb{S}_R^{q-1}$ is the sphere with radius R around 0 (i.e., $\mathbb{S}_R^{q-1} = \mathbb{S}_R^{q-1}(0)$).

$\mathbb{B}_R^q(y)$ denotes *the (open) ball in the Euclidean space* $\mathbb{R}^q$ *with center* $y \in \mathbb{R}^q$ *and radius* R:

$$\mathbb{B}_R^q(y) = \{x \in \mathbb{R}^q | \, |x - y| < R\}. \tag{2.38}$$

The closure of the ball $\mathbb{B}_R^q(y) \subset \mathbb{R}^q$ is given by

$$\overline{\mathbb{B}_R^q(y)} = \{x \in \mathbb{R}^q | \, |x - y| \le R\}. \tag{2.39}$$

We simply write $\mathbb{B}_R^q$ and $\overline{\mathbb{B}_R^q}$, respectively, for the open and closed ball with radius R around the origin 0.

By $\mathbb{B}_{\rho,R}^q(y)$, $0 \le \rho < R$, we denote the *ball ring in the Euclidean space* $\mathbb{R}^q$ *with center* $y \in \mathbb{R}^q$ *and radii* ρ *and* R given by

$$\mathbb{B}_{\rho,R}^q(y) = \{x \in \mathbb{R}^q | \, \rho < |x - y| < R\}, \tag{2.40}$$

i.e.,

$$\mathbb{B}_{\rho,R}^q(y) = \mathbb{B}_R^q(y) \backslash \overline{\mathbb{B}_\rho^q(y)}. \tag{2.41}$$

2.4 Radial and Angular Functions

A function $G : \overline{\mathbb{B}_{\rho,N}^q} \to \mathbb{C}$ is called *radial* in $\overline{\mathbb{B}_{\rho,N}^q}$, $0 \le \rho \le N$, if for all $x \in \overline{\mathbb{B}_{\rho,N}^q}$

$$G(x) = G(r\xi) = G(r), \quad x = r\xi, \, r = |x|. \tag{2.42}$$

A function $H : \overline{\mathbb{B}_{\rho,N}^q} \to \mathbb{C}$ is called *angular* in $\overline{\mathbb{B}_{\rho,N}^q}$, $0 < \rho \le N$, if for all $x \in \overline{\mathbb{B}_{\rho,N}^q}$

$$H(x) = H(r\xi) = H(\xi), \quad x = r\xi, \, r = |x|. \tag{2.43}$$

The Laplace derivative of a radial and angular function, respectively, is of particular significance for our later work

$$\Delta_x G(x) = r^{1-q} \frac{\partial}{\partial r} r^{q-1} \frac{\partial}{\partial r} G(r), \quad r \in [\rho, N], \tag{2.44}$$

$$\Delta_x H(x) = \frac{1}{r^2} \Delta_\xi^* H(\xi), \quad \xi \in \mathbb{S}^{q-1}. \tag{2.45}$$

3

One-Dimensional Auxiliary Material

CONTENTS

3.1 Gamma Function and Its Properties 17
        Definition and Functional Equation 18
        Euler's Beta Function 20
        Stirling's Formula 25
        Pochhammer's Factorial 28
3.2 Riemann–Lebesgue Limits 31
        Riemann–Lebesgue Theorem 31
        Extended Riemann–Lebesgue Theorem 32
3.3 Fourier Boundary and Stationary Point Asymptotics 34
        Boundary Point Asymptotics 35
        Stationary Point Asymptotics 35
3.4 Abel–Poisson and Gauß–Weierstraß Limits 38
        Gauß–Weierstraß Means 39
        Abel–Poisson Means 42

In this chapter we provide well known one-dimensional tools and methods of basic importance for this work. The point of departure is the Gamma function. A central topic is the Stirling formula. Particular attention is paid to generalizations of the Riemann–Lebesgue theorem known from the Fourier theory. We continue with some procedures of the stationary phase that turn out to be extremely helpful to secure the convergence of weighted lattice point sums including Fourier integrals (as discussed, for example, in Subsection 13.2). Finally, our considerations are dedicated to Abel–Poisson and Gauß–Weierstraß limit relations as canonical preparations for the Abel–Poisson and Gauß–Weierstraß transforms in multi-dimensional Euclidean spaces $\mathbb{R}^q$ (as studied in Section 7.4 and applied to the lattice point theory in Section 12.1).

3.1   Gamma Function and Its Properties

First our purpose is to introduce the classical Gamma function. Its essential properties are explained (for a more detailed discussion the reader is referred, e.g., to N. Nielsen [1906], E.T. Whittaker, G.N. Watson [1948], N.N. Lebedev [1973], C. Müller [1998], and the references therein). In particular, the Stirling formula is verified. The extension of the Gamma function to complex values is studied.

Definition and Functional Equation

For real values $x > 0$ we consider the integrals

$$(\alpha) \qquad \int_0^1 e^{-t} t^{x-1} dt \tag{3.1}$$

and

$$(\beta) \qquad \int_1^\infty e^{-t} t^{x-1} dt. \tag{3.2}$$

In order to show the convergence of (α), we observe that $0 < e^{-t} t^{x-1} \le t^{x-1}$ holds true for all $t \in (0, 1]$. Therefore, for $\varepsilon > 0$ sufficiently small, we have

$$\int_\varepsilon^1 e^{-t} t^{x-1} dt \le \int_\varepsilon^1 t^{x-1} dt = \left. \frac{t^x}{x} \right|_\varepsilon^1 = \frac{1}{x} - \frac{\varepsilon^x}{x}. \tag{3.3}$$

Consequently, for all $x > 0$, the integral (α) is convergent. To guarantee the convergence of (β) we observe that

$$e^{-t} t^{x-1} \le \frac{n!}{t^{n-x+1}} \tag{3.4}$$

for all $n \in \mathbb{N}$ and $t \ge 1$. This shows us that

$$\int_1^A e^{-t} t^{x-1} dt \le n! \int_1^A \frac{1}{t^{n-x+1}} dt = n! \left. \frac{t^{-n+x}}{x-n} \right|_1^A = \frac{n!}{x-n} \left(\frac{1}{A^{n-x}} - 1 \right) \tag{3.5}$$

provided that A is sufficiently large and n is chosen such that $n \ge x+1$. Thus, the integral (β) is convergent.

The point of departure is the following integral representation.

Lemma 3.1. *For all $x > 0$, the integral*

$$\int_0^\infty e^{-t} t^{x-1} dt \tag{3.6}$$

is convergent.

By definition we let

$$\Gamma(x) = \int_0^\infty e^{-t} t^{x-1} dt. \tag{3.7}$$

Definition 3.1. *The function $x \mapsto \Gamma(x)$, $x > 0$, as defined by (3.7), is called the Gamma function.*

Obviously, we have the following properties:

(i) Γ *is positive for all* $x > 0$,

(ii) $\Gamma(1) = \int_0^\infty e^{-t}dt = 1.$

Integration by parts yields

$$
\begin{aligned}
\Gamma(x+1) &= \int_0^\infty e^{-t}\, t^x dt = -e^{-t}t^x\big|_0^\infty - \int_0^\infty (-e^{-t})\, xt^{x-1}dt \\
&= x \int_0^\infty e^{-t}t^{x-1}dt = x\,\Gamma(x).
\end{aligned}
\tag{3.8}
$$

Lemma 3.2. *The Gamma function* Γ *satisfies the functional equation*

$$\Gamma(x+1) = x\Gamma(x), \qquad x > 0. \tag{3.9}$$

As an immediate consequence we obtain

$$\Gamma(x+n) = (x+n-1)\cdots(x+1)x\,\Gamma(x) \tag{3.10}$$

for $x > 0$ and $n \in \mathbb{N}$. This gives us

Lemma 3.3. *For* $n \in \mathbb{N}_0$,

$$\Gamma(n+1) = n!\,. \tag{3.11}$$

Proof. The assertion is clear for $n = 0, 1$. For $n \geq 2$ we have

$$
\begin{aligned}
\Gamma(n+1) &= n\Gamma(n) & (3.12) \\
&= n(n-1)\Gamma(n-1) \\
&= n\cdot\ldots\cdot 1\underbrace{\Gamma(1)}_{=1} \\
&= n!\,,
\end{aligned}
$$

as required. □

Remark 3.1. *The Gamma function restricted to positive integers is the well known factorial function.*

Next we deal with the derivatives of the Gamma function.

Lemma 3.4. *The Gamma function* Γ *is differentiable for all* $x > 0$, *and we have*

$$\Gamma'(x) = \int_0^\infty e^{-t}(\ln(t))t^{x-1}dt. \tag{3.13}$$

Γ is infinitely often differentiable for all $x > 0$, and we have

$$\Gamma^{(k)}(x) = \int_0^\infty e^{-t}(\ln(t))^k t^{x-1} dt, \quad k \in \mathbb{N}. \tag{3.14}$$

An elementary calculation shows us that

$$(\Gamma'(x))^2 = \left(\int_0^\infty e^{-t}(\ln(t)) t^{x-1} dt \right)^2 \tag{3.15}$$

$$= \left(\int_0^\infty e^{-\frac{t}{2}} t^{\frac{x-1}{2}} (\ln(t)) e^{-\frac{t}{2}} t^{\frac{x-1}{2}} dt \right)^2.$$

The Cauchy-Schwarz inequality yields

$$(\Gamma'(x))^2 \leq \int_0^\infty \left(e^{-\frac{t}{2}} t^{\frac{x-1}{2}} \right)^2 dt \int_0^\infty \left(e^{-\frac{t}{2}} t^{\frac{x-1}{2}} (\ln(t)) \right)^2 dt \tag{3.16}$$

$$= \int_0^\infty e^{-t} t^{x-1} dt \int_0^\infty e^{-t} t^{x-1} (\ln(t))^2 dt$$

$$= \Gamma(x)\,\Gamma''(x).$$

Lemma 3.5. *(Gauß' Expression of the Second Order Logarithmic Derivative)*
For $x > 0$,

$$(\Gamma'(x))^2 \leq \Gamma(x)\,\Gamma''(x). \tag{3.17}$$

Equivalently, we have

$$\left(\frac{d}{dx} \right)^2 \ln(\Gamma(x)) = \frac{\Gamma''(x)}{\Gamma(x)} - \left(\frac{\Gamma'(x)}{\Gamma(x)} \right)^2 > 0. \tag{3.18}$$

In other words, $x \mapsto \ln(\Gamma(x))$, $x > 0$, is a convex function.

Euler's Beta Function

Next we notice that for $\gamma > 0$, $\delta > 0$, the integral

$$\int_0^1 t^{\gamma-1}(1-t)^{\delta-1} dt \tag{3.19}$$

is convergent.

Definition 3.2. *The function $(\gamma, \delta) \mapsto B(\gamma, \delta)$, $\gamma, \delta > 0$, defined by*

$$B(\gamma, \delta) = \int_0^1 t^{\gamma-1}(1-t)^{\delta-1} dt \tag{3.20}$$

is called the Euler Beta function.

For $\gamma, \delta > 0$ we see that

$$
\begin{aligned}
\Gamma(\gamma)\Gamma(\delta) &= \int_0^\infty e^{-t} t^{\gamma-1} dt \int_0^\infty e^{-s} s^{\delta-1}\, ds \qquad (3.21) \\
&= \iint_{\substack{0 \le t < \infty \\ 0 \le s < \infty}} e^{-(t+s)} t^{\gamma-1} s^{\delta-1} dt\, ds.
\end{aligned}
$$

Note that the transition from one-dimensional to two-dimensional integrals is permitted by Fubini's theorem.

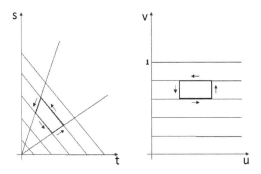

FIGURE 3.1
The illustration of the coordinate transformation relating the Beta and the Gamma functions.

We make a coordinate transformation (cf. Figure 3.1) as follows:

$$
\begin{aligned}
t &= u(1-v)\ , & 0 \le u < \infty, \\
s &= uv & ,\quad 0 \le v \le 1.
\end{aligned}
\qquad (3.22)
$$

It is not difficult to verify that the functional determinant of the coordinate transformation is given by

$$
\frac{\partial(t,s)}{\partial(u,v)} = \begin{vmatrix} 1-v & -u \\ v & u \end{vmatrix} = u(1-v) + uv = u \ge 0 \qquad (3.23)
$$

Thus we find

$$
\iint\limits_{\substack{0 \le t < \infty \\ 0 \le s < \infty}} e^{-(t+s)} t^{\gamma-1} s^{\delta-1} dt \, ds \;=\; \iint\limits_{\substack{0 \le v \le 1 \\ 0 \le u < \infty}} e^{-u} (u(1-v))^{\gamma-1} (uv)^{\delta-1} u \, du \, dv
$$

$$
=\; \iint\limits_{\substack{0 \le v \le 1 \\ 0 \le u < \infty}} e^{-u} u^{\gamma+\delta-2} (1-v)^{\gamma-1} v^{\delta-1} u \, du \, dv
$$

$$
=\; \int_0^\infty e^{-u} u^{\gamma+\delta-1} du \int_0^1 v^{\delta-1} (1-v)^{\gamma-1} dv.
$$
(3.24)

This leads to

Theorem 3.1. *For $\gamma, \delta > 0$*

$$
B(\gamma, \delta) = \frac{\Gamma(\gamma)\Gamma(\delta)}{\Gamma(\gamma+\delta)}.
$$
(3.25)

In particular,

$$
B\left(\frac{1}{2}, \frac{1}{2}\right) \;=\; \int_0^1 t^{-\frac{1}{2}} (1-t)^{-\frac{1}{2}} \, dt
$$
(3.26)

$$
=\; 2 \int_0^1 (1-u^2)^{-\frac{1}{2}} \, du
$$

$$
=\; 2 \arcsin(1) = 2\frac{\pi}{2} = \pi.
$$

Therefore we have

$$
\frac{\Gamma^2\left(\frac{1}{2}\right)}{\Gamma(1)} = \pi.
$$
(3.27)

This shows that

$$
\Gamma\left(\frac{1}{2}\right) = \sqrt{\pi} = \int_0^\infty e^{-t} t^{-\frac{1}{2}} \, dt.
$$
(3.28)

Other types of integrals can be derived from

$$
\int_0^\infty e^{-t^\alpha} dt \;\overset{u=t^\alpha}{=\!=}\; \frac{1}{\alpha} \int_0^\infty e^{-u} u^{\frac{1}{\alpha}-1} \, du
$$
(3.29)

$$
=\; \frac{1}{\alpha} \Gamma\left(\frac{1}{\alpha}\right), \quad \alpha > 0.
$$

Lemma 3.6. *For $\alpha > 0$,*

$$
\int_0^\infty e^{-t^\alpha} dt = \Gamma\left(\frac{\alpha+1}{\alpha}\right).
$$
(3.30)

In particular,

$$\int_0^\infty e^{-t^2} dt = \Gamma\left(\frac{3}{2}\right) = \frac{1}{2}\Gamma\left(\frac{1}{2}\right) = \frac{\sqrt{\pi}}{2}. \tag{3.31}$$

Moreover, we have

$$\int_0^\infty t^{\gamma-1} e^{-t^\alpha} dt = \frac{1}{\alpha}\Gamma\left(\frac{\gamma}{\alpha}\right), \quad \gamma, \alpha > 0 \tag{3.32}$$

and

$$\int_0^\infty t^{\gamma-1} e^{-\alpha t^2} dt = \frac{1}{2}\alpha^{-\frac{\gamma}{2}}\Gamma\left(\frac{\gamma}{2}\right), \quad \gamma, \alpha > 0. \tag{3.33}$$

q=1	$\|\mathbb{S}^0\| = 2$
q=2	$\|\mathbb{S}^1\| = 2\pi$
q=3	$\|\mathbb{S}^2\| = 4\pi$

TABLE 3.1
The area of the unit sphere $\mathbb{S}^{q-1}$ for $q = 1, 2, 3$.

Within the notational framework of polar coordinates (2.29), (2.30) we give the well known calculation of the area $\|\mathbb{S}^{q-1}\|$ of the unit sphere $\mathbb{S}^{q-1}$ in $\mathbb{R}^q$: By definition (see Table 3.1), we set

$$\|\mathbb{S}^0\| = 2, \tag{3.34}$$

$\mathbb{S}^1$ is the unit circle in $\mathbb{R}^2$; hence, its area is equal to

$$\|\mathbb{S}^1\| = 2\pi. \tag{3.35}$$

Furthermore, $\mathbb{S}^2$ is the unit sphere in $\mathbb{R}^3$; hence, its area is known to be equal to

$$\|\mathbb{S}^2\| = 4\pi. \tag{3.36}$$

We are interested in deriving the area of the sphere $\mathbb{S}^{q-1}$ in $\mathbb{R}^q$ $(q > 3)$:

$$\|\mathbb{S}^{q-1}\| = \int_{\mathbb{S}^{q-1}} dS_{(q-1)}(\xi_{(q)}). \tag{3.37}$$

In terms of *spherical coordinates* (2.29) and (2.30) in $\mathbb{R}^q$ the surface element admits the representation

$$dS_{(q-1)}\left(\xi_{(q)}\right) \;=\; dS_{(q-2)}\left(\sqrt{1-t^2}\xi_{(q-1)}\right) dt \tag{3.38}$$
$$+ (-1)^{q-1}t \; dV_{(q-1)}\left(\sqrt{1-t^2}\xi_{(q-1)}\right).$$

Now, we notice that

$$dV_{(q-1)}\left(\sqrt{1-t^2}\xi_{(q-1)}\right) \;=\; -t(1-t^2)^{\frac{q-3}{2}} \; dt \; dS_{(q-2)}\left(\xi_{(q-1)}\right) \tag{3.39}$$
$$= \; (-1)^{q-1}t(1-t^2)^{\frac{q-3}{2}} dS_{(q-2)}\left(\xi_{(q-1)}\right) \; dt.$$

In addition, it is not difficult to see that

$$dS_{(q-2)}\left(\sqrt{1-t^2}\xi_{(q-1)}\right) = (1-t^2)^{\frac{q-1}{2}} \; dS_{(q-2)}\left(\xi_{(q-1)}\right). \tag{3.40}$$

Combining our results we are led to the identity

$$dS_{(q-1)}\left(t\epsilon^q + \sqrt{1-t^2}\xi_{(q-1)}\right) \tag{3.41}$$
$$= \; (1-t^2)^{\frac{q-3}{2}}\left(1-t^2+t^2\right) \; dS_{(q-2)}\left(\xi_{(q-1)}\right) \; dt$$

such that

$$dS_{(q-1)}\left(\xi_{(q)}\right) = (1-t^2)^{\frac{q-3}{2}} dS_{(q-2)}\left(\xi_{(q-1)}\right) \; dt. \tag{3.42}$$

Thus we find

$$\|\mathbb{S}^{q-1}\| \;=\; \int_{-1}^{1}\int_{\mathbb{S}^{q-2}}(1-t^2)^{\frac{q-3}{2}}dS_{(q-2)}(\xi_{(q-1)}) \; dt$$
$$= \; \|\mathbb{S}^{q-2}\| \int_{-1}^{1}(1-t^2)^{\frac{q-3}{2}} \; dt.$$

For the computation of the remaining integral it is helpful to use some facts known from the Gamma function. More explicitly,

$$\int_{-1}^{1}(1-t^2)^{\frac{q-3}{2}}dt \;=\; 2\int_{0}^{1}(1-t^2)^{\frac{q-3}{2}}dt \tag{3.43}$$
$$\overset{t^2=v}{=}\; \int_{0}^{1}v^{-\frac{1}{2}}(1-v)^{\frac{q-3}{2}}dv$$
$$= \; B\left(\frac{1}{2},\frac{q-1}{2}\right)$$
$$= \; \frac{\Gamma\left(\frac{1}{2}\right)\Gamma\left(\frac{q-1}{2}\right)}{\Gamma\left(\frac{q}{2}\right)}$$
$$= \; \frac{\sqrt{\pi}\Gamma\left(\frac{q-1}{2}\right)}{\Gamma\left(\frac{q}{2}\right)}.$$

By recursion we get from (3.43)

Lemma 3.7. For $q \geq 2$,

$$||\mathbb{S}^{q-1}|| = 2\frac{\pi^{\frac{q}{2}}}{\Gamma\left(\frac{q}{2}\right)}. \tag{3.44}$$

The area of the sphere $\mathbb{S}_R^{q-1}(y)$ with center $y \in \mathbb{R}^q$ and radius $R > 0$ is given by

$$||\mathbb{S}_R^{q-1}(y)|| = ||\mathbb{S}^{q-1}|| R^{q-1} = 2\frac{\pi^{\frac{q}{2}}}{\Gamma\left(\frac{q}{2}\right)} R^{q-1}. \tag{3.45}$$

Furthermore, the volume of the ball $\mathbb{B}_R^q(y)$ with center $y \in \mathbb{R}^q$ and radius $R > 0$ is given by

$$
\begin{aligned}
||\mathbb{B}_R^q(y)|| = \int_{\mathbb{B}_R^q(y)} dV_{(q)}(x) &= \int_{r=0}^{R} \left(\int_{\mathbb{S}_r^{q-1}(y)} dS_{(q-1)}(x) \right) dr \\
&= 2\frac{\pi^{\frac{q}{2}}}{\Gamma\left(\frac{q}{2}\right)} \int_0^R r^{q-1} dr \\
&= \frac{\pi^{\frac{q}{2}}}{\Gamma\left(\frac{q}{2}+1\right)} R^q. \tag{3.46}
\end{aligned}
$$

Stirling's Formula

Next we are interested in the behavior of the Gamma function Γ for large positive values x. We therefore study the integral as x goes to infinity.

We first regard x as fixed. Substituting

$$t = x(1+s), \quad -1 \leq s < \infty, \tag{3.47}$$

we get

$$\Gamma(x) = x^x e^{-x} \int_{-1}^{\infty} (1+s)^{x-1} e^{-xs} \, ds. \tag{3.48}$$

For brevity we set

$$\Gamma(x) = x^x e^{-x} I(x). \tag{3.49}$$

Our aim is to verify that $I(x)$ satisfies

$$\left| I(x) - \sqrt{\frac{2\pi}{x}} \right| \leq \frac{1}{x}. \tag{3.50}$$

For that purpose we write

$$(1+s)^x e^{-xs} = e^{-x(s-\ln(1+s))} = e^{-xu^2(s)} \tag{3.51}$$

where

$$u(s) = \begin{cases} |s - \ln(1+s)|^{\frac{1}{2}}, & s \in [0, \infty) \\ -|s - \ln(1+s)|^{\frac{1}{2}}, & s \in (-1, 0). \end{cases} \tag{3.52}$$

By Taylor's formula we get for $s \in (-1, \infty)$

$$u^2(s) - u^2(0) = s - \ln(1 + s) = \frac{s^2}{2} \frac{1}{(1 + s\vartheta)^2} = u^2(s) \quad (3.53)$$

with $0 < \vartheta < 1$, where $u(0) = 0$. We interpret ϑ as a uniquely defined function of s; i.e., $\vartheta : s \mapsto \vartheta(s)$, so that

$$\frac{u(s)}{s} = \frac{1}{\sqrt{2}} \frac{1}{(1 + s\vartheta(s))} \quad (3.54)$$

is a positive continuous function for $s \in (-1, \infty)$ showing the property

$$\left| \frac{u(s)}{s} - \frac{1}{\sqrt{2}} \right| = \frac{1}{\sqrt{2}} \left| \frac{s\vartheta(s)}{1 + s\vartheta(s)} \right| \leq |u(s)|. \quad (3.55)$$

From $u^2(s) = s - \ln(1 + s)$ we are immediately able to see that

$$2u \frac{du}{ds} = \frac{s}{1 + s}. \quad (3.56)$$

Obviously, $s : u \mapsto s(u)$, $u \in \mathbb{R}$, is a member of class $C^{(1)}(\mathbb{R})$, and we get

$$\int_{-1}^{\infty} (1 + s)^{x-1} e^{-xs} \, ds = 2 \int_{-\infty}^{+\infty} e^{-xu^2} \frac{u}{s(u)} \, du. \quad (3.57)$$

Thus we are able to deduce that

$$\left| 2 \int_{-\infty}^{\infty} e^{-xu^2} \frac{u}{s(u)} \, du - \sqrt{2} \int_{-\infty}^{+\infty} e^{-xu^2} \, du \right| \leq 4 \int_{0}^{\infty} e^{-xu^2} du. \quad (3.58)$$

In connection with (3.33) and (3.55) this yields for $x > 0$

$$\left| \int_{-1}^{+\infty} (1 + s)^{x-1} e^{-xs} \, dx - \sqrt{\frac{2\pi}{x}} \right| \leq \frac{1}{x}. \quad (3.59)$$

This leads to

Theorem 3.2. *(Stirling's Formula). For $x > 0$*

$$\left| \frac{\Gamma(x)}{\sqrt{2\pi} \, x^{x-\frac{1}{2}} \, e^{-x}} - 1 \right| \leq \sqrt{\frac{2}{\pi x}}. \quad (3.60)$$

Remark 3.2. *Stirling's formula can be rewritten in the form*

$$\lim_{x \to \infty} \frac{\Gamma(x)}{\sqrt{2\pi} \, x^{x-\frac{1}{2}} e^{-x}} = 1. \quad (3.61)$$

The limit relation (3.61) may be equivalently written in the form

$$\Gamma(x) \sim \sqrt{2\pi} \, x^{x-\frac{1}{2}} \, e^{-x}, \quad x \to \infty, \quad (3.62)$$

where the symbol "$\sim$" means "asymptotically equal" (in the sense that the quotient of both sides tends to 1).

An immediate application is the limit relation

$$\lim_{x \to \infty} \frac{\Gamma(x+a)}{x^a \Gamma(x)} = 1, \quad a > 0. \tag{3.63}$$

This can be seen from Stirling's formula by

$$\lim_{x \to \infty} \frac{\Gamma(x+a)}{\sqrt{2\pi}(x+a)^{x+a-\frac{1}{2}} e^{-x-a}} = 1 \tag{3.64}$$

due to the relation

$$(x+a)^{x+a-\frac{1}{2}} = x^{x+a-\frac{1}{2}} \left(1 + \frac{a}{x}\right)^{x+a-\frac{1}{2}} \tag{3.65}$$

and the limits

$$\lim_{x \to \infty} \frac{(1 + \frac{a}{x})^x}{e^a} = 1, \tag{3.66}$$

$$\lim_{x \to \infty} \left(1 + \frac{a}{x}\right)^{a-\frac{1}{2}} = 1. \tag{3.67}$$

Next we prove the so–called *Legendre relation ("duplicator formula")*.

Lemma 3.8. *For $x > 0$ we have*

$$2^{x-1} \Gamma\left(\frac{x}{2}\right) \Gamma\left(\frac{x+1}{2}\right) = \sqrt{\pi}\, \Gamma(x). \tag{3.68}$$

Proof. We consider the function $x \mapsto \phi(x), x > 0$, defined by

$$\phi(x) = \frac{2^{x-1} \Gamma(\frac{x}{2}) \Gamma(\frac{x+1}{2})}{\Gamma(x)} \tag{3.69}$$

for $x > 0$. Setting $x + 1$ instead of x, we find with the functional equation (3.9) for the numerator

$$2^x \Gamma\left(\frac{x+1}{2}\right) \Gamma\left(\frac{x}{2} + 1\right) = 2^{x-1} x \Gamma\left(\frac{x}{2}\right) \Gamma\left(\frac{x+1}{2}\right), \tag{3.70}$$

so that the numerator satisfies the same functional equation as the denominator. This means $\phi(x+1) = \phi(x), x > 0$. By repetition we get for all $n \in \mathbb{N}$ and x fixed $\phi(x+n) = \phi(x)$. We let n tend toward ∞. For the numerator of $\phi(x+n)$ we then find by use of (3.63)

$$\lim_{n \to \infty} \frac{2^{x+n-1} \Gamma(\frac{x+n}{2}) \Gamma(\frac{x+n+1}{2})}{2^{x+n-1}(2\pi) \left(\frac{n}{2}\right)^{\frac{x}{2}} \left(\frac{n}{2}\right)^{\frac{x+1}{2}} \Gamma^2(\frac{n}{2})} = 1. \tag{3.71}$$

For the denominator we get

$$\lim_{n \to \infty} \frac{2\Gamma(x+n)}{2\sqrt{2\pi} n^{n+x-\frac{1}{2}} e^{-n}} = 1. \tag{3.72}$$

We therefore get for every $x > 0$ and all $n \in \mathbb{N}$

$$\phi(x) = \phi(x+n) = \lim_{n \to \infty} \phi(x+n) = \sqrt{\pi}. \qquad (3.73)$$

A periodical function with this property must be constant. This proves the desired Lemma 3.8.   □

A generalization of the Legendre relation ("duplicator formula") is the *Gauß multiplicator formula*.

Lemma 3.9. *For $x > 0$ and $n \geq 2$*

$$\Gamma\left(\frac{x}{n}\right)\Gamma\left(\frac{x+1}{n}\right) \cdot \ldots \cdot \Gamma\left(\frac{x+n-1}{n}\right) n^x = (2\pi)^{\frac{n-1}{2}} \sqrt{n}\, \Gamma(x). \qquad (3.74)$$

Pochhammer's Factorial

Thus far, the Gamma function Γ is defined for positive values, i.e., $x \in \mathbb{R}^+$ (cf. Figure 3.2). We are interested in an extension of Γ to the real line $\mathbb{R}$ (or even to the complex plane $\mathbb{C}$) if possible.

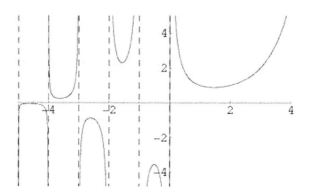

FIGURE 3.2
The Gamma function on the real line $\mathbb{R}$.

To this end we first consider the so-called *Pochhammer factorial* $(x)_n$ with $x \in \mathbb{R}$, $n \in \mathbb{N}$, which is defined by

$$(x)_n = x(x+1)\ldots(x+n-1). \qquad (3.75)$$

For $x > 0$ it is clear that

$$(x)_n = \frac{\Gamma(x+n)}{\Gamma(x)} \qquad (3.76)$$

or

$$\frac{(x)_n}{\Gamma(x+n)} = \frac{1}{\Gamma(x)}. \qquad (3.77)$$

The left-hand side is defined for $x > -n$ and gives the same value for all $n \in \mathbb{N}$ with $n > -x$. We may use this relation to define $\frac{1}{\Gamma(x)}$ for all $x \in \mathbb{R}$, and we see that this function vanishes for $x = 0, -1, -2, \ldots$ (cf. Figure 3.2).

This leads to the following conclusion: The Gamma integral (3.7) is absolutely convergent for $x \in \mathbb{C}$ with real part $\Re(x) > 0$, and represents a holomorphic function for all $x \in \mathbb{C}$ with $\Re(x) > 0$. Moreover, the Pochhammer factorial $(x)_n$ can be defined for all complex numbers x. Therefore we have a definition of $\frac{1}{\Gamma(x)}$ for all complex values x.

Lemma 3.10. *The Γ-function is a meromorphic function that has simple poles in $0, -1, -2, \ldots$. The reciprocal function $x \mapsto \frac{1}{\Gamma(x)}$, $x \in \mathbb{C}$, is an entire function.*

The identity

$$\frac{(x)_n}{\Gamma(n)} \frac{\Gamma(n)}{\Gamma(x+n)} = \frac{1}{\Gamma(x)} \tag{3.78}$$

is valid for all $x \in \mathbb{C}$ and all $n > \Re(x)$. Furthermore it is easy to see that

$$\frac{(x)_n}{\Gamma(n)} = x \frac{(x+1)(x+2)\ldots(x+n-1)}{1 \cdot 2 \ldots (n-1)} = x \prod_{k=1}^{n-1}\left(1 + \frac{x}{k}\right). \tag{3.79}$$

Stirling's formula tells us that

$$\lim_{n \to \infty} \frac{1}{n^{-x}} \frac{\Gamma(n)}{\Gamma(x+n)} = 1. \tag{3.80}$$

Thus we find with $n \to \infty$ from (3.79) the following lemma.

Lemma 3.11. *For $x \in \mathbb{C}$,*

$$\frac{1}{\Gamma(x+1)} = \lim_{n \to \infty} n^{-x} \prod_{k=1}^{n-1}\left(1 + \frac{x}{k}\right). \tag{3.81}$$

This limit relation can be expressed as an infinite product by use of *Euler's constant* (sometimes also called the *Euler–Mascheroni constant*)

$$C = \lim_{m \to \infty}\left(\sum_{k=1}^{m} \frac{1}{k} - \ln(m)\right) \tag{3.82}$$

(note that $C = 0,577\,215\,664\,\ldots$). In fact, for $n \to \infty$ we have

$$\lim_{n \to \infty} \frac{e^{Cx} \prod_{k=1}^{n-1}(1 + \frac{x}{k})e^{-\frac{x}{k}}}{n^{-x} \prod_{k=1}^{n-1}(1 + \frac{x}{k})} = 1. \tag{3.83}$$

Summarizing our results we therefore obtain the following lemma.

Lemma 3.12. *For $x \in \mathbb{C}$,*

$$\frac{1}{\Gamma(x)} = xe^{Cx} \prod_{k=1}^{\infty} \left(1 + \frac{x}{k}\right) e^{-\frac{x}{k}}. \tag{3.84}$$

Let us consider the expression

$$Q(x) = \frac{1}{\pi}\Gamma(x)\Gamma(1-x)\sin(\pi x) \tag{3.85}$$

which has no singularities and is holomorphic for all $x \in \mathbb{C}$. It is not difficult
to show that

$$
\begin{aligned}
Q(x) &= \frac{1}{\pi x}\Gamma(1+x)\Gamma(1-x)\sin(\pi x) \\
&= \frac{1}{\pi(1-x)}\Gamma(x)\Gamma(2-x)\sin(\pi x).
\end{aligned} \tag{3.86}
$$

Obviously,

$$Q(0) = 1, \quad Q(1) = 1. \tag{3.87}$$

In the interval $[0,1]$ the function Q is positive and twice continuously differentiable. With the duplicator formula (Lemma 3.8) we get

$$Q\left(\frac{x}{2}\right) Q\left(\frac{x+1}{2}\right) = Q(x), \tag{3.88}$$

which is easily verified. Letting

$$R(x) = \ln(Q(x)) \tag{3.89}$$

we see that

$$R\left(\frac{x}{2}\right) + R\left(\frac{x+1}{2}\right) = R(x). \tag{3.90}$$

By differentiation we obtain

$$\frac{1}{4}R''\left(\frac{x}{2}\right) + \frac{1}{4}R''\left(\frac{x+1}{2}\right) = R''(x). \tag{3.91}$$

As the second order derivative R'' is continuous on the compact interval $[0,1]$,
there is a value $\xi \in [0,1]$ such that

$$|R''(\xi)| \geq |R''(x)|, \quad x \in [0,1]. \tag{3.92}$$

Therefore we obtain from (3.91)

$$|R''(\xi)| \leq \frac{1}{4}\left|R''\left(\frac{\xi}{2}\right)\right| + \frac{1}{4}\left|R''\left(\frac{\xi+1}{2}\right)\right| \leq \frac{1}{2}|R''(\xi)|, \tag{3.93}$$

which implies $|R''(\xi)| = 0$, that is, $R''(x) = 0$. From $R(1) = R(0) = 0$ we then deduce $Q(x) = 1$. This result can be written in the form

$$\Gamma(x)\Gamma(1-x) = \frac{\pi}{\sin(\pi x)}. \tag{3.94}$$

It establishes an identity between the meromorphic functions $\Gamma(\cdot)$, $\Gamma(1-\cdot)$ and $(\sin \pi \cdot)^{-1}$. Altogether we have

$$\frac{1}{\Gamma(x)}\frac{1}{\Gamma(1-x)} = x\prod_{k=1}^{\infty}\left(1 - \frac{x^2}{k^2}\right). \tag{3.95}$$

In connection with (3.94) we therefore obtain

Lemma 3.13. *For $x \in \mathbb{C}$,*

$$\sin(\pi x) = \pi x \prod_{k=1}^{\infty}\left(1 - \frac{x^2}{k^2}\right). \tag{3.96}$$

3.2 Riemann–Lebesgue Limits

The following considerations can be regarded as certain realizations of the *Riemann–Lebesgue theorem* of the one-dimensional classical Fourier theory.

Riemann–Lebesgue Theorem

Lemma 3.14. *Let F be continuous in the interval $(-1, 1)$ such that*

$$\int_{-1}^{1} |F(t)| \, dt < \infty. \tag{3.97}$$

Then

$$\lim_{r\to\infty} \int_{-1}^{1} e^{irt} F(t) \, dt = 0. \tag{3.98}$$

Proof. We form

$$\xi_\mu = \frac{2\pi}{r}\mu \tag{3.99}$$

with

$$\mu = -\left\lfloor\frac{r}{2\pi}\right\rfloor, -\left\lfloor\frac{r}{2\pi}\right\rfloor + 1, \ldots, \left\lfloor\frac{r}{2\pi}\right\rfloor. \tag{3.100}$$

Hence, we obtain a set of points which divides the interval $[-1, 1]$ of integration into compartments of length not larger than $\frac{2\pi}{r}$. For $r \to \infty$, we have

$$\int_{-1}^{1} e^{irt} F(t) \, dt = \sum_{\mu = -\left\lfloor\frac{r}{2\pi}\right\rfloor}^{\left\lfloor\frac{r}{2\pi}\right\rfloor - 1} \int_{\xi_\mu}^{\xi_{\mu+1}} e^{irt} F(t) \, dt + o(1). \tag{3.101}$$

An elementary calculation shows that

$$\int_{\xi_\mu}^{\xi_{\mu+1}} e^{irt} \, dt = \frac{1}{ir} \left(e^{ir\xi_{\mu+1}} - e^{ir\xi_\mu} \right) = 0. \tag{3.102}$$

Thus we get

$$\int_{-1}^{1} e^{irt} F(t) \, dt = \sum_{\mu=-\lfloor \frac{r}{2\pi} \rfloor}^{\lfloor \frac{r}{2\pi} \rfloor - 1} \int_{\xi_\mu}^{\xi_{\mu+1}} e^{irt} \left(F(t) - F(\xi_\mu) \right) \, dt + o(1), \tag{3.103}$$

where F is uniformly continuous on $[\xi_\mu, \xi_{\mu+1}]$. Together with

$$\xi_{\mu+1} - \xi_\mu = \frac{2\pi}{r} \tag{3.104}$$

this implies the result stated in Lemma 3.14.   □

An immediate consequence is the following asymptotic relation.

Lemma 3.15. *Let F be continuous in $[-1, 1]$ and continuously differentiable in $(-1, 1)$ such that*

$$\int_{-1}^{1} |F'(t)| \, dt < \infty. \tag{3.105}$$

Then

$$\lim_{r \to \infty} \int_{-1}^{1} e^{irt} F'(t) \, dt = 0. \tag{3.106}$$

Of particular interest in the metaharmonic theory (see Section 6.7) is the following limit relation.

Extended Riemann–Lebesgue Theorem

Lemma 3.16. *Let F be continuous in $[-1, 1]$ and continuously differentiable in $(-1, 1)$ such that*

$$\int_{-1}^{1} |F'(t)| \, dt < \infty. \tag{3.107}$$

Then we have for $r \to \infty$

$$\sqrt{\frac{2r}{\pi}} \int_{-1}^{1} e^{irt} F(t)(1 - t^2)^{-\frac{1}{2}} \, dt = i^{-\frac{1}{2}} e^{ir} F(1) + i^{\frac{1}{2}} e^{-ir} F(-1) + o(1). \tag{3.108}$$

Proof. For real values $t \in [-1, 1]$ we consider the auxiliary function

$$X(r, t) = -\int_{t+0i}^{t+\infty i} \frac{e^{irz}}{\sqrt{1 - z^2}} \, dz. \tag{3.109}$$

By application of the Cauchy integral theorem we get at t (for fixed r)

$$\frac{\partial X}{\partial t}(r, t) = \frac{e^{irt}}{\sqrt{1 - t^2}}. \tag{3.110}$$

Letting $z = t + is$ for $|t| \leq 1$, $s \geq 0$ we find

$$\left|\sqrt{1 - z^2}\right| = \sqrt[4]{(1 - t^2 + s^2)^2 + 4t^2 s^2} \geq \sqrt[4]{(1 - t^2)^2} \tag{3.111}$$

such that

$$\left|\sqrt{1 - z^2}\right| \geq \sqrt{1 - t^2}. \tag{3.112}$$

Moreover, for $t \in [-1, 0]$ as well as for $t \in [0, 1]$ we have

$$\left|\sqrt{1 - z^2}\right| = \sqrt[4]{(1 + t)^2 + s^2} \sqrt[4]{(1 - t)^2 + s^2} \geq \sqrt{1}\sqrt{s}, \tag{3.113}$$

such that

$$\left|\sqrt{1 - z^2}\right| \geq \sqrt{s}. \tag{3.114}$$

Therefore we obtain

$$|X(r, t)| \leq \frac{1}{\sqrt{1 - t^2}} \int_0^\infty e^{-rs} \, ds = \frac{1}{r} \frac{1}{\sqrt{1 - t^2}}, \tag{3.115}$$

and

$$|X(t, r)| \leq \int_0^\infty \frac{e^{-rs}}{\sqrt{s}} \, ds = \sqrt{\frac{\pi}{r}}. \tag{3.116}$$

By partial integration we find

$$\int_{-1}^1 e^{irt} F(t)(1 - t^2)^{-\frac{1}{2}} \, dt \tag{3.117}$$

$$= X(r, 1)F(1) - X(r, -1)F(-1) - \int_{-1}^1 X(r, t)F'(t) \, dt.$$

From (3.115) and (3.116) it is not hard to verify that

$$\sqrt{r} \int_{-1}^1 X(r, t)F'(t) \, dt = o(1). \tag{3.118}$$

Furthermore we have with $z = 1 + \frac{iu}{r}$

$$X(r, 1) = -\int_{1+0i}^{1+\infty i} \frac{e^{irz}}{\sqrt{1 - z^2}} dz = \frac{e^{ir}}{\sqrt{2ir}} \int_0^\infty \frac{e^{-u}}{\sqrt{u}} \frac{du}{\sqrt{1 + \frac{iu}{2r}}} \tag{3.119}$$

so that

$$X(r, 1) = \frac{e^{ir}}{\sqrt{2ir}} \left(\int_0^\infty \frac{e^{-u}}{u} \, du + \int_0^\infty \frac{e^{-u}}{\sqrt{u}} \left(\frac{1}{\sqrt{1 + \frac{iu}{2r}}} - 1 \right) du \right). \tag{3.120}$$

From our definition of the root function we get

$$\left| 1 + \sqrt{1 + \frac{iu}{2r}} \right| \geq 1 \tag{3.121}$$

such that

$$
\begin{aligned}
\left| \frac{1}{\sqrt{1 + \frac{iu}{2r}}} - 1 \right| &= \frac{\left| 1 - \sqrt{1 + \frac{iu}{2r}} \right| \left| 1 + \sqrt{1 + \frac{iu}{2r}} \right|}{\left| \sqrt{1 + \frac{iu}{2r}} \right| \left| 1 + \sqrt{1 + \frac{iu}{2r}} \right|} \\
&\leq \frac{\frac{u}{2r}}{\left| 1 + \sqrt{1 + \frac{iu}{2r}} \right|} \leq \frac{u}{2r}.
\end{aligned}
\tag{3.122}
$$

This yields

$$\left| \frac{e^{ir}}{\sqrt{2ir}} \int_0^\infty \frac{e^{-u}}{\sqrt{u}} \left(\frac{1}{\sqrt{1 + \frac{iu}{2r}}} - 1 \right) du \right| \leq \frac{1}{\sqrt{2r}} \int_0^\infty \frac{e^{-u}}{\sqrt{u}} \frac{u}{2r} \, du. \tag{3.123}$$

Therefore we get

$$\frac{e^{ix}}{\sqrt{2ir}} \int_0^\infty \frac{e^{-u}}{\sqrt{u}} \left(\frac{1}{\sqrt{1 + \frac{iu}{2r}}} - 1 \right) du = O\left(r^{-\frac{3}{2}} \right). \tag{3.124}$$

Summarizing our results we obtain

$$X(r, 1) = \frac{e^{ir}}{\sqrt{ir}} \sqrt{\frac{\pi}{2}} + O\left(r^{-\frac{3}{2}} \right). \tag{3.125}$$

In the same way we find

$$X(r, -1) = \sqrt{\frac{i}{r}} e^{-ir} \sqrt{\frac{\pi}{2}} + O\left(r^{-\frac{3}{2}} \right). \tag{3.126}$$

Combining (3.125) and (3.126) we get the desired result.   □

3.3   Fourier Boundary and Stationary Point Asymptotics

In one-dimensional Fourier theory we find a large number of extensions of Lemma 3.15 and Lemma 3.16.

Boundary Point Asymptotics

From P. I. Natanson [1961] we borrow (without proof) the following asymptotic relation.

Theorem 3.3. *Let G be of class $C^{(m)}([a,b])$, $a, b \in \mathbb{R}$ with $a < b$. Suppose that F is given in the form*

$$F(t) = G(t)(t-a)^{\lambda-1}(b-t)^{\mu-1}, \quad t \in [a,b], \tag{3.127}$$

$0 < \lambda, \mu \leq 1$. *Then, for $r \to \infty$,*

$$\int_a^b F(t)e^{irt}\, dt = -A_m(r) + B_m(r) + O\left(\frac{1}{r^m}\right), \tag{3.128}$$

where

$$A_m(r) = \sum_{n=0}^{m-1} \frac{\Gamma(n+\lambda)}{n!\, r^{n+\lambda}} e^{ira+i\frac{\pi}{2}(n+\lambda-2)} \left(\frac{\partial}{\partial u}\right)^n ((b-u)^{\mu-1}G(u))\, \big|_{u=a} \tag{3.129}$$

and

$$B_m(r) = \sum_{n=0}^{m-1} \frac{\Gamma(n+\mu)}{n!\, r^{n+\mu}} e^{irb+i\frac{\pi}{2}(n-\mu)} \left(\frac{\partial}{\partial v}\right)^n ((v-a)^{\lambda-1}G(v))\, \big|_{v=b}. \tag{3.130}$$

Remark 3.3. *For $\lambda = \mu = 1$ the term $O(r^{-m})$ can be replaced by $o(r^{-m})$.*

An easy consequence is the following asymptotic expansion.

Corollary 3.1. *Suppose that $F \in C^{(1)}([a,b])$, $H \in C^{(2)}([a,b])$ such that $H'(t) > 0$ for $t \in [a,b]$. Then, for $r \to \infty$,*

$$\int_a^b F(t)e^{irH(t)}\, dt = \frac{1}{ir}\left(\frac{F(b)}{H'(b)}e^{irH(b)} - \frac{F(a)}{H'(a)}e^{irH(a)}\right) + o\left(\frac{1}{r}\right). \tag{3.131}$$

Remark 3.4. *Corollary 3.1 is based on the assumption that the function H does not possess a stationary point, i.e., $\tau \in [a,b]$ such that $H'(\tau) = 0$; hence, the asymptotic relation (3.131) depends only on the values of the endpoints of the interval $[a,b]$.*

Stationary Point Asymptotics

The following asymptotic relation (3.134) includes influences from stationary points (see P. I. Natanson [1961] for the proof).

Theorem 3.4. *Let G be of class $C^{(m)}([a,b])$. Suppose that $\tilde{H}$ is of class $C^{(m)}([a,b])$ with $\tilde{H}(t) > 0$ for all $t \in [a,b]$. Furthermore, we let*

$$F(t) = G(t)(t-a)^{\lambda-1}(b-t)^{\mu-1}, \quad t \in [a,b], \tag{3.132}$$

$0 \le \lambda,\ \mu \le 1$, *and* $H \in C^{(1)}([a,b])$ *with*

$$H'(t) = \tilde{H}(t)(t-a)^{\rho-1}(b-t)^{\sigma-1}, \quad t \in [a,b], \tag{3.133}$$

$\rho, \sigma \ge 1$. *Then, for* $r \to \infty$,

$$\int_a^b F(t)e^{ikH(t)}\, dt = -A_m(r) + B_m(r) + O\left(\frac{1}{r^{\frac{m}{\rho}}}\right) + O\left(\frac{1}{r^{\frac{m}{\sigma}}}\right), \tag{3.134}$$

where

$$A_m(r) = -e^{irH(a)}\sum_{n=0}^{m-1} \frac{\psi^{(n)}(0)}{\rho}\frac{\Gamma(\frac{n+\lambda}{\rho})}{n!}e^{i\frac{\pi}{2}\frac{n+\lambda}{\rho}}\frac{1}{r^{\frac{n+\lambda}{\rho}}} \tag{3.135}$$

and

$$B_m(r) = -e^{irH(b)}\sum_{n=0}^{m-1} \frac{\chi^{(n)}(0)}{\sigma}\frac{\Gamma(\frac{n+\mu}{\sigma})}{n!}e^{-i\frac{\pi}{2}\frac{n+\mu}{\sigma}}\frac{1}{r^{\frac{n+\mu}{\sigma}}} \tag{3.136}$$

with

$$\psi(u) = F(t)u^{1-\lambda}\frac{dt}{du}, \quad u^\rho(t) = H(t) - H(a), \tag{3.137}$$

and

$$\chi(v) = F(t)v^{1-\mu}\frac{dt}{dv}, \quad v^\sigma(t) = H(b) - H(t). \tag{3.138}$$

Remark 3.5. *The following statements hold true:*

(i) If $\lambda = \mu = 1$, *then the "O" symbol can be replaced by the "o" symbol.*

(ii) If $\rho = \sigma = 1$, *then Theorem 3.4 reduces to Theorem 3.3 (by the use of the substitution* $H(t) = t$*).*

Definition 3.3. *A point* $\tau \in [a,b]$ *is called a stationary point of order n in* $[a,b]$, *if the following conditions are satisfied:*

(i) $H \in C^{(n+1)}([a,b])$,

(ii) $H(\tau) = \ldots = H^{(n)}(\tau) = 0$,

(iii) $H^{(n+1)}(\tau) \ne 0$.

Note that a stationary point of order 1 is simply called a stationary point.

From Theorem 3.4 we are able to formulate some corollaries.

Corollary 3.2. *Suppose that* $F \in C^{(1)}([a,b])$ *and* $H \in C^{(3)}([a,b])$. *If there is a stationary point* $\tau \in (a,b)$ *such that* $H'(\tau) = 0$ *and* $H''(\tau) > 0$, *then for* $r \to \infty$

$$\int_a^b F(t)e^{irH(t)}\, dt = F(\tau)\sqrt{\frac{2\pi}{H''(\tau)r}}\, e^{irH(\tau)+i\frac{\pi}{4}} + o\left(\frac{1}{r^{\frac{1}{2}}}\right). \tag{3.139}$$

Proof. The properties $H \in C^{(3)}([a,b])$ and $H'(\tau) = 0$ imply that $H_0 : t \to H_0(t) = H'(t)(t-\tau)^{-1}$ is continuously differentiable in the interval $[a,b]$ with $H_0(\tau) = H''(\tau)$ and $H_0'(\tau) = \frac{1}{2}H'''(\tau)$. We split the integral into two parts

$$\int_a^b \ldots = \int_a^\tau \ldots + \int_\tau^b \ldots . \tag{3.140}$$

The integral $J_1 = \int_a^\tau F(t)e^{irH(t)} \, dt$ can be recognized in the notation of Theorem 3.4 by choosing $\lambda = \mu = \rho = 1$ and $\sigma = 2$. An easy calculation shows that

$$\int_a^\tau F(t)e^{irH(t)} \, dt = B_1(r) - A_1(r) + o\left(\frac{1}{r}\right) + o\left(\frac{1}{r^{\frac{1}{2}}}\right), \tag{3.141}$$

where (note that $\rho = 1$)

$$A_1(r) = O\left(\frac{1}{r}\right) \tag{3.142}$$

and

$$B_1(r) = \frac{1}{2}e^{irH(\tau)+i\frac{\pi}{4}F(\tau)}\sqrt{\frac{2\pi}{H''(\tau)r}}. \tag{3.143}$$

This gives

$$J_1 = \frac{1}{2}F(\tau)\sqrt{\frac{2\pi}{H''(\tau)r}}\, e^{irH(\tau)+i\frac{\pi}{4}} + o\left(\frac{1}{r^{\frac{1}{2}}}\right). \tag{3.144}$$

The integral $J_2 = \int_\tau^b F(t)e^{irH(t)} \, dt$ can be evaluated in the same way. We get

$$J_2 = \frac{1}{2}F(\tau)\sqrt{\frac{2\pi}{H''(\tau)r}}\, e^{irH(\tau)+i\frac{\pi}{4}} + o\left(\frac{1}{r^{\frac{1}{2}}}\right). \tag{3.145}$$

The sum of J_1 and J_2 yields the desired result. □

Remark 3.6. *If $\tau = b$ and/or $\tau = a$ we obtain the same result, however, equipped with a factor $\frac{1}{2}$.*

Corollary 3.3. *Suppose that $F \in C^{(1)}([a,b])$ and $H \in C^{(3)}([a,b])$. If there is a stationary point $\tau \in (a,b)$ such that $H'(\tau) = 0$ and $H''(\tau) < 0$, then for $r \to \infty$*

$$\int_a^b F(t)e^{irH(t)} \, dt = F(\tau)\sqrt{\frac{-2\pi}{H''(\tau)r}}\, e^{irH(\tau)-i\frac{\pi}{4}} + o\left(\frac{1}{r^{\frac{1}{2}}}\right). \tag{3.146}$$

3.4 Abel–Poisson and Gauß–Weierstraß Limits

In this section some limit relations are collected that are needed as auxiliary tools in the theory of Gauß–Weierstraß and Abel–Poisson transforms. The considerations are well known (see, e.g., C. Müller [1998]); they are listed here for a better understanding of the multi-dimensional integral transforms of Section 7.4.

We start our consideration with an elementary asymptotic relation in the Abel–Poisson summation of infinite integrals.

Lemma 3.17. *Suppose that* $\varphi : r \mapsto \varphi(r)$, $r \geq 0$, *is continuous with*

$$\lim_{r \to \infty} \varphi(r) = 0. \tag{3.147}$$

Then the integral $\int_0^\infty e^{-tr} \varphi(r) \, dr$ *exists for all* $t > 0$, *and we have*

$$\lim_{\substack{t \to 0 \\ t > 0}} \int_0^\infty e^{-tr} \varphi(r) \, dr = \int_0^\infty \varphi(r) \, dr \tag{3.148}$$

provided that the last integral exists in the sense

$$\int_0^\infty \cdots = \lim_{T \to \infty} \int_0^T \cdots . \tag{3.149}$$

Proof. For $t > 0$ we get by partial integration

$$\int_0^\infty e^{-tr} \varphi(r) \, dr = \int_0^\infty \varphi(r) \, dr - t \int_0^\infty e^{-tr} \left(\int_r^\infty \varphi(s) \, ds \right) dr. \tag{3.150}$$

If $\int_0^\infty \varphi(r) \, dr$ exists in the indicated way, then it clearly follows that

$$\lim_{r \to \infty} \int_r^\infty \varphi(s) \, ds = 0. \tag{3.151}$$

Thus, given $\varepsilon > 0$, we are able to find $R(= R(\varepsilon))$ such that

$$\left| \int_r^\infty \varphi(s) \, ds \right| \leq \frac{\varepsilon}{2} \tag{3.152}$$

for all $r \geq R$. This shows us that, for all $t > 0$,

$$\left| \int_0^\infty e^{-tr} \varphi(r) \, dr - \int_0^\infty \varphi(s) \, ds \right| \leq t \int_0^R \left| \int_r^\infty \varphi(s) \, ds \right| dr + \frac{\varepsilon}{2} \tag{3.153}$$

$$\leq tC + \frac{\varepsilon}{2}$$

with

$$C = \int_0^R \left| \int_r^\infty \varphi(s)ds \right| \, dr. \tag{3.154}$$

Thus we arrive at the limit relation

$$\lim_{\substack{t \to 0 \\ t>0}} \int_0^\infty e^{-tr} \varphi(r) \, dr = \int_0^\infty \varphi(r) \, dr, \tag{3.155}$$

as desired. $\qquad\qquad\qquad\qquad\qquad\qquad\qquad\qquad\qquad\qquad\qquad\square$

Remark 3.7. *Note that (3.155) may be used to interpret the integral on the right side by its limit on the left side.*

Gauß–Weierstraß Means

Lemma 3.18. *Suppose that $\varphi : r \mapsto \varphi(r)$, $r \geq 0$ is continuous with*

$$\lim_{r \to \infty} \varphi(r) = 0. \tag{3.156}$$

Then the integral $\int_0^\infty e^{-\pi tr^2} \varphi(r) r^{l-1} \, dr$ exists for all $t > 0$, $l \in \mathbb{N}$, and we have

$$\lim_{\substack{t \to 0 \\ t>0}} \int_0^\infty e^{-\pi tr^2} \varphi(r) r^{l-1} \, dr = \int_0^\infty \varphi(r) r^{l-1} \, dr \tag{3.157}$$

if the integral on the right side exists in the sense $\int_0^\infty \cdots = \lim_{T \to \infty} \int_0^T \cdots$.

Proof. By use of the substitution $r = \sqrt{\frac{s}{\pi}}$ we are able to transform the integral

$$\int_0^\infty e^{-\pi tr^2} \varphi(r) r^{l-1} \, dr \tag{3.158}$$

into

$$\frac{1}{2} \pi^{-\frac{l}{2}} \int_0^\infty e^{-ts} \varphi\left(\sqrt{\frac{s}{\pi}}\right) s^{\frac{l-2}{2}} \, ds. \tag{3.159}$$

Consequently, we obtain from the same arguments leading to Lemma 3.17

$$\lim_{\substack{t \to 0 \\ t>0}} \int_0^\infty e^{-\pi tr^2} \varphi(r) \, r^{l-1} \, dr = \int_0^\infty \varphi(r) \, r^{l-1} \, dr \tag{3.160}$$

$$= \lim_{T \to \infty} \int_0^T \varphi(r) \, r^{l-1} \, dr,$$

provided that the integral on the right side exists. $\qquad\qquad\qquad\square$

The integral of Lemma 3.17, which provides a limit for $t \to 0$, should be discussed for the limit $t \to \infty$, too.

Lemma 3.19. *Suppose that $\psi : r \mapsto \psi(r)$, $r \geq 0$, is continuous such that, for fixed $n \geq 1$,*

$$\psi(r) = a_0 + \ldots + a_n r^n + O\left(r^{n+1}\right) \tag{3.161}$$

for all $r \in [0, 1)$ and

$$\psi(r) = O\left(r^n\right) \tag{3.162}$$

for $r \to \infty$. Then, for $l \geq 2$ and $t \to \infty$, we have

$$\int_0^\infty e^{-tr} \, r^{\frac{l-3}{2}} \, \psi(r) \, dr \tag{3.163}$$

$$= \left(\frac{1}{t}\right)^{\frac{l-1}{2}} \left(b_0 + \ldots + \frac{b_n}{t^n}\right) + O\left(\frac{1}{t^{n+1}}\right),$$

where the coefficients a_k and b_k satisfy the following relationship

$$b_k = \Gamma\left(k + \frac{l-1}{2}\right) a_k, \quad k = 1, \ldots, n. \tag{3.164}$$

Proof. We split the integral (3.163) into two parts

$$\int_0^\infty \ldots = \int_0^1 \ldots + \int_1^\infty \ldots . \tag{3.165}$$

The integrand of the integral $\int_1^\infty \ldots$ admits the estimate

$$\left| e^{-tr} r^{\frac{l-3}{2}} \psi(r) \right| = O\left(e^{-r} \left| e^{-t(r-1)} r^{\frac{l-3}{2}} r^n \right|\right). \tag{3.166}$$

Therefore, $\int_1^\infty \ldots$ decays exponentially with t.

For the integral $\int_0^1 \ldots$ we observe the expansion in terms of the coefficients a_k, where we use, for $k = 0, \ldots, n+1, l \in \mathbb{N}$, the identity

$$\int_0^1 e^{-tr} r^{\frac{l-3}{2}} r^k \, dr = \left(\frac{1}{t}\right)^{k+\frac{l-1}{2}} \int_0^t e^{-s} s^{k+\frac{l-3}{2}} \, ds \tag{3.167}$$

$$= \left(\frac{1}{t}\right)^{\frac{l-1}{2}} \left(\frac{\Gamma\left(k + \frac{l-1}{2}\right)}{t^k} + O\left(\frac{1}{t^{n+1}}\right)\right).$$

This yields the coefficients b_k together with the error estimate stated in Lemma 3.19. $\square$

The limit relation (i.e., Lemma 3.18) and its consequences are used in the Fourier analysis of the Gauß–Weierstraß integral transform (see Subsection 7.4).

Lemma 3.20. *Suppose that $\varphi : r \mapsto \varphi(r)$, $r \geq 0$, is integrable on $[0, \infty)$ and continuous at the origin. Moreover, assume that $\int_0^\infty |\varphi(r)|\, r^{l-1}\, dr$, $l \in \mathbb{N}$, exists. Then*

$$\lim_{\substack{t \to 0 \\ t > 0}} \left(\frac{1}{t}\right)^{\frac{l}{2}} \int_0^\infty e^{-\frac{\pi}{t}r^2} \varphi(r)\, r^{l-1}\, dr = \frac{\varphi(0)}{\|\mathbb{S}^{l-1}\|}. \tag{3.168}$$

Proof. Since φ is assumed to be continuous at the origin, there is a function $\mu : [0, 1] \to \mathbb{R}^+$ with $\lim_{\substack{\tau \to 0 \\ \tau > 0}} \mu(\tau) = 0$ and

$$|\varphi(r) - \varphi(0)| \leq \mu(\tau), \quad 0 \leq r \leq \tau. \tag{3.169}$$

We have

$$\left(\frac{1}{t}\right)^{\frac{l}{2}} \int_0^\infty e^{-\frac{\pi}{t}r^2} r^{l-1}\, dr = \frac{1}{2}\pi^{-\frac{l}{2}} \int_0^\infty e^{-r} r^{\frac{l-3}{2}}\, dr = \frac{1}{\|\mathbb{S}^{l-1}\|}. \tag{3.170}$$

Moreover, for $\delta > 0$, we find

$$\left| \left(\frac{1}{t}\right)^{\frac{l}{2}} \int_\delta^\infty e^{-\frac{\pi}{t}r^2} \varphi(r)\, r^{l-1}\, dr \right| \leq t^{-\frac{l}{2}} e^{-\frac{\pi}{t}\delta^2} \int_0^\infty |\varphi(r)|\, r^{l-1}\, dr. \tag{3.171}$$

We discuss the expressions

$$I_1(t, \delta) = \left(\frac{1}{t}\right)^{\frac{l}{2}} \int_0^\delta e^{-\frac{\pi}{t}r^2} \varphi(r)\, r^{l-1}\, dr, \tag{3.172}$$

$$I_2(t, \delta) = \left(\frac{1}{t}\right)^{\frac{l}{2}} \int_\delta^\infty e^{-\frac{\pi}{t}r^2} \varphi(r)\, r^{l-1}\, dr \tag{3.173}$$

separately. In connection with (3.169) and (3.170) we have

$$\left| I_1(t, \delta) - \varphi(0) \left(\frac{1}{t}\right)^{\frac{l}{2}} \int_0^\delta e^{-\frac{\pi}{t}r^2} r^{l-1}\, dr \right| \leq \frac{\mu(\delta)}{\|\mathbb{S}^{l-1}\|}. \tag{3.174}$$

Furthermore, observing (3.170) we get

$$\left(\frac{1}{t}\right)^{\frac{l}{2}} \int_0^\delta e^{-\frac{\pi}{t}r^2} r^{l-1}\, dr = \frac{1}{2}\pi^{-\frac{l}{2}} \int_0^{\frac{\pi}{t}\delta^2} e^{-r} r^{\frac{l-2}{2}}\, dr. \tag{3.175}$$

We now set $\delta = t^{\frac{1}{3}}$. By passing to the limit $t \to 0$ the integral (3.175) tends to $\|\mathbb{S}^{l-1}\|^{-1}$. From (3.174) it follows that

$$\lim_{\substack{t \to 0 \\ t > 0}} I_1(t, t^{\frac{1}{3}}) = \frac{\varphi(0)}{\|\mathbb{S}^{l-1}\|}. \tag{3.176}$$

From (3.171) and (3.173) we get

$$\lim_{\substack{t \to 0 \\ t > 0}} I_2(t, t^{\frac{1}{3}}) = 0. \tag{3.177}$$

This is the required result. ∎

Abel–Poisson Means

Finally we come to an auxiliary result, which is of importance in the context of the Abel–Poisson integral transform (see Subsection 7.4).

Lemma 3.21. *Under the assumptions of Lemma 3.20*

$$\lim_{\substack{t \to 0 \\ t > 0}} \frac{2}{\|\mathbb{S}^l\|} \int_0^\infty \frac{t\varphi(r)r^{l-1}}{(t^2 + r^2)^{\frac{l+1}{2}}} \, dr = \frac{\varphi(0)}{\|\mathbb{S}^{l-1}\|}. \tag{3.178}$$

Proof. An easy calculation, with $r = st$, shows that

$$\int_0^\infty \frac{tr^{l-1} \, dr}{(t^2 + r^2)^{\frac{l+1}{2}}} = \int_0^\infty \frac{s^{l-1} \, ds}{(1 + s^2)^{\frac{l+1}{2}}} = \frac{1}{2} \int_0^\infty \frac{s^{l-2} d(s^2)}{(1 + s^2)^{\frac{l+1}{2}}}. \tag{3.179}$$

Substituting $v = (1 + s^2)^{-1}$ we obtain

$$\int_0^\infty \frac{tr^{l-1} \, dr}{(t^2 + r^2)^{\frac{l+1}{2}}} \tag{3.180}$$

$$= \frac{1}{2} \int_0^1 (1 - v)^{\frac{l-2}{2}} v^{-\frac{1}{2}} \, dv = \frac{1}{2} \frac{\Gamma(\frac{l}{2})\sqrt{\pi}}{\Gamma\left(\frac{l+1}{2}\right)} = \frac{1}{2} \frac{\|\mathbb{S}^l\|}{\|\mathbb{S}^{l-1}\|}.$$

We separate the integral into two parts, depending on a parameter $\delta > 0$,

$$I_1(t, \delta) = 2 \frac{\|\mathbb{S}^{l-1}\|}{\|\mathbb{S}^l\|} \int_0^\delta \frac{t\varphi(r)r^{l-1}}{(t^2 + r^2)^{\frac{l+1}{2}}} \, dr, \tag{3.181}$$

$$I_2(t, \delta) = 2 \frac{\|\mathbb{S}^{l-1}\|}{\|\mathbb{S}^l\|} \int_\delta^\infty \frac{t\varphi(r)r^{l-1}}{(t^2 + r^2)^{\frac{l+1}{2}}} \, dr. \tag{3.182}$$

First we obtain

$$\left| I_1(t, \delta) - 2\varphi(0) \frac{\|\mathbb{S}^{l-1}\|}{\|\mathbb{S}^l\|} \int_0^\delta \frac{tr^{l-1}}{(t^2 + r^2)^{\frac{l+1}{2}}} \, dr \right| \tag{3.183}$$

$$\leq 2 \frac{\|\mathbb{S}^{l-1}\|}{\|\mathbb{S}^l\|} \int_0^\delta \frac{t|\varphi(r) - \varphi(0)|}{(t^2 + r^2)^{\frac{l+1}{2}}} \, dr \leq \mu(\delta).$$

Second we have

$$|I_2(t, \delta)| \leq 2 \frac{\|\mathbb{S}^{l-1}\|}{\|\mathbb{S}^l\|} \frac{t}{\delta^{l+1}} \int_0^\infty r^{l-1} |\varphi(r)| \, dr. \tag{3.184}$$

The substitution (3.179) supplies the identity

$$\int_0^\delta \frac{tr^{l-1}}{(t^2 + r^2)^{\frac{l+1}{2}}} \, dr = \int_0^{\delta/t} \frac{s^{l-1}}{(1 + s^2)^{\frac{l+1}{2}}} \, ds. \tag{3.185}$$

We set $\delta = t^{\frac{1}{l+2}}$ and get $\frac{\delta}{t} = t^{-\frac{l+1}{l+2}}$, which tends to ∞ as $t \to 0$, while δ and $\frac{t}{\delta^{l+1}} = t^{\frac{1}{l+2}}$ tend to 0 as $t \to 0$. Hence, we obtain with $\delta(t) = t^{\frac{1}{l+2}}$

$$\lim_{\substack{t \to 0 \\ t > 0}} 2 \frac{\|\mathbb{S}^{l-1}\|}{\|\mathbb{S}^l\|} \int_0^\infty \frac{t\varphi(r)r^{l-1}}{(t^2 + r^2)^{\frac{l+1}{2}}}\, dr \qquad (3.186)$$

$$= \lim_{\substack{t \to 0 \\ t > 0}} \Big(I_1(t, \delta(t)) + I_2(t, \delta(t)) \Big) = \varphi(0).$$

This proves our assertion. $\qquad\qquad\qquad\qquad\qquad\qquad\qquad\qquad$ $\square$

4

One-Dimensional Euler and Poisson Summation Formulas

CONTENTS

4.1 Lattice Function 46
    Periodical Polynomials 47
    Eigenspectrum of the Laplace Operator 48
    Eigenfunction Expansions 49
    Properties of the Lattice Function 50
4.2 Euler Summation Formula for the Laplace Operator 54
    Extended Stirling's Formula Involving the Lattice Function 57
    Euler Summation to Periodical Boundary Conditions 59
4.3 Riemann Zeta Function and Lattice Function 63
    Functional Equation of the Zeta Function 63
    Kronecker's Limit Formula 66
    Euler's Product Representation 67
4.4 Poisson Summation Formula for the Laplace Operator 68
    Elementary Examples 73
    Theta Functions 74
4.5 Euler Summation Formula for Helmholtz Operators 76
    Lattice Function for the Helmholtz Operator 77
    Summation Formula for the Helmholtz Operator 80
4.6 Poisson Summation Formula for Helmholtz Operators 82
    Sufficient Criteria 82
    Hardy–Landau Identity 85

Our considerations in this chapter are concerned with different types of summation formulas for the one-dimensional Euclidean space $\mathbb{R}$ based on the work of L.J. Mordell [1928a,b, 1929]. We start with the classical Euler summation formula for the operator of the first and second order derivative, i.e., the gradient and the Laplace operator in one dimension, respectively, and "periodical boundary conditions". Moreover, we recapitulate the close relationship between the classical Euler summation formula and the one-dimensional Riemann Zeta function. Kronecker's formula is mentioned, and the Theta function and its functional equation are presented. In addition, we are interested in the intimate relationship between the Euler summation formula and the Poisson summation formula interconnected by properties of the Laplacian.

The particular goal of this chapter, i.e., the non-standard part of our investigation, is the transition of the one-dimensional Euler summation formula for the one-dimensional Laplace operator to Euler summation formulas for one-dimensional Helmholtz operators such that alternative (sufficient) conditions can be provided for the validity of the Poisson summation formula in one dimension. These conditions turn out to be of tremendous significance for the discussion of one-dimensional alternating sums involving discontinuous integrals such as the one-dimensional counterpart of the Hardy–Landau identity (expressing the number of lattice points inside an interval as series in terms of sinc-functions).

All one-dimensional results are formulated in such a way that their extensions to the multi-variate theory become obvious.

4.1   Lattice Function

Let $\mathbb{Z}$ denote the one-dimensional lattice of integral points, i.e., the additive group of points in $\mathbb{R}$ having integral coordinates (the addition being, of course, the one derived from the vector structure of $\mathbb{R}$).

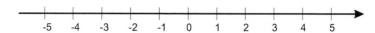

FIGURE 4.1
The integer lattice $\mathbb{Z}$.

The fundamental cell $\mathcal{F}$ of the integer lattice $\mathbb{Z}$ is given by

$$\mathcal{F} = \left\{ x \in \mathbb{R} \,\middle|\, -\frac{1}{2} \le x < \frac{1}{2} \right\};$$ (4.1)

it is a half-open interval.

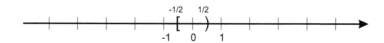

FIGURE 4.2
The fundamental cell $\mathcal{F}$ of the integer lattice $\mathbb{Z}$.

Definition 4.1. *A function $F : \mathbb{R} \to \mathbb{C}$ is called $\mathbb{Z}$-periodical if*

$$F(x + g) = F(x)$$ (4.2)

holds for all $x \in \mathcal{F}$ and $g \in \mathbb{Z}$.

Periodical Polynomials

Example 4.1. *The function* $\Phi_h : \mathbb{R} \to \mathbb{C}$, $h \in \mathbb{Z}$, *given by*

$$x \mapsto \Phi_h(x) = e^{2\pi i h x} \tag{4.3}$$

is $\mathbb{Z}$*-periodical:*

$$
\begin{aligned}
\Phi_h(x + g) &= e^{2\pi i h(x+g)} \\
&= e^{2\pi i h x} e^{2\pi i h g} \\
&= e^{2\pi i h x} \\
&= \Phi_h(x)
\end{aligned} \tag{4.4}
$$

for all $x \in \mathcal{F}$ *and all* $g \in \mathbb{Z}$.

Remark 4.1. *By convention,* $e(hx) = \exp(2\pi i h x) = e^{2\pi i h x}$, $x \in \mathbb{R}$, $h \in \mathbb{Z}$.

The space of all $F \in \mathrm{C}^{(m)}(\mathbb{R})$ that are $\mathbb{Z}$-periodical is denoted by $\mathrm{C}_{\mathbb{Z}}^{(m)}(\mathbb{R})$, $0 \le m \le \infty$. $\mathrm{L}_{\mathbb{Z}}^2(\mathbb{R})$ is the space of all $F : \mathbb{R} \to \mathbb{C}$ that are $\mathbb{Z}$-periodical and are Lebesgue–measurable on $\mathcal{F}$ with

$$\|F\|_{\mathrm{L}_{\mathbb{Z}}^2(\mathbb{R})} = \left(\int_{\mathcal{F}} |F(x)|^2 \, dx \right)^{\frac{1}{2}} < \infty. \tag{4.5}$$

Clearly, the space $\mathrm{L}_{\mathbb{Z}}^2(\mathbb{R})$ is the completion of $\mathrm{C}_{\mathbb{Z}}^{(0)}(\mathbb{R})$ with respect to the norm $\|\cdot\|_{\mathrm{L}_{\mathbb{Z}}^2(\mathbb{R})}$:

$$\mathrm{L}_{\mathbb{Z}}^2(\mathbb{R}) = \overline{\mathrm{C}_{\mathbb{Z}}^{(0)}(\mathbb{R})}^{\|\cdot\|_{\mathrm{L}_{\mathbb{Z}}^2(\mathbb{R})}}. \tag{4.6}$$

An easy calculation shows that the system $\{\Phi_h\}_{h \in \mathbb{Z}}$ is orthonormal with respect to the $\mathrm{L}_{\mathbb{Z}}^2(\mathbb{R})$-inner product

$$
\begin{aligned}
(\Phi_h, \Phi_{h'})_{\mathrm{L}_{\mathbb{Z}}^2(\mathbb{R})} &= \int_{\mathcal{F}} \Phi_h(x) \overline{\Phi_{h'}(x)} \, dx \\
&= \delta_{hh'} = \begin{cases} 1 & , \quad h = h' \\ 0 & , \quad h \neq h'. \end{cases}
\end{aligned} \tag{4.7}
$$

In more detail,

$$
\begin{aligned}
\int_{\mathcal{F}} \Phi_h(x) \overline{\Phi_{h'}(x)} \, dx &= \int_{-\frac{1}{2}}^{\frac{1}{2}} e^{2\pi i h x} e^{-2\pi i h' x} \, dx \\
&= \int_{-\frac{1}{2}}^{\frac{1}{2}} e^{2\pi i (h - h') x} \, dx \\
&= \begin{cases} 1 & , \quad h = h' \\ 0 & , \quad h \neq h'. \end{cases}
\end{aligned} \tag{4.8}
$$

An elementary calculation yields (with $\nabla_x = \frac{d}{dx}$ as the one-dimensional gradient and $\Delta_x = \left(\frac{d}{dx}\right)^2$ as the one-dimensional Laplacian)

$$\nabla_x \Phi_h(x) = \frac{d}{dx}\Phi_h(x) = \frac{d}{dx}e^{2\pi i h x} = 2\pi i h \Phi_h(x) \tag{4.9}$$

such that

$$\Delta_x \Phi_h(x) = \left(\frac{d}{dx}\right)^2 \Phi_h(x) = (2\pi i h)^2 \Phi_h(x) = \underbrace{-4\pi^2 h^2}_{=-\Delta^\wedge(h)} \Phi_h(x), \quad h \in \mathbb{Z},\ x \in \mathbb{R}. \tag{4.10}$$

Eigenspectrum of the Laplace Operator

By convention we say that λ is an *eigenvalue of the lattice $\mathbb{Z}$ with respect to the operator Δ* of the second order derivative (i.e., the one-dimensional Laplace operator), if there is a non-trivial solution U of the differential equation

$$(\Delta + \lambda)U = 0 \tag{4.11}$$

satisfying the "boundary condition" of periodicity

$$U(x + g) = U(x) \tag{4.12}$$

for all $x \in \mathcal{F}$ and $g \in \mathbb{Z}$. From classical Fourier analysis (see, e.g., R. Courant, D. Hilbert [1924]) we know that the operator Δ has a half-bounded and discrete eigenspectrum $\{\Delta^\wedge(h)\}_{h \in \mathbb{Z}} \subset \mathbb{R}$ such that

$$(\Delta_x + \Delta^\wedge(h))\, \Phi_h(x) = 0, \quad x \in \mathcal{F}, \tag{4.13}$$

with eigenvalues $\Delta^\wedge(h)$ given by

$$\Delta^\wedge(h) = 4\pi^2 h^2, \quad h \in \mathbb{Z}, \tag{4.14}$$

and eigenfunctions

$$\Phi_h(x) = e^{2\pi i h x}, \quad h \in \mathbb{Z}, x \in \mathcal{F}. \tag{4.15}$$

Remark 4.2. *We consistently write $\Delta^\wedge(h)$ instead of $\Delta_{\mathbb{Z}}^\wedge(h)$, $h \in \mathbb{Z}$, if no confusion is likely to arise.*

Consequently, the *eigenspectrum of the operator Δ* (with respect to $\mathbb{Z}$) is given by

$$\mathrm{Spect}_\Delta(\mathbb{Z}) = \{\Delta^\wedge(h) \mid \Delta^\wedge(h) = 4\pi^2 h^2,\ h \in \mathbb{Z}\}. \tag{4.16}$$

The orthonormal system $\{\Phi_h\}_{h \in \mathbb{Z}}$ of (eigen)functions $\Phi_h : x \mapsto \Phi_h(x) =$

$e^{2\pi i h x}$, $x \in \mathbb{R}$, is closed in the space $C_{\mathbb{Z}}^{(0)}(\mathbb{R})$; i.e., for every $\varepsilon > 0$ and every $F \in C_{\mathbb{Z}}^{(0)}(\mathbb{R})$ there exist an integer $N(= N(\varepsilon))$ and a linear combination

$$\sum_{\substack{|h| \leq N \\ h \in \mathbb{Z}}} a_h \Phi_h \tag{4.17}$$

such that

$$\sup_{x \in \mathcal{F}} \left| F(x) - \sum_{\substack{|h| \leq N \\ h \in \mathbb{Z}}} a_h \Phi_h(x) \right| \leq \varepsilon. \tag{4.18}$$

By virtue of the norm estimate

$$\|F\|_{L_{\mathbb{Z}}^2(\mathbb{R})} = \left(\int_{\mathcal{F}} |F(x)|^2 \, dx \right)^{\frac{1}{2}} \leq \sup_{x \in \mathcal{F}} |F(x)| = \|F\|_{C_{\mathbb{Z}}^{(0)}(\mathbb{R})}, \quad F \in C_{\mathbb{Z}}^{(0)}(\mathbb{R}), \tag{4.19}$$

the closure of the system $\{\Phi_h\}_{h \in \mathbb{Z}}$ in $\left(C_{\mathbb{Z}}^{(0)}(\mathbb{R}), \| \cdot \|_{C_{\mathbb{Z}}^{(0)}(\mathbb{R})} \right)$ implies the closure in $\left(C_{\mathbb{Z}}^{(0)}(\mathbb{R}), \| \cdot \|_{L_{\mathbb{Z}}^2(\mathbb{R})} \right)$.

Eigenfunction Expansions

Since $C_{\mathbb{Z}}^{(0)}(\mathbb{R})$ is dense in $L_{\mathbb{Z}}^2(\mathbb{R})$ with respect to the norm $\| \cdot \|_{L_{\mathbb{Z}}^2(\mathbb{R})}$, the validity of the following equivalences is finally implied.

Theorem 4.1. *The following statements are equivalent:*

(i) The system $\{\Phi_h\}_{h \in \mathbb{Z}}$ is closed in $L_{\mathbb{Z}}^2(\mathbb{R})$; i.e., for every $\varepsilon > 0$ and every $F \in L_{\mathbb{Z}}^2(\mathbb{R})$ there exist an index $N(= N(\varepsilon))$ and coefficients $a_h \in \mathbb{C}$ such that

$$\left(\int_{\mathcal{F}} \left| F(x) - \sum_{\substack{|h| \leq N \\ h \in \mathbb{Z}}} a_h \Phi_h(x) \right|^2 \, dx \right)^{\frac{1}{2}} \leq \varepsilon. \tag{4.20}$$

(ii) The Fourier series of $F \in L_{\mathbb{Z}}^2(\mathbb{R})$

$$\sum_{h \in \mathbb{Z}} F_{\mathbb{Z}}^{\wedge}(h) \Phi_h(x) \tag{4.21}$$

with the "Fourier coefficients"

$$F_{\mathbb{Z}}^{\wedge}(h) = (F, \Phi_h)_{L_{\mathbb{Z}}^2(\mathbb{R})} = \int_{\mathcal{F}} F(x) \overline{\Phi_h(x)} \, dx, \ h \in \mathbb{Z}, \tag{4.22}$$

converges in the $||\cdot||_{L^2_{\mathbb{Z}}(\mathbb{R})}$ *-norm:*

$$\lim_{N\to\infty}\left(\int_{\mathcal{F}}\left|F(x)-\sum_{\substack{|h|\leq N \\ h\in\mathbb{Z}}}F^{\wedge}_{\mathbb{Z}}(h)\Phi_h(x)\right|^2 dx\right)^{\frac{1}{2}}=0. \qquad (4.23)$$

(iii) Parseval's identity holds. That is, for any $F\in L^2_{\mathbb{Z}}(\mathbb{R})$,

$$||F||^2_{L^2_{\mathbb{Z}}(\mathbb{R})}=\sum_{h\in\mathbb{Z}}|F^{\wedge}_{\mathbb{Z}}(h)|^2. \qquad (4.24)$$

(iv) The extended Parseval identity holds. That is, for any $F,H\in L^2_{\mathbb{Z}}(\mathbb{R})$,

$$(F,H)_{L^2_{\mathbb{Z}}(\mathbb{R})}=\sum_{h\in\mathbb{Z}}F^{\wedge}_{\mathbb{Z}}(h)\overline{H^{\wedge}_{\mathbb{Z}}(h)}. \qquad (4.25)$$

(v) There is no strictly larger orthonormal system containing the orthonormal system $\{\Phi_h\}_{h\in\mathbb{Z}}$.

(vi) The system $\{\Phi_h\}_{h\in\mathbb{Z}}$ *has the completeness property. That is,* $F\in L^2_{\mathbb{Z}}(\mathbb{R})$ *and* $F^{\wedge}_{\mathbb{Z}}(h)=0$ *for all* $h\in\mathbb{Z}$ *implies* $F=0$.

(vii) An element $F\in L^2_{\mathbb{Z}}(\mathbb{R})$ *is uniquely determined by its orthogonal coefficients. That is, if* $F^{\wedge}_{\mathbb{Z}}(h)=H^{\wedge}_{\mathbb{Z}}(h)$ *for all* $h\in\mathbb{Z}$, *then* $F=H$.

For more details on the system $\{\Phi_h\}_{h\in\mathbb{Z}}$ the reader is referred to monographs on the Fourier theory in Euclidean spaces, for example, P.L. Butzer, R. Nessel [1971], and E.M. Stein, G. Weiss [1971]. The proof of Theorem 4.1 can be found in every textbook on constructive approximation (for example, P.J. Davis [1963]).

After these preliminaries on $\mathbb{Z}$-periodical functions (i.e., functions with period 1) we now come to the definition of the $\mathbb{Z}$-lattice function with respect to the one-dimensional Laplacian $\Delta=\nabla^2$, $\nabla=\frac{d}{dx}$, i.e., Green's function with respect to Δ corresponding to $\mathbb{Z}$-periodical "boundary conditions". Based on the constituting properties of this function the Euler summation formula can be developed by integration by parts. Finally, we derive some variants of the Poisson summation formula.

Properties of the Lattice Function

We start with the definition of the $\mathbb{Z}$-lattice function for the Laplacian (see the remarks in our Introduction).

Definition 4.2. *A function $G(\Delta; \cdot) : \mathbb{R} \to \mathbb{R}$ is called the Green function for the operator Δ with respect to the $\mathbb{Z}$-periodicity (in brief, $\mathbb{Z}$-lattice function for Δ) if it satisfies the following properties:*

(i) (Periodicity) $G(\Delta; \cdot)$ is continuous in $\mathbb{R}$, and

$$G(\Delta; x + g) = G(\Delta; x) \tag{4.26}$$

for all $x \in \mathbb{R}$ and $g \in \mathbb{Z}$.

(ii) (Differential equation) G is twice continuously differentiable with

$$\Delta G(\Delta; x) = -1 \tag{4.27}$$

for all $x \notin \mathbb{Z}$.

(iii) (Characteristic singularity)

$$x \mapsto G(\Delta; x) - \frac{1}{2} x \, \mathrm{sign}(x) \tag{4.28}$$

is continuously differentiable for all $x \in \mathcal{F}$.

(iv) (Normalization)

$$\int_{\mathcal{F}} G(\Delta; x) \, dx = 0. \tag{4.29}$$

First we prove that the $\mathbb{Z}$-*lattice function for the operator of the second order derivative*, i.e., for the one-dimensional Laplacian, is uniquely determined by its constituting properties.

Lemma 4.1. *$G(\Delta; \cdot)$ is uniquely determined by the properties (i)–(iv).*

Proof. Denote by $D(\Delta; \cdot)$ the difference between two $\mathbb{Z}$-lattice functions for Δ. Then we have the following properties:

(i) D is continuous in $\mathbb{R}$, and for all $x \in \mathbb{R}$ and $g \in \mathbb{Z}$

$$D(\Delta; x + g) = D(\Delta; x). \tag{4.30}$$

(ii) $D(\Delta; \cdot)$ is twice continuously differentiable for all $x \notin \mathbb{Z}$ with

$$\Delta D(\Delta; \cdot) = 0. \tag{4.31}$$

(iii) $D(\Delta; \cdot)$ is continuously differentiable in $\mathbb{R}$.

(iv)

$$\int_{\mathcal{F}} D(\Delta; x) \, dx = 0. \tag{4.32}$$

The properties *(i)–(iii)* show that $D(\Delta; \cdot)$ is a constant function. The last condition *(iv)* shows us that the constant must be zero. Thus, G is uniquely determined.   □

It is not hard to see (see, e.g., W. Magnus et al. [1966]) that $G(\Delta; \cdot)$ is explicitly available in elementary form. In fact, the function (cf.(1.4))

$$x \mapsto G(\Delta; x) = -\frac{(x - \lfloor x \rfloor)^2}{2} + \frac{x - \lfloor x \rfloor}{2} - \frac{1}{12}, \quad x \in \mathbb{R} \qquad (4.33)$$

satisfies all defining properties *(i)–(iv)* of the $\mathbb{Z}$-lattice function for the one-dimensional Laplace operator Δ. Apart from a multiplicative constant, $G(\Delta; \cdot)$ therefore coincides with the Bernoulli function B_2 of degree 2 given by

$$x \mapsto B_2(x) = (x - \lfloor x \rfloor)^2 - (x - \lfloor x \rfloor) + \frac{1}{6}. \qquad (4.34)$$

Remark 4.3. $\lfloor x \rfloor$ *means that integer in* $\mathbb{Z}$ *for which* $\lfloor x \rfloor \leq x < \lfloor x \rfloor + 1$.

Theorem 4.2. *The $\mathbb{Z}$-lattice function $G(\Delta; \cdot)$ for Δ possesses the explicit representation*

$$G(\Delta; x) = -\frac{(x - \lfloor x \rfloor)^2}{2} + \frac{x - \lfloor x \rfloor}{2} - \frac{1}{12}, \quad x \in \mathbb{R}. \qquad (4.35)$$

Observing the $\mathbb{Z}$-periodicity *(i)* and the characteristic singularity *(iii)* of the $\mathbb{Z}$-lattice function for Δ we obtain by applying integration by parts

$$(G(\Delta; \cdot), \Phi_h)_{\mathrm{L}^2_{\mathbb{Z}}(\mathbb{R})} = \int_{\mathcal{F}} G(\Delta; x) \overline{\Phi_h(x)} \, dx = \frac{1}{-4\pi^2 h^2} = \frac{1}{-\Delta^\wedge(h)} \qquad (4.36)$$

provided that $h \neq 0$. Thus, the classical representation theorem of one-dimensional Fourier theory gives us the Fourier series representation

$$G(\Delta; x) = \sum_{\substack{h \neq 0 \\ h \in \mathbb{Z}}} \frac{1}{-\Delta^\wedge(h)} \Phi_h(x), \quad x \in \mathbb{R}. \qquad (4.37)$$

Lemma 4.2. *For all* $x \in \mathbb{R}$

$$G(\Delta; x) = \sum_{\substack{h \neq 0 \\ h \in \mathbb{Z}}} \frac{1}{-\Delta^\wedge(h)} \Phi_h(x) = \sum_{\substack{h \neq 0 \\ h \in \mathbb{Z}}} \frac{1}{-\Delta^\wedge(h)} \overline{\Phi_h(x)}, \qquad (4.38)$$

and the series on the right side is absolutely and uniformly convergent on each compact interval $I \subset \mathbb{R}$.

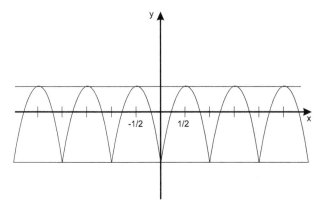

FIGURE 4.3
The illustration of the $\mathbb{Z}$-lattice function $G(\Delta; \cdot)$ for Δ.

Figure 4.3 gives an illustration of the $\mathbb{Z}$-lattice function for Δ. Obviously, for all $x \in \mathbb{R}$ we have

$$G(\Delta; x) = \sum_{n=1}^{\infty} \frac{1}{-4\pi^2 n^2} \left(e^{2\pi i n x} + e^{-2\pi i n x} \right) \tag{4.39}$$

$$= -\frac{1}{2\pi^2} \sum_{n=1}^{\infty} \frac{1}{n^2} \cos(2\pi n x).$$

Moreover, the following property of $G(\Delta; \cdot)$ should be noted: $G(\Delta; \cdot)$ is a piecewise polynomial of degree 2, and we have

$$G(\Delta; g) = -\frac{1}{12}, \quad g \in \mathbb{Z}. \tag{4.40}$$

Since the Fourier series of $G(\Delta; \cdot)$ converges absolutely and uniformly for all $x \in \mathbb{R}$, elementary differentiation yields

$$\nabla_x G(\Delta; x) = G(\nabla; x) = -(x - \lfloor x \rfloor) + \frac{1}{2}, \quad x \in \mathbb{R} \setminus \mathbb{Z}. \tag{4.41}$$

Remark 4.4. *The function* $-G(\nabla; \cdot)$ *is called the Bernoulli function* B_1 *of degree 1 given by* $x \mapsto B_1(x) = x - \lfloor x \rfloor - \frac{1}{2}, \quad x \in \mathbb{R}.$

The Fourier series of $G(\nabla; \cdot)$ reads as follows

$$G(\nabla; x) = \sum_{\substack{h \neq 0 \\ h \in \mathbb{Z}}} \frac{1}{2\pi i h} \Phi_h(x), \tag{4.42}$$

where the equality is understood in the $L_{\mathbb{Z}}^2(\mathbb{R})$-sense. Using the well known representation of the sin-function we find

$$\sum_{\substack{h\neq 0 \\ h\in\mathbb{Z}}} \frac{1}{2\pi i h}\Phi_h(x) = \sum_{n=1}^{\infty}\frac{1}{2i}\left(\frac{e^{2\pi i n x}}{\pi n} - \frac{e^{-2\pi i n x}}{\pi n}\right) = \sum_{n=1}^{\infty}\frac{\sin(2\pi n x)}{\pi n}. \qquad (4.43)$$

We mention the following result concerning the convergence of the series (4.43) and its explicit representation (see Figure 4.4).

Lemma 4.3. *The series*

$$\sum_{n=1}^{\infty}\frac{\sin(2\pi n x)}{\pi n} \qquad (4.44)$$

converges uniformly in each compact interval $I \subset (g, g+1)$, $g \in \mathbb{Z}$.
Moreover, for $x \in I$ *with* I *a compact subset of* $(g, g+1)$, $g \in \mathbb{Z}$, *we have*

$$\underbrace{-(x - \lfloor x\rfloor - \frac{1}{2})}_{=B_1(x)} = \frac{d}{dx}\left(-\sum_{n=1}^{\infty}\frac{\cos(2\pi n x)}{2\pi^2 n^2}\right) = \sum_{n=1}^{\infty}\frac{\sin(2\pi n x)}{\pi n}. \qquad (4.45)$$

The proof is well known in the literature (see, e.g., H. Rademacher [1973]).

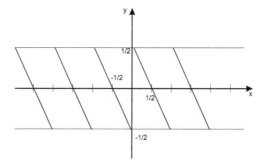

FIGURE 4.4
The derivative $G(\nabla; x) = \nabla G(\Delta; x)$ of the $\mathbb{Z}$-lattice function $G(\Delta; x)$.

4.2   Euler Summation Formula for the Laplace Operator

First we want to verify the *Euler summation formula* in its classical form related to the one-dimensional operator of the first order derivative, i.e., the gradient ∇, and the one-dimensional operator of the second order derivative, i.e., the Laplace operator Δ, respectively. A particular role is played by the $\mathbb{Z}$-lattice function for Δ as introduced by Definition 4.2.

Theorem 4.3. *If* $F \in C^{(1)}([0,n]), n \in \mathbb{N}$, *then*

$$\sum_{\substack{g \in [0,n] \\ g \in \mathbb{Z}}} F(g) = \int_0^n F(x)dx + \frac{1}{2}\big(F(n) + F(0)\big) \tag{4.46}$$

$$- \int_0^n G(\nabla; x) \, \nabla F(x) \, dx.$$

If $F \in C^{(2)}([0,n]), n \in \mathbb{N}$, *then*

$$\sum_{\substack{g \in [0,n] \\ g \in \mathbb{Z}}} F(g) = \int_0^n F(x) \, dx + \frac{1}{2}\big(F(n) + F(0)\big) \tag{4.47}$$

$$+ \frac{1}{12}\big((\nabla F)(n) - (\nabla F)(0)\big)$$

$$+ \int_0^n G(\Delta; x) \, \Delta F(x) \, dx.$$

Let $F : [a,b] \to \mathbb{C}$, $a < b$, *be a twice continuously differentiable function, i.e.,* $F \in C^{(2)}([a,b])$. *Then*

$$\sideset{}{'}\sum_{\substack{g \in [a,b] \\ g \in \mathbb{Z}}} F(g) = \int_a^b F(x)dx + \int_a^b G(\Delta; x) \, \Delta F(x) \, dx \tag{4.48}$$

$$+ \{F(x)(\nabla G(\Delta; x)) - G(\Delta; x) \, (\nabla F(x))\}\big|_a^b,$$

where

$$\sideset{}{'}\sum_{\substack{g \in [a,b] \\ g \in \mathbb{Z}}} F(g) = \sum_{\substack{g \in (a,b) \\ g \in \mathbb{Z}}} F(g) + \frac{1}{2}\sum_{\substack{g=a \\ g=b \\ g \in \mathbb{Z}}} F(g), \tag{4.49}$$

and the last sum in (4.49) occurs only if a and/or b are lattice points. Otherwise it is assumed to be zero.

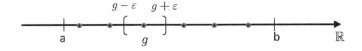

FIGURE 4.5
The illustration of the integration interval in Euler summation.

Proof. We restrict ourselves to the proof of the formula (4.48).

First we are concerned with the case that both endpoints a, b are non-integers (cf. Figure 4.5). By partial integration we get for every (sufficiently small) $\varepsilon > 0$

$$\int_{\substack{x \in [a,b] \\ x \notin \bigcup_{g \in \mathbb{Z}} \mathbb{B}^1_\varepsilon(g)}} \{G(\Delta; x) \Delta F(x) - F(x) \Delta G(\Delta; x)\} \; dx \qquad (4.50)$$

$$= \{G(\Delta; x) \left(\nabla F(x)\right) - F(x) \left(\nabla G(\Delta; x)\right)\} \Big|_a^b$$

$$+ \sum_{\substack{g \in (a,b) \\ g \in \mathbb{Z}}} \{G(\Delta; x) \left(\nabla F(x)\right) - F(x) \left(\nabla G(\Delta; x)\right)\} \Big|_{g+\varepsilon}^{g-\varepsilon} .$$

By virtue of the differential equation $\Delta G(\Delta; x) = -1$, $x \in \mathbb{R} \setminus \mathbb{Z}$, it follows that

$$\int_{\substack{x \in [a,b] \\ x \notin \bigcup_{g \in \mathbb{Z}} \mathbb{B}^1_\varepsilon(g)}} F(x) \, \Delta G(\Delta; x) \; dx = - \int_{\substack{x \in [a,b] \\ x \notin \bigcup_{g \in \mathbb{Z}} \mathbb{B}^1_\varepsilon(g)}} F(x) \; dx. \qquad (4.51)$$

By letting $\varepsilon \to 0$ and observing the (limit) values of the $\mathbb{Z}$-lattice function for Δ and its derivatives in the lattice points, we obtain

$$\sum_{\substack{g \in (a,b) \\ g \in \mathbb{Z}}} F(g) = \int_a^b F(x) \; dx + \int_a^b G(\Delta; x) \, \Delta F(x) \; dx \qquad (4.52)$$

$$+ \{F(x) \left(\nabla G(\Delta; x)\right) - G(\Delta; x) \left(\nabla F(x)\right)\} \Big|_a^b.$$

Second, the cases where a and/or b are integers, i.e., members of the lattice $\mathbb{Z}$, follow by obvious modifications, where $\{G(\nabla; x) F(x)\}\big|_a^b$ is understood in the usual sense

$$G(\nabla; b - 0) F(b) - G(\nabla; a + 0) F(a). \qquad (4.53)$$

This is the desired result.

$\square$

Lemma 4.4. *For $a, b \in \mathbb{R}$, $a < b$,*

$$\sideset{}{'}\sum_{\substack{g \in [a,b] \\ g \in \mathbb{Z}}} 1 = (b - a) + \{G(\nabla; x)\}\big|_a^b . \qquad (4.54)$$

In other words, the total number of integer points in the interval $[a, b]$ can be compared with the length $b - a$ of the interval under explicit knowledge of the remainder term, i.e., the *"lattice point discrepancy"*

$$G(\nabla; x)\big|_a^b = \nabla G(\Delta; x)\big|_a^b. \qquad (4.55)$$

Extended Stirling's Formula Involving the Lattice Function

Next we come to a generalization of Stirling's formula by use of Euler summation: Let F be of class $C^{(1)}([0, n])$. Then we know from Theorem 4.3 that

$$\sum_{k=0}^{n} F(k) - \frac{1}{2}(F(0) + F(n)) = \int_0^n F(x)\,dx - \int_0^n (\nabla G(\Delta; x))\,(\nabla F(x))\,dx,$$

$$(4.56)$$

i.e., the Euler summation formula on the interval $[0, n]$ with respect to the operator $\nabla_x = \left(\frac{d}{dx}\right)$. In particular, with $F(x) = (1+x)^{-1}, x \in [0, n]$, we get

$$\sum_{k=0}^{n}(1 + k)^{-1} = \frac{1}{2(n+1)} + \frac{1}{2} + \ln(n+1) + \int_0^n G(\nabla; x)\,(1 + x)^{-2}\,dx. \quad (4.57)$$

An immediate consequence is the following representation of Euler's constant in terms of the function $G(\nabla; \cdot)$

$$C = \lim_{n\to\infty} \left(\sum_{k=1}^{n} \frac{1}{k} - \ln(n) \right) = \frac{1}{2} + \int_0^\infty G(\nabla; x)\,(1 + x)^{-2}\,dx. \quad (4.58)$$

For $x > 0$ we know from Lemma 3.11 that

$$\ln\left(\frac{1}{\Gamma(x)}\right) = \lim_{n\to\infty} \left(\ln\left(n^{-x} x \prod_{k=1}^{n} \left(\frac{k + x}{k}\right) \right) \right). \quad (4.59)$$

This is equivalent to

$$\ln(\Gamma(x)) = \lim_{n\to\infty} \left(x\ln(n) + \sum_{k=0}^{n-1} \ln(1 + k) - \sum_{k=0}^{n} \ln(x + k) \right). \quad (4.60)$$

Now, again with the above variant (4.56) of the Euler summation formula, we obtain

$$\sum_{k=0}^{n} \ln(x + k) \;=\; \int_0^n \ln(x + t)\,dt + \frac{1}{2}\,(\ln(x + n) + \ln(x))$$

$$- \int_0^n G(\nabla; t)(x + t)^{-1}\,dt. \quad (4.61)$$

Thus, elementary calculations yield for $x > 0$

$$\ln(\Gamma(x)) \;=\; 1 - \int_0^\infty G(\nabla; t)(1 + t)^{-1}\,dt - x + \left(x - \frac{1}{2} \right) \ln(x)$$

$$+ \int_0^\infty G(\nabla; t)(x + t)^{-1}\,dt. \quad (4.62)$$

The integral $\int_0^\infty G(\nabla; t)(x+t)^{-1}\, dt$ exists for $x > 0$. Even more, for $x > 0$, we get

$$\frac{d}{dx}\ln\left(\Gamma(x)\right) = \frac{\Gamma'(x)}{\Gamma(x)} = \ln(x) - \frac{1}{2x} - \int_0^\infty G(\nabla; t)\,(x+t)^{-2}\, dt. \quad (4.63)$$

In connection with (4.58) and Lemma 3.4 we therefore obtain

Lemma 4.5. *(Euler's constant)*

$$C = -\Gamma'(1) = -\int_0^\infty e^{-t}\left(\ln(t)\right)\, dt. \quad (4.64)$$

Furthermore, we have

$$\int_0^\infty G\left(\nabla; t\right)(x+t)^{-1}\, dt = O\left(\frac{1}{x}\right) \quad (4.65)$$

for $x \to \infty$. This leads us to the formula

$$\ln(\Gamma(x)) = 1 - \int_0^\infty G(\nabla; t)(1+t)^{-1}\, dt - x + \left(x - \frac{1}{2}\right)\ln(x)$$
$$+ O\left(\frac{1}{x}\right). \quad (4.66)$$

From Stirling's formula (Theorem 3.2) we get

$$\ln(\Gamma(x)) = \ln(\sqrt{2\pi}) + \left(x - \frac{1}{2}\right)\ln(x) - x + O\left(\frac{1}{x}\right). \quad (4.67)$$

Thus, by combination we find

$$1 - \int_0^\infty G(\nabla; t)(1+t)^{-1}\, dt = \ln(\sqrt{2\pi}). \quad (4.68)$$

All in all, we have

$$e^{\ln(\Gamma(x))} = e^{\ln(\sqrt{2\pi})} e^{-x} e^{\ln(x^{x-\frac{1}{2}})} e^{\int_0^\infty G(\nabla; t)(x+t)^{-1}\, dt}. \quad (4.69)$$

Consequently we get the following *extension of Stirling's formula* involving the derivative $G(\nabla; \cdot)$ of the lattice function $G(\Delta; \cdot)$.

Theorem 4.4. *For $x > 0$,*

$$\Gamma(x) = \sqrt{2\pi}\, x^{x-\frac{1}{2}}\, e^{-x}\, e^{\int_0^\infty G(\nabla; t)(x+t)^{-1}\, dt}. \quad (4.70)$$

Euler Summation to Periodical Boundary Conditions

In what follows we want to present an extension of the Euler summation formula. This can be achieved by replacing the lattice $\mathbb{Z}$ by a "translated lattice" $\mathbb{Z} + \{x\}$ based at $x \in \mathbb{R}$.

Corollary 4.1. *Let x be a point of $\mathbb{R}$. Suppose that F is of class $\mathrm{C}^{(2)}([a, b])$. Then*

$$\sum_{\substack{g+x \in [a,b] \\ g \in \mathbb{Z}}}{}' F(g + x) = \int_a^b F(y) \, dy \tag{4.71}$$

$$+ \int_a^b G(\Delta; x - y) \, \Delta_y F(y) \, dy$$
$$+ \{ F(y)(\nabla_y G(\Delta; x - y)) - (\nabla_y F(y)) G(\Delta; x - y)\} \big|_a^b.$$

An easy consequence is the following corollary (note that the operators $\nabla_y = \frac{d}{dy}$ and $\Delta_y = \nabla_y^2 = (\frac{d}{dy})^2$ apply to the y-variable).

Corollary 4.2. *Let x be a point of the fundamental cell $\mathcal{F}$ of the lattice $\mathbb{Z}$. Suppose that F is of class $\mathrm{C}^{(2)}(\mathcal{F})$. Then*

$$F(x) = \int_{\mathcal{F}} F(y) \, dy \tag{4.72}$$

$$+ \int_{\mathcal{F}} G(\Delta; x - y) \, \Delta_y F(y) \, dy$$
$$+ \{ F(y) \, (\nabla_y G(\Delta; x - y)) - (\nabla_y F(y)) G(\Delta; x - y)\} \big|_{-\frac{1}{2}}^{\frac{1}{2}}.$$

Next we mention the following result involving $\mathbb{Z}$-periodical functions.

Corollary 4.3. *Assume that $x \in \mathcal{F}$ and $F \in \mathrm{C}_{\mathbb{Z}}^{(2)}(\mathbb{R})$. Then*

$$F(x) = \int_{\mathcal{F}} F(y) \, dy + \int_{\mathcal{F}} G(\Delta; x - y) \, \Delta_y F(y) \, dy.$$

It follows from Corollary 4.3 that

$$F(x) = \int_{\mathcal{F}} F(y) \, dy + \int_{\mathcal{F}} (\Delta_y G(\Delta^2; x - y)) \, (\Delta_y F(y)) \, dy \tag{4.73}$$

can be used for both best approximate integration of $\mathbb{Z}$-periodical functions and periodical spline interpolation (see W. Freeden, R. Reuter [1981], W. Freeden, J. Fleck [1987], W. Freeden [1988], O. Schulte [2009], and the references therein).

By the way, Corollary 4.3 leads back to the (well known) solutions of the following differential equations corresponding to "periodical boundary conditions".

Corollary 4.4. *Assume that F is of class $C_{\mathbb{Z}}^{(2)}(\mathbb{R})$ satisfying $\Delta F(y) = 0$, $y \in \mathcal{F}$. Then F is a constant function. More explicitly,*

$$F(x) = \int_{\mathcal{F}} F(y) \, dy, \quad x \in \mathcal{F}. \tag{4.74}$$

Corollary 4.5. *Assume that H is of class $C_{\mathbb{Z}}^{(0)}(\mathbb{R})$ such that*

$$\int_{\mathcal{F}} H(y) \, dy = 0. \tag{4.75}$$

Let $F \in C_{\mathbb{Z}}^{(2)}(\mathbb{R})$ satisfy $\Delta F(y) = H(y)$, $y \in \mathcal{F}$, such that

$$\int_{\mathcal{F}} F(y) \, dy = 0. \tag{4.76}$$

Then

$$F(x) = \int_{\mathcal{F}} G(\Delta; x - y) H(y) \, dy, \quad x \in \mathcal{F}. \tag{4.77}$$

By virtue of the Cauchy–Schwarz inequality we get from Corollary 4.3

$$\left| F(x) - \int_{\mathcal{F}} F(y) \, dy \right| \le \left(G\left(\Delta^2; 0\right) \right)^{\frac{1}{2}} \left(\int_{\mathcal{F}} |\Delta F(y)|^2 \, dy \right)^{\frac{1}{2}}, \tag{4.78}$$

where $G(\Delta^2; \cdot) : x \mapsto G(\Delta^2; x)$, $x \in \mathbb{R}$, is defined by the convolution

$$G\left(\Delta^2; x - y\right) = \int_{\mathcal{F}} G(\Delta; x - z) G(\Delta; z - y) \, dz. \tag{4.79}$$

Obviously, the bilinear series reads as follows

$$G(\Delta^2; x - y) = \sum_{\substack{h \ne 0 \\ h \in \mathbb{Z}}} \frac{1}{-(\Delta^2)^\wedge(h)} \Phi_h(x - y) \tag{4.80}$$

$$= \sum_{\substack{k \ne 0 \\ k \in \mathbb{Z}}} \sum_{\substack{h \ne 0 \\ h \in \mathbb{Z}}} \frac{\Phi_h(x)}{-\Delta^\wedge(h)} \frac{\overline{\Phi_k(y)}}{-\Delta^\wedge(k)} \int_{\mathcal{F}} \overline{\Phi_h(z)} \Phi_k(z) \, dz$$

$$= \sum_{\substack{k \ne 0 \\ k \in \mathbb{Z}}} \sum_{\substack{h \ne 0 \\ h \in \mathbb{Z}}} \frac{\Phi_h(x)}{-\Delta^\wedge(h)} \frac{\overline{\Phi_k(y)}}{-\Delta^\wedge(k)} \delta_{hk}$$

$$= \sum_{\substack{h \ne 0 \\ h \in \mathbb{Z}}} \frac{1}{(-\Delta^\wedge(h))^2} \Phi_h(x - y).$$

Moreover, it follows that

$$\Delta_y G\left(\Delta^2; x - y\right) = G(\Delta; x - y). \tag{4.81}$$

In other words, $G(\Delta^2; \cdot)$ is the lattice function to the iterated Laplace operator $\Delta^2 = \Delta\Delta$.

Next we mention the explicit representation of the lattice function $G(\Delta^2; \cdot)$ for $\Delta^2 \colon x \mapsto G(\Delta^2; x)$, $x \in \mathbb{R}$. To this end, we observe that

$$G\left(\Delta^2; x\right) = \frac{1}{8\pi^4} \sum_{h=1}^{\infty} \frac{1}{h^4} \cos(2\pi h x), \tag{4.82}$$

such that (cf. (4.78))

$$G(\Delta^2; 0) = \frac{1}{720}. \tag{4.83}$$

Lemma 4.6. *The function $G\left(\Delta^2; \cdot\right)$ with respect to the operator $\Delta^2 = \nabla^4$ is $\mathbb{Z}$-periodical, twice continuously differentiable in $\mathbb{R}$ such that*

$$\Delta G\left(\Delta^2; x\right) = G(\Delta; x), \quad x \in \mathbb{R}, \tag{4.84}$$

and we have

$$G\left(\Delta^2; x\right) = -\frac{1}{24}(x - \lfloor x \rfloor)^4 + \frac{1}{12}(x - \lfloor x \rfloor)^3 - \frac{1}{24}(x - \lfloor x \rfloor)^2 + \frac{1}{720}. \tag{4.85}$$

In analogy to (4.79) we define the $\mathbb{Z}$-lattice functions $G(\Delta^l; \cdot)$, $l \in \mathbb{N}$, by

$$G\left(\Delta^{l+1}; x - y\right) = \int_{\mathcal{F}} G\left(\Delta^l; x - z\right) G(\Delta; z - y)\, dz, \quad x, y \in \mathbb{R}. \tag{4.86}$$

Clearly, we have

$$\Delta G\left(\Delta^{l+1}; x\right) = G(\Delta^l; x), \quad x \in \mathbb{R}, \tag{4.87}$$

so that the l-iterated $\mathbb{Z}$-lattice function is nothing else than the $\mathbb{Z}$-lattice function to the iterated operator Δ^l

$$G\left(\Delta^l; x\right) = \sum_{\substack{h \neq 0 \\ h \in \mathbb{Z}}} \frac{1}{-(\Delta^l)^{\wedge}(h)}\, \Phi_h(x) = \sum_{\substack{h \neq 0 \\ h \in \mathbb{Z}}} \frac{1}{(-\Delta^{\wedge}(h))^l}\, \Phi_h(x), \quad x \in \mathbb{R}. \tag{4.88}$$

Furthermore, it is clear that

$$\nabla G\left(\Delta^l; x\right) = G\left(\nabla^{2l-1}; x\right), \quad x \in \mathbb{R}, \tag{4.89}$$

for every $l \in \mathbb{N}$. An easy calculation shows that

$$G\left(\Delta^l; x\right) = 2(-1)^l \sum_{\substack{h > 0 \\ h \in \mathbb{Z}}} \frac{\cos(2\pi h x)}{(2\pi h)^{2l}} \tag{4.90}$$

and

$$\nabla G\left(\Delta^l; x\right) = 2(-1)^{l-1} \sum_{\substack{h > 0 \\ h \in \mathbb{Z}}} \frac{\sin(2\pi h x)}{(2\pi h)^{2l-1}} \tag{4.91}$$

such that

$$G\left(\nabla^k; x\right) = -2\frac{1}{(2\pi)^k} \sum_{m=1}^{\infty} \frac{1}{m^k} \cos\left(2\pi m x - k\frac{\pi}{2}\right). \tag{4.92}$$

Remark 4.5. *The functions B_k defined by $B_k(x) = -k!\, G\left(\nabla^k; x\right)$, $x \in \mathbb{R}$, are known as Bernoulli functions of degree k. In more detail, we have*

$$
\begin{aligned}
B_1(x) &= x - \lfloor x \rfloor - \frac{1}{2}, \\
B_2(x) &= (x - \lfloor x \rfloor)^2 - (x - \lfloor x \rfloor) + \frac{1}{6}, \\
B_3(x) &= (x - \lfloor x \rfloor)^3 - \frac{3}{2}(x - \lfloor x \rfloor)^2 + \frac{1}{2}(x - \lfloor x \rfloor), \\
B_4(x) &= (x - \lfloor x \rfloor)^4 - 2(x - \lfloor x \rfloor)^3 + (x - \lfloor x \rfloor)^2 - \frac{1}{30}.
\end{aligned}
$$

B_k is a piecewise polynomial of degree k, and it is $\mathbb{Z}$-periodical. Moreover, we have (see, e.g., W. Magnus et al. [1966], H. Rademacher [1973])

$$B'_{k+1}(x) = (k+1)B_k(x), \quad k = 0, 1, \ldots, \tag{4.93}$$

$$B_k(x+1) - B_k(x) = k\, x^{k-1}, \quad k = 0, 1, \ldots, \tag{4.94}$$

and

$$B_{2l+1}(0) = B_{2l+1}(1) = 0, \quad l = 1, 2, \ldots. \tag{4.95}$$

Remark 4.6. *The values B_k, $k = 2, 3, \ldots$, given by*

$$B_k = B_k(0) = B_k(1) \tag{4.96}$$

are called Bernoulli numbers. We have

$$B_1 = -\frac{1}{2}, B_2 = \frac{1}{6}, B_3 = 0, B_4 = -\frac{1}{30}, B_5 = 0, B_6 = \frac{1}{42}, \ldots \tag{4.97}$$

Obviously,

$$G\left(\nabla^k; 0\right) = -\frac{1}{k!}\, B_k = G\left(\nabla^k; 1\right), \tag{4.98}$$

and

$$B_{2l+1} = 0, \quad l = 1, 2, \ldots, \tag{4.99}$$

$$B_{2l} = 2(-1)^{l+1}\,(2l)! \sum_{k=1}^{\infty} \left(\frac{1}{2\pi k}\right)^{2l}, \quad l = 1, 2, \ldots. \tag{4.100}$$

Note that the integer values have to be exempted for the Bernoulli function of degree 1 because of the discontinuity, i.e., the characteristic singularity of $G(\nabla; \cdot) = \nabla G\left(\nabla^2; \cdot\right)$ for $x \in \mathbb{Z}$. For more details the reader is referred to, e.g., H. Rademacher [1973], P.L. Butzer, R.L. Stens [1983], and the literature therein.

4.3   Riemann Zeta Function and Lattice Function

In number theory, the Riemann Zeta function has become the basis of the whole theory of the distribution of primes. In addition, it has significant relations to the lattice function which are of our particular interest (for more details see, e.g., L.J. Mordell [1928a,b], E.C. Titchmarsh [1951], E. Krätzel [2000]). In the sequel, the one-dimensional Riemann Zeta function is introduced in such a way that its extension to the multi-variate case becomes transparent.

Functional Equation of the Zeta Function

For $s \in \mathbb{C}$ with $\Re(s) > 1$ we consider the *(Riemann) Zeta function* ζ given by the series

$$s \mapsto \zeta(s) = \sum_{n=1}^{\infty} \frac{1}{n^s}. \tag{4.101}$$

The series is absolutely convergent since it can be majorized by $\sum_{n=1}^{\infty} n^{-\Re(s)}$, which shows us that the convergence of (4.101) is uniform in every half-plane $\{s \in \mathbb{C} \mid \Re(s) \geq 1 + \delta, \ \delta > 0\}$. Thus, ζ is holomorphic in these half-planes (note that we define c^s, where $s \in \mathbb{C}$ with $\Re(s) > 0$, as $e^{s \ln(c)}$ with $-\frac{\pi}{2} < \Im(\ln(s)) < \frac{\pi}{2}$). For the function $x \mapsto \frac{1}{x^s}$, $x > 0$, we have

$$\Delta_x \frac{1}{x^s} = \left(\frac{d}{dx}\right)^2 \frac{1}{x^s} = s(s+1)\, x^{-s-2}, \qquad x > 0. \tag{4.102}$$

The Euler summation formula (Theorem 4.3) yields

$$\sideset{}{'}\sum_{\substack{\rho \leq n \leq N \\ n \in \mathbb{Z}}} \frac{1}{n^s} = \left(\frac{1}{1-s} x^{1-s}\right)\Big|_\rho^N \tag{4.103}$$

$$+ s(s+1) \int_\rho^N G(\Delta; x) \frac{1}{x^{s+2}}\, dx$$

$$+ \left\{\frac{1}{x^s}\left(\nabla G(\Delta; x)\right) + s\frac{1}{x^{s+1}} G(\Delta; x)\right\}\Big|_\rho^N$$

for all $\rho \in \mathbb{R}$ with $0 < \rho \leq 1$. The explicit representation of the lattice function gives

$$\frac{1}{x^s} \nabla G(\Delta; x)\Big|_\rho^N = \frac{1}{x^s}\left(-(x - \lfloor x \rfloor) + \frac{1}{2}\right)\Big|_\rho^N \tag{4.104}$$

and

$$s\frac{1}{x^{s+1}} G(\Delta; x)\Big|_\rho^N = s\frac{1}{x^{s+1}}\left(-\frac{(x-\lfloor x \rfloor)^2}{2} + \frac{x - \lfloor x \rfloor}{2} - \frac{1}{12}\right)\Big|_\rho^N. \tag{4.105}$$

$|G(\nabla;\cdot)|$ and $|G(\Delta;\cdot)|$ are bounded such that the terms (4.104) and (4.105) do not contribute as $N \to \infty$. For $s \in \mathbb{C}$ with $\Re(s) > 1$ we therefore obtain

$$\zeta(s) = \frac{1}{s-1}\rho^{1-s} + s(s+1)\int_\rho^\infty G(\Delta;x)\frac{1}{x^{s+2}}\,dx \qquad (4.106)$$

$$+ \frac{1}{\rho^s}\left(\rho - \frac{1}{2}\right) + s\frac{1}{\rho^{s+1}}\left(\frac{\rho^2}{2} - \frac{\rho}{2} + \frac{1}{12}\right).$$

With the help of the Fourier expansion of $G(\Delta;\cdot)$ (cf. Lemma 4.2) the integral

$$\int_\rho^N G(\Delta;x)\frac{1}{x^{s+2}}\,dx \qquad (4.107)$$

permits the representation

$$-\sum_{\substack{h\neq 0 \\ h\in\mathbb{Z}}} \frac{1}{4\pi^2 h^2}\int_\rho^N \frac{1}{x^{s+2}}e^{2\pi i h x}\,dx. \qquad (4.108)$$

Therefore, the integral

$$\int_\rho^\infty G(\Delta;x)\frac{1}{x^{s+2}}\,dx \qquad (4.109)$$

converges not only in the half-plane of all $s \in \mathbb{C}$ with $\Re(s) > 1$, but also for all $s \in \mathbb{C}$ with $\Re(s) > -1$. Thus (4.106) furnishes an analytic continuation of ζ into the half-plane $\{s \in \mathbb{C} \mid \Re(s) > -1\}$ showing as the only singularity the pole at $s = 1$. In addition, the expression (4.107) is convergent as $\rho \to 0$ provided that $\Re(s) < 0$; hence, ζ can be continued by (4.106) to any point in the s-plane, and ζ emerges a meromorphic function with the simple pole $s = 1$.

Even more, we are able to formulate the following lemma.

Lemma 4.7. *For $s \in \mathbb{C}$ with $-1 < \Re(s) < 0$ we have*

$$\zeta(s) = s(s+1)\int_0^\infty G(\Delta;x)\frac{1}{x^{s+2}}\,dx. \qquad (4.110)$$

Now, if $s \in \mathbb{C}$ with $-1 < \Re(s) < 0$, we get by partial integration

$$\zeta(s) = \sum_{\substack{h\neq 0 \\ h\in\mathbb{Z}}} \int_0^\infty e^{2\pi i h \cdot x}\frac{1}{x^s}\,dx. \qquad (4.111)$$

For $h \neq 0$ we have (see, e.g., L.J. Mordell [1928b])

$$\int_0^\infty e^{2\pi i h x}\frac{1}{x^s}\,dx = \begin{cases} \left(\frac{1}{2h\pi}\right)^{1-s}\Gamma(1-s)e^{\frac{\pi i}{2}(1-s)}, & h > 0 \\ \left(\frac{1}{2(-h)\pi}\right)^{1-s}\Gamma(1-s)e^{-\frac{\pi i}{2}(1-s)}, & h < 0. \end{cases} \qquad (4.112)$$

For all $s \in \mathbb{C}$ with $-1 < \Re(s) < 0$ it therefore follows that

$$\zeta(s) = 2^s \pi^{s-1} \, \Gamma(1-s) \, \sin\left(\frac{\pi s}{2}\right) \sum_{h=1}^{\infty} \frac{1}{h^{1-s}}. \tag{4.113}$$

Now, the left side of (4.113) is a meromorphic function (of s) with the only pole at $s = 1$, so that this equation provides an analytic continuation of the right hand member as a meromorphic function over the whole s-plane, and ζ appears as a meromorphic function only with the simple pole at $s = 1$.

Theorem 4.5. *(Functional Equation of the Riemann Zeta Function) The Zeta function ζ given by*

$$\zeta(s) = \sum_{n=1}^{\infty} \frac{1}{n^s}, \quad s \in \mathbb{C}, \ \Re(s) > 1, \tag{4.114}$$

can be extended analytically to a meromorphic function with the pole

$$\frac{1}{s-1} \tag{4.115}$$

to the whole complex plane $\mathbb{C}$. Moreover, ζ satisfies the functional equation

$$\zeta(s) = 2^s \pi^{s-1} \Gamma(1-s) \, \sin\left(\frac{\pi s}{2}\right) \zeta(1-s). \tag{4.116}$$

The functional equation can be put into a more illuminating form if we make use of the duplicator formula (Lemma 3.8) of the Gamma function

$$\Gamma(s)\Gamma\left(s + \frac{1}{2}\right) = 2\pi^{\frac{1}{2}} 2^{-2s} \Gamma(2s). \tag{4.117}$$

Replacing here $2s$ by $1 - s$ we obtain

$$\Gamma(1-s)\pi^{\frac{1}{2}} 2^s \sin\left(\frac{\pi s}{2}\right) = \Gamma\left(\frac{1}{2} - \frac{s}{2}\right)\Gamma\left(1 - \frac{s}{2}\right) \sin\left(\frac{\pi s}{2}\right) = \pi \frac{\Gamma(\frac{1-s}{2})}{\Gamma(\frac{s}{2})}. \tag{4.118}$$

Thus, Theorem 4.5 can be reformulated in the form of

Corollary 4.6. *Under the assumptions of Theorem 4.5*

$$\Gamma\left(\frac{s}{2}\right)\zeta(s) = \pi^{s-\frac{1}{2}} \, \Gamma\left(\frac{1-s}{2}\right) \zeta(1-s). \tag{4.119}$$

Remark 4.7. *Note that the function ξ given by*

$$\xi(s) = \pi^{-\frac{s}{2}} \, \Gamma\left(\frac{s}{2}\right) \zeta(s) \tag{4.120}$$

fulfills the functional equation

$$\xi(s) = \xi(1-s). \tag{4.121}$$

The identity (4.106) is also valid for $s \in \mathbb{C}$ with $\Re(s) > -1$, and we get from (4.106)

$$\zeta(0) = -\frac{1}{2}. \tag{4.122}$$

Even more, our considerations enable us to deduce that

$$\zeta(-n) = -\frac{B_{n+1}}{n+1}, \quad n = 1, 2, \ldots, \tag{4.123}$$

$$\zeta(-2n) = 0, \quad n = 1, 2, \ldots, \tag{4.124}$$

$$\zeta(-2n+1) = -\frac{B_{2n}}{2n}, \quad n = 1, 2, \ldots . \tag{4.125}$$

In addition, an easy calculation yields

$$\zeta'(0) = -\frac{1}{2}\ln(2\pi). \tag{4.126}$$

Clearly, $\zeta(2n)$ is proportional to the Bernoulli numbers $B_{2n} = -(2n)!\, G(\Delta^n; 0)$ (cf. (4.100), (4.123))

$$\zeta(2n) = (-1)^{n+1}\frac{(2\pi)^{2n}}{2(2n)!}B_{2n} = \frac{1}{2}(-4\pi^2)^n G(\Delta^n; 0). \tag{4.127}$$

Furthermore, we have

$$\zeta(-2n+1) = -\frac{1}{2n}B_{2n} = (2n-1)!\, G(\Delta^n; 0). \tag{4.128}$$

The zeros of ζ at $-2l$, $l = 1, 2, \ldots$ are often called the "trivial zeros" as they are easily found. The role of the "trivial zeros" of ζ is evident in (4.119). For $\Re(s) < 0$ we have $\Re(1-s) > 1$ so that the right member of (4.119) is regular. However, $\Gamma(\frac{s}{2})$ has poles for $s = -2n$, $n \in \mathbb{N}$, which are just neutralized by the zeros of ζ. Only for $s = 0$ we have a pole of first order on both sides of (4.119) since $\zeta(0) = -\frac{1}{2}$.

Kronecker's Limit Formula

Setting $s = 1 + t$ we obtain from Corollary 4.6

$$\zeta(1+t) = \frac{\pi^{t-\frac{1}{2}}\Gamma(-\frac{t}{2})}{t\Gamma(\frac{t-1}{2})}\zeta(-t). \tag{4.129}$$

In the neighborhood of $t = 0$ we have with (4.122) and (4.126)

$$\zeta(t) = -\frac{1}{2} - \frac{1}{2}\ln(2\pi)\,t + \ldots . \tag{4.130}$$

Furthermore,

$$\frac{\pi^{t-\frac{1}{2}}\Gamma\left(-\frac{t}{2}\right)}{\Gamma\left(\frac{t-1}{2}\right)} = -2 + (\ln(2\pi) - \Gamma'(1))\,t + \ldots \tag{4.131}$$

such that

$$\zeta(1+t) = \frac{1}{t} - \Gamma'(1) + \ldots \qquad (4.132)$$

With $-\Gamma'(1) = C$ (cf. Lemma 4.5) we therefore obtain the so–called *Kronecker limit formula*.

Lemma 4.8.

$$\lim_{s \to 1} \left(\zeta(s) - \frac{1}{s-1} \right) = C, \qquad (4.133)$$

where C is the Euler constant.

Combining Lemma 4.8 and the identity (4.106) we find with $\rho = 1$

$$C = \lim_{s \to 1} \left(s(s+1) \int_1^\infty G(\Delta; x) \frac{1}{x^{s+2}} \, dx + \frac{1}{2} - \frac{s}{12} \right) \qquad (4.134)$$

such that

$$C = 2 \int_1^\infty G(\Delta; x) \frac{1}{x^3} \, dx + \frac{7}{12}. \qquad (4.135)$$

From the Euler summation formula (Theorem 4.3) we get

$$\lim_{N \to \infty} \left(\sum_{\substack{1 \le n \le N \\ n \in \mathbb{Z}}} \frac{1}{n} - \int_1^N \frac{1}{x} \, dx \right) = 2 \int_1^\infty G(\Delta; x) \frac{1}{x^3} \, dx + \frac{7}{12}. \qquad (4.136)$$

By comparison of (4.135) and (4.136) we finally arrive at the already known limit relation (3.82) for the Euler constant.

Euler's Product Representation

The fundamental theorem of classical number theory, proved essentially by Euclid, states that every positive integer can be decomposed in only one manner into a product of powers of different primes (see, e.g., E. Hlawka et al. [1991], H. Rademacher [1973] for more details). For $s \in \mathbb{C}$ with $\Re(s) > 1$, we therefore have

$$\zeta(s) = \sum_{n=1}^\infty \frac{1}{n^s} = \prod_p \left(1 + \frac{1}{p^s} + \frac{1}{p^{2s}} + \cdots \right) = \prod_p \frac{1}{1 - p^{-s}}, \qquad (4.137)$$

where the product has to be extended over all prime numbers p. It is known that the product representation (4.137) was already used by Euler, at least for special values of s. Since in the product representation (4.137) all factors are different from 0 in the half-plane $\{s \in \mathbb{C} \mid \Re(s) > 1\}$, the function ζ can at most have zeros at the poles of $\Gamma(\frac{s}{2})$, i.e., 0 and at the negative even integers. Indeed, we know that the "trivial zeros" are $-2n$, $n \in \mathbb{N}$, of ζ. In addition, we have $\zeta(0) = -\frac{1}{2}$. All other zeros of ζ (and it is known that such zeros exist) must lie in the "critical strip" $\{s \in \mathbb{C} \mid 0 \leq \Re(s) \leq 1\}$.

The functional equation of the Zeta function immediately tells us that with s also $1 - s$ must be a zero in the "critical strip"; hence, zeros must pairwise be symmetrical to $s = \frac{1}{2}$. Furthermore, since the ζ-function is real on the real axis, it has conjugate complex values in conjugate complex points. Consequently its zeros must also be symmetrical with respect to the real axis. For a zero ρ on the line $\{s \in \mathbb{C} \mid \Re(s) = \frac{1}{2}\}$, these two symmetries lead to the same further zero $\widetilde{\rho} = 1 - \rho$. If, however, in the "critical strip" there should exist a zero with real part different from $\frac{1}{2}$, this zero would appear together with the further zeros $\widetilde{\rho}, 1 - \rho, 1 - \widetilde{\rho}$, which together with ρ form the vertices of a rectangle. For the location of the non-trivial zeros of ζ the *Riemann conjecture* (cf. B. Riemann [1859]) tells us that they all lie on the line $\{s \in \mathbb{C} \mid \Re(s) = \frac{1}{2}\}$ (for more details see, e.g., T. Christ et al. [2010] and the references therein).

4.4   Poisson Summation Formula for the Laplace Operator

Next we discuss some variants of the one-dimensional *Poisson summation formula* (due to L.J. Mordell [1928a,b, 1929]).

Theorem 4.6. *(Variant 1) Let F be twice continuously differentiable, i.e., $F \in C^{(2)}([a, b])$. Then*

$$\sum_{\substack{g \in [a,b] \\ g \in \mathbb{Z}}}{}' F(g) = \sum_{h \in \mathbb{Z}} \int_a^b F(x)\overline{\Phi_h(x)}\,dx. \tag{4.138}$$

Proof. Using the Fourier expansion of the $\mathbb{Z}$-lattice function $G(\Delta; \cdot)$ (see Lemma 4.2) we get

$$\int_a^b G(\Delta; x)\,\Delta F(x)\,dx = \sum_{\substack{h \neq 0 \\ h \in \mathbb{Z}}} \int_a^b (\Delta F(x))\,\frac{\overline{\Phi_h(x)}}{(2\pi i h)^2}\,dx \tag{4.139}$$

(note that the Fourier expansion is absolutely and uniformly convergent in $\mathbb{R}$; hence, summation and integration can be interchanged). As already pointed out in the Introduction, partial integration gives us (with $\nabla_x = \frac{d}{dx}$ and $\Delta_x =$

$\nabla_x^2 = \left(\frac{d}{dx}\right)^2$, and $(\Delta_x + \Delta^\wedge(h))\Phi_h(x) = 0$, $x \in \mathbb{R}$, $\Delta^\wedge(h) = 4\pi^2 h^2)$

$$\int_a^b (\Delta F(x))\,\overline{\Phi_h(x)}\,dx \;=\; \left.(\nabla F(x))\overline{\Phi_h(x)}\right|_a^b \qquad (4.140)$$

$$- (2\pi i h)\int_a^b (\nabla F(x))\,\overline{\Phi_h(x)}\,dx$$

$$= \left\{(\nabla F(x))\overline{\Phi_h(x)} - (2\pi i h)F(x)\overline{\Phi_h(x)}\right\}\Big|_a^b$$

$$+ (2\pi i h)^2 \int_a^b F(x)\overline{\Phi_h(x)}\,dx.$$

Thus we obtain from the Euler summation formula (cf. Theorem 4.3)

$$\int_a^b G(\Delta; x)\Delta F(x)dx \;=\; \left.\{(\nabla F(x))G(\Delta; x) - F(x)\nabla G(\Delta, x)\}\right|_a^b$$

$$+ \sum_{\substack{h\neq 0 \\ h\in\mathbb{Z}}} \int_a^b F(x)\overline{\Phi_h(x)}\,dx. \qquad (4.141)$$

Combining all results we get the Poisson summation formula

$$\sideset{}{'}\sum_{\substack{g\in[a,b] \\ g\in\mathbb{Z}}} F(g) \;=\; \int_a^b F(x)\,dx \;+\; \sum_{\substack{h\neq 0 \\ h\in\mathbb{Z}}} \int_a^b F(x)\overline{\Phi_h(x)}\,dx \qquad (4.142)$$

$$= \sum_{h\in\mathbb{Z}} \int_a^b F(x)\overline{\Phi_h(x)}\,dx,$$

as announced. $\qquad\qquad\qquad\qquad\qquad\qquad\qquad\qquad\qquad\qquad\qquad\qquad$ □

Example 4.2. *We know from Variant 1 of the Poisson summation formula (Theorem 4.6) that*

$$\sideset{}{'}\sum_{\substack{g\in[a,b] \\ g\in\mathbb{Z}}} 1 \;=\; \sum_{h\in\mathbb{Z}} \int_a^b \overline{\Phi_h(x)}\,dx. \qquad (4.143)$$

This gives us (cf. (1.19))

$$\sideset{}{'}\sum_{\substack{g\in[a,b] \\ g\in\mathbb{Z}}} 1 = \int_a^b dx \;+\; \sum_{\substack{h\neq 0 \\ h\in\mathbb{Z}}} \int_a^b \overline{\Phi_h(x)}\,dx. \qquad (4.144)$$

In particular, for $[a,b] = [-R, R]$*,* $R > 0$*, we obtain as a one-dimensional* *counterpart of the so–called Hardy–Landau identity (see Section 10.4)*

$$\sideset{}{'}\sum_{\substack{g\in[-R,R]\\ g\in\mathbb{Z}}} 1 = 2R + 2R \sum_{\substack{|h|\neq 0\\ h\in\mathbb{Z}}} \frac{\sin(2\pi hR)}{2\pi hR} \,. \tag{4.145}$$

In other words (see, e.g., A.M. Ostrowski [1969], F. Stenger [1976, 1981]), we have

$$\sideset{}{'}\sum_{\substack{g\in[-R,R]\\ g\in\mathbb{Z}}} 1 = 2R \sum_{h\in\mathbb{Z}} \operatorname{sinc}(2\pi hR), \tag{4.146}$$

where

$$\operatorname{sinc}(2\pi hR) = \frac{\sin(2\pi hR)}{2\pi hR}. \tag{4.147}$$

It should be noted that the Hardy–Landau series on the right side of (4.145) is alternating. Nevertheless, the (pointwise) convergence of the alternating series (4.145) can be readily seen from the derived Poisson summation formula for the interval $[-R, R]$. Moreover, it is uniformly convergent on each compact interval $I \subset (g, g + 1), g \in \mathbb{Z}$. Since the formulation of the Poisson summation formula (Theorem 4.6) on finite domains is not straightforward in higher dimensions (because of the behavior of the Fourier series of the lattice function), the convergence will bother us much more for the two- and certainly for higher-dimensional counterparts of the Hardy–Landau series. A way out will be found by formulating a Poisson summation formula corresponding to an adaptive "wave number" λ of a suitably chosen Helmholtz operator $\Delta + \lambda$ (for more details concerning the one-dimensional case the reader is referred to Section 4.5).

Theorem 4.7. *(Variant 2) Let F be twice continuously differentiable in $[0, \infty)$. Moreover, suppose that*

$$F(x) \to 0, \qquad x \to \infty,$$
$$\nabla F(x) \to 0, \qquad x \to \infty.$$

Furthermore, assume that the limits

$$\int_0^\infty F(x)dx = \lim_{N\to\infty} \int_0^N F(x)\, dx, \tag{4.148}$$

$$\int_0^\infty |\Delta F(x)|\, dx = \lim_{N\to\infty} \int_0^N |\Delta F(x)|\, dx \tag{4.149}$$

exist. Then

$$\sum_{g\in\mathbb{Z}} F(g) = \sum_{h\in\mathbb{Z}} \int_0^\infty F(x)\overline{\Phi_h(x)}\, dx. \tag{4.150}$$

Proof. For every $\varepsilon > 0$ there exists an $N(= N(\varepsilon))$ such that

$$\left| G(\Delta; x) - \sum_{\substack{0 < |h| \leq N \\ h \in \mathbb{Z}}} \frac{1}{-\Delta^{\wedge}(h)} \overline{\Phi_h(x)} \right| \leq \varepsilon \tag{4.151}$$

for all $x \in \mathbb{R}$. From the Euler summation formula it is not difficult to show that

$$\sum_{\substack{0 \leq g < \infty \\ g \in \mathbb{Z}}} F(g) = \int_0^\infty F(x)\, dx - \frac{1}{12}\,(\nabla F)\,(0) + \frac{1}{2}F(0) \tag{4.152}$$

$$+ \int_0^\infty G(\Delta; x)\, \Delta F(x)\, dx.$$

It is clear that

$$\int_0^\infty \left| G(\Delta; x) - \sum_{\substack{0 < |h| \leq N \\ h \in \mathbb{Z}}} \frac{1}{-\Delta^{\wedge}(h)} \overline{\Phi_h(x)} \right| |\Delta F(x)|\ dx \leq \varepsilon \int_0^\infty |\Delta F(x)|\ dx,$$

$$\tag{4.153}$$

where the integral on the right side of (4.153) is assumed to be convergent. Thus we find

$$\lim_{N \to \infty} \left(\int_0^\infty \left(G(\Delta; x) - \sum_{\substack{0 < |h| \leq N \\ h \in \mathbb{Z}}} \frac{1}{-\Delta^{\wedge}(h)} \overline{\Phi_h(x)} \right) \Delta F(x)\, dx \right) = 0. \tag{4.154}$$

Finally we note that

$$\int_0^\infty G(\Delta; x)\, \Delta F(x)\, dx = \sum_{\substack{0 < |h| < \infty \\ h \in \mathbb{Z}}} \int_0^\infty \frac{1}{-\Delta^{\wedge}(h)}\, (\Delta F(x))\, \overline{\Phi_h(x)}\, dx$$

$$= \frac{1}{12}\,(\nabla F)\,(0) - \frac{1}{2}F(0) \tag{4.155}$$

$$+ \sum_{\substack{0 < |h| < \infty \\ h \in \mathbb{Z}}} \int_0^\infty F(x)\overline{\Phi_h(x)}\, dx.$$

From (4.152) we therefore obtain the desired result. $\qquad\square$

From Theorem 4.7 we immediately obtain

Theorem 4.8. *(Variant 3) Let F be twice continuously differentiable in $\mathbb{R}$. Moreover, suppose that*

$$F(x) \to 0, \qquad x \to \pm\infty,$$
$$\nabla F(x) \to 0, \qquad x \to \pm\infty. \tag{4.156}$$

Furthermore, assume that the limits

$$\int_{\mathbb{R}} F(x) \, dx = \lim_{N \to \infty} \int_{-N}^{N} F(x) \, dx \tag{4.157}$$

and

$$\int_{\mathbb{R}} |\Delta F(x)| \, dx = \lim_{N \to \infty} \int_{-N}^{N} |\Delta F(x)| \, dx \tag{4.158}$$

exist. Then

$$\sum_{g \in \mathbb{Z}} F(g) = \sum_{h \in \mathbb{Z}} \int_{\mathbb{R}} F(x) \overline{\Phi_h(x)} \, dx. \tag{4.159}$$

Theorem 4.8 admits a canonical generalization to the multi-dimensional case, that unfortunately cannot be applied directly within the framework of Hardy–Landau identities of the analytic theory of numbers.

A modification of Variant 3 is formulated in the next corollary.

Corollary 4.7. *Let $y \mapsto F(x+y)$, $x \in \mathbb{R}$, satisfy the assumptions of Theorem 4.8. Then*

$$\sum_{g \in \mathbb{Z}} F(g + x) = \sum_{h \in \mathbb{Z}} \int_{\mathbb{R}} F(y) \overline{\Phi_h(y)} \, dy \, \Phi_h(x) \tag{4.160}$$

$$= \sum_{h \in \mathbb{Z}} \int_{\mathbb{R}} F(y) \Phi_h(y) \, dy \, \overline{\Phi_h(x)}. \tag{4.161}$$

Proof. From Theorem 4.8 it follows that

$$\sum_{g \in \mathbb{Z}} F(g + x) = \sum_{h \in \mathbb{Z}} \int_{\mathbb{R}} F(x + y) \overline{\Phi_h(y)} \, dy. \tag{4.162}$$

Clearly, under the assumptions of Theorem 4.8,

$$\int_{\mathbb{R}} F(x + y) \overline{\Phi_h(y)} \, dy = \Phi_h(x) \int_{\mathbb{R}} F(y) \overline{\Phi_h(y)} \, dy. \tag{4.163}$$

This is the desired result.   □

Elementary Examples

The following identities are well known in the literature (see, for example, M. Abramowitz, I.A. Stegun [1972], I.S. Gradshteyn, I.M. Ryzhik [1965]): they are presented here as one-dimensional manifestations of multi-variate lattice point sums which will play an important role in the context of Euler and Poisson summation formulas, later on. Of particular significance in this respect is the one-dimensional Theta series (4.182) and its functional equation (4.191).

Example 4.3. *For $\tau \in \mathbb{R}$ with $\tau > 0$ we consider the function $F : \mathbb{R} \to \mathbb{R}$ given by*

$$F(x) = \frac{1}{x^2 + \tau^2}. \tag{4.164}$$

It is easy to see that F satisfies all assumptions of Variant 3. Therefore we obtain

$$\sum_{g \in \mathbb{Z}} \frac{1}{g^2 + \tau^2} = \sum_{h \in \mathbb{Z}} \int_{\mathbb{R}} \frac{e^{-2\pi i h x}}{x^2 + \tau^2} \, dx. \tag{4.165}$$

The integral on the right side is explicitly calculable

$$\int_{\mathbb{R}} \frac{e^{-2\pi i h x}}{x^2 + \tau^2} \, dx = \frac{\pi}{\tau} e^{-|2\pi h \tau|}. \tag{4.166}$$

This yields

$$\sum_{g \in \mathbb{Z}} \frac{1}{g^2 + \tau^2} = \frac{\pi}{\tau} \sum_{h \in \mathbb{Z}} e^{-|2\pi h \tau|} \tag{4.167}$$

$$= \frac{\pi}{\tau} + 2\frac{\pi}{\tau} \sum_{n=1}^{\infty} e^{-2\pi \tau n}$$

$$= \frac{\pi}{\tau} + \frac{2\pi}{\tau} \frac{e^{-2\pi \tau}}{1 - e^{-2\pi \tau}}$$

$$= \frac{\pi}{\tau} \frac{1 + e^{-2\pi \tau}}{1 - e^{-2\pi \tau}}.$$

Lemma 4.9. *For positive real τ,*

$$\sum_{g \in \mathbb{Z}} \frac{1}{g^2 + \tau^2} = \frac{\pi}{\tau} \frac{e^{\pi \tau} + e^{-\pi \tau}}{e^{\pi \tau} - e^{-\pi \tau}} = \frac{\pi}{\tau} \coth(\pi \tau). \tag{4.168}$$

Example 4.4. *We deal with $F : \mathbb{R} \to \mathbb{C}$ given by*

$$F(x) = \frac{e^{-i\pi x}}{x^2 + \tau^2}, \quad \tau > 0. \tag{4.169}$$

Obviously, F satisfies all assumptions of Variant 3 (Theorem 4.8). Therefore, *the Poisson summation formula yields*

$$\sum_{g\in\mathbb{Z}} F(g) = \sum_{h\in\mathbb{Z}} \int_{\mathbb{R}} \frac{e^{-\pi i x}}{x^2 + \tau^2} e^{-2\pi i h x}\, dx \tag{4.170}$$

$$= \frac{\pi}{\tau} \sum_{h\in\mathbb{Z}} e^{-\pi\tau|2h+1|}$$

$$= \frac{2\pi}{\tau} \frac{1}{e^{\pi\tau} - e^{-\pi\tau}}.$$

Lemma 4.10. *For positive real τ,*

$$\sum_{g\in\mathbb{Z}} \frac{(-1)^g}{g^2 + \tau^2} = \frac{\pi}{\tau} \frac{1}{\sinh(\pi\tau)}. \tag{4.171}$$

For all $\tau \in \mathbb{C}$ with $\Re(\tau) > 0$ and all $y \in \mathbb{R}$, a standard calculation using tools of complex analysis yields

$$\int_{\mathbb{R}} \frac{1}{\cosh(\pi\tau x)} e^{-2\pi i y x}\, dx = \frac{1}{\tau} \frac{1}{\cosh\left(\pi\frac{y}{\tau}\right)}. \tag{4.172}$$

Thus the Poisson summation formula enables us to derive the following identity.

Lemma 4.11. *For all $\tau \in \mathbb{C}$ with $\Re(\tau) > 0$,*

$$\sum_{g\in\mathbb{Z}} \int_{\mathbb{R}} \frac{1}{\cosh(\pi\tau x)} e^{-2\pi i g x}\, dx = \frac{1}{\tau} \sum_{h\in\mathbb{Z}} \frac{1}{\cosh\left(\pi\frac{h}{\tau}\right)}. \tag{4.173}$$

In other words, the function β given by

$$\beta(\tau) = \sum_{g\in\mathbb{Z}} \frac{1}{\cosh(\pi\tau g)}, \quad \tau \in \mathbb{C}, \ \Re(\tau) > 0, \tag{4.174}$$

satisfies the functional equation

$$\beta(\tau) = \frac{1}{\tau} \beta\left(\frac{1}{\tau}\right), \quad \Re(\tau) > 0. \tag{4.175}$$

Theta Functions

Next we discuss $F : \mathbb{R} \to \mathbb{C}$ given by

$$F(x) = e^{-\pi\tau x^2}, \quad \tau \in \mathbb{C}, \ \Re(\tau) > 0. \tag{4.176}$$

Clearly, we have

$$\sum_{g\in\mathbb{Z}} e^{-\pi\tau g^2} = \sum_{h\in\mathbb{Z}} \int_{\mathbb{R}} e^{-\pi\tau x^2} e^{-2\pi i h x}\, dx. \tag{4.177}$$

The integral on the right side should be computed in more detail. To this end we observe that

$$\int_{\mathbb{R}} e^{-\pi\tau x^2} e^{-2\pi i h x}\,dx = e^{-\pi \frac{h^2}{\tau}} \int_{\mathbb{R}} e^{-\pi\tau\left(x+\frac{ih}{\tau}\right)^2}\,dx. \qquad (4.178)$$

Now by letting $u = x + \frac{ih}{\tau}$ we get

$$\int_{\mathbb{R}} e^{-\pi\tau\left(x+\frac{ih}{\tau}\right)^2}\,dx = \int_{\mathbb{R}} e^{-\pi\tau u^2}\,du. \qquad (4.179)$$

Combining our results we therefore find

$$\begin{aligned}
\int_{\mathbb{R}} e^{-\pi\tau x^2} e^{-2\pi i h x}\,dx &= e^{-\pi\frac{h^2}{\tau}} \int_{\mathbb{R}} e^{-\pi\tau x^2}\,dx \qquad (4.180)\\
&= e^{-\pi\frac{h^2}{\tau}} \frac{1}{\sqrt{\pi\tau}} \int_{\mathbb{R}} e^{-s^2}\,ds \\
&= e^{-\pi\frac{h^2}{\tau}} \frac{1}{\sqrt{\pi\tau}} \sqrt{\pi}.
\end{aligned}$$

This leads to

Lemma 4.12. *For all $\tau \in \mathbb{C}$ with $\Re(\tau) > 0$,*

$$\sum_{g\in\mathbb{Z}} e^{-\pi\tau g^2} = \frac{1}{\sqrt{\tau}} \sum_{h\in\mathbb{Z}} e^{-\pi\frac{1}{\tau}h^2}. \qquad (4.181)$$

In other words, the *(one-dimensional) Theta function* ϑ (see, e.g., W. Magnus et al. [1966] and the references therein) given by

$$\vartheta(\tau) = \sum_{g\in\mathbb{Z}} e^{-\pi\tau g^2}, \quad \tau \in \mathbb{C}, \ \Re(\tau) > 0, \qquad (4.182)$$

satisfies the functional equation

$$\vartheta(\tau) = \frac{1}{\sqrt{\tau}} \, \vartheta\left(\frac{1}{\tau}\right). \qquad (4.183)$$

It is not difficult to see that, for $x \in \mathbb{R}$, $\tau \in \mathbb{C}$, $\Re(\tau) > 0$,

$$\int_{\mathbb{R}} e^{-\frac{|y-x|^2}{\tau}} e^{-2\pi i h y}\,dy = e^{-2\pi i h x} \int_{\mathbb{R}} e^{-\frac{u^2}{\tau}} e^{-2\pi i h u}\,du. \qquad (4.184)$$

Thus it follows that

$$\int_{\mathbb{R}} e^{-\frac{|y-x|^2}{\tau}} e^{-2\pi i h y}\,dy = \sqrt{\pi\tau}\,e^{-2\pi i h x} e^{-2\tau\pi^2 h^2}. \qquad (4.185)$$

In particular,

$$\frac{1}{\sqrt{\pi\tau}} \int_{\mathbb{R}} e^{-\frac{|y-x|^2}{\tau}}\,dy = 1. \qquad (4.186)$$

Lemma 4.13. *For all real x and all $\tau \in \mathbb{C}$ with $\Re(\tau) > 0$,*

$$\frac{1}{\sqrt{\pi\tau}} \sum_{g \in \mathbb{Z}} e^{-\frac{|g-x|^2}{\tau}} = \sum_{h \in \mathbb{Z}} e^{-\tau\pi^2 h^2} e^{-2\pi i h x}. \tag{4.187}$$

Lemma 4.13 motivates us to introduce

Definition 4.3. *For $\tau \in \mathbb{C}$ with $\Re(\tau) > 0$ and all $x, y \in \mathbb{R}$, the function $\vartheta(\cdot; x, y; \mathbb{Z})$ given by (cf. (4.177))*

$$\vartheta(\tau; x, y; \mathbb{Z}) = \lim_{N \to \infty} \sum_{\substack{|g| \leq N \\ g \in \mathbb{Z}}} e^{-\pi\tau|g-x|^2} e^{2\pi i g y} \tag{4.188}$$

is called the Theta function (of degree 0 and dimension 1).

Obviously, for all $\tau \in \mathbb{C}$ with $\Re(\tau) > 0$, we have $\vartheta(\tau) = \vartheta(\tau; 0, 0; \mathbb{Z})$. Moreover, from Lemma 4.13, we are able to deduce the functional equation

$$\vartheta\left(\frac{1}{\pi\tau}; x, 0; \mathbb{Z}\right) = (\pi\tau)^{-\frac{1}{2}} \, \vartheta\left(\pi\tau; 0, x; \mathbb{Z}\right). \tag{4.189}$$

More generally, in accordance with our previous considerations, the one-dimensional Poisson summation formula delivers

$$\lim_{N \to \infty} \sum_{\substack{|g| \leq N \\ g \in \mathbb{Z}}} e^{-\pi\tau|g-x|^2} e^{2\pi i g y} = \frac{1}{\sqrt{\tau}} e^{2\pi i x \cdot y} \lim_{R \to \infty} \sum_{\substack{|h| \leq R \\ h \in \mathbb{Z}}} e^{-\frac{\pi}{\tau}|y+h|^2} e^{2\pi i h \cdot y}. \tag{4.190}$$

This leads us to the following *functional equation of the one-dimensional Theta function* (cf. Theorem 12.1).

Lemma 4.14. *The Theta function $\vartheta(\cdot; x, y; \mathbb{Z})$ is holomorphic for all $\tau \in \mathbb{C}$ with $\Re(\tau) > 0$. Furthermore, $\vartheta(\cdot; x, y; \mathbb{Z})$ satisfies the functional equation*

$$\vartheta(\tau; x, y; \mathbb{Z}) = \frac{e^{2\pi i x \cdot y}}{\sqrt{\tau}} \vartheta\left(\frac{1}{\tau}; -y, x; \mathbb{Z}\right). \tag{4.191}$$

4.5 Euler Summation Formula for Helmholtz Operators

In the sequel we make the attempt to apply the assumptions (4.156), (4.157), and (4.158) of Variant 3 (i.e., Theorem 4.8) to the function known from the (one-dimensional) Hardy–Landau identity

$$F(x) = \operatorname{sinc}(2\pi R x) = \frac{\sin(2\pi R x)}{2\pi R x}, \quad x \in \mathbb{R}, \ R > 0, \tag{4.192}$$

i.e., for the problem of determining the total number of lattice points inside an interval $[-R, R]$ (see (4.145)).

An elementary calculation yields the identities

$$\nabla F(x) = \frac{\cos(2\pi Rx)}{x} - \frac{\sin(2\pi Rx)}{2\pi Rx^2} \tag{4.193}$$

and

$$\Delta F(x) = -4\pi^2 R^2 \frac{\sin(2\pi Rx)}{2\pi Rx} - 2\frac{\cos(2\pi Rx)}{x^2} + 2\frac{\sin(2\pi Rx)}{2\pi Rx^3}. \tag{4.194}$$

In other words, we have

$$(\Delta + 4\pi^2 R^2) \frac{\sin(2\pi Rx)}{2\pi Rx} = O\left(\frac{1}{x^2}\right), \quad x \to \pm\infty. \tag{4.195}$$

Therefore, F is an arbitrarily often differentiable function in $\mathbb{R}$ satisfying (4.156). Furthermore, the integral

$$\int_{\mathbb{R}} F(x)\, dx = \lim_{R \to \infty} \int_{-R}^{R} F(x)\, dx \tag{4.196}$$

exists. Unfortunately, *the second order derivative (4.194) is not absolutely integrable over* $\mathbb{R}$. Thus, the assumption (4.158) is not satisfied so that the Poisson summation formula for the function F given by (4.192) can not be based on the Variant 3. From (4.194), however, we see that $(\Delta + 4\pi R^2)F$ is absolutely integrable over $\mathbb{R}$, i.e.,

$$\int_{\mathbb{R}} \left| (\Delta + 4\pi^2 R^2)\, F(x) \right|\, dx < \infty. \tag{4.197}$$

Thus, the resulting question of our observation can be put into the following words: *can we base the Poisson summation formula on assumptions involving the "Helmholtz operator"* $\Delta + 4\pi^2 R^2$ *instead of the "Laplace operator"* Δ? The answer leads to the metaharmonic theory, viz. the $\mathbb{Z}$-lattice functions for the one-dimensional "Helmholtz operator" $\Delta + \lambda$, $\lambda \in \mathbb{R}$ (note that, in our context, the most interesting properties of the Helmholtz operator appear for real "wave numbers" λ).

Lattice Function for the Helmholtz Operator

We begin with a heuristic argument to demonstrate that the $\mathbb{Z}$-lattice function $G(\Delta + \lambda; \cdot)$ for the Helmholtz operator $\Delta + \lambda$ and the "boundary condition" of $\mathbb{Z}$-periodicity exists in two variants dependent on the choice of the parameter $\lambda \in \mathbb{R}$:

(1) For $\lambda \notin \mathrm{Spect}_\Delta(\mathbb{Z})$, i.e., $\lambda \neq \Delta^\wedge(h) = 4\pi^2 h^2$ for all $h \in \mathbb{Z}$, the $\mathbb{Z}$-lattice

function shows the essential ingredients of a classical Green function, viz. homogeneous Helmholtz differential equation, characteristic singularity (i.e., a finite jump of its derivative in lattice points), and ($\mathbb{Z}$-periodical) boundary condition. Following mathematical physics we have to describe $G(\Delta + \lambda; \cdot)$ in formal consistency with the following identities (of course, to be understood more precisely in a distributional sense)

$$
\begin{aligned}
(\Delta_x + \lambda)G(\Delta + \lambda; x) &= (\Delta_x + \lambda) \sum_{h \in \mathbb{Z}} \frac{1}{-(\Delta + \lambda)^\wedge(h)} \Phi_h(x) \\
&= \sum_{h \in \mathbb{Z}} \Phi_h(x) \qquad (4.198) \\
&= \delta_{\mathbb{Z}}(x),
\end{aligned}
$$

where $\delta_{\mathbb{Z}}$ is the *Dirac function(al)* and $(\Delta + \lambda)^\wedge(h) = 4\pi^2 h^2 - \lambda$, $h \in \mathbb{Z}$.

(2) For $\lambda \in \mathrm{Spect}_\Delta(\mathbb{Z})$, i.e., $\lambda = \Delta^\wedge(h) = 4\pi^2 h^2$ for some $h \in \mathbb{Z}$, in the terminology of D. Hilbert [1912], the $\mathbb{Z}$-lattice function is a Green function in an enlarged sense; i.e., the right side of the differential equation (4.198) must be modified which implies an additional normalization condition to ensure the uniqueness. Again, in accordance with mathematical physics, we have to specify $G(\Delta + \lambda; \cdot)$ in such a way that

$$
\begin{aligned}
(\Delta_x + \lambda)G(\Delta + \lambda; x) &= (\Delta_x + \lambda) \sum_{\substack{(\Delta+\lambda)^\wedge(h)\neq 0 \\ h \in \mathbb{Z}}} \frac{1}{-(\Delta + \lambda)^\wedge(h)} \Phi_h(x) \\
&= \sum_{\substack{(\Delta+\lambda)^\wedge(h)\neq 0 \\ h \in \mathbb{Z}}} \Phi_h(x) \qquad (4.199)
\end{aligned}
$$

holds true (more precisely in a distributional sense) such that

$$
(\Delta_x + \lambda)G(\Delta + \lambda; x) = \delta_{\mathbb{Z}}(x) - \sum_{\substack{(\Delta+\lambda)^\wedge(h)= 0 \\ h \in \mathbb{Z}}} \Phi_h(x). \qquad (4.200)
$$

In other words, for $\lambda \in \mathrm{Spect}_\Delta(\mathbb{Z})$, the differential equation becomes inhomogeneous with the right side indicated by (4.200).

Remark 4.8. *It should be noted that the classical Bernoulli function B_2 of degree 2 is (apart from a factor) the one-dimensional $\mathbb{Z}$-lattice function corresponding to the eigenvalue $0 \in \mathrm{Spect}_\Delta(\mathbb{Z})$. In consequence, in the jargon of our approach, the standard one-dimensional analytic theory of numbers usually deals with a Green function corresponding to an eigenvalue, namely $\lambda = 0$.*

Altogether, from our formal considerations presented in (4.198), (4.199), and (4.200) we are led to the following introduction of $\mathbb{Z}$-lattice functions with respect to Helmholtz operators $\Delta + \lambda$, $\lambda \in \mathbb{R}$.

Definition 4.4. *A function $G(\Delta + \lambda; \cdot) : \mathbb{R} \to \mathbb{R}$ is called the Green function for the Helmholtz operator $\Delta + \lambda$, $\lambda \in \mathbb{R}$ with respect to the lattice $\mathbb{Z}$ (in brief, $\mathbb{Z}$-lattice function for $\Delta + \lambda$), if it fulfills the following properties:*

(i) (Periodicity) $G(\Delta + \lambda; \cdot)$ is continuous in $\mathbb{R}$, and

$$G(\Delta + \lambda; x + g) = G(\Delta + \lambda; x) \tag{4.201}$$

for all $x \in \mathbb{R}$ and $g \in \mathbb{Z}$.

(ii) (Differential equation) $G(\Delta + \lambda; \cdot)$ is twice continuously differentiable for all $x \notin \mathbb{Z}$ with

$$(\Delta + \lambda)G(\Delta + \lambda; x) = 0 \tag{4.202}$$

provided that $\lambda \notin \operatorname{Spect}_\Delta(\mathbb{Z})$,
  $G(\Delta + \lambda; \cdot)$ is twice continuously differentiable for all $x \notin \mathbb{Z}$ with

$$(\Delta + \lambda)G(\Delta + \lambda; x) = - \sum_{\substack{(\Delta+\lambda)^\wedge(h)=0 \\ h \in \mathbb{Z}}} \Phi_h(x) \tag{4.203}$$

provided that $\lambda \in \operatorname{Spect}_\Delta(\mathbb{Z})$ (note that the summation on the right side of (4.203) is to be taken over all lattice points $h \in \mathbb{Z}$ satisfying $(\Delta + \lambda)^\wedge(h) = 0$, i.e., $4\pi^2 h^2 - \lambda = 0$).

(iii) (Characteristic singularity)

$$x \mapsto G(\Delta + \lambda; \cdot) - \frac{1}{2}x \operatorname{sign}(x) \tag{4.204}$$

is continuously differentiable for all $x \in \mathcal{F}$.

(iv) (Normalization) For all $h \in \mathbb{Z}$ with $(\Delta + \lambda)^\wedge(h) = 0$,

$$\int_{\mathcal{F}} G(\Delta + \lambda; x)\overline{\Phi_h(x)} \, dx = 0. \tag{4.205}$$

It is not difficult to show that the difference $D(\Delta + \lambda; \cdot)$ between two $\mathbb{Z}$-lattice functions for $\Delta + \lambda$, $\lambda \in \mathbb{R}$, is a twice continuously differentiable function in $\mathbb{R}$ with

$$(\Delta + \lambda)D(\Delta + \lambda; x) = 0 \tag{4.206}$$

and

$$D(\Delta + \lambda; x + g) = D(\Delta + \lambda; x) \tag{4.207}$$

for all $g \in \mathbb{Z}$. Moreover, by partial integration it follows that

$$\int_{\mathcal{F}} D(\Delta + \lambda; x)\overline{\Phi_h(x)} \, dx = 0 \tag{4.208}$$

for all $h \in \mathbb{Z}$ with $(\Delta + \lambda)^\wedge(h) \neq 0$. In addition, the normalization condition (4.205) yields

$$\int_{\mathcal{F}} D(\Delta + \lambda; x)\overline{\Phi_h(x)} \, dx = 0 \tag{4.209}$$

for all $h \in \mathbb{Z}$ with $(\Delta + \lambda)^\wedge(h) = 0$. Hence, because of the completeness of the system $\{\Phi_h\}_{h\in\mathbb{Z}}$ in $L^2_\mathbb{Z}(\mathbb{R})$, we find $D(\Delta + \lambda; \cdot) = 0$.

Lemma 4.15. *(Uniqueness) For* $\lambda \in \mathbb{R}$, *the* $\mathbb{Z}$-*lattice function* $G(\Delta + \lambda; \cdot)$ *is uniquely determined by its constituting properties (as stated in Definition 4.4).*

Remark 4.9. *With the help of its Fourier series*

$$G(\Delta + \lambda; x) \quad = \sum_{\substack{(\Delta+\lambda)^\wedge(h)\neq 0 \\ h\in\mathbb{Z}}} \frac{1}{-(\Delta + \lambda)^\wedge(h)} \Phi_h(x) \qquad (4.210)$$

$$= \sum_{\substack{4\pi^2 h^2 \neq \lambda \\ h\in\mathbb{Z}}} \frac{1}{\lambda - 4\pi^2 h^2} \Phi_h(x),$$

$G(\Delta + \lambda; \cdot)$ *can be shown to be representable for all numbers* $\lambda \in \mathbb{R}$ *in closed form by means of elementary functions (see F. Oberhettinger [1973], W. Freeden, J. Fleck [1987]).*

Summation Formula for the Helmholtz Operator

In analogy to the Euler summation formula for the Laplace operator we are able to prove the following Euler summation formula for Helmholtz operators $\Delta + \lambda$, $\lambda \in \mathbb{R}$ (see Theorem 9.13 for the proof in the multi-variate theory).

Theorem 4.9. *(Euler Summation Formula for the Helmholtz Operator* $\Delta+\lambda$, $\lambda \in \mathbb{R}$) *Let* F *be of class* $C^{(2)}([a, b])$, $a < b$. *Then, the following identities hold true:*

If $\lambda \notin \mathrm{Spect}_\Delta(\mathbb{Z})$, *then*

$$\sideset{}{'}\sum_{\substack{g\in[a,b] \\ g\in\mathbb{Z}}} F(g) \qquad\qquad\qquad\qquad (4.211)$$

$$= \int_a^b G(\Delta + \lambda; x)(\Delta + \lambda)F(x) \, dx$$

$$+ \left\{ F(x) \left(\nabla G(\Delta + \lambda; x)\right) - (\nabla F(x)) \, G(\Delta + \lambda; x) \right\}\big|_a^b.$$

If $\lambda \in \mathrm{Spect}_\Delta(\mathbb{Z})$, *then*

$$\sideset{}{'}\sum_{\substack{g\in[a,b] \\ g\in\mathbb{Z}}} F(g) \quad = \sum_{\substack{(\Delta+\lambda)^\wedge(h)=0 \\ h\in\mathbb{Z}}} \int_a^b F(x)\overline{\Phi_h(x)} \, dx \qquad (4.212)$$

$$+ \int_a^b G(\Delta + \lambda; x)(\Delta + \lambda)F(x) \, dx$$

$$+ \left\{ F(x) \, \left(\nabla G(\Delta + \lambda; x)\right) - (\nabla F(x)) \, G(\Delta + \lambda; x) \right\}\big|_a^b.$$

The Euler summation formulas for the Helmholtz operators $\Delta + \lambda, \lambda \in \mathbb{R}$, show a striking difference: the first integral on the right side occurs only if λ is an eigenvalue, i.e., $\lambda \in \mathrm{Spect}_\Delta(\mathbb{Z})$. The sum in (4.212) is extended over all lattice points $h \in \mathbb{Z}$ with $(\Delta + \lambda)^\wedge(h) = 0$, i.e., $\Delta^\wedge(h) = 4\pi^2 h^2 = \lambda$. It seems that the extension of the Euler summation formula to Helmholtz operators (i.e., $\lambda \neq 0$) has never been used in more detail in classical number theory (even in the one-dimensional case).

Putting the information together and changing the lattice $\mathbb{Z}$ to a translated lattice $\mathbb{Z} + \{x\}$, we obtain the following variant of the Euler summation formula (Theorem 4.9).

Corollary 4.8. *Let x be an arbitrary point of $\mathbb{R}$. Suppose that F is of class* $\mathrm{C}^{(2)}([a, b])$. *Then*

$$
\sideset{}{'}\sum_{\substack{g+x \in [a,b] \\ g \in \mathbb{Z}}} F(g + x) = \sum_{\substack{(\Delta+\lambda)^\wedge(h)=0 \\ h \in \mathbb{Z}}} \int_a^b F(y) \overline{\Phi_h(y)} \; dy \; \Phi_h(x) \tag{4.213}
$$

$$
+ \int_a^b G(\Delta + \lambda; x - y)(\Delta_y + \lambda) F(y) \; dy
$$

$$
+ \left\{ F(y)(\nabla_y G(\Delta + \lambda; x - y)) - (\nabla_y F(y)) G(\Delta + \lambda; x - y) \right\} \big|_a^b,
$$

where the sum on the right side is to be taken over all points $h \in \mathbb{Z}$ for which $(\Delta + \lambda)^\wedge(h) = 4\pi^2 h^2 - \lambda = 0$. *In case of $(\Delta + \lambda)^\wedge(h) = 4\pi^2 h^2 - \lambda \neq 0$ for all* $h \in \mathbb{Z}$, *this sum is understood to be zero.*

Remark 4.10. *The Euler summation can be generalized by use of lattice functions to iterated Helmholtz operators (cf. Chapter 11 for the multi-dimensional theory). The details are omitted here; they can be taken immediately from our multi-dimensional theory.*

Easy to handle for practical purposes are summation formulas for which the boundary terms vanish.

Corollary 4.9. *Let $F \in \mathrm{C}^{(2)}([a, b])$ satisfy the homogeneous boundary conditions*

$$
F(a) = \nabla F(a) = F(b) = \nabla F(b) = 0. \tag{4.214}
$$

Then

$$
\sideset{}{'}\sum_{\substack{g+x \in [a,b] \\ g \in \mathbb{Z}}} F(g + x) = \sum_{\substack{(\Delta+\lambda)^\wedge(h)=0 \\ h \in \mathbb{Z}}} \int_a^b F(y) \overline{\Phi_h(y)} \; dy \; \Phi_h(x) \tag{4.215}
$$

$$
+ \int_a^b G(\Delta + \lambda; x - y)(\Delta_y + \lambda) F(y) \; dy.
$$

For $\mathbb{Z}$-periodical functions we obtain

Corollary 4.10. *Let F be of the class* $C_{\mathbb{Z}}^{(2)}(\mathbb{R})$. *Then, for each* $x \in \mathcal{F}$,

$$F(x) \quad = \quad \sum_{\substack{(\Delta+\lambda)^{\wedge}(h)=0 \\ h \in \mathbb{Z}}} \int_{\mathcal{F}} F(y)\overline{\Phi_h(y)} \, dy \, \Phi_h(x) \tag{4.216}$$

$$+ \int_{\mathcal{F}} G(\Delta + \lambda; x - y)(\Delta_y + \lambda)F(y) \, dy.$$

In other words, assume that H is of class $C_{\mathbb{Z}}^{(0)}(\mathbb{R})$ with

$$\int_{\mathcal{F}} H(y)\overline{\Phi_h(y)} \, dy = 0 \tag{4.217}$$

for all $h \in \mathbb{Z}$ with $(\Delta + \lambda)^{\wedge}(h) = 0$. Let $F \in C_{\mathbb{Z}}^{(2)}(\mathbb{R})$ satisfy

$$(\Delta + \lambda)F(y) = H(y), \tag{4.218}$$

such that

$$\int_{\mathcal{F}} F(y)\overline{\Phi_h(y)} \, dy = 0 \tag{4.219}$$

for all $h \in \mathbb{Z}$ with $(\Delta + \lambda)^{\wedge}(h) = 0$. Then

$$F(x) = \int_{\mathcal{F}} G(\Delta + \lambda; x - y)H(y) \, dy, \quad x \in \mathcal{F}. \tag{4.220}$$

4.6   Poisson Summation Formula for Helmholtz Operators

Next we are interested in a variant of the Poisson summation formula involving the (one-dimensional) Helmholtz differential operator $\Delta + \lambda$, $\lambda \in \mathbb{R}$. Our approach as presented in Theorem 4.10 is formulated in such a way that its generalization to higher dimensional Euclidean spaces $\mathbb{R}^q$, $q \geq 2$, becomes immediately obvious (for the details the reader is referred to Chapter 10).

Sufficient Criteria

Theorem 4.10. *(Modified Variant 3) For given values* $\varepsilon > 0$, $\lambda \in \mathbb{R}$, *let F be a member of class* $C^{(2)}(\mathbb{R})$ *satisfying the following properties:*

(i) F obeys the asymptotic relations

$$\begin{aligned} F(x) &\to 0, & |x| &\to \infty, \\ \nabla F(x) &\to 0, & |x| &\to \infty. \end{aligned} \tag{4.221}$$

(ii) $(\Delta + \lambda)F$ *fulfills the asymptotic relation*

$$(\Delta + \lambda)H(x) = O\left(|x|^{-(1+\varepsilon)}\right), \qquad |x| \to \infty. \tag{4.222}$$

(iii) for all $h \in \mathbb{Z}$ with $(\Delta + \lambda)^{\wedge}(h) = 0$, i.e., $4\pi^2 h^2 = \lambda$, the integrals

$$\int_{\mathbb{R}} H(x)\overline{\Phi_h(x)}\ dx < \infty \tag{4.223}$$

exist in the sense

$$\int_{\mathbb{R}} \ldots = \lim_{N \to \infty} \int_{-N}^{N} \ldots \ . \tag{4.224}$$

Then the Poisson summation formula

$$\lim_{N \to \infty} \sum_{\substack{|g| \leq N \\ g \in \mathbb{Z}}} F(g) = \sum_{h \in \mathbb{Z}} \int_{\mathbb{R}} F(x)\overline{\Phi_h(x)}\ dx. \tag{4.225}$$

holds true.

Proof. First, from Theorem 4.9, we obtain for all $N > 0$

$$\sum_{\substack{|g| \leq N \\ g \in \mathbb{Z}}}{}' F(g) = \sum_{\substack{(\Delta + \lambda)^{\wedge}(h) = 0 \\ h \in \mathbb{Z}}} \int_{-N}^{N} F(x)\overline{\Phi_h(x)}\ dx$$

$$+ \int_{-N}^{N} G(\Delta + \lambda; x)(\Delta + \lambda)F(x)\ dx$$

$$+ \left\{ F(x)\ (\nabla G(\Delta + \lambda; x)) - (\nabla F(x))\ G(\Delta + \lambda; x) \right\}\Big|_{-N}^{N}.$$

Assume that F satisfies (4.221). Then it follows that

$$\sum_{\substack{|g| = N \\ g \in \mathbb{Z}}} F(g) = o(1), \ N \to \infty. \tag{4.226}$$

For each $\lambda \in \mathbb{R}$, the $\mathbb{Z}$-lattice function $G(\Delta + \lambda; \cdot)$ and its first order derivative are bounded in $\mathbb{R}$; i.e., there exists a constant C such that

$$|G(\Delta + \lambda; x)| \leq C, \tag{4.227}$$
$$|\nabla G(\Delta + \lambda; x)| \leq C \tag{4.228}$$

for all $x \in \mathbb{R}$. By virtue of (4.221) we therefore are able to verify, for $N \to \infty$,

$$\left\{ F(x)\ (\nabla G(\Delta + \lambda; x)) - (\nabla F(x))\ G(\Delta + \lambda; x) \right\}\Big|_{-N}^{N} = o(1). \tag{4.229}$$

For the specified parameters $\varepsilon > 0$, $\lambda \in \mathbb{R}$, and for all $N > 0$, the asymptotic estimate (4.222) leads to

$$\left| \int_{-N}^{N} G(\Delta + \lambda; x)(\Delta + \lambda)F(x) \, dx \right| \tag{4.230}$$

$$\leq \ C \int_{-N}^{N} |(\Delta + \lambda)F(x)| \ dx$$

$$\leq C \int_{\mathbb{R}} |(\Delta + \lambda)F(x)| \ dx < \infty \ .$$

This shows that, for a function $F \in C^{(2)}(\mathbb{R})$ satisfying (4.221) and (4.222), we have

$$\lim_{N \to \infty} \left(\sum_{\substack{|g| \leq N \\ g \in \mathbb{Z}}} F(g) \ - \sum_{\substack{(\Delta+\lambda)^{\wedge}(h)=0 \\ h \in \mathbb{Z}}} \int_{-N}^{N} F(x)\overline{\Phi_h(x)} \, dx \right) \tag{4.231}$$

$$= \int_{\mathbb{R}} G(\Delta + \lambda; x)(\Delta + \lambda)F(x) \, dx.$$

Observing the absolute and uniform convergence of the Fourier expansion

$$G(\Delta + \lambda; x) \ = \sum_{\substack{(\Delta+\lambda)^{\wedge}(h)\neq 0 \\ h \in \mathbb{Z}}} \frac{1}{-(\Delta + \lambda)^{\wedge}(h)} \ \overline{\Phi_h(x)} \tag{4.232}$$

in $\mathbb{R}$ we find by Lebesgue's theorem

$$\int_{\mathbb{R}} G(\Delta + \lambda; x)(\Delta + \lambda)F(x) \, dx \tag{4.233}$$

$$= \sum_{\substack{(\Delta+\lambda)^{\wedge}(h)\neq 0 \\ h \in \mathbb{Z}}} \frac{1}{-(\Delta + \lambda)^{\wedge}(h)} \int_{\mathbb{R}} \overline{\Phi_h(x)}(\Delta + \lambda)F(x) \, dx.$$

For all $h \in \mathbb{Z}$ with $(\Delta+\lambda)^{\wedge}(h) \neq 0$ partial integration, in connection with the asymptotic relations (4.221), gives

$$\int_{\mathbb{R}} \overline{\Phi_h(x)} \left((\Delta + \lambda)F(x) \right) \, dx = -(\Delta + \lambda)^{\wedge}(h) \int_{\mathbb{R}} F(x)\overline{\Phi_h(x)} \, dx. \tag{4.234}$$

This yields

$$\lim_{N \to \infty} \left(\sum_{\substack{|g| \leq N \\ g \in \mathbb{Z}}} F(g) - \sum_{\substack{(\Delta+\lambda)^{\wedge}(h)=0 \\ h \in \mathbb{Z}}} \int_{-N}^{N} F(x)\overline{\Phi_h(x)} \, dx \right) \tag{4.235}$$

$$= \sum_{\substack{(\Delta+\lambda)^{\wedge}(h)\neq 0 \\ h \in \mathbb{Z}}} \int_{\mathbb{R}} F(x)\overline{\Phi_h(x)} \, dx.$$

We treat two cases:

(i) if $\lambda \notin \mathrm{Spect}_\Delta(\mathbb{Z})$ and $F \in C^{(2)}(\mathbb{R})$ satisfies (4.221) and (4.222), then we have

$$\lim_{N \to \infty} \sum_{\substack{|g| \leq N \\ g \in \mathbb{Z}}} F(g) = \sum_{h \in \mathbb{Z}} \int_{\mathbb{R}} F(x)\overline{\Phi_h(x)}\, dx. \tag{4.236}$$

(ii) if $\lambda \in \mathrm{Spect}_\Delta(\mathbb{Z})$ and and $F \in C^{(2)}(\mathbb{R})$ satisfies (4.221), (4.222), and (4.223), then we find

$$\lim_{N \to \infty} \sum_{\substack{|g| \leq N \\ g \in \mathbb{Z}}} F(g) = \sum_{\substack{(\Delta+\lambda)^\wedge(h)=0 \\ h \in \mathbb{Z}}} \int_{\mathbb{R}} F(x)\overline{\Phi_h(x)}\, dx \tag{4.237}$$

$$+ \sum_{\substack{(\Delta+\lambda)^\wedge(h)\neq 0 \\ h \in \mathbb{Z}}} \int_{\mathbb{R}} F(x)\overline{\Phi_h(x)}\, dx.$$

Summing up the expressions on the right side of (4.237) we obtain the desired result. $\qquad\qquad\square$

Remark 4.11. *The Poisson summation formula (Theorem 4.10) can be extended by use of iterated operators $(\Delta+\lambda)^m$, $\lambda \in \mathbb{R}$, $m \in \mathbb{N}$, in straightforward way (cf. Theorem 10.5 for the multi-dimensional case).*

Hardy–Landau Identity

Finally we come back to the one-dimensional Hardy–Landau identity on the interval $[-R, R]$, $R > 0$. Under the particular choice

$$\varepsilon = 1, \quad \lambda = 4\pi^2 R^2, \tag{4.238}$$

the sinc-function F, given by (4.192),

$$F(x) = \mathrm{sinc}(2\pi Rx) = \frac{\sin(2\pi Rx)}{2\pi Rx}, \quad x \in \mathbb{R},\ R > 0, \tag{4.239}$$

satisfies all properties (4.221), (4.222), and (4.223) listed in Theorem 4.10. Therefore we are allowed to deduce that

$$\lim_{N \to \infty} \sum_{\substack{|g| \leq N \\ g \in \mathbb{Z}}} \frac{\sin(2\pi Rg)}{2\pi Rg} = \sum_{h \in \mathbb{Z}} \int_{\mathbb{R}} \frac{\sin(2\pi Rx)}{2\pi Rx} e^{-2\pi ihx}\, dx. \tag{4.240}$$

The (discontinuous) integral on the right side of (4.240) can be calculated explicitly by elementary manipulations. Indeed, we have

$$\lim_{N \to \infty} \int_{-N}^{N} \frac{\sin(2\pi Rx)}{\pi x} e^{-2\pi ihx}\, dx = \begin{cases} 1 & , \quad |h| < R \\ \frac{1}{2} & , \quad |h| = R \\ 0 & , \quad |h| > R. \end{cases} \tag{4.241}$$

Thus, we end up with the already known (one-dimensional) Hardy–Landau identity (see (4.145))

$$\sum_{\substack{|h|\leq R \\ h\in\mathbb{Z}}}{}' 1 \;=\; 2R \;+\; 2R \sum_{\substack{g\neq 0 \\ g\in\mathbb{Z}}} \operatorname{sinc}(2\pi g R), \qquad (4.242)$$

but now as an immediate consequence of the Poisson summation formula, i.e., Theorem 4.10.

5

Preparatory Tools of Analytic Theory of Numbers

CONTENTS

5.1 Lattices in Euclidean Spaces 88
    Inverse Lattices 89
    Ingredients of Two-Dimensional Lattices 91
5.2 Basic Results of the Geometry of Numbers 92
    Covering and Filling Lattices 93
    Blichfeldt's and Minkowski's Lattice Point Theorem 95
5.3 Lattice Points Inside Circles 98
5.4 Lattice Points on Circles 105
5.5 Lattice Points Inside Spheres 113
5.6 Lattice Points on Spheres 118

In this chapter we recapitulate basic results of geometric and analytic theory of numbers (for more details the reader is referred to G.H. Hardy, E.M. Wright [1958], F. Fricker [1982], E. Hlawka et al. [1991], E. Krätzel [2000], and many others). Our intent is twofold: on the one hand, known key information of analytic number theory should be provided as preparation; on the other hand, the material should serve as an appropriate reference for our later work on lattice point summation.

The organization of this chapter is as follows: Section 5.1 introduces the concept of lattices and their inverse counterparts in Euclidean spaces $\mathbb{R}^q$. A closer look is directed at the two-dimensional case. Section 5.2 is devoted to some fundamentals of the geometry of numbers. Figure lattices are geometrically specified, from which ball lattices play a particular role in our context. The character of covering and filling lattices is explained. Blichfeldt's Theorem and Minkowski's Theorem are formulated within the standard framework of the geometry of numbers. Some procedures for representing an integer as the sum of finitely many integral squares are discussed. The sum of two integral squares is investigated in more detail. Moreover, as key results, the Lagrange Theorem and the Fermat–Euler Theorem are mentioned for later use. Section 5.3 lists the (classical) asymptotic relations for the total number of lattice points inside circles starting from the fundamental work due to C.F. Gauß [1801]. Our central feature is to explain that the ideas of counting lattice points inside circles have surprisingly deep connections to the sum of squares function, which occurs frequently in number theory. A bound on a solution, therefore, is a difficult task. In fact, the discussion of upper bounds leads to the famous Hardy conjecture. Its description demands a deeper study of the number of lattice points on circles which is provided in Section 5.4. Finally, Section 5.5 gives some basic information about lattice points inside and on spheres in Euclidean spaces $\mathbb{R}^q$, while Section 5.6 discusses lattice points on spheres in $\mathbb{R}^q$.

5.1   Lattices in Euclidean Spaces

Let $g_1, \ldots, g_q$ be linearly independent vectors in the q-dimensional Euclidean space $\mathbb{R}^q$.

Definition 5.1. *The set Λ (more precisely, $\Lambda_{(q)}$) of all points*

$$g = n_1 g_1 + \ldots + n_q g_q \tag{5.1}$$

$n_i \in \mathbb{Z}$, $i = 1, \ldots, q$, *is called a lattice in $\mathbb{R}^q$ with basis $g_1, \ldots, g_q$.*

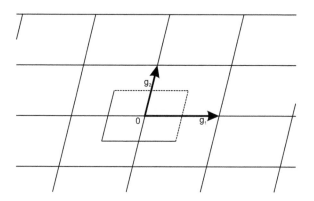

FIGURE 5.1
Two-dimensional lattice and its fundamental cell $\mathcal{F}$.

Clearly, the vectors $\epsilon^1, \ldots, \epsilon^q$ form a lattice basis of $\mathbb{Z}^q$. Trivially, a lattice basis $\{g_1, \ldots, g_q\}$ is related to the canonical basis $\{\epsilon^1, \ldots, \epsilon^q\}$ in $\mathbb{R}^q$ via the formula

$$g_i = \sum_{r=1}^{q} (g_i \cdot \epsilon^r) \, \epsilon^r. \tag{5.2}$$

Definition 5.2. *The half-open parallelotope $\mathcal{F}$ (more precisely, $\mathcal{F}_{(q)}$) consisting of the points $x \in \mathbb{R}^q$ with*

$$x = t_1 g_1 + \ldots + t_q g_q, \qquad -\frac{1}{2} \leq t_i < \frac{1}{2}, \tag{5.3}$$

$i = 1, \ldots, q$, *is called the fundamental cell of the lattice* Λ *(cf. Figure 5.1 for the two-dimensional case).*

Remark 5.1. *Obviously, there are infinitely many cells of* Λ *reflecting the* Λ*-periodicity.* $\mathcal{F}$*, as specified by (5.3), is both simple and appropriate for our purposes.*

From linear algebra (see, e.g., P.J. Davis [1963]) it is well known that the volume of $\mathcal{F}$ is equal to the quantity

$$\|\mathcal{F}\| = \int_{\mathcal{F}} dV = \sqrt{\det \left((g_i \cdot g_j)_{i,j=1,\ldots,q} \right)}. \tag{5.4}$$

For each $g \in \Lambda$ we have $\mathcal{F} + \{g\} = \{y + g \mid y \in \mathcal{F}\}$, such that

$$\|\mathcal{F}\| = \|\mathcal{F} + \{g\}\|. \tag{5.5}$$

Clearly, because of $(\mathcal{F} + \{g\}) \cap (\mathcal{F} + \{g'\}) = \emptyset$ for $g \neq g'$, $g, g' \in \Lambda$, we have

$$\mathbb{R}^q = \bigcup_{g \in \Lambda} (\mathcal{F} + \{g\}) = \bigcup_{g \in \Lambda} (\mathcal{F} - \{g\}). \tag{5.6}$$

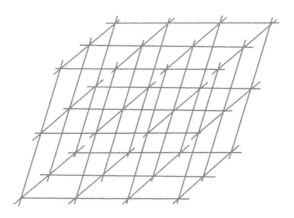

FIGURE 5.2
An illustration of a three-dimensional lattice.

Inverse Lattices

Since the vectors $g_1, \ldots, g_q$ are assumed to be linearly independent, there exists a system of vectors $h_1, \ldots, h_q$ in $\mathbb{R}^q$ such that

$$h_j \cdot g_i = \delta_{ij} = \begin{cases} 0 & , \quad i \neq j \\ 1 & , \quad i = j \end{cases} \tag{5.7}$$

(δ_{ij} is the Kronecker symbol). In more detail, for $i, j = 1, \ldots, q$, we let

$$g_i \cdot g_j = \gamma_{ij}. \tag{5.8}$$

The scalars γ^{ij}, $i, j = 1, \ldots, q$, are defined by

$$\sum_{j=1}^{q} \gamma^{ij} \gamma_{jk} = \delta_{ik}. \tag{5.9}$$

The vectors h_j, $j = 1, \ldots, q$, given by

$$h_j = \sum_{k=1}^{q} \gamma^{jk} g_k, \quad j = 1, \ldots, q, \tag{5.10}$$

satisfy the equations

$$h_j \cdot g_i = \sum_{k=1}^{q} \gamma^{jk} g_k \cdot g_i = \sum_{k=1}^{q} \gamma^{jk} \gamma_{ki} = \delta_{ji}, \tag{5.11}$$

$i, j = 1, \ldots, q$. Moreover, we find

$$
\begin{aligned}
h_i \cdot h_j &= \sum_{k=1}^{q} \gamma^{ik} g_k \cdot \sum_{l=1}^{q} \gamma^{jl} g_l \\
&= \sum_{l=1}^{q} \gamma^{jl} \sum_{k=1}^{q} \gamma^{ik} \gamma_{kl} \\
&= \gamma^{ji}, \quad i, j = 1, \ldots, q.
\end{aligned}
\tag{5.12}
$$

Definition 5.3. *The lattice with basis $h_1, \ldots, h_q$ given by (5.10) is called the inverse (or dual) lattice Λ^{-1} to Λ.*

The inverse lattice Λ^{-1} consists of all vectors $h \in \mathbb{R}^q$ such that the inner product $h \cdot g$ is an integer for all $g \in \Lambda$. Obviously,

$$\Lambda = (\Lambda^{-1})^{-1}. \tag{5.13}$$

Moreover, for the fundamental cell $\mathcal{F}_{\Lambda^{-1}}$ of the inverse lattice Λ^{-1} (throughout this work, denoted by $\mathcal{F}^{-1}$) we have

$$\|\mathcal{F}^{-1}\| = \|\mathcal{F}\|^{-1}. \tag{5.14}$$

Example 5.1. *Let $\Lambda = \tau \mathbb{Z}^q$ be the lattice generated by the "dilated" basis $\tau \epsilon^1, \ldots, \tau \epsilon^q$, where τ is a positive number and $\epsilon^1, \ldots, \epsilon^q$ forms the canonical orthonormal basis in $\mathbb{R}^q$. Then, the volume of the fundamental cell of $\tau \mathbb{Z}^q$ is $\|\mathcal{F}\| = \tau^q$. Generating vectors of the inverse lattice Λ^{-1} are $\tau^{-1} \epsilon^1, \ldots, \tau^{-1} \epsilon^q$. The volume of the fundamental cell of the inverse lattice is given by*

$$\|\mathcal{F}^{-1}\| = \tau^{-q} = \|\mathcal{F}\|^{-1}. \tag{5.15}$$

In particular, for $\tau = 1$, i.e., the lattice $\Lambda = \mathbb{Z}^q$, we have $\Lambda^{-1} = \mathbb{Z}^q = \Lambda$ such that

$$\|\mathcal{F}^{-1}\| = 1 = \|\mathcal{F}\|. \tag{5.16}$$

This fact has been used permanently in our one-dimensional theory (see Chapter 4), and it is always obvious throughout this work whenever $\Lambda = \mathbb{Z}^q$.

Remark 5.2. *Suppose that $(\mu_{ij})_{i,j=1,...,q}$ is a "unimodular" matrix; i.e., $(\mu_{ij})_{i,j=1,...,q}$ satisfies $\mu_{ij} \in \mathbb{Z}$, $i, j = 1, \ldots, q$, and $\det(\mu_{ij}) = \pm 1$. Assume $g_1, \ldots, g_q$ are the generating vectors of Λ. Then, $\sum_{j=1}^{q} \mu_{1j} g_j, \ldots, \sum_{j=1}^{q} \mu_{qj} g_j$ are also generating vectors of Λ since Λ is invariant with respect to unimodular transformations (for more details see, e.g., J.W.S. Cassels [1968], C.G. Lekkerkerker [1969]).*

Ingredients of Two-Dimensional Lattices

We conclude our brief introduction of lattices with some more detailed information for the planar case: assume that the two-dimensional lattice Λ is generated by the linearly independent vectors $g_1, g_2 \in \mathbb{R}^2$ with $\gamma_{ij} = g_i \cdot g_j$, $i, j = 1, 2$ (see Section 5.1). According to our construction, the scalars γ^{ij}, $i, j = 1, 2$, are related to γ_{ij}, $i, j = 1, 2$, by

$$\gamma^{11}\gamma_{11} + \gamma^{12}\gamma_{21} = 1 \quad , \quad \gamma^{11}\gamma_{12} + \gamma^{12}\gamma_{22} = 0, \tag{5.17}$$
$$\gamma^{21}\gamma_{11} + \gamma^{22}\gamma_{21} = 0 \quad , \quad \gamma^{21}\gamma_{12} + \gamma^{22}\gamma_{22} = 1, \tag{5.18}$$

where

$$h_j = \gamma^{j1} g_1 + \gamma^{j2} g_2, \quad j = 1, 2. \tag{5.19}$$

Furthermore

$$\|\mathcal{F}\|^2 = \gamma_{11}\gamma_{22} - \gamma_{12}\gamma_{21}, \tag{5.20}$$
$$\|\mathcal{F}\|^{-2} = \gamma^{11}\gamma^{22} - \gamma^{12}\gamma^{21}.$$

From (5.17) and (5.18) we readily obtain

$$\frac{\gamma^{12}}{\gamma^{11}} = -\frac{\gamma^{11}\gamma_{12}}{\gamma_{22}} \frac{1}{\gamma^{11}} = -\frac{\gamma_{12}}{\gamma_{22}} \tag{5.21}$$

and

$$\frac{\gamma^{21}}{\gamma^{22}} = -\frac{\gamma^{22}\gamma_{21}}{\gamma_{11}} \frac{1}{\gamma^{22}} = -\frac{\gamma_{21}}{\gamma_{11}}. \tag{5.22}$$

Moreover, in connection with (5.17) and (5.21), it follows that

$$\frac{1}{\gamma^{11}} = \frac{\|\mathcal{F}\|^2}{\gamma^{11}(\gamma_{11}\gamma_{22} - \gamma_{12}\gamma_{21})} = \frac{\|\mathcal{F}\|^2}{\gamma_{22}\left(\gamma^{11}\gamma_{11} - \frac{\gamma^{11}}{\gamma_{22}}\gamma_{12}\gamma_{21}\right)} \tag{5.23}$$

$$= \frac{\|\mathcal{F}\|^2}{\gamma_{22}\left(1 - \gamma^{12}\gamma_{21} - \frac{\gamma^{11}}{\gamma_{22}}\gamma_{12}\gamma_{21}\right)}$$

$$= \frac{\|\mathcal{F}\|^2}{\gamma_{22}\left(1 - \gamma^{12}\gamma_{21} + \gamma^{12}\gamma_{21}\right)} = \frac{\|\mathcal{F}\|^2}{\gamma_{22}}.$$

In an analogous way we find

$$\gamma^{22} = \frac{\gamma_{11}}{\|\mathcal{F}\|^2}. \tag{5.24}$$

5.2 Basic Results of the Geometry of Numbers

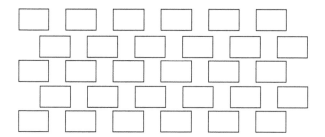

FIGURE 5.3
A two-dimensional figure lattice associated to a rectangle.

If $\Lambda \subset \mathbb{R}^q$ is a lattice and $x \in \mathbb{R}^q$, then $\Lambda + \{x\} \subset \mathbb{R}^q$ is called the *(point) lattice based at* x. If $\Lambda \subset \mathbb{R}^q$ is a lattice and $\mathcal{M} \subset \mathbb{R}^q$ is a set, then the sum $\mathcal{M} + \Lambda$ is called the *figure lattice* associated to $\mathcal{M}$ by Λ (see Figure 5.3). If $\mathcal{M}$ is a ball around the origin 0, then the figure lattice $\mathcal{M} + \Lambda$ is called a *ball lattice* (see Figure 5.4).

For a set $\mathcal{M} \subset \mathbb{R}^q$ we define *the characteristic function* $\mathcal{X}_{\mathcal{M}}$ by

$$\mathcal{X}_{\mathcal{M}}(x) = \begin{cases} 1 & , \quad x \in \mathcal{M} \\ 0 & , \quad x \notin \mathcal{M}. \end{cases} \tag{5.25}$$

From the considerations above it follows that a point $x \in \mathbb{R}^q$ belongs to the figure lattice $\mathcal{M} + \Lambda$ if and only if

$$\sum_{g \in \Lambda} \mathcal{X}_\mathcal{M}(x + g) \geq 1. \tag{5.26}$$

The point $x \in \mathbb{R}^q$ does not belong to the figure lattice $\mathcal{M} + \Lambda$ if and only if

$$\sum_{g \in \Lambda} \mathcal{X}_\mathcal{M}(x + g) = 0. \tag{5.27}$$

FIGURE 5.4
An example of a two-dimensional ball lattice (i.e., circle lattice).

Covering and Filling Lattices

If for all points from $\mathbb{R}^q$ the formula

$$\sum_{g \in \Lambda} \mathcal{X}_\mathcal{M}(x + g) \geq 1 \tag{5.28}$$

is valid, then we call Λ a *covering lattice* for $\mathcal{M}$ (see Figure 5.5) because in this case $\mathbb{R}^q = \mathcal{M} + \Lambda$. If, on the other hand, for all points $x \in \mathbb{R}^q$ the formula

$$\sum_{g \in \Lambda} \mathcal{X}_\mathcal{M}(x + g) \leq 1 \tag{5.29}$$

holds true, the lattice Λ is called a *filling lattice* for $\mathcal{M}$ (see Figure 5.6). In this case an arbitrary point $x \in \mathbb{R}^q$ can belong to at most one of the sets $\mathcal{M} + \{g\}$ with $g \in \Lambda$.

FIGURE 5.5
An example of a two-dimensional covering lattice.

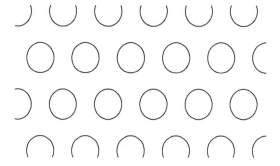

FIGURE 5.6
An example of a two-dimensional filling lattice.

Clearly, for the fundamental cell $\mathcal{F}$ of Λ, the lattice Λ simultaneously is a covering and filling lattice for $\mathcal{F}$. Furthermore it is clear that

$$
\int_{\mathbb{R}^q} F(x)\, dV_{(q)}(x) \;=\; \sum_{g \in \Lambda} \int_{\mathcal{F}-\{g\}} F(x)\, dV_{(q)}(x) \tag{5.30}
$$

$$
=\; \sum_{g \in \Lambda} \int_{\mathcal{F}} F(x+g)\, dV_{(q)}(x)
$$

$$
=\; \int_{\mathcal{F}} \left(\sum_{g \in \Lambda} F(x+g) \right) dV_{(q)}(x)
$$

holds for an integrable function F in $\mathbb{R}^q$. Consequently, if for some positive constant C we have

$$\int_{\mathbb{R}^q} F(x)\, dV_{(q)}(x) \geq C\, \|\mathcal{F}\|, \tag{5.31}$$

then for at least one point x of the fundamental cell $\mathcal{F}$ we must have

$$\sum_{g \in \Lambda} F(x+g) \geq C. \tag{5.32}$$

If conversely

$$\sum_{g \in \Lambda} F(x+g) < C \tag{5.33}$$

holds for all points $x \in \mathcal{F}$, the calculation

$$
\begin{aligned}
\int_{\mathbb{R}^q} F(x)\, dV_{(q)}(x) &= \int_{\mathcal{F}} \left(\sum_{g \in \Lambda} F(x+g) \right) dV_{(q)}(x) \\
&< C \int_{\mathcal{F}} dV_{(q)}(x) = C\, \|\mathcal{F}\|
\end{aligned}
$$

would give a contradiction to the assumption.

Blichfeldt's and Minkowski's Lattice Point Theorem

If we restrict ourselves to the function $F = \mathcal{X}_{\mathcal{M}}$, where $\mathcal{M}$ is a bounded measurable set, the volume $\|\mathcal{M}\|$ of $\mathcal{M}$ is given by

$$\int_{\mathbb{R}^q} \mathcal{X}_{\mathcal{M}}(x)\, dV_{(q)}(x) = \|\mathcal{M}\|. \tag{5.34}$$

The relation

$$\sum_{g \in \Lambda} \mathcal{X}_{\mathcal{M}}(x+g) \geq \frac{\|\mathcal{M}\|}{\|\mathcal{F}\|} \tag{5.35}$$

says that at least $\lceil \|\mathcal{M}\| / \|\mathcal{F}\| \rceil$ points of the form $x+g$, $g \in \Lambda$ lie in the set $\mathcal{M}$. This leads us to the following statement.

Theorem 5.1. *(Blichfeldt's Theorem) For each lattice Λ and each non-empty, bounded, measurable set $\mathcal{M}$ there exists a point $x \in \mathbb{R}^q$ such that the number N of lattice points of the lattice $\Lambda + \{x\}$ based at $x \in \mathbb{R}^q$ lying inside the set $\mathcal{M}$ satisfies the relation*

$$N \geq \frac{\|\mathcal{M}\|}{\|\mathcal{F}\|}. \tag{5.36}$$

In Theorem 5.1 little is assumed about the set $\mathcal{M}$. In particular, the point $x \in \mathbb{R}^q$ serving as origin for the lattice $\Lambda + \{x\}$ remains completely unspecified. If one wishes to avoid this lack of information, additional assumptions on the set $\mathcal{M}$ are necessary.

A set $\mathcal{M}$ is said to be *symmetrical* with respect to a point $z \in \mathbb{R}^q$ if $x = z + a$ belonging to the set implies $x' = z - a$ also does. The point z is called the center of symmetry of this set.

A set $\mathcal{M}$ is said to be *star-shaped* with respect to a point z if $x = z + a$ belonging to the set implies that $x_t = z + ta$ does also for all $t \in [0, 1]$.

A set $\mathcal{M}$ is called *convex* (see Figure 5.7) if it is star-shaped with respect to each of its points. A criterion for convex sets reads: a set $\mathcal{M}$ is convex if and only if, for each two points x and z belonging to it, so too does the line segment consisting of all points $y_t = x + t(z - x)$, $t \in [0, 1]$.

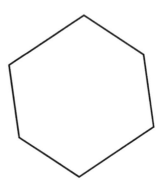

FIGURE 5.7
An example of a convex set in $\mathbb{R}^2$.

With these preparations we are able to formulate *Minkowski's lattice point theorem* that is classical in the geometry of numbers. For more details the reader is referred, e.g., to L.J. Mordell [1935], E. Hlawka [1954], E. Hlawka et al. [1991], and the references therein.

Theorem 5.2. *If x denotes the center of symmetry of the symmetrical, bounded, and convex region $\mathcal{G} \subset \mathbb{R}^q$, and $\Lambda \subset \mathbb{R}^q$ is a lattice with*

$$\|\mathcal{G}\| > 2^q \, \|\mathcal{F}\|, \tag{5.37}$$

then $\mathcal{G}$ contains at least one lattice point $z = x+g$, distinct from $x \in \mathbb{R}^q, g \in \Lambda$, from the lattice $\Lambda + \{x\}$ based at x.

Proof. The idea is to apply Blichfeldt's Theorem (Theorem 5.1) for the set $\mathcal{M}$ of all $x + \frac{1}{2}a \in \mathbb{R}^q$, as $x + a$ runs through all points of $\mathcal{G}$, i.e.,

$$\mathcal{M} = \left\{ x + \frac{1}{2}a \in \mathbb{R}^q \ \middle| \ x + a \in \mathcal{G} \right\}. \tag{5.38}$$

Since we have

$$\|\mathcal{M}\| = \frac{1}{2^q}\|\mathcal{G}\| > \frac{1}{2^q}2^q\|\mathcal{F}\| = \|\mathcal{F}\|, \tag{5.39}$$

by Blichfeldt's Theorem, at least two points of the form $w+g'$, $w+g''$ lie in $\mathcal{M}$ for distinct g', g'' from Λ. Since $x+(w-x+g')$, $x+(w-x+g'')$ lie in $\mathcal{M}$, by our assumption the points $x+2(w-x+g')$ and $x+2(w-x+g'')$ are in $\mathcal{G}$. Because $\mathcal{G}$ is symmetrical, the points $z' = x + 2(w - x + g')$, $z'' = x + 2(x - w - g'')$ lie in $\mathcal{G}$ and, because $\mathcal{G}$ is convex, their mid-point

$$z = z' + \frac{1}{2}(z'' - z') \tag{5.40}$$

is also in $\mathcal{G}$. Hence,

$$
\begin{aligned}
z &= x + 2(w - x + g') + \frac{1}{2}\left(2(x - w) - g''\right) - 2(w - x + g')) \\
  &= x + (g' - g'') \tag{5.41}
\end{aligned}
$$

is also in $\mathcal{G}$. Since $g = g' - g'' \neq 0$ this point is certainly a distinct lattice point from x, as required. □

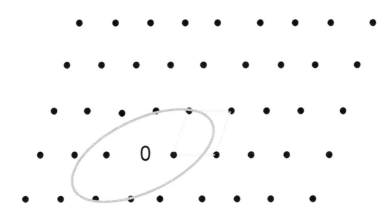

FIGURE 5.8
The geometric situation of Minkowski's theorem.

From the example of the cube

$$\mathcal{G} = \left\{ x + \sum_{i=1}^{q} t_i \epsilon^i \in \mathbb{R}^q \;\middle|\; t_i \in (-1, 1), \; i = 1, \ldots, q \right\} \tag{5.42}$$

we see that the condition $\|\mathcal{G}\| > 2^q \, \|\mathcal{F}\|$ cannot be weakened for Λ being the lattice $\mathbb{Z}^q$. However, we are able to formulate a variant of Minkowski's theorem (see Figure 5.8) showing the not-so-restrictive inequality $\|\mathcal{G}\| \geq 2^q \, \|\mathcal{F}\|$ (see, e.g., E. Hlawka et al. [1991]).

Corollary 5.1. *If x denotes the center of symmetry of the symmetrical, compact, convex set $\mathcal{K}$, and if $\Lambda \subset \mathbb{R}^q$ is a lattice with*

$$\|\mathcal{K}\| \geq 2^q \, \|\mathcal{F}\|, \tag{5.43}$$

then $\mathcal{K}$ contains at least one lattice point $y = x+g$ distinct from $x \in \mathbb{R}^q, g \in \Lambda$, from the lattice $\Lambda + \{x\}$ based at x.

Remark 5.3. *Our (standard) proof of Minkowski's Theorem (Theorem 5.2) is based on Mordell's approach (cf. L.J. Mordell [1935]; for more details the reader is referred to L.J. Mordell [1969], P. Erdös et al. [1989], E. Hlawka et al. [1991], and many others. For recent aspects in computer aided geometric design see R. Ait-Haddou et al. [2000]).*

In Chapter 12, we give a different proof of Minkowski's Theorem completely based on an analytic proposition, namely a multi-dimensional variant of the Poisson summation formula for regular regions. This concept is motivated by the two-dimensional approach presented in C. Müller [1956].

5.3   Lattice Points Inside Circles

The branch of analytic theory of numbers concerned with lattice point summation has a long history, which reaches back to L. Euler [1736a] and C.F. Gauß [1801]. Enlightening accounts of the developments within this theory are, e.g., due to G.H. Hardy [1915], E. Landau [1927], A. Walfisz [1927], to mention just a few.

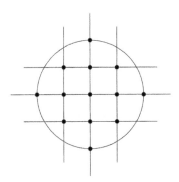

FIGURE 5.9
Lattice points inside a circle.

In what follows we are aiming to recapitulate some results on the number of lattice points inside circles $\mathbb{S}_N^1$ of radii $N > \frac{\sqrt{2}}{2}$ around the origin 0; more accurately, we deal with closed disks $\overline{\mathbb{B}_N^2}$ of radii $N > \frac{\sqrt{2}}{2}$ (for more background material the reader is referred, e.g., to the monographs F. Fricker [1982], E. Krätzel [1988], and the more recent survey paper A. Ivic et al. [2004]).

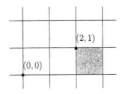

FIGURE 5.10
Northwest edges.

The problem of determining the total number of lattice points of $\mathbb{Z}^2$ inside and on a circle with radius N, i.e., the determination of the quantity

$$\#_{\mathbb{Z}^2}\left(\overline{\mathbb{B}_N^2}\right) = \#\left\{(n_1, n_2)^T \in \mathbb{Z}^2 \,\middle|\, n_1^2 + n_2^2 \le N^2\right\} \tag{5.44}$$

is very old. In fact, for the sum

$$\#_{\mathbb{Z}^2}\left(\overline{\mathbb{B}_N^2}\right) = \sum_{\substack{n_1^2 + n_2^2 \le N^2 \\ (n_1, n_2)^T \in \mathbb{Z}^2}} 1, \tag{5.45}$$

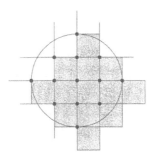

FIGURE 5.11
The polyhedral set $\mathbb{P}_N^2$.

C.F. Gauß [1801] found a simple, but efficient, method for its estimation: associate to every square (cf. Figure 5.10) the Northwest edge as lattice point. The union of all squares with lattice points inside $\overline{\mathbb{B}_N^2}$ defines a polyhedral set $\mathbb{P}_N^2$ with area $\|\mathbb{P}_N^2\| = \#_{\mathbb{Z}^2}\left(\overline{\mathbb{B}_N^2}\right)$ (cf. Figure 5.11). Since the diagonal of each square is $\sqrt{2}$, the geometry of Figure 5.11 tells us that

$$\pi \left(N - \frac{\sqrt{2}}{2}\right)^2 \leq \#_{\mathbb{Z}^2}\left(\overline{\mathbb{B}_N^2}\right) \leq \pi \left(N + \frac{\sqrt{2}}{2}\right)^2. \tag{5.46}$$

Therefore, $\#_{\mathbb{Z}^2}\left(\overline{\mathbb{B}_N^2}\right) - \pi N^2$ after division by N is bounded for $N \to \infty$, which is usually written with the O-symbol as

$$\#_{\mathbb{Z}^2}\left(\overline{\mathbb{B}_N^2}\right) = \pi N^2 + O(N). \tag{5.47}$$

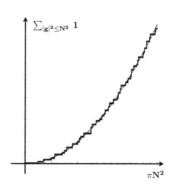

FIGURE 5.12
Two-dimensional lattice point sum versus circle area.

In other words, the number of lattice points in $\overline{\mathbb{B}_N^2}$ is equal to the area of that circle plus a remainder of the order of the boundary. In particular (cf. Figure 5.12),

$$\#_{\mathbb{Z}^2}\left(\overline{\mathbb{B}_N^2}\right) \sim \pi N^2 \tag{5.48}$$

i.e.,

$$\lim_{N \to \infty} \frac{\#_{\mathbb{Z}^2}\left(\overline{\mathbb{B}_N^2}\right)}{N^2} = \pi. \tag{5.49}$$

C.F. Gauß [1801] illustrated his result by taking $N^2 = 100\,000$. In this case he calculated

$$\sum_{\substack{|g|^2 \leq 100\,000 \\ g \in \mathbb{Z}^2}} 1 = 314\,197. \tag{5.50}$$

This determines the number π up to three decimals after the comma.

Altogether, in our nomenclature, the formula (5.47) of C.F. Gauß [1801] allows the representation

$$\#_{\mathbb{Z}^2}\left(\overline{\mathbb{B}_N^2}\right) = \pi N^2 + O(N). \tag{5.51}$$

The so–called *circle problem* is concerned with the question of determining the bound

$$\alpha_2 = \inf \left\{ \gamma \,\Big|\, \#_{\mathbb{Z}^2}\left(\overline{\mathbb{B}^2_{\sqrt{N}}}\right) = \pi N + O(N^\gamma) \right\}. \tag{5.52}$$

Until now, we knew from (5.51) that $\alpha_2 \leq \frac{1}{2}$. An improvement of the situation, however, turns out to be very laborious, in fact, requiring a great effort. A first remarkable result is due to W. Sierpinski [1906], who proved by use of a method of his teacher G. Voronoi [1903] that

$$\#_{\mathbb{Z}^2}\left(\overline{\mathbb{B}^2_{\sqrt{N}}}\right) = \pi N + O\left(N^{\frac{1}{3}}\right), \tag{5.53}$$

i.e., $\alpha_2 \leq \frac{1}{3}$. The proof of Sierpinski is elementary; it is a link between geometry and number theory. Today, his proof can be shortened by far, and a sketch of the proof should be presented (see, e.g., F. Fricker [1975]). Elementary manipulations give us the identities

$$\#_{\mathbb{Z}^2}\left(\overline{\mathbb{B}^2_{\sqrt{N}}}\right) = 1 + \sum_{\substack{0 < n_1^2 + n_2^2 \leq N \\ (n_1, n_2)^T \in \mathbb{Z}^2}} 1 \tag{5.54}$$

$$= 1 + \sum_{0 < n_1^2 \leq N} 1 + \sum_{0 < n_2^2 \leq N} 1 + \sum_{\substack{0 < n_1^2 + n_2^2 \leq N \\ n_1 \neq 0, n_2 \neq 0}} 1$$

$$= 1 + 2 \sum_{0 < n_1^2 \leq N} 1 + 4 \sum_{\substack{0 < n_1^2 + n_2^2 \leq N \\ n_1 > 0, n_2 > 0}} 1.$$

We are able to continue as follows

$$\#_{\mathbb{Z}^2}\left(\overline{\mathbb{B}^2_{\sqrt{N}}}\right) = 1 + 4\left\lfloor \sqrt{N} \right\rfloor \tag{5.55}$$

$$+ 4\left(\sum_{\substack{0 < n_1^2 + n_2^2 \leq N \\ 0 < n_1 \leq \sqrt{\frac{N}{2}} \\ n_2 > 0}} 1 + \sum_{\substack{0 < n_1^2 + n_2^2 \leq N \\ n_1 > 0 \\ 0 < n_2 \leq \sqrt{\frac{N}{2}}}} 1 - \sum_{\substack{0 < n_1^2 + n_2^2 \leq N \\ 0 < n_1 \leq \sqrt{\frac{N}{2}} \\ 0 < n_2 \leq \sqrt{\frac{N}{2}}}} 1 \right)$$

such that

$$\#_{\mathbb{Z}^2}\left(\overline{\mathbb{B}^2_{\sqrt{N}}}\right) = 1 + 4\left\lfloor \sqrt{N} \right\rfloor$$

$$+ 4\left(2 \sum_{\substack{0 < n_1^2 + n_2^2 \leq N \\ 0 < n_1 \leq \sqrt{\frac{N}{2}} \\ n_2 > 0}} 1 - \sum_{\substack{0 < n_1 \leq \sqrt{\frac{N}{2}} \\ 0 < n_2 \leq \sqrt{\frac{N}{2}}}} 1 \right)$$

and

$$\#_{\mathbb{Z}^2}\left(\overline{\mathbb{B}^2_{\sqrt{N}}}\right) = 1 + 4\left\lfloor\sqrt{N}\right\rfloor \tag{5.56}$$

$$+ 8 \sum_{0<n_1\leq\sqrt{\frac{N}{2}}} \sum_{0<n_2\leq\sqrt{N-n_1^2}} 1 - 4\left\lfloor\sqrt{\frac{N}{2}}\right\rfloor^2.$$

Finally, we arrive at

$$\#_{\mathbb{Z}^2}\left(\overline{\mathbb{B}^2_{\sqrt{N}}}\right) = 1 + 4\left\lfloor\sqrt{N}\right\rfloor + 8\sum_{n\leq\sqrt{\frac{N}{2}}}\left\lfloor\sqrt{N-n^2}\right\rfloor - 4\left\lfloor\sqrt{\frac{N}{2}}\right\rfloor^2. \tag{5.57}$$

Expressing in (5.57) $\lfloor x\rfloor$ in terms of $G(\nabla;x)$ and applying the Euler summation formula to the sum $\sum_{n\leq\sqrt{\frac{N}{2}}}\sqrt{N-n^2}$, we obtain after some elementary manipulations

$$\#_{\mathbb{Z}^2}\left(\overline{\mathbb{B}^2_{\sqrt{N}}}\right) = \pi N + 8\sum_{n\leq\sqrt{\frac{N}{2}}} G\left(\nabla;\sqrt{N-n^2}\right) + O(1). \tag{5.58}$$

The rough estimate $|G(\nabla;x)| \leq \frac{1}{2}$ in (5.58) provides us with the Gaussian result (5.47).

By a "tour de force", van der Corput [1928] was able to formulate the following analytic tool (which will not be proved within this context).

Lemma 5.1. *Suppose that, for $F \in C^{(2)}([a,b])$, either $\Delta F(x) \geq \beta$ or $\Delta F(x) \leq -\beta$ for all $x \in [a,b]$ with some constant $\beta \in (0,1]$. Then*

$$\left|\sum_{a\leq l\leq b} G\left(\nabla;F(l)\right)\right| \leq C\left(|\nabla F(b) - \nabla F(a)|\ \beta^{-\frac{2}{3}} + \beta^{-\frac{1}{2}}\right). \tag{5.59}$$

Now, Sierpinski's asymptotic expansion (5.53) is an immediate consequence of Lemma 5.1: choose, especially, $F(x) = \sqrt{N-x^2}$. This function is twice continuously differentiable on the interval $\left[1,\sqrt{\frac{N}{2}}\right]$. Its derivatives can be calculated easily, namely

$$\nabla F(x) = F'(x) = -\frac{x}{\sqrt{N-x^2}} \tag{5.60}$$

and

$$\Delta F(x) = F''(x) = -\frac{N}{\sqrt{(N-x^2)^3}}. \tag{5.61}$$

It follows that

$$\Delta F(x) = F''(x) \leq -\frac{N}{\sqrt{N^3}} = -\beta \tag{5.62}$$

with $\beta = N^{-\frac{1}{2}}$. Since $0 < \beta < 1$ for $N > 2$ the lemma of van der Corput [1928] (i.e., Lemma 5.1) can be applied such that

$$\left| \sum_{n \leq \sqrt{\frac{N}{2}}} G(\nabla; \sqrt{N - n^2}) \right| = \left| \sum_{1 \leq l \leq \sqrt{\frac{N}{2}}} G(\nabla; F(l)) \right| \tag{5.63}$$

$$\leq C \left(\left| -1 + \frac{1}{\sqrt{N - 1}} \right| N^{\frac{1}{3}} + N^{\frac{1}{4}} \right)$$

$$\leq C \left(N^{\frac{1}{3}} + N^{\frac{1}{4}} \right)$$

$$\leq 2CN^{\frac{1}{3}}.$$

Consequently, we arrive at the estimate

$$\#_{\mathbb{Z}^2} \left(\overline{\mathbb{B}^2_{\sqrt{N}}} \right) = \pi N + O \left(N^{\frac{1}{3}} \right). \tag{5.64}$$

By use of advanced methods on exponential sums (based on the work by, e.g., H. Weyl [1916], L.-K. Hua [1959], H.R. Chen [1963], and many others) the estimate $\frac{1}{3}$ could be strengthened to some extent. It culminated in the publication by G. Kolesnik [1985], who had as his sharpest result with these techniques

$$\#_{\mathbb{Z}^2} \left(\overline{\mathbb{B}^2_{\sqrt{N}}} \right) - \pi N = O \left(N^{\frac{139}{429}} \right). \tag{5.65}$$

M.N. Huxley [2003] devised a substantially new approach (not discussed here); his strongest result was the estimate

$$\#_{\mathbb{Z}^2} \left(\overline{\mathbb{B}^2_{\sqrt{N}}} \right) - \pi N = O \left(N^{\frac{131}{416}} \right). \tag{5.66}$$

Note that $\frac{139}{429} = 0.324009...$, while $\frac{131}{416} = 0.315068...$.

Remark 5.4. *Hardy's conjecture tells us that*

$$\#_{\mathbb{Z}^2} \left(\overline{\mathbb{B}^2_{\sqrt{N}}} \right) - \pi N = O \left(N^{\frac{1}{4}+\varepsilon} \right) \tag{5.67}$$

*for **every** $\varepsilon > 0$. This conjecture seems to be still a challenge for future work. However, in the year 2007, S. Cappell and J. Shaneson deposited a paper entitled "Some Problems in Number Theory I: The Circle Problem" in the arXiv:math/0702613 claiming to prove the bound of $O \left(N^{\frac{1}{4}+\varepsilon} \right)$ for $\varepsilon > 0$.*

G.H. Hardy [1916] also contributed significantly to the understanding of the circle problem by detecting a lower bound,

$$\#_{\mathbb{Z}^2} \left(\overline{\mathbb{B}^2_{\sqrt{N}}} \right) - \pi N = \Omega \left(N^{\frac{1}{4}} \right); \tag{5.68}$$

i.e., there is a constant C and arbitrarily large values of N so that

$$\left| \#_{\mathbb{Z}^2} \left(\overline{\mathbb{B}^2_{\sqrt{N}}} \right) - \pi N \right| > CN^{\frac{1}{4}} \tag{5.69}$$

is satisfied. For the circle problem this means $\alpha_2 \geq \frac{1}{4}$. Stronger $\Omega-$bounds have been developed in many papers; the note A. Ivic et al. [2004] provides an excellent overview.

0.250000	Gauß (1801)
0.083333...	G. Voronoi (1903), Sierpinski (1906)
0.080357...	J.E. Littlewood, A. Walfisz (1924)
0.079268...	J.G. van der Corput (1928)
0.074324...	J.-R. Chen (1963)
0.074009...	G. Kolesnik (1985)
0.064903...	M.N. Huxley (2003)

TABLE 5.1
Incremental improvements for the values ε_2 in the estimate (5.70).

Table 5.1 summarizes incomplete incremental improvements for the quantities ε_2 of the upper limit for the circle problem

$$\#_{\mathbb{Z}^2}\left(\overline{\mathbb{B}^2_{\sqrt{N}}}\right) - \pi N = O\left(N^{\frac{1}{4}+\varepsilon_2}\right). \tag{5.70}$$

For all recent improvements, the proofs became rather long and made use of some of the more heavy machinery in analysis.

Summarizing our results about lattice points inside circles we are confronted with the following situation:

$$\frac{1}{4} \leq \alpha_2 \leq \frac{1}{4} + \varepsilon_2 \tag{5.71}$$

and

$$\#_{\mathbb{Z}^2}\left(\overline{\mathbb{B}^2_{\sqrt{N}}}\right) - \pi N \neq O\left(N^{\frac{1}{4}}\right), \tag{5.72}$$

$$\#_{\mathbb{Z}^2}\left(\overline{\mathbb{B}^2_{\sqrt{N}}}\right) - \pi N = O\left(N^{\frac{1}{4}+\varepsilon_2}\right), \tag{5.73}$$

where $0 < \varepsilon_2 \leq \frac{1}{4}$ (e.g., Huxley's bound $\varepsilon_2 = 0.064903\ldots$).

Remark 5.5. *There are many perspectives to formulate variants of lattice point problems for the circle. It already was the merit of E. Landau [1924] to point out particularly interesting areas, such as*

- *general lattices can be used instead of the unit lattice $\mathbb{Z}^2$,*

- *lattice points can be affected by non-constant weights,*

- *multi-dimensional generalizations can be described, e.g., the lattice point problem for spheres in $\mathbb{R}^q$.*

5.4 Lattice Points on Circles

Before we come to a result on the representation of an integer as the sum of two integral squares we give some preparatory information: as usual, the number of representations of n by two squares; i.e., the number of integral solutions of $n_1^2 + n_2^2 = n$ is denoted by $r(n)$ (sometimes more explicitly, $r_2(n)$):

$$r(n) = \sum_{\substack{n_1^2+n_2^2=n \\ (n_1,n_2)^T \in \mathbb{Z}^2}} 1 \ = \sum_{\substack{|g|^2=n \\ g \in \mathbb{Z}^2}} 1. \tag{5.74}$$

We have to pay attention to the sign and order of $n_1, n_2 \in \mathbb{Z}$. For example, we have

$$1 = (\pm 1)^2 + 0^2 = 0^2 + (\pm 1)^2, \tag{5.75}$$
$$5 = (\pm 2)^2 + (\pm 1)^2 = (\pm 1)^2 + (\pm 2)^2; \tag{5.76}$$

hence, $r(1) = 4$, $r(5) = 8$. If $n_1^2 + n_2^2 = n$ possesses no solution, we set $r(n) = 0$. For example, $r(n) = 0$ whenever n is a member of the arithmetical sequence $\{3, 7, 11, \ldots\}$.

The problem of determining $r(n)$ in terms of simpler arithmetical functions of n can be attacked by use of the Theta function (see, e.g., G.H. Hardy [1940], E. Landau [1962], H. Rademacher [1973], and the references therein). For $\sigma \in \mathbb{C}$ with $\Re(\sigma) > 0$ we have the expression (see Lemma 4.14)

$$\alpha(\sigma) = (\vartheta(\sigma; 0, 0; \mathbb{Z}))^2 = \left(\sum_{g \in \mathbb{Z}} e^{-\pi \sigma g^2}\right)^2. \tag{5.77}$$

It is not hard to see that

$$\alpha(\sigma) = (\vartheta(\sigma; 0, 0; \mathbb{Z}))^2 = \sum_{n=0}^{\infty} r(n)\, e^{-\pi\sigma n}. \tag{5.78}$$

Moreover, elementary manipulations show that α is holomorphic for all $\sigma \in \mathbb{C}$ with $\Re(\sigma) > 0$ such that the following properties are satisfied:

$$(P1) \qquad \alpha(\sigma + i) = \frac{4}{\sigma} \left(e^{-\frac{\pi}{2\sigma}} + o\left(e^{-\frac{\pi}{2\sigma}} \right) \right), \ \sigma \to 0, \tag{5.79}$$

$$(P2) \qquad \lim_{\Re(\sigma)\to\infty} \alpha(\sigma) = 1, \tag{5.80}$$

$$(P3) \qquad \alpha(\sigma) = \frac{1}{\sigma}\, \alpha\left(\frac{1}{\sigma} \right), \tag{5.81}$$

$$(P4) \qquad \alpha(\sigma + 2i) = \alpha(\sigma). \tag{5.82}$$

In addition, the function β (cf. Lemma 4.11) given by

$$\beta(\sigma) = \sum_{g\in\mathbb{Z}} \frac{1}{\cosh(\pi\sigma g)} \tag{5.83}$$

is a holomorphic function for all $\sigma \in \mathbb{C}$ with $\Re(\sigma) > 0$ such that β satisfies the same properties *(P1), (P2), (P3), (P4)*.

In consequence, the function $w : \overline{D} \to \mathbb{C}$, $\overline{D} = D \cup \partial D$, given by

$$w(\sigma) = \frac{\beta(\sigma)}{\alpha(\sigma)}, \qquad \sigma \in \overline{D}, \tag{5.84}$$

is holomorphic in the region D (see Figure 5.13), bounded and continuous on $\overline{D}$ such that

$$w(\tau) = w\left(\frac{1}{\tau} \right) \tag{5.85}$$

and

$$w(\tau) = w\left(\tau + 2\, i \right). \tag{5.86}$$

Then the theory of elliptic functions (see, e.g., K. Knopp [1971]) implies that

$$w = const, \tag{5.87}$$

from which we know by virtue of (P1) that

$$\lim_{\sigma\to 0} w(\sigma) = 1. \tag{5.88}$$

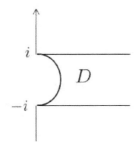

FIGURE 5.13
The region D implying Lemma 5.2.

Summarizing our results we therefore obtain

Lemma 5.2. *For $\sigma \in \mathbb{C}$ with $\Re(\sigma) > 0$*

$$(\vartheta(\sigma; 0, 0; \mathbb{Z}))^2 = \sum_{n=0}^{\infty} r(n)\, e^{-\pi\sigma n} = \sum_{n=-\infty}^{\infty} \frac{1}{\cosh(\pi\sigma n)}. \qquad (5.89)$$

An easy manipulation yields for all $\sigma \in \mathbb{R}$ with $\sigma > 0$

$$\sum_{n=-\infty}^{\infty} \frac{1}{\cosh(\pi\sigma n)} = 1 + 4 \sum_{n=1}^{\infty} \frac{e^{-\pi\sigma n}}{1 + e^{-2\pi\sigma n}} \qquad (5.90)$$

$$= 1 + 4 \sum_{n=1}^{\infty} \frac{e^{-\pi\sigma n}}{1 - e^{-4\pi\sigma n}} - 4 \sum_{n=1}^{\infty} \frac{e^{-3\pi\sigma n}}{1 - e^{-4\pi\sigma n}}$$

$$= 1 + 4 \sum_{n=1}^{\infty} \sum_{k=0}^{\infty} e^{-(4k+1)\pi\sigma n} - 4 \sum_{n=1}^{\infty} \sum_{k=0}^{\infty} e^{-(4k+3)\pi\sigma n}.$$

Thus it follows by equating the coefficients that the problem of determining $r(n)$ by simpler arithmetical functions of n reduces to a problem of calculating a sum of its divisors.

Theorem 5.3. *For $n \in \mathbb{N}$, the solution number $r(n)$ of the equation*

$$n_1^2 + n_2^2 = n, \quad n_1, n_2 \in \mathbb{Z}, \qquad (5.91)$$

is equal to

$$r(n) = 4(d_1(n) - d_3(n)), \qquad (5.92)$$

where $d_1(n)$ and $d_3(n)$ are the numbers of divisors of n of the form $4m + 1$ and $4m + 3$, respectively.

As examples we mention $d_1(7) = d_3(7) = 1$; hence, $r(7) = 0$ and $d_1(5^m) = m + 1$, $d_3(5^m) = 0$, hence, $r(5^m) = 4(m + 1)$.

For all $x \in \mathbb{R}$ with $|x| < 1$ we find (cf. G.H. Hardy, E.M. Wright [1958])

$$\left(\sum_{n=-\infty}^{\infty} x^{n^2}\right)^2 = \left(1 + 2\sum_{m=1}^{\infty} x^{m^2}\right)^2 \tag{5.93}$$

$$= 1 + 4\sum_{k=0}^{\infty}(-1)^k \frac{x^{2k+1}}{1 - x^{2k+1}}$$

and

$$\left(\sum_{n=-\infty}^{\infty} x^{n^2}\right)^2 = 1 + \sum_{n=1}^{\infty} r(n)x^n. \tag{5.94}$$

Now it is clear that

$$\frac{1}{1 - x^{2k+1}} = \sum_{l=0}^{\infty} x^{(2k+1)l}. \tag{5.95}$$

Hence, we get

$$\left(\sum_{n=-\infty}^{\infty} x^{n^2}\right)^2 = 1 + 4\sum_{l=0}^{\infty}\sum_{k=0}^{\infty}(-1)^k x^{(2k+1)(l+1)} \tag{5.96}$$

$$= 1 + 4\sum_{n=1}^{\infty}\left(\sum_{\substack{(2k+1)|n \\ k \in \mathbb{N}_0}} (-1)^k\right) x^n.$$

Consequently, by comparison of the coefficients of the sums (5.94) and (5.96), we obtain

Lemma 5.3. *For $n \in \mathbb{N}$*

$$r(n) = 4\sum_{\substack{(2k+1)|n \\ k \in \mathbb{N}_0}} (-1)^k, \tag{5.97}$$

where the sum is extended over all non-negative integers k, for which the odd numbers $2k+1$ are divisors of n.

Example 5.2. *929 is a prime number, and we have*

$$r\left(929^2\right) = 4\sum_{(2k+1)|929^2} (-1)^k \tag{5.98}$$

$$= 4\left((-1)^0 + (-1)^{464} + (-1)^{930 \cdot 464}\right)$$
$$= 4(1 + 1 + 1)$$
$$= 12.$$

Moreover, from (5.97) it is not difficult to see that

$$r(2n) = \sum_{\substack{(2k+1)|2n \\ k \in \mathbb{N}_0}} (-1)^k = \sum_{\substack{(2k+1)|n \\ k \in \mathbb{N}_0}} (-1)^k = r(n). \tag{5.99}$$

Further consequences are obtainable in the following way (see, e.g., G.H. Hardy, E.M. Wright [1958]):

Lemma 5.4. *Let p be a prime number of the form $4n + 1$. Suppose that P is an arbitrary positive integer with $(P, p) = 1$. Then*

$$r(p^\alpha P) = (1 + \alpha)\, r(P) \tag{5.100}$$

holds for all positive integers α.

Proof. It can be readily seen that

$$r\left(p^\alpha P\right) = 4 \sum_{\substack{(2k+1)|p^\alpha P \\ k \geq 0}} (-1)^k = 4 \sum_{\substack{p^i(2k'+1)|p^\alpha P \\ (2k'+1,p)=1 \\ k' \geq 0,\ \alpha \geq i \geq 0}} (-1)^{\frac{p^i(2k'+1)-1}{2}}, \tag{5.101}$$

such that

$$r\left(p^\alpha P\right) = 4 \sum_{i=0}^{\alpha} \sum_{\substack{(2k'+1)|P \\ k' \geq 0}} (-1)^{k'} = (1 + \alpha)\, r(P), \tag{5.102}$$

since $p = 4n + 1$ implies $p^i = 4N + 1$; hence,

$$\frac{p^i(2k'+1) - 1}{2} = 2N\,(2k'+1) + k' \tag{5.103}$$

(note that the assumption $(P, p) = 1$ is essential in our consideration). $\qquad\square$

Lemma 5.5. *Let q be a prime number of the form $4n - 1$. Suppose that Q is an arbitrary positive integer with $(Q, q) = 1$. Then*

$$r(q^\beta Q) = \frac{1 + (-1)^\beta}{2}\, r(Q) \tag{5.104}$$

holds for all non-negative integers β.

Proof. We notice that

$$r\left(q^\beta Q\right) = 4 \sum_{\substack{(2k+1)|q^\beta Q \\ k \geq 0}} (-1)^k = 4 \sum_{\substack{q^i(2k'+1)|q^\beta Q \\ (q,2k'+1)=1 \\ k' \geq 0,\ \beta \geq i \geq 0}} (-1)^{\frac{q^i(2k'+1)-1}{2}}. \tag{5.105}$$

In other words, we find

$$r\left(q^\beta Q\right) \quad = \quad 4\sum_{i=0}^{\beta} \sum_{\substack{(2k'+1)\mid Q \\ k'\geq 0}} (-1)^{k'+i}. \tag{5.106}$$

Note that, because of $q^i = 4M + (-1)^i$, we have

$$(-1)^{\frac{q^i(2k'+1)-1}{2}} = (-1)^{(-1)^i k' + \frac{(-1)^i-1}{2}} = (-1)^{k'+i}, \tag{5.107}$$

so that

$$r\left(q^\beta Q\right) \quad = \quad \sum_{\substack{(2k'+1)\mid Q \\ k'\geq 0}} (-1)^{k'} \left(4\sum_{i=0}^{\beta}(-1)^i\right) = r(Q) \sum_{i=0}^{\beta}(-1)^i. \tag{5.108}$$

Observing the identity

$$\sum_{i=0}^{\beta}(-1)^i \quad = \quad \frac{1+(-1)^\beta}{2}. \tag{5.109}$$

we get the announced result from (5.108) (again, the assumption $(Q,q) = 1$ is essential). $\qquad\qquad\square$

Summarizing our results we obtain (by observing $r(1) = 4$)

Theorem 5.4. *(Fermat–Euler Theorem) Let p_k and q_l be prime numbers of the form $4m+1$ and $4m-1$, respectively. If $n \in \mathbb{N}$ is of the form*

$$n = 2^\gamma \prod_{k=1}^{i} p_k^{\alpha_k} \prod_{l=1}^{j} q_l^{\beta_l}, \tag{5.110}$$

then

$$r\left(2^\gamma \prod_{k=1}^{i} p_k^{\alpha_k} \prod_{l=1}^{j} q_l^{\beta_l}\right) = 4\prod_{k=1}^{i}(1+\alpha_k)\prod_{l=1}^{j} \frac{1+(-1)^{\beta_l}}{2}. \tag{5.111}$$

From Theorem 5.4 it follows that $r(n) \neq 0$ if and only if the prime numbers q_l in n are of even power. In particular, we have

$$r\left(n^2\right) \geq 4. \tag{5.112}$$

All in all, for n of the form (5.110) we are able to write

$$d_1(n) = \frac{1}{2} \prod_{k=1}^{i}(1+\alpha_k) \left(\prod_{l=1}^{j}(1+\beta_l) + \prod_{l=1}^{j}\frac{1+(-1)^{\beta_l}}{2}\right) \tag{5.113}$$

and

$$d_3(n) = \frac{1}{2} \prod_{k=1}^{i} (1 + \alpha_k) \left(\prod_{l=1}^{j} (1 + \beta_l) - \prod_{l=1}^{j} \frac{1 + (-1)^{\beta_l}}{2} \right) \tag{5.114}$$

hence,

$$r(n) = 4(d_1(n) - d_3(n)) \tag{5.115}$$

and

$$\prod_{k=1}^{i} (1 + \alpha_k) \prod_{l=1}^{j} (1 + \beta_l) = d_1(n) + d_3(n). \tag{5.116}$$

Next we are interested in the asymptotic behavior of $r(n)$ for $n \to \infty$. On the one hand,

$$\liminf_{n \to \infty} r(n) = 0 \tag{5.117}$$

since $r\left(2^k m\right) = 0$ for $m \equiv 3 \mod 4$. On the other hand,

$$\limsup_{n \to \infty} r(n) = \infty \tag{5.118}$$

since $r\left(p^k\right) = 4(k + 1)$. More precise results (see, e.g., F. Fricker [1982]) can be obtained by observing $r(n) \le 4\,d(n)$, where $d(n)$ is the number of positive divisors of n. In fact, the following O-estimate of $r(n)$ is motivated by this observation together with the decomposition (5.110).

Lemma 5.6. *For $n \to \infty$ and arbitrary $\varepsilon > 0$*

$$\frac{r(n)}{n^\varepsilon} = O(1). \tag{5.119}$$

Proof. We start from the decomposition (5.110)

$$n = 2^\gamma \prod_{k=1}^{i} p_k^{\alpha_k} \prod_{l=1}^{j} q_l^{\beta_l}, \tag{5.120}$$

where p_k and q_l is the k-th prime number of the form $4m + 1$ and the l-th prime number of the form $4m + 3$, respectively. It follows that

$$\frac{r(n)}{n^\varepsilon} \le 4 \prod_{k=1}^{i} \frac{1 + \alpha_k}{p_k^{\varepsilon \alpha_k}}. \tag{5.121}$$

The denominator of the factors can be estimated by

$$p_k^{\varepsilon \alpha_k} \ge 2^{\varepsilon \alpha_k} = e^{\varepsilon \alpha_k \ln(2)} > \varepsilon \alpha_k \ln(2); \tag{5.122}$$

hence, the factors themselves admit the estimates

$$\frac{1 + \alpha_k}{p_k^{\varepsilon \alpha_k}} < \frac{2 \alpha_k}{\varepsilon \, \alpha_k \, \ln(2)} < \frac{4}{\varepsilon}. \tag{5.123}$$

We use these estimates only in the case that $p_k \le 2^{\frac{1}{\varepsilon}}$ (note that there are at most $2^{\frac{1}{\varepsilon}}$ numbers p_k such that $p_k \le 2^{\frac{1}{\varepsilon}}$). Even better, for the remaining factors we have

$$\frac{1+\alpha_k}{p_k^{\varepsilon \alpha_k}} < \frac{1+\alpha_k}{(p_k^\varepsilon)^{\alpha_k}} < \frac{1+\alpha_k}{2^{\alpha_k}} \le 1. \tag{5.124}$$

Combining all results we finally get

$$\frac{r(n)}{n^\varepsilon} \le 4 \left(\frac{4}{\varepsilon}\right)^{2^{\frac{1}{\varepsilon}}}, \tag{5.125}$$

where the term on the right side is independent of n. This yields Lemma 5.6. $\qquad\square$

We summarize our results: the Gaussian approach to π via the Equation (5.51) initiated a number of theoretical investigations of the diophantine inequality $n_1^2 + n_2^2 \le N^2$, where, as usual, "diophantine" means that the numbers n_1, n_2 are integers. From (5.49) we readily obtain that $r(n)$ is "equidistributed" in the sense

$$\lim_{l \to \infty} \frac{r(1) + \ldots + r(l)}{l} = \pi. \tag{5.126}$$

In more detail, the average order of $r(n)$ is in terms of the sum (1.1)

$$\lim_{l \to \infty} \frac{r(1) + \ldots + r(l)}{l} = \pi = 4 \sum_{g=0}^{\infty} \frac{(-1)^g}{2g+1} \tag{5.127}$$

(note that the sum in (5.97) has to be extended only over odd positive divisors). This result bridging the theories of numbers and series was already pointed out by D. Hilbert, S. Cohn-Vossen [1932]. It is remarkable since $r(n)$ is not "very regular" (note that it certainly becomes 0 after a few integers, for example, $r(124) = 0, r(125) = 16, r(126) = 0, \ldots$).

The naive approach to find the order of the magnitude of $r(n)$ by using the area of a circle of radius $\sqrt{n}$, i.e., πn, divided by the number n of circles would lead us to expect a constant number of lattice points on circles. But we already know that

$$r(n) = 4\left(d_1(n) - d_3(n)\right) = 4 \sum_{\substack{u|n \\ u \text{ odd}}} (-1)^{\frac{u-1}{2}} = O\left(n^\varepsilon\right) \tag{5.128}$$

holds for every $\varepsilon > 0$. There are infinite families of circles with "very few" lattice points; radii that are a power of 2 yield 4 points, for instance. Radii that are the square root of a prime of the form $4k+1$ have exactly 8 points. Even more, there are circles with large numbers of lattice points of $\mathbb{Z}^2$ that are "poorly distributed". For more details see, e.g., the work due to J. Cilleruelo [1993] who showed that, for any $\varepsilon > 0$, and for any integer k, there exists a circle around the origin with radius $\sqrt{n}$ possessing more than k lattice points such that all the lattice points are on arcs $\sqrt{n}e^{\frac{\pi}{2}(l+\theta)i}$ with $|\theta| < \varepsilon$ and $l = 0, 1, 2, 3$.

It was already known to E. Landau [1909] that

$$\sum_{n \leq N} (r(n))^2 = O(N \ln(N)), \ N \to \infty, \tag{5.129}$$

and

$$\sum_{\substack{1 \leq n \leq N \\ n \notin B}} 1 \sim \frac{C\,N}{\sqrt{\ln(N)}}, \ N \to \infty, \tag{5.130}$$

for some constant C, where $B = \{n \in \mathbb{Z} \mid r(n) > 0\}$. Roughly speaking, this suggests that apart from a constant factor the elements of B behave like the sequence $\left\{n\sqrt{\ln(n)}\right\}_{n \in B}$.

5.5 Lattice Points Inside Spheres

Just as in the planar case, we may discuss the problem of counting the number of points $(n_1, \ldots, n_q)^T$ with all the n_i integers inside spheres $\mathbb{S}_N^{q-1}$ of radii N centered at the origin 0. More precisely, by

$$\#_{\mathbb{Z}^q}\left(\overline{\mathbb{B}_N^q}\right) = \sum_{\substack{|g| \leq N \\ g \in \mathbb{Z}^q}} 1 \tag{5.131}$$

we denote the *total number of lattice points* g of the lattice $\mathbb{Z}^q$ such that $|g| \leq N$; i.e., g is a member of $\overline{\mathbb{B}_N^q}$. Following the idea of C.F. Gauß [1801] we consider the polyhedron

$$\mathbb{P}_N^q = \bigcup_{\substack{g \in \overline{\mathbb{B}_N^q} \\ g \in \mathbb{Z}^q}} (\mathcal{F} + \{g\}), \tag{5.132}$$

where, as usual, $\mathcal{F} + \{g\}$ is the translate of $\mathcal{F}$ by the lattice point $g \in \mathbb{Z}^q$. Clearly, the volume $\|\mathbb{P}_N^q\|$ of $\mathbb{P}_N^q$ is equal to $\#_\Lambda\left(\overline{\mathbb{B}_N^q}\right)$. There exists a constant d such that the estimate $|x| \leq d$ holds true for all $x \in \mathcal{F}$ (for example, $d = \frac{\sqrt{q}}{2}$), so that for $\mathbb{P}_N^q$ the following inclusion is valid $(N > d)$:

$$\overline{\mathbb{B}_{N-d}^q} \subset \mathbb{P}_N^q \subset \overline{\mathbb{B}_{N+d}^q}. \tag{5.133}$$

This leads to the following inequality

$$\left\|\overline{\mathbb{B}_{N-d}^q}\right\| \leq \underbrace{\#_{\mathbb{Z}^q}\left(\overline{\mathbb{B}_N^q}\right)}_{=\|\mathbb{P}_N^q\|} \leq \left\|\overline{\mathbb{B}_{N+d}^q}\right\|, \tag{5.134}$$

which is equivalent by (3.46) to

$$\frac{\pi^{\frac{q}{2}}}{\Gamma(\frac{q}{2}+1)}(N-d)^q \leq \underbrace{\#_{\mathbb{Z}^q}\left(\overline{\mathbb{B}_N^q}\right)}_{=\|\mathbb{P}_N^q\|} \leq \frac{\pi^{\frac{q}{2}}}{\Gamma(\frac{q}{2}+1)}(N+d)^q. \tag{5.135}$$

Following the Gaussian approach we therefore obtain the following fundamental result in the theory of lattice points inside spheres.

Theorem 5.5. *For $N \to \infty$,*

$$\#_{\mathbb{Z}^q}\left(\overline{\mathbb{B}_N^q}\right) = \frac{\pi^{\frac{q}{2}}}{\Gamma(\frac{q}{2}+1)}N^q + O\left(N^{q-1}\right). \tag{5.136}$$

In analogy to the two-dimensional case, the problem of determining

$$\alpha_q = \inf\left\{\gamma \mid \#_{\mathbb{Z}^q}\left(\overline{\mathbb{B}_{\sqrt{N}}^q}\right) = \left\|\overline{\mathbb{B}_{\sqrt{N}}^q}\right\| + O(N^\gamma)\right\} \tag{5.137}$$

is known as the *lattice point problem for the sphere* (for $q = 2$, as we know, *lattice point problem for the circle*).

It is clear from the Gaussian approach that

$$\alpha_q \leq \frac{q-1}{2}. \tag{5.138}$$

Moreover, it is not hard (see, e.g., F. Fricker [1975, 1982] and the references therein) to conclude that

$$\alpha_q \geq \frac{q-2}{2}. \tag{5.139}$$

In more detail,

$$\#_{\mathbb{Z}^q}\left(\overline{\mathbb{B}_{\sqrt{N}}^q}\right) = \left\|\overline{\mathbb{B}_{\sqrt{N}}^q}\right\| + \Omega\left(N^{\frac{q-2}{2}}\right), \quad N \to \infty, \tag{5.140}$$

where, as usual, Ω means that there is a constant C and arbitrarily large values of N so that

$$\left|\#_{\mathbb{Z}^q}\left(\overline{\mathbb{B}_{\sqrt{N}}^q}\right) - \left\|\overline{\mathbb{B}_{\sqrt{N}}^q}\right\|\right| \geq CN^{\frac{q-2}{2}}. \tag{5.141}$$

It is advisable to discuss the dimensions $q = 3, 4$ separately. For *the case* $q = 3$ we already know that

$$\#_{\mathbb{Z}^3}\left(\overline{\mathbb{B}_{\sqrt{N}}^3}\right) - \left\|\overline{\mathbb{B}_{\sqrt{N}}^3}\right\| = O(N). \tag{5.142}$$

In other words, for the *three-dimensional lattice point problem*

$$\alpha_3 = \inf\left\{\gamma \ \Big| \ \#_{\mathbb{Z}^3}\left(\overline{\mathbb{B}^3_{\sqrt{N}}}\right) - \left\|\overline{\mathbb{B}^3_{\sqrt{N}}}\right\| = O(N^\gamma)\right\} \tag{5.143}$$

it is known that $\alpha_3 \leq 1$. In order to improve this result great efforts must be made (similar to the two-dimensional case). E. Landau [1927] proved $\alpha_3 \leq \frac{3}{4}$, I.M. Vinogradov [1955] with $\alpha_3 \leq \frac{11}{16}$ and O.M. Fomenko [1961] with $\alpha_3 \leq \frac{701}{1020}$ gave essential improvements; more recent progress is due to F. Chamizo, H. Iwaniec [1995] with $\alpha_3 \leq \frac{29}{44}$ and D.R. Heath-Brown [1999] with $\alpha_3 \leq \frac{21}{32}$.

0.250000	E. Landau (1927)
0.187500	I. M. Winogradow (1955)
0.187254...	O.M. Fomenko (1961)
0.159090...	F. Chamizo, H. Iwaniec (1995)
0.156250	D. R. Heath-Brown (1999)

TABLE 5.2
Incremental improvements for the quantities ε_3 in the estimate (5.144).

Table 5.2 summarizes incomplete incremental improvements for the quantities ε_3 within the estimate

$$\#_{\mathbb{Z}^3}\left(\overline{\mathbb{B}^3_{\sqrt{N}}}\right) - \left\|\overline{\mathbb{B}^3_{\sqrt{N}}}\right\| = O\left(N^{\frac{1}{2}+\varepsilon_3}\right) \tag{5.144}$$

for the three-dimensional problem for the sphere. Again, the proofs of the improvements are rather long and difficult.

K. Chandrasekharan, R. Narasimhan [1961] proved that

$$\limsup_{N\to\infty} \frac{\#_{\mathbb{Z}^3}\left(\overline{\mathbb{B}^3_{\sqrt{N}}}\right) - \left\|\overline{\mathbb{B}^3_{\sqrt{N}}}\right\|}{N^{\frac{1}{2}}} = \infty. \tag{5.145}$$

For an overview of Ω-bounds the reader is referred to, e.g., A. Ivic et al. [2004] and the references therein. Altogether it can be stated that

$$\frac{1}{2} \leq \alpha_3 \leq \frac{1}{2} + \varepsilon_3 \tag{5.146}$$

and

$$\#_{\mathbb{Z}^3}\left(\overline{\mathbb{B}^3_{\sqrt{N}}}\right) - \left\|\overline{\mathbb{B}^3_{\sqrt{N}}}\right\| \neq O\left(N^{\frac{1}{2}}\right), \tag{5.147}$$

where $0 < \varepsilon_3 \leq \frac{1}{2}$ (e.g., Heath-Brown's bound $\varepsilon_3 = 0.156250$).

For *the dimension* $q = 4$ the classical result due to Lagrange (that each positive integer can be expressed as the sum of at most four integral squares) helps us to verify an estimate (for more details see, e.g., F. Fricker [1975, 1982], E. Krätzel [2000]) of the form

$$\#_{\mathbb{Z}^4}\left(\overline{\mathbb{B}^4_{\sqrt{N}}}\right) - \left\|\overline{\mathbb{B}^4_{\sqrt{N}}}\right\| = O\left(N\ln(N)\right), \tag{5.148}$$

and it is known that the ln-term in (5.148) cannot be omitted. In fact, A. Walfisz [1960b] used some extremely deep analysis to show an estimate of the form

$$\#_{\mathbb{Z}^4}\left(\overline{\mathbb{B}^4_{\sqrt{N}}}\right) - \left\|\overline{\mathbb{B}^4_{\sqrt{N}}}\right\| = O\left(N(\ln(N))^{\frac{2}{3}}\right). \tag{5.149}$$

In other words, we are confronted with the situation that

$$\alpha_4 = 1, \tag{5.150}$$

however,

$$\#_{\mathbb{Z}^4}\left(\overline{\mathbb{B}^4_{\sqrt{N}}}\right) - \left\|\overline{\mathbb{B}^4_{\sqrt{N}}}\right\| \neq O\left(N\right). \tag{5.151}$$

For more details including Ω-bounds the reader is referred to S.D. Adhikari, Y.-F.S. Petermann [1991].

For *higher dimensions* $q \geq 5$, the situation of the *lattice point problem for the sphere* changes drastically (see, e.g., E. Grosswald [1985] for more details). The asymptotic growth of the number of points has been known for a long time. In fact, we have the following relations, leaving no room for progress.

Theorem 5.6. *For* $q \geq 5$,

$$\alpha_q = \frac{q-2}{2}, \tag{5.152}$$

and

$$\#_{\mathbb{Z}^q}\left(\overline{\mathbb{B}^q_{\sqrt{N}}}\right) - \left\|\overline{\mathbb{B}^q_{\sqrt{N}}}\right\| = \Omega\left(N^{\alpha_q}\right), \tag{5.153}$$

$$\#_{\mathbb{Z}^q}\left(\overline{\mathbb{B}^q_{\sqrt{N}}}\right) - \left\|\overline{\mathbb{B}^q_{\sqrt{N}}}\right\| = O\left(N^{\alpha_q}\right). \tag{5.154}$$

We give only a motivation of the proof of the upper bound (for a more detailed discussion of the higher dimensional lattice point discrepancies, the reader is referred, e.g., to E. Landau [1925], A. Walfisz [1957], F. Fricker [1982], E. Krätzel [1988, 2000]).

The essential idea in Theorem 5.6 is the transfer of a recursion relation for the sums $\#_{\mathbb{Z}^q}\left(\overline{\mathbb{B}^q_{\sqrt{N}}}\right)$ to the so–called *lattice point discrepancies*

$$\#_{\mathbb{Z}^q}\left(\overline{\mathbb{B}^q_{\sqrt{N}}}\right) - \left\|\mathbb{B}^q_{\sqrt{N}}\right\| = \sum_{\substack{|g|\leq\sqrt{N}\\ g\in\mathbb{Z}^q}} 1 - \left\|\mathbb{B}^q_{\sqrt{N}}\right\| \tag{5.155}$$

by induction over the dimension q. To be more concrete,

$$\#_{\mathbb{Z}^{q+1}}\left(\overline{\mathbb{B}^{q+1}_{\sqrt{N}}}\right) = \sum_{\substack{n_1^2+\cdots+n_{q+1}^2\leq N\\ (n_1,\ldots,n_{q+1})^T\in\mathbb{Z}^{q+1}}} 1 \tag{5.156}$$

$$= \sum_{\substack{-\sqrt{N}\leq n_{q+1}\leq\sqrt{N}\\ n_{q+1}\in\mathbb{Z}}} \sum_{\substack{n_1^2+\cdots+n_q^2\leq N-n_{q+1}^2\\ (n_1,\ldots,n_q)^T\in\mathbb{Z}^q}} 1$$

$$= \sum_{\substack{-\sqrt{N}\leq n_{q+1}\leq\sqrt{N}\\ n_{q+1}\in\mathbb{Z}}} \#_{\mathbb{Z}^q}\left(\overline{\mathbb{B}^q_{\sqrt{N-n_{q+1}^2}}}\right).$$

We observe that

$$\#_{\mathbb{Z}^q}\left(\overline{\mathbb{B}^q_{\sqrt{N-n_{q+1}^2}}}\right) = \|\mathbb{B}^q_1\|\ (N-n_{q+1}^2)^{\frac{q}{2}} \tag{5.157}$$

$$+ \#_{\mathbb{Z}^q}\left(\overline{\mathbb{B}^q_{\sqrt{N-n_{q+1}^2}}}\right) - \left\|\overline{\mathbb{B}^q_{\sqrt{N-n_{q+1}^2}}}\right\|.$$

The one-dimensional Euler summation formula yields the estimate

$$\|\mathbb{B}^q_1\| \sum_{\substack{-\sqrt{N}\leq n_{q+1}\leq\sqrt{N}\\ n_{q+1}\in\mathbb{Z}}} (N-n_{q+1}^2)^{\frac{q}{2}} = \|\mathbb{B}^{q+1}_1\|\ N^{\frac{q+1}{2}} + O\left(N^{\frac{q+1}{2}-1}\right), \tag{5.158}$$

such that

$$\#_{\mathbb{Z}^{q+1}}\left(\overline{\mathbb{B}^{q+1}_{\sqrt{N}}}\right) - \left\|\overline{\mathbb{B}^{q+1}_{\sqrt{N}}}\right\| \tag{5.159}$$

$$= \sum_{\substack{-\sqrt{N}\leq n_{q+1}\leq\sqrt{N}\\ n_{q+1}\in\mathbb{Z}}} \#_{\mathbb{Z}^q}\left(\overline{\mathbb{B}^q_{\sqrt{N-n_{q+1}^2}}}\right) - \left\|\overline{\mathbb{B}^q_{\sqrt{N-n_{q+1}^2}}}\right\|$$

$$+ O\left(N^{\frac{q+1}{2}-1}\right).$$

In other words, from (5.148), we obtain for $q \geq 4$

$$\#_{\mathbb{Z}^{q+1}}\left(\overline{\mathbb{B}^{q+1}_{\sqrt{N}}}\right) - \left\|\overline{\mathbb{B}^{q+1}_{\sqrt{N}}}\right\| = O\left(N^{\frac{q+1}{2}-1}\log(N)\right), \tag{5.160}$$

which can be improved (especially for the recursion step $q = 4$ to $q = 5$) by a
suitable use of Lemma 5.1 to

$$\#_{\mathbb{Z}^{q+1}}\left(\overline{\mathbb{B}^{q+1}_{\sqrt{N}}}\right) - \left\|\overline{\mathbb{B}^{q+1}_{\sqrt{N}}}\right\| = \sum_{\substack{|g| \leq \sqrt{N} \\ g \in \mathbb{Z}^{q+1}}} 1 - \left\|\overline{\mathbb{B}^{q+1}_{\sqrt{N}}}\right\| = O\left(N^{\frac{q+1}{2}-1}\right) \qquad (5.161)$$

(for more details see A. Walfisz [1924]). Therefore, by repeated application of
the recursion for $\#_{\mathbb{Z}^q}\left(\overline{\mathbb{B}^q_{\sqrt{N}}}\right)$, the validity of (5.154) for all dimensions $q \geq 5$
becomes obvious.

Remark 5.6. *The impression that the lattice point problem for the higher-
dimensional spheres becomes more and more problematic is wrong. The essential difficulties arising in the two- and three-dimensional cases do not occur for
dimensions $q \geq 5$. The dimension $q = 4$ is an intermediate case. The reason
seems to be that the decomposition of a positive integer as a sum of integral
squares for dimensions $q \geq 5$ is not as "pathologic" as, for instance, for $q = 2$
and $q = 3$.*

5.6   Lattice Points on Spheres

The sum $\#_{\mathbb{Z}^q}\left(\overline{\mathbb{B}^q_{\sqrt{N}}}\right)$, as discussed in Theorem 5.5, can be written in the form

$$\#_{\mathbb{Z}^q}\left(\overline{\mathbb{B}^q_{\sqrt{N}}}\right) = \sum_{\substack{|g| \leq \sqrt{N} \\ g \in \mathbb{Z}^q}} 1 = \sum_{n \leq N} r_q(n), \qquad (5.162)$$

where $r_q(n)$ (more precisely, $r_q(\mathbb{Z}^q; n)$) is the number of solutions of the diophantine equation $n_1^2 + \ldots + n_q^2 = n$, $(n_1, \ldots, n_q)^T \in \mathbb{Z}^q$, i.e.,

$$r_q(n) = \#_{\mathbb{Z}^q}\left\{(n_1, \ldots, n_q)^T \in \mathbb{Z}^q \mid n_1^2 + \ldots + n_q^2 = n\right\}. \qquad (5.163)$$

From the estimate due to C.F. Gauß [1801] we know that

$$\lim_{n \to \infty} \frac{r_q(1) + \ldots + r_q(n)}{n^{\frac{q}{2}}} = \frac{\pi^{\frac{q}{2}}}{\Gamma\left(\frac{q+2}{2}\right)}. \qquad (5.164)$$

The term (5.163) has a large pertinent literature (see, e.g., E. Grosswald [1985],
Y. L. Linnik [1968]). Again, the answers depend very much on the dimensionality of the problem.

The naive approach to determine the order of magnitude for a given dimension q is to use the volume of a ball, divided by the number of spheres

contained in the ball. The volume of a ball of radius $\sqrt{n}$ grows as $n^{\frac{q}{2}}$, while the number of spheres is n. However, for small q, this approach is misleading. The dimension $q = 2$ is already known to us (see Section 5.4). Here, from Lemma 5.6 we already have $r(n) = O(n^\varepsilon)$ for every $\varepsilon > 0$.

For the dimension $q = 3$, e.g., we have $r_3(n) = 0$ if n is a member of the sequence $\{7, 15, 23, \ldots\}$. Based on a slightly weaker result by C.F. Gauß [1826], C.L. Siegel [1935] proved the estimate: if $n = 4^s m$, where $4 \nmid m$ and $m \not\equiv 7 \pmod 8$, then

$$C_1(\varepsilon) m^{\frac{1-\varepsilon}{2}} \le r(n) \le C_2(\varepsilon) m^{\frac{1-\varepsilon}{2}} \tag{5.165}$$

for any $\epsilon > 0$ and positive constants C_1, C_2. Furthermore, C.L. Siegel [1935] pointed out that $n_1^2 + n_2^2 + n_3^2 = n$ and n_1, n_2, n_3 even holds true if and only if $n \equiv 0 \pmod 4$. It has already been noted by C.F. Gauß that n is a sum of three squares of integers if and only if $n \ne 4^s(8k + 7)$.

The higher-dimensional cases behave in more regular fashion: for the dimension $q = 4$, *Jacobi's formula* (see C.G. Jacobi [1829]) tells us that

$$r_4(2^l u) = \begin{cases} 8\,\sigma(u), & l = 0, \\ 24\,\sigma(u), & l \ge 1, \end{cases} \tag{5.166}$$

where

$$\sigma(n) = \sum_{\substack{d|n \\ d>0}} d. \tag{5.167}$$

In other words, the calculation of $r_4(n)$, $n = 2^l u$, with u being a positive odd integer amounts (apart from a multiplicative constant) to the sum of all positive odd divisors of n. For illustration, we determine $r_4(30)$. The positive divisors of 30 are given by $\{1, 2, 3, 5, 6, 10, 15, 30\}$. Consequently, $r_4(30) = 24(1 + 3 + 5 + 15) = 576$. Clearly, it follows from (5.167) that $\sigma(n) \ge 1 + n$. Jacobi's formula therefore includes the aforementioned result due to Lagrange that every positive integer can be written as the sum of four squares. Thus, for the dimension $q = 4$, every sphere whose squared radius is an integer admits lattice points, i.e., $r_4(n) > 0$ for all n. Moreover, it is known (cf. F. Fricker [1982]) that

$$\limsup_{n \to \infty} \frac{r_4(n)}{n \, \ln(\ln(n))} \ge 1. \tag{5.168}$$

Nevertheless, for $q = 4$, the number of lattice points still oscillates rather wildly (e.g., spheres with radius equal to a power of 2 possess 24 points).

For $q \ge 5$, questions of both number of solutions and repartition of them possess satisfactory answers (cf. Y. L. Linnik [1968]). A. Walfisz [1957] pointed out that the series

$$\sum_{n=1}^{\infty} \frac{r_q(n)}{n^\sigma} \tag{5.169}$$

is convergent for every $\sigma \in \mathbb{R}$ with $\sigma > \frac{q}{2}$. For all dimensions $q \geq 5$, indeed, the "naive" estimate gives the correct asymptotic growth of the number of lattice points on spheres, i.e.,

$$\sum_{\substack{|g|^2=n \\ g \in \mathbb{Z}^q}} 1 = r_q(n) = O\left(n^{\frac{q-2}{2}}\right), \qquad n \to \infty. \qquad (5.170)$$

Remark 5.7. *The goal of our further chapters is to develop lattice point identities specifically imbedded in the metaharmonic framework such as classical Bessel function expansions or certain (generalized) counterparts originated from the expansion theory of solutions for the Helmholtz wave equation. Within this concept the need of summability techniques (such as Gauß–Weierstraß or Abel–Poisson summability) in lattice point identities (arising from calamities of the convergence of the occurring series expansions) becomes successively stronger with the growth of their dimensionality. In consequence, we are confronted with the amazing situation that lattice point identities become harder to verify with increasing dimensions, while their asymptotic laws with radius N going to infinity simply allow an exact control.*

6

<hr>

Preparatory Tools of Mathematical Physics

CONTENTS

6.1 Integral Theorems for the Laplace Operator 122
   Integral Theorems 122
   Harmonic and Metaharmonic Functions 124
   Fundamental Solutions of the Laplacian 124
   Fundamental Solutions for Iterated Laplacians 129
6.2 Integral Theorems for the Laplace–Beltrami Operator 133
   Sphere Function for the Laplace–Beltrami Operator 134
   Integral Formulas for the Laplace–Beltrami Operator 136
   Laplace–Beltrami Differential Equation 138
6.3 Tools Involving the Laplace Operator 139
   Integral Estimates 139
   Multi-Dimensional Angles 142
6.4 Radial and Angular Decomposition of Harmonics 144
   Homogeneous Harmonic Polynomials 144
   Legendre Polynomials 153
   Spherical Harmonics 158
   Associated Legendre Polynomials 171
   Pointwise Expansion Theorem 173
   Asymptotic Relations for the Spherical Harmonic Coefficients ... 178
6.5 Integral Theorems for the Helmholtz–Beltrami Operator 180
   Sphere Function for the Helmholtz–Beltrami Operator 180
   Integral Formulas for the Helmholtz–Beltrami Operator 183
   Helmholtz–Beltrami Differential Equation 186
   Spherical Harmonics as Eigenfunctions 186
   Lattice Point Generated Spherical Equidistribution 187
6.6 Radial and Angular Decomposition of Metaharmonics 192
   Bessel Functions 193
   Modified Bessel Functions 200
   Hankel Functions 202
   Kelvin Functions 207
   Expansion Theorems 212
6.7 Tools Involving Helmholtz Operators 215
   An Asymptotic Expansion for an Entire Integral Solution 217
   Canonical Extensions 221

This chapter deals with harmonic and metaharmonic functions in Euclidean spaces $\mathbb{R}^q$. A particular role is played by entire metaharmonic functions, i.e., solutions of the Helmholtz equation which are valid for the whole Euclidean space $\mathbb{R}^q$.

The layout of the chapter is as follows: in Section 6.1 we are concerned with some material of the (poly)harmonic theory in Euclidean spaces $\mathbb{R}^q$ such as the fundamental solution for iterated Laplacians and its role in Green's theorems (see O.D. Kellogg [1929], N. Aronszajn et al. [1983], E. Wienholtz et al. [2009] to mention a few textbooks). Section 6.2 contains corresponding results on the (unit) sphere for the Beltrami operator Δ^* and its iterations (cf. W. Freeden [1978, 1979, 1980b], W. Freeden, R. Reuter [1982]). After these preparations we are interested in establishing some integral estimates involving fundamental solutions (see W. Freeden [1975, 1978a] for the two-dimensional case, C. Müller, W. Freeden [1980] for the q-dimensional case), which play a significant role in specifying the asymptotic behavior of the multi-dimensional lattice functions (as introduced later in Chapter 8). In Section 6.4, we turn to homogeneous harmonic polynomials in Euclidean spaces $\mathbb{R}^q$, $q \geq 2$, and their restrictions to the unit sphere $\mathbb{S}^{q-1}$, i.e., spherical harmonics. Their theory is self-contained. In fact, our approach to spherical harmonics is intended to avoid any singularity that arises from the special choice of a local coordinate system. Section 6.6 contains a short approach to the theory of Bessel functions. Asymptotic estimates for Bessel, Hankel, and Kelvin functions are developed, as far as they are needed for our lattice point concerns. Finally, Section 6.7 presents results on the metaharmonic theory, namely asymptotic relations for entire solutions of the Helmholtz wave equation. These asymptotics take a decisive role in the estimation of integral expressions occurring in multi-dimensional Euler and Poisson summation formulas.

6.1   Integral Theorems for the Laplace Operator

First we replicate some key results of classical vector analysis and potential theory, which are useful for our work.

Integral Theorems

A regular region $\mathcal{G} \subset \mathbb{R}^q$ (cf. Section 2.1) allows the *Gauss theorem*

$$\int_{\mathcal{G}} \nabla \cdot f(x) \, dV(x) = \int_{\partial \mathcal{G}} f(x) \cdot \nu(x) \, dS(x) \qquad (6.1)$$

for all continuously differentiable vector fields f on $\overline{\mathcal{G}}, \overline{\mathcal{G}} = \mathcal{G} \cup \partial \mathcal{G}$ (throughout this work, ν is the (unit) normal field on $\partial \mathcal{G}$ directed to the exterior of $\overline{\mathcal{G}}$). By letting $f = \nabla F, F \in C^{(2)}(\overline{\mathcal{G}}), \mathcal{G} \subset \mathbb{R}^q$ regular, we obtain from (6.1)

$$\int_{\mathcal{G}} \Delta F(x) \, dV(x) = \int_{\partial \mathcal{G}} \frac{\partial F}{\partial \nu}(x) \, dS(x), \qquad (6.2)$$

where $\frac{\partial}{\partial\nu}$ denotes the derivative in the direction of the outer (unit) normal field ν. Consequently, for all functions $F \in C^{(2)}(\overline{\mathcal{G}})$ satisfying the Laplace equation $\Delta F = 0$ in $\mathcal{G}$, we have

$$\int_{\partial\mathcal{G}} \frac{\partial F}{\partial\nu}(x) \, dS(x) = 0. \tag{6.3}$$

For all vector fields $f = F\nabla G$, $F \in C^{(1)}(\overline{\mathcal{G}})$, $G \in C^{(2)}(\overline{\mathcal{G}})$, we get from the Gauss theorem

Theorem 6.1. *(First Green Theorem) Suppose that $\mathcal{G} \subset \mathbb{R}^q$ is a regular region. For $F \in C^{(1)}(\overline{\mathcal{G}}), G \in C^{(2)}(\overline{\mathcal{G}})$ we have*

$$\int_{\mathcal{G}} \{F(x)\Delta G(x) + \nabla F(x) \cdot \nabla G(x)\} \, dV(x) = \int_{\partial\mathcal{G}} F(x)\frac{\partial G}{\partial\nu}(x) \, dS(x).$$

Taking $f = F\nabla G - G\nabla F$ with $F, G \in C^{(2)}(\overline{\mathcal{G}})$ we obtain

Theorem 6.2. *(Second Green Theorem) Suppose that $\mathcal{G} \subset \mathbb{R}^q$ is a regular region. For $F, G \in C^{(2)}(\overline{\mathcal{G}})$ we have*

$$\int_{\mathcal{G}} \{G(x)\Delta F(x) - F(x)\Delta G(x)\} \, dV(x)$$

$$= \int_{\partial\mathcal{G}} \left\{G(x)\frac{\partial F}{\partial\nu}(x) - F(x)\frac{\partial G}{\partial\nu}(x)\right\} \, dS(x).$$

Remark 6.1. *Note that, for $q = 1$ and $\overline{\mathcal{G}} = [a, b]$, $a < b$, we adopt the conventions*

$$\int_{\partial\mathcal{G}} F(x) \, dS(x) = F(a) + F(b) \tag{6.4}$$

and

$$\left.\frac{\partial}{\partial\nu}\right|_a = -\left.\frac{d}{dx}\right|_a \, , \quad \left.\frac{\partial}{\partial\nu}\right|_b = \left.\frac{d}{dx}\right|_b. \tag{6.5}$$

Next we mention an extension of the Second Green Theorem.

Theorem 6.3. *(Extended Second Green Theorem) For any number $\lambda \in \mathbb{R}$ and any regular region $\mathcal{G} \subset \mathbb{R}^q$, $q \geq 2$, and for $F \in C^{(2m)}(\overline{\mathcal{G}})$, $m \in \mathbb{N}$, we have*

$$\int_{\mathcal{G}} G(x)(\Delta + \lambda)^m F(x) \, dV(x) = \int_{\mathcal{G}} F(x)(\Delta + \lambda)^m G(x) \, dV(x) \tag{6.6}$$

$$+ \sum_{r=0}^{m-1} \int_{\partial\mathcal{G}} \left(\frac{\partial}{\partial\nu}(\Delta + \lambda)^r F(x)\right)\left((\Delta + \lambda)^{m-(r+1)}G(x)\right) dS(x)$$

$$- \sum_{r=0}^{m-1} \int_{\partial\mathcal{G}} \left((\Delta + \lambda)^r F(x)\right)\left(\frac{\partial}{\partial\nu}(\Delta + \lambda)^{m-(r+1)}G(x)\right) dS(x).$$

Its simplest form is the case with vanishing boundary terms, i.e.,

$$\sum_{r=0}^{m-1} \int_{\partial\mathcal{G}} \left(\frac{\partial}{\partial\nu}(\Delta+\lambda)^r F(x)\right)\left((\Delta+\lambda)^{m-(r+1)}G(x)\right)dS(x) \qquad (6.7)$$

$$-\sum_{r=0}^{m-1}\int_{\partial\mathcal{G}}\left((\Delta+\lambda)^r F(x)\right)\left(\frac{\partial}{\partial\nu}(\Delta+\lambda)^{m-(r+1)}G(x)\right)dS(x)$$

$$=\ 0,$$

such that (6.6) reduces to the formula

$$\int_{\mathcal{G}} \{G(x)(\Delta+\lambda)^m F(x) - F(x)(\Delta+\lambda)^m G(x)\}\ dV(x) = 0. \qquad (6.8)$$

Harmonic and Metaharmonic Functions

Next we come to the well known definition of harmonic and metaharmonic functions.

Definition 6.1. $U \in C^{(2)}(\mathcal{G})$ *is called a harmonic function (sometimes also called a harmonic) in a region $\mathcal{G} \subset \mathbb{R}^q$ if it satisfies the Laplace equation*

$$\Delta U(x) = \sum_{i=1}^{q}\left(\frac{\partial}{\partial x_i}\right)^2 U(x_1,\ldots,x_q) = 0\ ,\ x = (x_1,\ldots,x_q)^T \in \mathcal{G}. \qquad (6.9)$$

$U \in C^{(2m)}(\mathcal{G})$, $m \in \mathbb{N}$, *is called a polyharmonic function of degree m in $\mathcal{G} \subset \mathbb{R}^q$ if*

$$\Delta^m U(x) = 0, \quad x \in \mathcal{G}. \qquad (6.10)$$

$U \in C^{(2)}(\mathcal{G})$ *is called a metaharmonic function (sometimes also called a metaharmonic) with respect to the Helmholtz operator $\Delta + \lambda$, $\lambda \in \mathbb{R}$, in a region $\mathcal{G} \subset \mathbb{R}^q$ if it satisfies the Helmholtz equation*

$$(\Delta + \lambda)\, U(x) = 0\ ,\ x \in \mathcal{G}. \qquad (6.11)$$

$U \in C^{(2m)}(\mathcal{G})$, $m \in \mathbb{N}$, *is called a polymetaharmonic function of degree m in $\mathcal{G}$ if*

$$(\Delta + \lambda)^m\, U(x) = 0, \quad x \in \mathcal{G}. \qquad (6.12)$$

Fundamental Solutions of the Laplacian

Let $y \in \mathcal{G}$ be fixed, where $\mathcal{G}$ is a region in $\mathbb{R}^q$. We are looking for a harmonic function U in $\mathcal{G} \setminus \{y\}$ such that

$$U(x) = F(|x - y|), \quad x \in \mathcal{G} \setminus \{y\}; \qquad (6.13)$$

i.e., U depends only on the mutual distance of x and y. From the identities

$$\frac{\partial}{\partial x_i} F(|x-y|) = F'(|x-y|)\frac{x_i - y_i}{|x-y|}, \tag{6.14}$$

$$\left(\frac{\partial}{\partial x_i}\right)^2 F(|x-y|) = F''(|x-y|)\frac{(x_i - y_i)^2}{|x-y|^2} \tag{6.15}$$

$$+ F'(|x-y|)\left(\frac{1}{|x-y|} - \frac{(x_i - y_i)^2}{|x-y|^3}\right),$$

we easily obtain

$$\Delta_x F(|x-y|) = F''(|x-y|) + \frac{q-1}{|x-y|} F'(|x-y|) = 0, \tag{6.16}$$

In other words, $F(|x-y|)$ can be written in the form

$$F(|x-y|) = \begin{cases} C_1 \ln(|x-y|) + C_2 & , \quad q = 2 \\ C_1 |x-y|^{2-q} + C_2 & , \quad q \geq 3 \end{cases} \tag{6.17}$$

with some constants C_1, C_2. By convention, the function

$$x \mapsto F_q(|x-y|) = \begin{cases} -\frac{1}{2\pi}\ln(|x-y|) & , \quad q = 2 \\ \frac{|x-y|^{2-q}}{(q-2)||\mathbb{S}^{q-1}||} & , \quad q \geq 3 \end{cases} \tag{6.18}$$

is called the *fundamental solution in* $\mathbb{R}^q$ *for the Laplace operator* Δ (later on, we will see that the choice (6.18) for the coefficients C_1, C_2 is very useful in our context).

Remark 6.2. *As already known, the one-dimensional fundamental solution for the Laplacian, i.e., the operator of the second order derivative, is given by the continuous function*

$$x \mapsto F_1(|x-y|) = -\frac{1}{2}|x-y| = -\frac{1}{2}(x-y)\,\mathrm{sign}(x-y), \tag{6.19}$$

$x, y \in \mathbb{R}$. *This is a wonderful feature with nice applications - unfortunately only in the one-dimensional lattice point theory.*

The fundamental solution of the Laplace operator possesses the following property.

Lemma 6.1. *For continuous functions* G, H *in the ball* $\mathbb{B}_R^q(y), y \in \mathbb{R}^q, q \geq 2$, $R > r > 0$, *we have*

$$\lim_{\substack{r \to 0 \\ r > 0}} \int_{|x-y|=r} G(x)\frac{\partial}{\partial \nu_x} F_q(|x-y|)\, dS(x) = -G(y), \tag{6.20}$$

$$\lim_{\substack{r \to 0 \\ r > 0}} \int_{|x-y|=r} H(x) F_q(|x-y|)\, dS(x) = 0, \tag{6.21}$$

where the (unit) normal field ν *is directed to the exterior of* $\mathbb{B}_r^q(y)$.

Proof. We restrict ourselves to dimensions $q \geq 3$. The case $q = 2$ easily follows by analogous arguments.

Because of the continuity of the function H in each ball $\overline{\mathbb{B}_r^q}(y)$, $r < R$, we find

$$\left| \int_{|x-y|=r} H(x) \frac{|x-y|^{2-q}}{(q-2)||\mathbb{S}^{q-1}||} \, dS(x) \right| \leq \frac{C}{||\mathbb{S}^{q-1}||} \int_{|x-y|=r} |x-y|^{2-q} \, dS(x)$$

$$= \frac{C}{||\mathbb{S}^{q-1}||} r^{2-q} ||\mathbb{S}^{q-1}|| r^{q-1}$$

$$= Cr \qquad\qquad (6.22)$$

for some positive constant C. This shows the second limit relation (6.21).

For the first limit relation we observe that the normal derivative can be understood as the radial derivative. From the mean value theorem we therefore obtain

$$\int_{|x-y|=r} G(x) \frac{\partial}{\partial \nu_x} \frac{|x-y|^{2-q}}{(q-2)||\mathbb{S}^{q-1}||} \, dS(x) = -\frac{r^{1-q}}{||\mathbb{S}^{q-1}||} \int_{|x-y|=r} G(x) \, dS(x)$$

$$= -\frac{r^{1-q}}{||\mathbb{S}^{q-1}||} r^{q-1} ||\mathbb{S}^{q-1}|| \, G(x_r)$$

$$(6.23)$$

for certain points $x_r \in \mathbb{S}_r^{q-1}(y)$. The limit $r \to 0$ implies $x_r \to y$, such that the continuity of G yields

$$\lim_{\substack{r \to 0 \\ r > 0}} G(x_r) = G(y). \qquad\qquad (6.24)$$

This is the desired result.   □

Next we want to apply the Second Green Theorem (for a regular region $\mathcal{G}$ with continuously differentiable boundary $\partial\mathcal{G}$) especially to the functions

$$F : x \mapsto \qquad F(x) = 1 \qquad , \; x \in \overline{\mathcal{G}}, \qquad\qquad (6.25)$$

$$G : x \mapsto \qquad G(x) = F_q(|x-y|) \qquad , \; x \in \overline{\mathcal{G}} \backslash \{y\} \qquad\qquad (6.26)$$

where $y \in \mathbb{R}^q$ is positioned in accordance with the following three cases:
Case $y \in \mathcal{G}$: For sufficiently small $\varepsilon > 0$ we obtain by integration by parts, i.e., the Second Green Theorem

$$\int_{\substack{x \in \mathcal{G} \\ |x-y| \geq \varepsilon}} \underbrace{\Delta_x F_q(|x-y|)}_{=0} \, dV(x) = \int_{x \in \partial\mathcal{G}} \frac{\partial}{\partial \nu_x} F_q(|x-y|) \, dS(x) \qquad (6.27)$$

$$+ \int_{\substack{x \in \mathcal{G} \\ |x-y|=\varepsilon}} \frac{\partial}{\partial \nu_x} F_q(|x-y|) \, dS(x).$$

In connection with Lemma 6.1 we therefore obtain by letting $\varepsilon \to 0$

$$\int_{\partial\mathcal{G}} \frac{\partial}{\partial \nu_x} F_q(|x-y|) \, dS(x) \; = \; -1. \tag{6.28}$$

Case $y \in \partial\mathcal{G}$**:** Again, by Green's theorem we obtain for $\varepsilon > 0$

$$-\int_{\substack{x\in\mathcal{G} \\ |x-y|=\varepsilon}} \frac{\partial}{\partial \nu_x} F_q(|x-y|) \, dS(x) = \int_{x\in\partial\mathcal{G}} \frac{\partial}{\partial \nu_x} F_q(|x-y|) \, dS(x). \tag{6.29}$$

By letting $\varepsilon \to 0$ we now find in case of a continuously differentiable surface $\partial\mathcal{G}$

$$\int_{\partial\mathcal{G}} \frac{\partial}{\partial \nu_x} F_q(|x-y|) \, dS(x) = \; -\frac{1}{2}. \tag{6.30}$$

Case $y \notin \overline{\mathcal{G}}$**:** The second Green theorem now yields

$$\int_{\mathcal{G}} \underbrace{\Delta_x F_q(|x-y|)}_{=0} \, dV(x) = \int_{\partial\mathcal{G}} \frac{\partial}{\partial \nu_x} F_q(|x-y|) \, dS(x). \tag{6.31}$$

Summarizing all our results we obtain the following identity from (6.28), (6.30), and (6.31).

Lemma 6.2. *Let* $\mathcal{G} \subset \mathbb{R}^q$ *be a regular region with continuously differentiable boundary* $\partial\mathcal{G}$. *Then*

$$-\int_{\partial\mathcal{G}} \frac{\partial}{\partial \nu_x} F_q(|x-y|) \, dS(x) = \begin{cases} 1 & , \quad y \in \mathcal{G} \\ \frac{1}{2} & , \quad y \in \partial\mathcal{G} \\ 0 & , \quad y \notin \overline{\mathcal{G}}. \end{cases} \tag{6.32}$$

In other words, the integral is a measure for the "solid angle" $\alpha(y)$ subtended by the boundary $\partial\mathcal{G}$ at the point $y \in \mathbb{R}^q$.

Remark 6.3. *From potential theory (see, e.g., O.D. Kellogg [1929]) it is known that Lemma 6.2 may be extended to regular regions* $\mathcal{G}$ *such as cube, simplex, polyhedron, more concretely, to regular regions with solid angle* $\alpha(y)$ *at* $y \in \mathbb{R}^q$ *subtended by the surface* $\partial\mathcal{G}$:

$$-\int_{\partial\mathcal{G}} \frac{\partial}{\partial \nu_x} F_q(|x-y|) \, dS(x) = \alpha(y), \quad y \in \mathbb{R}^q. \tag{6.33}$$

Example 6.1. *For the cube (see Figure 6.1)*

$$\mathcal{G} \; = \; (-1,1)^3 \subset \mathbb{R}^3 \tag{6.34}$$

we have

(i) $\alpha(y) = 1$ *if* y *is located in the open cube* $\mathcal{G}$

(ii) $\alpha(y) = \frac{1}{2}$ *if* y *is located on one of the six faces of the boundary* $\partial \mathcal{G}$ *of the cube* $\mathcal{G}$ *but not on an edge or in a vertex,*

(iii) $\alpha(y) = \frac{1}{4}$ *if* y *is located on one of the eight edges of* $\partial \mathcal{G}$ *but not in a vertex,*

(iv) $\alpha(y) = \frac{1}{8}$ *if* y *is located in one of the eight vertices of* $\partial \mathcal{G}$.

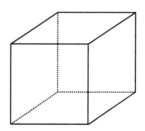

FIGURE 6.1
Three-dimensional cube $\mathcal{G} = (-1, 1)^3$.

Lemma 6.2 is a special case of the third Green theorem in $\mathbb{R}^q$ (see, e.g., O.D. Kellogg [1929], N. Aronszaijn et al. [1983]) that will be mentioned next.

Theorem 6.4. *(Third Green Theorem)*
(i) Let $\mathcal{G}$ *be a regular region with continuously differentiable boundary* $\partial \mathcal{G}$. *Suppose that* $U : \overline{\mathcal{G}} \to \mathbb{R}$ *is twice continuously differentiable, i.e.,* $U \in C^{(2)}\left(\overline{\mathcal{G}}\right)$. *Then we have*

$$
\int_{\partial \mathcal{G}} \left\{ F_q(|x-y|) \left(\frac{\partial U}{\partial \nu}(x) \right) - U(x) \left(\frac{\partial}{\partial \nu_x} F_q(|x-y|) \right) \right\} dS(x)
$$

$$
- \int_{\mathcal{G}} F_q(|x-y|)\, \Delta U(x)\, dV(x) = \begin{cases} U(y) & , \quad y \in \mathcal{G} \\ \frac{1}{2} U(y) & , \quad y \in \partial \mathcal{G} \\ 0 & , \quad y \in \mathbb{R}^q \setminus \overline{\mathcal{G}}. \end{cases} \qquad (6.35)
$$

(ii) Let $\mathcal{G}$ *be a regular region. Suppose that* $U : \overline{\mathcal{G}} \to \mathbb{R}$ *is twice continuously differentiable, i.e.,* $U \in C^{(2)}\left(\overline{\mathcal{G}}\right)$. *Then we have*

$$
\int_{\partial \mathcal{G}} \left\{ F_q(|x-y|) \left(\frac{\partial U}{\partial \nu}(x) \right) - U(x) \left(\frac{\partial}{\partial \nu_x} F_q(|x-y|) \right) \right\} dS(x)
$$

$$
- \int_{\mathcal{G}} F_q(|x-y|)\, \Delta U(x)\, dV(x) = \alpha(y)\, U(y), \qquad (6.36)
$$

where $\alpha(y)$, $y \in \mathbb{R}^q$, *is the solid angle at* y *subtended by the surface* $\partial \mathcal{G}$.

Proof. We consider only the case $y \in \mathcal{G}$. For every $\varepsilon > 0$, Green's formula tells us that

$$- \int\limits_{\substack{|x-y| \geq \varepsilon \\ x \in \mathcal{G}}} F_q(|y-x|) \, \Delta U(x) \, dV(x) \tag{6.37}$$

$$= \int_{\partial \mathcal{G}} \left\{ U(x) \left(\frac{\partial}{\partial \nu_x} F_q(|y-x|) \right) - F_q(|y-x|) \left(\frac{\partial U}{\partial \nu}(x) \right) \right\} \, dS(x)$$

$$+ \int\limits_{\substack{|x|=\varepsilon \\ x \in \mathcal{G}}} \left\{ U(x) \left(\frac{\partial}{\partial \nu_x} F_q(|y-x|) \right) - F_q(|y-x|) \left(\frac{\partial U}{\partial \nu}(x) \right) \right\} \, dS(x).$$

Letting $\varepsilon \to 0$ the theorem follows immediately from Lemma 6.1. $\square$

Finally we mention the *Poisson differential equation*, which is a classical result in potential theory (see, for example, S.G. Michlin [1970] or W. Freeden, M. Schreiner [2009] for an alternative proof in $\mathbb{R}^3$). We restrict our formulation to the case $q \geq 3$. The case $q = 2$ follows by obvious modification.

Lemma 6.3. *Let F be of class $C^{(0)}\left(\overline{\mathcal{G}}\right)$, $\overline{\mathcal{G}} = \mathcal{G} \cup \partial \mathcal{G} \subset \mathbb{R}^q$. Then $U : \overline{\mathcal{G}} \to \mathbb{R}$ given by*

$$U(x) = \int_{\mathcal{G}} F(y) \, F_q(|x-y|) \, dV(y) \tag{6.38}$$

is of class $C^{(1)}\left(\overline{\mathcal{G}}\right)$, and we have

$$\nabla U(x) = \int_{\mathcal{G}} F(y) \, \nabla_x F_q(|x-y|) \, dV(y). \tag{6.39}$$

If F is bounded in $\overline{\mathcal{G}} = \mathcal{G} \cup \partial \mathcal{G}$ satisfying a Lipschitz-condition in the neighborhood of the point $x \in \mathcal{G}$ (more precisely, $|F(y) - F(z)| \leq C_F |y-z|$ for all y, z in the neighborhood of the point $x \in \mathcal{G}$), then U as given by (6.38) is twice continuously differentiable in $x \in \mathcal{G}$, and we have

$$\Delta U(x) = - \, F(x). \tag{6.40}$$

Fundamental Solutions for Iterated Laplacians

By induction on m we obtain (see N. Aronszaijn et al. [1983])

$$\Delta^m |x|^k = A_{m,k} |x|^{k-2m}, \tag{6.41}$$

where we have used the abbreviations

$$\begin{aligned} A_{0,1} &= 1, \\ A_{m,k} &= k(k-2) \ldots (k-2m+2) \\ &\quad \times (k-2+q)(k-4+q) \ldots (k-2m+q). \end{aligned} \tag{6.42}$$

An elementary calculation yields

$$A_{m,k} = 2^{2m} \frac{\Gamma\left(\frac{k}{2} + 1\right)\Gamma\left(\frac{k+q}{2}\right)}{\Gamma\left(\frac{k}{2} - m + 1\right)\Gamma\left(\frac{k+q}{2} - m\right)} \tag{6.43}$$

except when k is even and $k \leq -2$ or $k+q$ is even and $k+q \leq 0$. Moreover,

$$\begin{aligned} \Delta\left(|x|^k \ln(|x|)\right) &= k(k+q-2)|x|^{k-2}\ln(|x|) \\ &\quad + (2k+q-2)|x|^{k-2}. \end{aligned} \tag{6.44}$$

By induction on m we find

$$\Delta^m\left(|x|^k \ln(|x|)\right) = A_{m,k}|x|^{k-2m}\ln(|x|) \tag{6.45}$$
$$+ |x|^{k-2m}\sum_{i=1}^{m} A_{i-1,k}A_{m-i,k-2i}(2k+q+2-4i).$$

By virtue of (6.43) the preceding formula can be transformed into the identity

$$\Delta^m\left(|x|^k \ln(|x|)\right) \tag{6.46}$$
$$= A_{m,k}|x|^{k-2m}\left(\ln(|x|) + \sum_{i=1}^{m}\left(\frac{1}{k-2i+2} + \frac{1}{k+q-2i}\right)\right).$$

For each operator Δ^l there corresponds a *characteristic function* $S_q^{(l)}$, $q \geq 2$, solving the equation $\Delta^l S_q^{(l)} = 0$ in $\mathbb{R}^q\backslash\{0\}$. To be more concrete, $S_q^{(l)}$ is given by

$$S_q^{(l)}(x) = \begin{cases} \dfrac{|x|^{2l-q}}{\gamma_{l-1}}, & q \text{ odd}, \quad l = 1, 2, \ldots, \\[2mm] \dfrac{|x|^{2l-q}}{\tilde{\gamma}_{l-1}}, & q \text{ even}, \quad l = 1, 2, \ldots, \frac{q}{2} - 1, \\[2mm] \dfrac{-|x|^{2l-q}\ln(|x|)}{\tilde{\gamma}_{l-1}}, & q \text{ even}, \quad l = \frac{q}{2}, \frac{q}{2} + 1, \ldots. \end{cases} \tag{6.47}$$

The constants γ_l and $\tilde{\gamma}_l$ (in the notation of (6.42)) are given as follows:

q odd:

$$\gamma_l = A_{l,2l+2-q} = 2^{2l}l!\frac{\Gamma\left(l+2-\frac{q}{2}\right)}{\Gamma\left(2-\frac{q}{2}\right)}, \tag{6.48}$$

$l = 0, 1, \ldots,$

q even:

$q = 2r \geq 4$

$$\tilde{\gamma}_l = A_{l,2l+2-q} = (-1)^l 2^{2l}l!\frac{(r-2)!}{(r-l-2)!}, \tag{6.49}$$

$l = 0, \ldots, r-2,$

$$\tilde{\gamma}_l = (-1)^{r+1}2^{2l-1}(r-2)!\,l!\,(l+1-r)!, \tag{6.50}$$

$$l = r - 1, r, \ldots,$$

$q = 2$:

$$\tilde{\gamma}_l = 2^{2l}(l!)^2, \quad l \in \mathbb{N}_0. \tag{6.51}$$

For brevity we let

$$\Omega_q = \begin{cases} \frac{1}{(q-2)\|\mathbb{S}^{q-1}\|} & , \quad q \neq 2, \\ \frac{1}{\|\mathbb{S}^{q-1}\|} & , \quad q = 2. \end{cases} \tag{6.52}$$

The *fundamental solution* $F_q^{(l)} : \mathbb{R}^q \backslash \{0\} \to \mathbb{R}$ *for the equation* $\Delta^l F_q^{(l)} = 0$ is defined by

$$F_q^{(l)}(|x|) = \begin{cases} \Omega_q \frac{|x|^{2l-q}}{\gamma_{l-1}}, & q \text{ odd}, \quad l = 1, 2, \ldots, \\ \Omega_q \frac{|x|^{2l-q}}{\tilde{\gamma}_{l-1}}, & q = 2r, \quad l = 1, \ldots, r - 1 \\ \Omega_q \left(\frac{-|x|^{2l-q} \ln(|x|)}{\tilde{\gamma}_{l-1}} + \frac{C_{l-\frac{q}{2}}}{\tilde{\gamma}_{l-1}} |x|^{2l-q} \right), & q = 2r, \quad l = r, r+1, \ldots, \end{cases} \tag{6.53}$$

where the constants C_l, $l = 0, 1, \ldots,$ are given by

$$C_0 = 0, \tag{6.54}$$

$$C_l = \sum_{k=1}^{l} \frac{1}{2k} + \sum_{k=r}^{l+r-1} \frac{1}{2k}, \quad l > 0. \tag{6.55}$$

With this choice of the constants C_l it immediately follows that for all dimensions $q = 2, 3, \ldots$ and all degrees $l = 1, 2, \ldots$ we have

$$\Delta^m F_q^{(l)} = F_q^{(l-m)}, \quad m < l. \tag{6.56}$$

Remark 6.4. *The fundamental solution (6.53) plays an important role in the characterization of the singularity behavior of the lattice function (see Section 8.1).*

Now, suppose that $\mathcal{G}$ is a regular region with continuously differentiable boundary $\partial\mathcal{G}$. By standard arguments we obtain from the Extended Second Green Theorem the following extension of the Third Green Theorem.

Theorem 6.5. *(Extended Third Green Theorem)*
(i) Let $\mathcal{G} \subset \mathbb{R}^q$, $q \geq 2$, be a regular region with continuously differentiable

boundary $\partial \mathcal{G}$. Let $F : \overline{\mathcal{G}} \to \mathbb{R}$ be of class $\mathrm{C}^{(2m)} \left(\overline{\mathcal{G}} \right), \overline{\mathcal{G}} = \mathcal{G} \cup \partial \mathcal{G}, m \geq 1$. Then

$$\sum_{l=0}^{m-1} \int_{\partial \mathcal{G}} F_q^{(l+1)}(|x-y|) \left(\frac{\partial}{\partial \nu_y} \Delta_y^l F(y) \right) \, dS(y)$$

$$- \sum_{l=0}^{m-1} \int_{\partial \mathcal{G}} \left(\frac{\partial F_q^{(l+1)}}{\partial \nu_y} (|x-y|) \right) \Delta_y^l F(y) \, dS(y)$$

$$- \int_{\mathcal{G}} \left(F_q^{(m)}(|x-y|) \right) \left(\Delta_y^m F(y) \right) \, dV(y)$$

$$= \begin{cases} F(x) & , \quad x \in \mathcal{G} \\ \frac{1}{2}F(x) & , \quad x \in \partial \mathcal{G} \\ 0 & , \quad x \notin \overline{\mathcal{G}}. \end{cases}$$

(ii) Let $\mathcal{G} \subset \mathbb{R}^q$, $q \geq 2$, be a regular region. Let $F : \overline{\mathcal{G}} \to \mathbb{R}$ be of the class $\mathrm{C}^{(2m)} \left(\overline{\mathcal{G}} \right), \overline{\mathcal{G}} = \mathcal{G} \cup \partial \mathcal{G}, m \geq 1$. Then

$$\sum_{l=0}^{m-1} \int_{\partial \mathcal{G}} F_q^{(l+1)}(|x-y|) \left(\frac{\partial}{\partial \nu_y} \Delta_y^l F(y) \right) \, dS(y)$$

$$- \sum_{l=0}^{m-1} \int_{\partial \mathcal{G}} \left(\frac{\partial F_q^{(l+1)}}{\partial \nu_y} (|x-y|) \right) \Delta_y^l F(y) \, dS(y)$$

$$- \int_{\mathcal{G}} F_q^{(m)}(|x-y|) \, \Delta_y^m F(y) \, dV(y)$$

$$= \alpha(x) \, F(x),$$

where $\alpha(x)$, $x \in \mathbb{R}^q$, is the solid angle at x subtended by the surface $\partial \mathcal{G}$.

Remark 6.5. *For $m = 1$, Theorem 6.5 is the usual Third Green Theorem for the Laplacian.*

Observing the differential equation (6.56) we get the following reformulation of Theorem 6.5.

Corollary 6.1. *Under the assumptions of Theorem 6.5 (ii) we have*

$$\sum_{l=0}^{m-1} \int_{\partial \mathcal{G}} \left(\Delta_y^{2m-l-1} F_q^{(2m)}(|x-y|) \right) \left(\frac{\partial}{\partial \nu_y} \Delta_y^l F(y) \right) \, dS(y) \qquad (6.57)$$

$$- \sum_{l=0}^{m-1} \int_{\partial \mathcal{G}} \left(\frac{\partial}{\partial \nu_y} \Delta_y^{2m-l-1} F_q^{(2m)}(|x-y|) \right) \left(\Delta_y^l F(y) \right) \, dS(y)$$

$$- \int_{\mathcal{G}} \left(\Delta_y^m F_q^{(2m)}(|x-y|) \right) \left(\Delta_y^m F(y) \right) \, dV(y)$$

$$= \alpha(x) \, F(x),$$

where $\alpha(x)$, $x \in \mathbb{R}^q$, is the solid angle at x subtended by the surface $\partial \mathcal{G}$.

6.2 Integral Theorems for the Laplace–Beltrami Operator

The set of scalar functions $F : \mathbb{S}^{q-1} \to \mathbb{C}$ which are measurable and for which

$$\|F\|_{\mathrm{L}^p(\mathbb{S}^{q-1})} = \left(\int_{\mathbb{S}^{q-1}} |F(\xi)|^p \, dS_{(q-1)}(\xi) \right)^{\frac{1}{p}} < \infty, \quad 1 \le p < \infty, \quad (6.58)$$

is known as $\mathrm{L}^p(\mathbb{S}^{q-1})$. Clearly, $\mathrm{L}^p(\mathbb{S}^{q-1}) \subset \mathrm{L}^q(\mathbb{S}^{q-1})$ for $1 \le q < p$. A function $F : \mathbb{S}^{q-1} \to \mathbb{C}$ possessing k continuous derivatives on the unit sphere $\mathbb{S}^{q-1}$ is said to be of class $\mathrm{C}^{(k)}(\mathbb{S}^{q-1})$, $0 \le k \le \infty$. $\mathrm{C}(\mathbb{S}^{q-1})$ ($= \mathrm{C}^{(0)}(\mathbb{S}^{q-1})$) denotes the class of continuous scalar-valued functions on $\mathbb{S}^{q-1}$. $\mathrm{C}(\mathbb{S}^{q-1})$ is the complete normed space endowed with

$$\|F\|_{\mathrm{C}(\mathbb{S}^{q-1})} = \sup_{\xi \in \mathbb{S}^{q-1}} |F(\xi)|. \quad (6.59)$$

By $\mu(F; \delta)$, we denote the *modulus of continuity* of a function $F \in \mathrm{C}(\mathbb{S}^{q-1})$

$$\mu(F; \delta) = \max_{\substack{\xi, \zeta \in \mathbb{S}^{q-1} \\ 1 - \xi \cdot \zeta \le \delta}} |F(\xi) - F(\zeta)|, \quad 0 < \delta < 2. \quad (6.60)$$

A function $F : \mathbb{S}^{q-1} \to \mathbb{C}$ is said to be *Lipschitz-continuous* if there exists a (Lipschitz) constant $C_F > 0$ such that the inequality

$$|F(\xi) - F(\eta)| \le C_F \, |\xi - \eta| = \sqrt{2} \, C_F \sqrt{1 - \xi \cdot \eta} \quad (6.61)$$

holds for all $\xi, \eta \in \mathbb{S}^{q-1}$. The class of all Lipschitz-continuous functions on $\mathbb{S}^{q-1}$ is denoted by $\mathrm{Lip}(\mathbb{S}^{q-1})$. Clearly, $\mathrm{C}^{(1)}(\mathbb{S}^{q-1}) \subset \mathrm{Lip}(\mathbb{S}^{q-1})$.

$\mathrm{L}^2(\mathbb{S}^{q-1})$ is the Hilbert space with respect to the inner product $(\cdot, \cdot)_{\mathrm{L}^2(\mathbb{S}^{q-1})}$ defined by

$$(F, G)_{\mathrm{L}^2(\mathbb{S}^{q-1})} = \int_{\mathbb{S}^{q-1}} F(\xi) \overline{G(\xi)} \, dS_{(q-1)}(\xi), \quad F, G \in \mathrm{L}^2(\mathbb{S}^{q-1}). \quad (6.62)$$

In connection with $(\cdot, \cdot)_{\mathrm{L}^2(\mathbb{S}^{q-1})}$, $\mathrm{C}(\mathbb{S}^{q-1})$ is a pre-Hilbert space. For each $F \in \mathrm{C}(\mathbb{S}^{q-1})$ we have the norm estimate

$$\|F\|_{\mathrm{L}^2(\mathbb{S}^{q-1})} \le \sqrt{\|\mathbb{S}^{q-1}\|} \, \|F\|_{\mathrm{C}(\mathbb{S}^{q-1})}. \quad (6.63)$$

$\mathrm{L}^2(\mathbb{S}^{q-1})$ is the completion of $\mathrm{C}(\mathbb{S}^{q-1})$ with respect to the norm $\| \cdot \|_{\mathrm{L}^2(\mathbb{S}^{q-1})}$, i.e.,

$$\mathrm{L}^2(\mathbb{S}^{q-1}) = \overline{\mathrm{C}(\mathbb{S}^{q-1})}^{\| \cdot \|_{\mathrm{L}^2(\mathbb{S}^{q-1})}}. \quad (6.64)$$

Any function of the form

$$G_\xi : \mathbb{S}^{q-1} \to \mathbb{R}, \eta \mapsto G_\xi(\eta) = G(\xi \cdot \eta), \quad \eta \in \mathbb{S}^{q-1}, \quad (6.65)$$

is called a ξ-*zonal function* on $\mathbb{S}^{q-1}$ (or ξ-*axial radial basis function*). Zonal functions are constant on the sets $\mathbb{S}^{q-1}(\xi;\tau) = \{\eta \in \mathbb{S}^{q-1} | \xi \cdot \eta = \tau\}$, where $\tau \in [-1,1]$. The set of all ξ-zonal functions is isomorphic to the set of functions $G : [-1,1] \to \mathbb{R}$. This allows us to interpret $C^{(0)}([-1,1])$ and $L^p([-1,1])$ (with norms defined correspondingly) as subspaces of $C^{(0)}(\mathbb{S}^{q-1})$ and $L^p(\mathbb{S}^{q-1})$.

Sphere Function for the Laplace–Beltrami Operator

It is natural to restrict the Green theorems known from $\mathbb{R}^q$ to $\mathbb{S}^{q-1}$. In case of the *First Green Surface Theorem* we find

$$
\begin{aligned}
\int_{\mathbb{S}^{q-1}} F(\xi)\Delta^*G(\xi)\, dS_{(q-1)}(\xi) &= \int_{\mathbb{S}^{q-1}} F(\xi)\,(\nabla^* \cdot \nabla^*G(\xi))\, dS_{(q-1)}(\xi) \\
&= -\int_{\mathbb{S}^{q-1}} \nabla^*F(\xi)\cdot\nabla^*G(\xi)\, dS_{(q-1)}(\xi),
\end{aligned}
$$
$$(6.66)$$

provided that $F \in C^{(1)}(\mathbb{S}^{q-1})$, $G \in C^{(2)}(\mathbb{S}^{q-1})$. In case of the *Second Green Surface Theorem* this leads to the identity (describing partial integration in terms of Δ^*)

$$
\begin{aligned}
\int_{\mathbb{S}^{q-1}} F(\xi)\,\Delta^*G(\xi)\, dS_{(q-1)}(\xi) &= -\int_{\mathbb{S}^{q-1}} \nabla^*F(\xi)\cdot\nabla^*G(\xi)\, dS_{(q-1)}(\xi) \\
&= \int_{\mathbb{S}^{q-1}} G(\xi)\,\Delta^*F(\xi)\, dS_{(q-1)}(\xi),
\end{aligned} \quad (6.67)
$$

provided that $F, G \in C^{(2)}(\mathbb{S}^{q-1})$.

Next we introduce the Green function of the unit sphere $\mathbb{S}^{q-1}$, $q \geq 3$, for the Beltrami operator Δ^* (cf. W. Freeden [1979], W. Freeden [1980b] for the case $q = 3$, and W. Freeden, R. Reuter [1982] for the cases $q \geq 3$).

Definition 6.2. $G(\Delta^*; \cdot, \cdot) : (\xi, \eta) \mapsto G(\Delta^*; \xi, \eta)$, $-1 \leq \xi \cdot \eta < 1$, *is called the Green function for the Beltrami operator* Δ^* *on the unit sphere* $\mathbb{S}^{q-1}$ *(in brief,* $\mathbb{S}^{q-1}$*-sphere function for* Δ^**) if it satisfies the following properties:*

(i) For each fixed $\xi \in \mathbb{S}^{q-1}$, $\eta \mapsto G(\Delta^*; \xi, \eta)$ *is twice continuously differentiable with respect to (the variable)* $\eta \in \mathbb{S}^{q-1}$, $1 - \xi \cdot \eta \neq 0$, *with*

$$
\Delta^*_\eta G(\Delta^*; \xi, \eta) = -\frac{1}{\|\mathbb{S}^{q-1}\|}. \tag{6.68}
$$

(ii) In the neighborhood of the point $\xi \in \mathbb{S}^{q-1}$ *the estimates*

$$
\begin{aligned}
G(\Delta^*; \xi, \eta) - \tfrac{1}{\|\mathbb{S}^2\|}\ln(1 - \xi \cdot \eta) &= O(1) \\
\nabla^*_\eta G(\Delta^*; \xi, \eta) - \tfrac{1}{\|\mathbb{S}^2\|}\nabla^*_\eta \ln(1 - \xi \cdot \eta) &= O(1)
\end{aligned} \quad (q = 3) \qquad (6.69)
$$

and

$$G(\Delta^*; \xi, \eta) + \frac{1}{q-3} \frac{2^{\frac{3-q}{2}}}{\|\mathbb{S}^{q-2}\|} (1 - \xi \cdot \eta)^{\frac{3-q}{2}} \tag{6.70}$$

$$= \begin{cases} O(\ln(1 - \xi \cdot \eta)), & q = 5 \\ O\left((1 - \xi \cdot \eta)^{\frac{5-q}{2}}\right), & q = 4, q \geq 6 \end{cases}$$

$$\nabla_\eta^* G(\Delta^*; \xi, \eta) + \frac{1}{q-3} \frac{2^{\frac{3-q}{2}}}{\|\mathbb{S}^{q-2}\|} \nabla_\eta^* (1 - \xi \cdot \eta)^{\frac{3-q}{2}} \tag{6.71}$$

$$= \begin{cases} O\left(\nabla_\eta^* \ln(1 - \xi \cdot \eta)\right), & q = 5 \\ O\left(\nabla_\eta^* (1 - \xi \cdot \eta)^{\frac{5-q}{2}}\right), & q = 4, q \geq 6 \end{cases}$$

are valid.

(iii) For all orthogonal transformations **t**

$$G(\Delta^*; \xi, \eta) = G(\Delta^*, \mathbf{t}\xi, \mathbf{t}\eta). \tag{6.72}$$

(iv) For all $\xi \in \mathbb{S}^{q-1}$

$$\int_{\mathbb{S}^{q-1}} G(\Delta^*; \xi, \eta) \, dS_{(q-1)}(\eta) = 0. \tag{6.73}$$

Because of Condition *(iii)* of Definition 6.2, $G(\Delta^*; \cdot)$ depends only on the scalar product of ξ and η (i.e., the function $G(\Delta^*; \cdot)$ is a zonal function); hence, it may be understood as a function defined on the (one-dimensional) interval $[-1, 1)$.

Remark 6.6. *Throughout this work, we consistently write* $G(\Delta^*; \xi \cdot \eta)$ *instead of* $G(\Delta^*; \xi, \eta)$ *(more generally,* $G(\Delta_{(q)}^*; \xi_{(q)}, \eta_{(q)})$*).*

By the defining properties as stated in Definition 6.2, the $\mathbb{S}^{q-1}$-sphere function for Δ^* is uniquely determined (cf. W. Freeden [1979, 1980b]).

Lemma 6.4. *For* $\xi, \eta \in \mathbb{S}^{q-1}$ *with* $-1 \leq \xi \cdot \eta < 1$

$$G(\Delta^*; \xi \cdot \eta) \tag{6.74}$$

$$= -\frac{1}{(q-2)\|\mathbb{S}^{q-1}\|} \int_0^1 \frac{1 + r^{q-2}}{r} \left[\frac{1}{(1 - 2r(\xi \cdot \eta) + r^2)^{\frac{q-2}{2}}} - 1 \right] dr.$$

We list the representation of $G(\Delta^*; \xi \cdot \eta)$ for the dimensions $q = 3, 4, 5$.

$q = 3$:

$$G(\Delta^*; \xi \cdot \eta) = -\frac{1}{4\pi} \int_0^1 \frac{1 + r}{r} \left[\frac{1}{(1 - 2r(\xi \cdot \eta) + r^2)^{\frac{1}{2}}} - 1 \right] dr. \tag{6.75}$$

Note that, with $t = \xi \cdot \eta$, we find

$$
\begin{aligned}
G(\Delta^*; \xi \cdot \eta) &= \frac{1}{4\pi} \left[r + \ln r + \ln \left(\frac{2}{r} \left(\sqrt{1 - 2rt + r^2} + 1 - tr \right) \right) \right. \\
&\quad \left. - \ln \left(2 \left(\sqrt{1 - 2rt + r^2} + r - t \right) \right) \right]_0^1 \\
&= \frac{1}{4\pi} \left[r + \ln \left(\frac{\sqrt{1 - 2rt + r^2} + 1 - rt}{\sqrt{1 - 2rt + r^2} + r - t} \right) \right]_0^1 \\
&= -\frac{1}{4\pi} \left(-1 + \ln 2 - \ln(1 - t) \right).
\end{aligned} \tag{6.76}
$$

$q = 4$:

$$
\begin{aligned}
&G(\Delta^*; \xi \cdot \eta) \\
&= -\frac{1}{4\pi^2} \int_0^1 \frac{1 + r^2}{r} \left[\frac{1}{1 - 2rt + r^2} - 1 \right] dr \\
&= -\frac{1}{2\pi^2} \left(-\frac{1}{4} + \frac{t}{\sqrt{1 - t^2}} \arctan \frac{\sqrt{1 + t}}{\sqrt{1 - t}} \right).
\end{aligned} \tag{6.77}
$$

$q = 5$:

$$
\begin{aligned}
G(\Delta^*; \xi \cdot \eta) &= -\frac{1}{8\pi^2} \int_0^1 \frac{1 + r^3}{r} \left[\frac{1}{(1 - 2r(\xi \cdot \eta) + r^2)^{\frac{3}{2}}} - 1 \right] dr \\
&= -\frac{1}{8\pi^2} \left[\frac{1}{1 - t} - \ln(1 - t) + \ln 2 - \frac{7}{3} \right].
\end{aligned} \tag{6.78}
$$

Integral Formulas for the Laplace–Beltrami Operator

The purpose now is to formulate a counterpart to the Third Green Theorem on the unit sphere $\mathbb{S}^{q-1}$ for the Beltrami operator Δ^* (see W. Freeden [1979, 1981], R. Reuter [1982]).

Suppose that F is a twice continuously differentiable function on $\mathbb{S}^{q-1}$, i.e., $F \in C^{(2)}(\mathbb{S}^{q-1})$. Then, for each sufficiently small $\varepsilon > 0$, the Second Green Surface Theorem gives

$$
\int_{\substack{\sqrt{1 - \xi \cdot \eta} \geq \varepsilon \\ \eta \in \mathbb{S}^{q-1}}} \{ G(\Delta^*; \xi \cdot \eta) \Delta_\eta^* F(\eta) \tag{6.79}
$$

$$
- F(\eta) \Delta_\eta^* G(\Delta^*; \xi \cdot \eta) \} \, dS_{(q-1)}(\eta)
$$

$$
= \int_{\substack{\sqrt{1 - \xi \cdot \eta} = \varepsilon \\ \eta \in \mathbb{S}^{q-1}}} \left\{ G(\Delta^*; \xi \cdot \eta) \left(\frac{\partial}{\partial \nu_\eta} F(\eta) \right) \right.
$$

$$
\left. - F(\eta) \left(\frac{\partial}{\partial \nu_\eta} G(\Delta^*; \xi \cdot \eta) \right) \right\} \, dS_{(q-2)}(\eta),
$$

where $dS_{(q-2)}$ denotes the surface element in $\mathbb{R}^{q-1}$, while ν is the (unit) vector normal to $\{\eta \in \mathbb{S}^{q-1} \mid \sqrt{1 - \xi \cdot \eta} = \varepsilon\}$ and tangential on $\mathbb{S}^{q-1}$ and directed into the exterior of $\{\eta \in \mathbb{S}^{q-1} \mid \sqrt{1 - \xi \cdot \eta} \geq \varepsilon\}$. Inserting the differential equations of the $\mathbb{S}^{q-1}$-sphere function for Δ^*, we obtain

$$\int_{\substack{\sqrt{1-\xi\cdot\eta}\geq\varepsilon \\ \eta\in\mathbb{S}^{q-1}}} F(\eta)\, \Delta_\eta^* G(\Delta^*; \xi \cdot \eta)\, dS_{(q-1)}(\eta) \tag{6.80}$$

$$= -\frac{1}{\|\mathbb{S}^{q-1}\|} \int_{\substack{\sqrt{1-\xi\cdot\eta}\geq\varepsilon \\ \eta\in\mathbb{S}^{q-1}}} F(\eta)\, dS_{(q-1)}(\eta).$$

Observing the characteristic singularity of the $\mathbb{S}^{q-1}$-sphere function for Δ^*, we are able to prove by analogous conclusions as known in potential theory

$$\int_{\substack{\sqrt{1-\xi\cdot\eta}=\varepsilon \\ \eta\in\mathbb{S}^{q-1}}} G(\Delta^*; \xi \cdot \eta) \frac{\partial}{\partial\nu_\eta} F(\eta)\, dS_{(q-2)}(\eta) = o(1), \quad \varepsilon \to 0, \tag{6.81}$$

and

$$\int_{\substack{\sqrt{1-\xi\cdot\eta}=\varepsilon \\ \eta\in\mathbb{S}^{q-1}}} F(\eta) \frac{\partial}{\partial\nu_\eta} G(\Delta^*; \xi \cdot \eta)\, dS_{(q-2)}(\eta) = -F(\xi) + o(1), \quad \varepsilon \to 0. \tag{6.82}$$

Summarizing our results we therefore get the following analogue to the Third Green Theorem for the unit sphere $\mathbb{S}^{q-1}$.

Theorem 6.6. *(Integral Formula for the Operator Δ^*) If $\xi \in \mathbb{S}^{q-1}$ and $F \in C^{(2)}(\mathbb{S}^{q-1})$, then*

$$\begin{aligned} F(\xi) &= \frac{1}{\|\mathbb{S}^{q-1}\|} \int_{\mathbb{S}^{q-1}} F(\eta)\, dS_{(q-1)}(\eta) \\ &\quad + \int_{\mathbb{S}^{q-1}} G(\Delta^*; \xi \cdot \eta) \left(\Delta_\eta^* F(\eta) \right) dS_{(q-1)}(\eta). \end{aligned} \tag{6.83}$$

This formula compares the integral mean over $\mathbb{S}^{q-1}$ with the functional value at $\xi \in \mathbb{S}^{q-1}$ under explicit knowledge of the remainder term in integral form. It also serves as a point of departure for the spherical spline theory and best approximate integration formulas (as proposed, e.g., by W. Freeden [1981], G. Wahba [1981], W. Freeden, P. Hermann [1986] and the references therein).

Next we are interested in extending Theorem 6.6 to iterated operators.

Definition 6.3. *Let $G\left((\Delta^*)^m; \xi \cdot \eta\right)$, $m = 2, 3, \ldots$, be recursively defined by the convolution integrals*

$$G\left((\Delta^*)^m; \xi \cdot \eta\right) = \int_{\mathbb{S}^{q-1}} G\left((\Delta^*)^{m-1}; \xi \cdot \zeta\right) G\left(\Delta^*; \zeta \cdot \eta\right) dS_{(q-1)}(\zeta), \tag{6.84}$$

where

$$G\left((\Delta^*)^m; \xi \cdot \eta\right) = G\left(\Delta^*; \xi \cdot \eta\right), \quad m = 1. \tag{6.85}$$

Then, $G\left((\Delta^*)^m; \cdot\right)$ is called the $\mathbb{S}^{q-1}$-sphere function for $(\Delta^*)^m$.

In analogy to techniques known in potential theory it can be shown that

$$
G\left((\Delta^*)^m, \xi \cdot \eta\right) =
\begin{cases}
O\left((1 - \xi \cdot \eta)^{m - \frac{q-1}{2}} \ln(1 - \xi \cdot \eta)\right), \\
\qquad\qquad\qquad\qquad\qquad 2m \leq q - 1, \quad q \text{ odd} \\
O\left((1 - \xi \cdot \eta)^{m - \frac{q-1}{2}}\right), \qquad\qquad \text{otherwise.}
\end{cases}
$$

Observing the differential equation

$$
\Delta^* G\left((\Delta^*)^m; \xi \cdot \eta\right) = G\left((\Delta^*)^{m-1}, \xi \cdot \eta\right), \tag{6.86}
$$

$-1 \leq \xi \cdot \eta < 1, m \geq 2$, we obtain by successive integration by parts the following extension of Theorem 6.6.

Theorem 6.7. *(Integral Formula for the Iterated Beltrami Operator). Suppose that $\xi \in \mathbb{S}^{q-1}$ and $F \in C^{(2m)}(\mathbb{S}^{q-1})$, $m \in \mathbb{N}$. Then*

$$
\begin{aligned}
F(\xi) \;=\; & \frac{1}{\|\mathbb{S}^{q-1}\|} \int_{\mathbb{S}^{q-1}} F(\eta)\, dS_{q-1}(\eta) \\
& + \int_{\mathbb{S}^{q-1}} G\left((\Delta^*)^m; \xi \cdot \eta\right)\, (\Delta^*)^m_\eta F(\eta)\, dS_{(q-1)}(\eta). \tag{6.87}
\end{aligned}
$$

This formula compares the integral mean of F over $\mathbb{S}^{q-1}$ with the functional value of F taken at the point $\xi \in \mathbb{S}^{q-1}$.

An immediate consequence of Theorem 6.7 is the following corollary.

Corollary 6.2. *Under the assumptions of Theorem 6.7 we have*

$$
\begin{aligned}
F(\xi) \;=\; & \frac{1}{\|\mathbb{S}^{q-1}\|} \int_{\mathbb{S}^{q-1}} F(\eta)\, dS_{(q-1)}(\eta) \\
& + \int_{\mathbb{S}^{q-1}} (\Delta^*)^m_\eta G\left((\Delta^*)^{2m}; \xi \cdot \eta\right)\, (\Delta^*)^m_\eta F(\eta)\, dS_{(q-1)}(\eta).
\end{aligned}
$$

Laplace–Beltrami Differential Equation

Theorem 6.7 can be used to discuss the differential equation

$$
(\Delta^*)^m V = W, \quad V \in C^{(2m)}(\mathbb{S}^{q-1}), \quad m \in \mathbb{N}. \tag{6.88}
$$

Theorem 6.8. *Let W be a function of class $C^{(0)}(\mathbb{S}^{q-1})$ with vanishing integral mean, i.e.,*

$$
\frac{1}{\|\mathbb{S}^{q-1}\|} \int_{\mathbb{S}^{q-1}} W(\eta)\, dS_{(q-1)}(\eta) = 0. \tag{6.89}
$$

Then the function V given by

$$
V(\xi) = \int_{\mathbb{S}^{q-1}} G\left((\Delta^*)^m; \xi \cdot \eta\right)\, W(\eta)\, dS_{(q-1)}(\eta) \tag{6.90}
$$

represents the only $(2m)$-times continuously differentiable solution of the differential equation $(\Delta^*)^m V = W$ on $\mathbb{S}^{q-1}$, which satisfies

$$\frac{1}{\|\mathbb{S}^{q-1}\|} \int_{\mathbb{S}^{q-1}} V(\eta)\, dS_{(q-1)}(\eta) = 0. \tag{6.91}$$

6.3 Tools Involving the Laplace Operator

Next we come to integral estimates involving fundamental solutions which play an essential role in asymptotic relations for Euler and Poisson summation in Euclidean spaces $\mathbb{R}^q$. Our considerations are based on the two-dimensional work by W. Freeden [1975, 1978a] and the q-dimensional generalization by C. Müller, W. Freeden [1980].

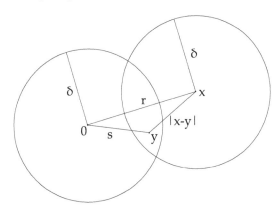

FIGURE 6.2
The geometric situation as discussed by Lemma 6.5.

Integral Estimates

For $x, y \in \mathbb{R}^q$, we introduce the polar coordinates

$$
\begin{array}{cccc}
x = r\xi & , & \xi^2 = 1 & , & r = |x|, \\
y = s\eta & , & \eta^2 = 1 & , & s = |y|
\end{array}
\tag{6.92}
$$

characterizing the geometric situation as illustrated by Figure 6.2. We are interested in an estimate for the scalar product $\xi \cdot \eta$ if x and y are assumed to satisfy $|x-y| \leq \delta$ with (sufficiently small) fixed positive δ. Clearly, $|x-y|^2 \leq \delta^2$ is equivalent to $r^2 + s^2 - 2rs\xi \cdot \eta \leq \delta^2$, such that

$$\left(1 - \frac{\delta^2}{r^2}\right) \leq -\left(\frac{s}{r} - \xi \cdot \eta\right)^2 + (\xi \cdot \eta)^2 \leq (\xi \cdot \eta)^2. \tag{6.93}$$

Thus, we are led to the following estimate.

Lemma 6.5. *For $x \in \mathbb{R}^q$ fixed with $|x| \geq \delta > 0$, the inequality $|x - y| \leq \delta$ implies in the polar coordinates (6.92)*

$$\xi \cdot \eta \geq \sqrt{1 - \frac{\delta^2}{r^2}}. \tag{6.94}$$

Based on Lemma 6.5 we prove the following lemma.

Lemma 6.6. *Suppose that δ is a fixed real number with $\delta \in (0,1)$. Then*

$$\int_{\substack{|x-y| \leq \delta \\ |x|=N \\ x \in \mathbb{R}^q}} \frac{1}{|x - y|^k} \, dS_{(q-1)}(x) \leq \frac{\|\mathbb{S}^{q-2}\|}{q - 1 - k} \frac{\delta^{q-1-k}}{\sqrt{1 - \frac{\delta^2}{N^2}}} \tag{6.95}$$

holds true for all $N \geq 1$ and all integers k with $0 \leq k < q - 1$.

Proof. We already know that the surface element of the sphere $\mathbb{S}_N^{q-1}$ with radius N around 0 is equal to $dS_{(q-1)}(x) = N^{q-1} dS_{(q-1)}(\xi)$, $x = N\xi$, where $dS_{(q-1)}(\xi)$ is the surface element of the unit sphere $\mathbb{S}^{q-1}$. By virtue of Lemma 6.5 we get

$$\int_{\substack{|x-y| \leq \delta \\ |x|=N \\ x \in \mathbb{R}^q}} \frac{dS_{(q-1)}(x)}{|x - y|^k} = \int_{\substack{\xi \cdot \eta \geq \sqrt{1 - \frac{\delta^2}{N^2}} \\ \xi \in \mathbb{S}^{q-1}}} \frac{N^{(q-1)-k} \, dS_{(q-1)}(\xi)}{(1 + \rho^2 - 2\rho(\xi \cdot \eta))^{\frac{k}{2}}}, \tag{6.96}$$

where we use the notation $\rho = \frac{|y|}{N} = \frac{s}{N}$. Now we have

$$\int_{\substack{\xi \cdot \eta \geq \sqrt{1 - \frac{\delta^2}{N^2}} \\ \xi \in \mathbb{S}^{q-1}}} \frac{N^{(q-1)-k}}{(1 + \rho^2 - 2\rho(\xi \cdot \eta))^{\frac{k}{2}}} \, dS_{(q-1)}(\xi) \tag{6.97}$$

$$= \|\mathbb{S}^{q-2}\| N^{(q-1)-k} \int_{\sqrt{1 - \frac{\delta^2}{N^2}}}^{1} \frac{(1 - t^2)^{\frac{q-3}{2}}}{(1 + \rho^2 - 2\rho t)^{\frac{k}{2}}} \, dt.$$

We observe the inequality

$$1 + \rho^2 - 2\rho t = (\rho - t)^2 + 1 - t^2 \geq 1 - t^2. \tag{6.98}$$

Because of $\sqrt{1 - \frac{\delta^2}{N^2}} \leq t \leq 1$ we therefore obtain

$$N^{(q-1)-k} \int_{\sqrt{1 - \frac{\delta^2}{N^2}}}^{1} \frac{(1 - t^2)^{\frac{q-3}{2}}}{(1 + \rho^2 - 2\rho t)^{\frac{k}{2}}} \, dt \tag{6.99}$$

$$\leq N^{(q-1)-k} \int_{\sqrt{1 - \frac{\delta^2}{N^2}}}^{1} (1 - t^2)^{\frac{q-3-k}{2}} \, dt$$

$$\leq N^{(q-1)-k} \int_{\sqrt{1 - \frac{\delta^2}{N^2}}}^{1} (1 - t^2)^{\frac{q-3-k}{2}} \frac{t}{\sqrt{1 - \frac{\delta^2}{N^2}}} \, dt.$$

Now, an elementary calculation gives

$$\int_{\sqrt{1-\frac{\delta^2}{N^2}}}^{1} (1-t^2)^{\frac{q-3-k}{2}} t\, dt = -\frac{1}{2} \frac{1}{\frac{q-1-k}{2}} (1-t^2)^{\frac{q-1-k}{2}} \Big|_{\sqrt{1-\frac{\delta^2}{N^2}}}^{1}$$

$$= \frac{1}{q-1-k} \frac{\delta^{q-1-k}}{N^{q-1-k}}. \tag{6.100}$$

Combining (6.97) and (6.100) we obtain the desired result. $\qquad\square$

For the two-dimensional case (i.e., $q = 2$) we find the following lemma.

Lemma 6.7. *Suppose that* $\delta \in (0,1)$*. Then*

$$\int_{\substack{|x-y|\leq\delta \\ |x|=N \\ x\in\mathbb{R}^2}} |\ln(|x-y|)|\, dS_{(1)}(x) \leq -\frac{2\delta\ln(\delta)}{\sqrt{1-\frac{\delta^2}{N^2}}} \tag{6.101}$$

holds true for all $N \geq 1$.

Proof. If $\delta \in (0,1)$, then it follows for $x, y \in \mathbb{R}^2$ with $|x-y| \leq \delta$ that

$$|\ln(|x-y|)| = -\ln(|x-y|) = -\frac{1}{2}\ln\left(N^2 + s^2 - 2Ns\xi\cdot\eta\right). \tag{6.102}$$

From Lemma 6.5 we know that

$$\xi\cdot\eta \geq \sqrt{1-\frac{\delta^2}{N^2}}. \tag{6.103}$$

Thus, we are able to conclude that

$$\int_{\substack{|x-y|\leq\delta \\ |x|=N \\ x\in\mathbb{R}^2}} |\ln(|x-y|)|\, dS_{(1)}(x) \tag{6.104}$$

$$\leq -N \int_{\sqrt{1-\frac{\delta^2}{N^2}}}^{1} (1-t^2)^{-\frac{1}{2}} \ln(N^2 + s^2 - 2Nst)\, dt.$$

We now use for $t \in \left[\sqrt{1-\frac{\delta^2}{N^2}}, 1\right]$ the identity

$$N^2 + s^2 - 2Nst = (s-Nt)^2 + N^2(1-t^2). \tag{6.105}$$

This leads us to the relation

$$N^2 + s^2 - 2Nst \geq N^2(1-t^2) \geq \delta^2. \tag{6.106}$$

Observing $\delta \in (0,1)$ we therefore obtain

$$
\begin{aligned}
\int_{\substack{|x-y|\le\delta \\ |x|=N \\ x\in\mathbb{R}^2}} |\ln(|x-y|)|\,dS_{(1)}(x) \;&\le\; -2N\ln(\delta)\int_{\sqrt{1-\frac{\delta^2}{N^2}}}^{1} \frac{t}{\sqrt{1-t^2}}\,\frac{dt}{\sqrt{1-\frac{\delta^2}{N^2}}} \\[2mm]
&\le\; \frac{-2N\ln(\delta)}{\sqrt{1-\frac{\delta^2}{N^2}}}\,\frac{\delta}{N} \\[2mm]
&=\; -\frac{2\delta\ln(\delta)}{\sqrt{1-\frac{\delta^2}{N^2}}}.
\end{aligned}
\tag{6.107}
$$

This is the result stated in Lemma 6.7. $\square$

Multi-Dimensional Angles

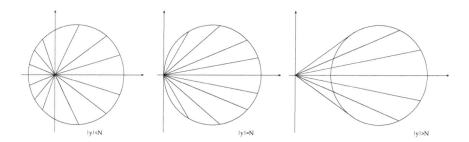

FIGURE 6.3
The three cases under consideration in Lemma 6.8.

Next we evaluate integrals which can be interpreted as the "q-dimensional angles" under which the sphere of radius N around a point $y \in \mathbb{R}^q$ is seen from the origin (see Figure 6.3). Three cases (illustrated by Figure 6.3) must be distinguished. Indeed, the three cases require three different calculations:

Lemma 6.8. *For $q \ge 2$, $y \in \mathbb{R}^q$, and $N > 0$*

$$
\int_{\substack{|x-y|=N \\ x\in\mathbb{R}^q}} \frac{|x\cdot\nu(x)|}{|x|^q}\,dS_{(q-1)}(x) \le \|\mathbb{S}^{q-1}\|.
\tag{6.108}
$$

Proof. We introduce polar coordinates $x = r(\xi)\xi$, $\xi \in \mathbb{S}^{q-1}$, to represent the sphere $\mathbb{S}_N^{q-1}(y) = \{x \in \mathbb{R}^q \mid |x-y| = N\}$ by its projection onto the unit sphere $\mathbb{S}^{q-1} = \{\xi \in \mathbb{R}^q \mid |\xi| = 1\}$. By means of the polar coordinates $x = r(\xi)\xi$, $\xi \in \mathbb{S}^{q-1}$, the surface element on $\mathbb{S}_N^{q-1}(y)$ can then be represented in the form

$$
dS_{(q-1)}(x) = dS_{(q-1)}(r(\xi)\xi) = (r(\xi))^{q-1}\,dS_{(q-1)}(\xi)
\tag{6.109}
$$

such that

$$|x \cdot \nu(x)| \, dS_{(q-1)}(x) \;=\; \frac{r(\xi)\xi \cdot r(\xi)\xi}{|r(\xi)|} \, dS_{(q-1)}(r(\xi)\xi) \qquad (6.110)$$

$$=\; r(\xi)(r(\xi))^{q-1} \, dS_{(q-1)}(\xi)$$

$$=\; (r(\xi))^{q} \, dS_{(q-1)}(\xi),$$

which relates the surface element $dS_{(q-1)}(x)$ on $\mathbb{S}_{N}^{q-1}(y)$ to its projection $dS_{(q-1)}(\xi)$ on $\mathbb{S}^{q-1}$. We understand $Z : \mathbb{S}^{q-1} \to \mathbb{R}$ to be the number of positive solutions $r : \xi \mapsto r(\xi)$, $\xi \in \mathbb{S}^{q-1}$, of the equation

$$|r(\xi)\xi - y| = N \qquad (6.111)$$

such that

$$\int_{\substack{|x-y|=N \\ x \in \mathbb{R}^{q}}} \frac{|x \cdot \nu(x)|}{|x|^{q}} \, dS(x) = \int_{\xi \in \mathbb{S}^{q-1}} Z(\xi) \, dS(\xi). \qquad (6.112)$$

Case $|y| < N$: We have exactly one positive r to every direction $\xi \in \mathbb{S}^{q-1}$ such that

$$|r(\xi)\xi - y|^{2} = (r(\xi) - y \cdot \xi)^{2} - (y \cdot \xi)^{2} + y^{2} = N^{2}. \qquad (6.113)$$

Consequently, $Z(\xi) = 1$ for all $\xi \in \mathbb{S}^{q-1}$. This shows that

$$\int_{\substack{|x-y|=N \\ x \in \mathbb{R}^{q}}} \frac{|x \cdot \nu(x)|}{|x|^{q}} \, dS(x) = \|\mathbb{S}^{q-1}\|. \qquad (6.114)$$

Case $|y| = N$: Now the equation

$$|r(\xi)\xi - y|^{2} = (r(\xi) - y \cdot \xi)^{2} - (y \cdot \xi)^{2} + y^{2} = N^{2} \qquad (6.115)$$

has just one positive solution for $y \cdot \xi > 0$ and no positive solution for $y \cdot \xi \leq 0$. This leads to

$$\int_{\substack{|x-y|=N \\ x \in \mathbb{R}^{q}}} \frac{|x \cdot \nu(x)|}{|x|^{q}} \, dS(x) = \frac{1}{2}\|\mathbb{S}^{q-1}\|. \qquad (6.116)$$

Case $|y| > N$: We have the two positive solutions

$$r_{\frac{1}{2}}(\xi) = y \cdot \xi \pm \sqrt{(y \cdot \xi)^{2} - (y^{2} - N^{2})} \qquad (6.117)$$

for $y \cdot \xi \geq \sqrt{y^{2} - N^{2}}$. Accordingly we get in this case

$$\int_{\substack{|x-y|=N \\ x \in \mathbb{R}^{q}}} \frac{|x \cdot \nu(x)|}{|x|^{q}} \, dS(x) = 2 \int_{\substack{\xi \cdot \eta \geq \sqrt{1 - \frac{N^{2}}{\delta^{2}}} \\ \xi \in \mathbb{S}^{q-1}}} dS(\xi). \qquad (6.118)$$

Therefore this yields the estimate

$$\int_{\substack{|x-y|=N \\ x \in \mathbb{R}^{q}}} \frac{|x \cdot \nu(x)|}{|x|^{q}} \, dS(x) = 2\|\mathbb{S}^{q-2}\| \int_{\sqrt{1-\frac{N^{2}}{\delta^{2}}}}^{1} (1-t^{2})^{\frac{q-3}{2}} \, dt < \|\mathbb{S}^{q-1}\|. \qquad (6.119)$$

Altogether, Lemma 6.8 follows from (6.114), (6.116), and (6.119). $\qquad\square$

A change of variables in Lemma 6.8 (more precisely, x is substituted by $x - y$, and y is replaced by $-y$) leads us to the following result.

Corollary 6.3. *For $q \geq 2$, $y \in \mathbb{R}^q$, and $N > 0$*

$$\int_{\substack{|x|=N \\ x \in \mathbb{R}^q}} \left| \frac{(x - y) \cdot \nu(x)}{|x - y|^q} \right| \, dS(x) \leq \|\mathbb{S}^{q-1}\|. \tag{6.120}$$

6.4 Radial and Angular Decomposition of Harmonics

This section gives a survey of the theory of homogeneous harmonic polynomials and spherical harmonics of dimension q including the addition theorem, the Funk–Hecke formula, the closure and completeness theorems and the characterization of spherical harmonics as eigensolutions of the Beltrami operator. Our approach is self-contained. It consistently contains all essential properties. Evidently - in the form as presented here - most of the material is known from the literature (see, e.g., C. Müller [1966] and the references therein). However, together with the comprehensive concept of Green functions for (Helmholtz) operators $\Delta^* + \lambda$, $\lambda \in \mathbb{R}$, the theory of spherical harmonics of dimension q seems to be not available in textbooks yet.

Finally, as link to approximate integration on the (unit) sphere, equidistributed point sets generated by projection of $\mathbb{Z}^q$-lattice points to $\mathbb{S}^{q-1}$ are investigated by use of the so–called Hlawka–Koksma formula involving the Green function for the Laplace–Beltrami operator Δ^*.

Homogeneous Harmonic Polynomials

Next we consider the theory on homogeneous harmonic polynomials in the Euclidean space $\mathbb{R}^q$ based on the work by W. Freeden [1979], R. Reuter [1982], W. Freeden et al. [1998], and C. Müller [1998].

Let $\operatorname{Hom}_n(\mathbb{R}^q)$ consist of all polynomials $H_n(q; \cdot)$ in q variables which are homogeneous of degree n (i.e., $H_n(q; \tau x) = \tau^n H_n(q; x)$ for all $\tau \in \mathbb{R}$ and all $x \in \mathbb{R}^q$). If $H_n(q; \cdot)$ is a member of the class $\operatorname{Hom}_n(\mathbb{R}^q)$, then there exist complex numbers $C_\alpha = C_{\alpha_1, \dots, \alpha_q}$ such that

$$H_n(q; x) = \sum_{[\alpha]=n} C_\alpha x^\alpha. \tag{6.121}$$

In Cartesian coordinates $x = (x_1, \dots, x_q)^T$,

$$H_n(q; x_1, \dots, x_q) = \sum_{\alpha_1 + \dots + \alpha_q = n} C_{\alpha_1, \dots, \alpha_q} x_1^{\alpha_1} \dots x_q^{\alpha_q}. \tag{6.122}$$

It is obvious that the set of monomials $x \mapsto x^\alpha$, $[\alpha] = n$, $x \in \mathbb{R}^q$, is a basis of the space $\text{Hom}_n(\mathbb{R}^q)$. The number of such monomials is precisely the number of ways a q-tuple can be chosen so that we have $[\alpha] = n$, i.e., the number of ways selecting $q - 1$ elements out of a collection of $n + q - 1$. This means that the dimension $\dim(\text{Hom}_n(\mathbb{R}^q))$ of $\text{Hom}_n(\mathbb{R}^q)$ is equal to

$$M(q; n) = \dim(\text{Hom}_n(\mathbb{R}^q)) = \binom{n + q - 1}{q - 1} = \binom{n + q - 1}{n}, \quad (6.123)$$

i.e.,

$$M(q; n) = \dim(\text{Hom}_n(\mathbb{R}^q)) = \frac{(n + q - 1)!}{(q - 1)! \, n!} = \frac{\Gamma(n + q)}{\Gamma(q)\Gamma(n + 1)}. \quad (6.124)$$

Let $H_n(q; \nabla_x)$ be the differential operator associated to $H_n(q; x)$ (i.e., replace x^α formally by $(\nabla_x)^\alpha$ in the expression of $H_n(q; x)$):

$$H_n(q; \nabla_x) = \sum_{[\alpha]=n} C_\alpha (\nabla_x)^\alpha. \quad (6.125)$$

If such an operator is applied to a homogeneous polynomial $U_n(q; \cdot)$ of the same degree

$$U_n(q; x) = \sum_{[\beta]=n} D_\beta x^\beta, \quad (6.126)$$

we obtain as result a number

$$H_n(q; \nabla_x)\overline{U_n(q; x)} = \sum_{[\alpha]=n} C_\alpha \overline{D_\alpha} \, \alpha!. \quad (6.127)$$

Clearly, we have

$$H_n(q; \nabla_x)\overline{U_n(q; x)} = \overline{U_n(q; \nabla_x)\overline{H_n(q; x)}}, \quad (6.128)$$

$$H_n(q; \nabla_x)\overline{H_n(q; x)} \geq 0. \quad (6.129)$$

This enables us (see W. Freeden [1979] for the three-dimensional case, R. Reuter [1982] for the q-dimensional case) to introduce an inner product $(\cdot, \cdot)_{\text{Hom}_n(\mathbb{R}^q)}$ on the space $\text{Hom}_n(\mathbb{R}^q)$ by letting

$$(H_n(q; \cdot), U_n(q; \cdot))_{\text{Hom}_n(\mathbb{R}^q)} = H_n(q; \nabla_x)\overline{U_n(q; x)}. \quad (6.130)$$

The space $\text{Hom}_n(\mathbb{R}^q)$ equipped with the inner product $(\cdot, \cdot)_{\text{Hom}_n(\mathbb{R}^q)}$ is a finite-dimensional Hilbert space. The set of monomials

$$\{x \mapsto (\alpha!)^{-\frac{1}{2}} x^\alpha | \, [\alpha] = n\} \quad (6.131)$$

forms an orthonormal system in the space $\text{Hom}_n(\mathbb{R}^q)$. For each $H_n(q; \cdot) \in$

$\mathrm{Hom}_n(\mathbb{R}^q)$ it follows in connection with the binomial theorem (see (2.22)) that

$$
\begin{aligned}
H_n(q;x) &= \sum_{[\alpha]=n} \frac{1}{\alpha!}\left(H_n(q;\nabla_y)\right) y^\alpha x^\alpha && (6.132)\\
&= \left(H_n(q;\nabla_y)\right)\frac{1}{n!}\sum_{[\alpha]=n}\frac{n!}{\alpha!}\, y^\alpha x^\alpha\\
&= \left(H_n(q;\nabla_y)\right)\frac{(x\cdot y)^n}{n!}\\
&= \frac{1}{n!}\,(x\cdot\nabla_y)^\alpha\,\overline{H_n(q;y)}.
\end{aligned}
$$

This shows that $\mathrm{Hom}_n(\mathbb{R}^q)$ equipped with the inner product $(\cdot,\cdot)_{\mathrm{Hom}_n(\mathbb{R}^q)}$ is an $M(q;n)$-dimensional Hilbert space with the reproducing kernel

$$
K_{\mathrm{Hom}_n(\mathbb{R}^q)}(x,y) = \frac{(x\cdot y)^n}{\Gamma(n+1)},\quad x,y\in\mathbb{R}^q, \tag{6.133}
$$

i.e., $K_{\mathrm{Hom}_n(\mathbb{R}^q)}(x,\cdot)\in\mathrm{Hom}_n(\mathbb{R}^q)$ for each $x\in\mathbb{R}^q$ and

$$
\left(K_{\mathrm{Hom}_n(\mathbb{R}^q)}(\cdot,x),H_n(q;\cdot)\right)_{\mathrm{Hom}_n(\mathbb{R}^q)} = H_n(q;x) \tag{6.134}
$$

for each $x\in\mathbb{R}^q$ and each $H_n(q;\cdot)\in\mathrm{Hom}_n(\mathbb{R}^q)$.

Let $H_{n,1}(q;\cdot),\ldots,H_{n,M(q;n)}(q;\cdot)$ be an orthonormal system in $\mathrm{Hom}_n(\mathbb{R}^q)$ with respect to $(\cdot,\cdot)_{\mathrm{Hom}_n(\mathbb{R}^q)}$. Then each $H_n(q;\cdot)\in\mathrm{Hom}_n(\mathbb{R}^q)$ can be represented in the form

$$
H_n(q;\cdot) = \sum_{j=1}^{M(q;n)}\left(H_n(q;\cdot),H_{n,j}(q;\cdot)\right)_{\mathrm{Hom}_n(\mathbb{R}^q)} H_{n,j}(q;\cdot), \tag{6.135}
$$

where

$$
K_{\mathrm{Hom}_n(\mathbb{R}^q)}(x,y) = \sum_{j=1}^{M(q;n)} H_{n,j}(q;x)\overline{H_{n,j}(q;y)},\quad x,y\in\mathbb{R}^q. \tag{6.136}
$$

In other words, we are led to the *addition theorem for homogeneous polynomials*.

Theorem 6.9. *Let $H_{n,1}(q;\cdot),\ldots,H_{n,M(q;n)}(q;\cdot)$ be an orthonormal system in $\mathrm{Hom}_n(\mathbb{R}^q)$ with respect to $(\cdot,\cdot)_{\mathrm{Hom}_n(\mathbb{R}^q)}$. Then, for all $x,y\in\mathbb{R}^q$,*

$$
\frac{(x\cdot y)^n}{n!} = \sum_{j=1}^{M(q;n)} H_{n,j}(q;x)\overline{H_{n,j}(q;y)}. \tag{6.137}
$$

Suppose that there are given $M(q;n)$ points $x_1, \ldots, x_{M(q;n)} \in \mathbb{R}^q$ such that

$$\sum_{k=1}^{M(q;n)} b_k \overline{H_{n,j}(q; x_k)} = \Big(H_n(q; \cdot), H_{n,j}(q; \cdot) \Big)_{\mathrm{Hom}_n(\mathbb{R}^q)}, \tag{6.138}$$

for $H_n(q; \cdot) \in \mathrm{Hom}_n(\mathbb{R}^q)$ and $j = 1, \ldots, M(q;n)$. Then

$$H_n(q; x) = \sum_{k=1}^{M(q;n)} \sum_{j=1}^{M(q;n)} b_k \overline{H_{n,j}(q; x_k)} \, H_{n,j}(q; x), \quad x \in \mathbb{R}^q. \tag{6.139}$$

This yields

$$H_n(q; x) = \sum_{k=1}^{M(q;n)} b_k \frac{(x \cdot x_k)^n}{n!}, \quad x \in \mathbb{R}^q. \tag{6.140}$$

Let $\mathrm{Harm}_n(\mathbb{R}^q)$ be the class of all polynomials in $\mathrm{Hom}_n(\mathbb{R}^q)$ that are harmonic:

$$\mathrm{Harm}_n(\mathbb{R}^q) = \{H_n(q; \cdot) \in \mathrm{Hom}_n(\mathbb{R}^q) \mid \Delta_x H_n(q; x) = 0, \ x \in \mathbb{R}^q \}. \tag{6.141}$$

For $n < 2$, of course, all homogeneous polynomials are harmonic. Assume that n is an integer with $n \geq 2$. Let $H_{n-2}(q; \cdot)$ be a homogeneous polynomial of degree $n - 2$, i.e., $H_{n-2}(q; \cdot) \in \mathrm{Hom}_{n-2}(\mathbb{R}^q)$. Then, for each homogeneous harmonic polynomial $K_n(q; \cdot)$, we have

$$\Big(|\cdot|^2 H_{n-2}(q; \cdot), K_n(q, \cdot) \Big)_{\mathrm{Hom}_n(\mathbb{R}^q)} = H_{n-2}(\nabla_x) \Delta_x \overline{K_n(q; x)} = 0. \tag{6.142}$$

This means $|\cdot|^2 H_{n-2}(q; \cdot)$ is orthogonal to $K_n(q; \cdot)$ in the sense of the inner product $(\cdot, \cdot)_{\mathrm{Hom}_n(\mathbb{R}^q)}$. Conversely, suppose that $K_n(q; \cdot) \in \mathrm{Harm}_n(\mathbb{R}^q)$ is orthogonal to all L_n of the form

$$L_n(x) = |x|^2 H_{n-2}(q; x), \ H_{n-2}(q; \cdot) \in \mathrm{Hom}_{n-2}(\mathbb{R}^q), \ x \in \mathbb{R}^q. \tag{6.143}$$

Then it follows that

$$0 = \Big(|\cdot|^2 H_{n-2}(q; \cdot), K_n(q; \cdot) \Big)_{\mathrm{Hom}_n(\mathbb{R}^q)} = \Big(H_{n-2}(q; \cdot), \Delta K_n(q; \cdot) \Big)_{\mathrm{Hom}_{n-2}(\mathbb{R}^q)} \tag{6.144}$$

for all $H_{n-2}(q; \cdot) \in \mathrm{Hom}_{n-2}(\mathbb{R}^q)$. This is true if $\Delta K_n(q; \cdot) = 0$; i.e, $K_n(q; \cdot)$ is a homogeneous harmonic polynomial. In other words, $\mathrm{Hom}_n(\mathbb{R}^q)$, $n \geq 2$, is the orthogonal direct sum of $\mathrm{Harm}_n(\mathbb{R}^q)$ and $\mathrm{Harm}_n^\perp(\mathbb{R}^q)$, where

$$\mathrm{Harm}_n^\perp(\mathbb{R}^q) = |\cdot|^2 \mathrm{Hom}_{n-2}(\mathbb{R}^q). \tag{6.145}$$

Furthermore, the dimension $N(q;n)$ of $\mathrm{Harm}_n(\mathbb{R}^q)$ can be determined as follows:

$$\begin{aligned} N(q;n) &= \dim(\mathrm{Harm}_n(\mathbb{R}^q)) \\ &= \dim(\mathrm{Hom}_n(\mathbb{R}^q)) - \dim(\mathrm{Hom}_{n-2}(\mathbb{R}^q)) \end{aligned} \tag{6.146}$$

such that

$$N(q;n) = \frac{\Gamma(n+q)}{\Gamma(q)\Gamma(n+1)} - \frac{\Gamma(n+q-2)}{\Gamma(q)\Gamma(n-1)} \tag{6.147}$$

$$= \frac{\Gamma(n+q-2)}{\Gamma(q)\Gamma(n+1)}((n+q-1)(n+q-2)-n(n-1))$$

$$= \frac{\Gamma(n+q-2)}{\Gamma(q)\Gamma(n+1)}((q-1)(2n+q-2))$$

$$= \frac{(2n+q-2)\Gamma(n+q-2)}{\Gamma(n+1)\Gamma(q-1)}.$$

Consequently,

$$N(q;0) = 1, \tag{6.148}$$

$$N(q;n) = \frac{(2n+q-2)\Gamma(n+q-2)}{\Gamma(n+1)\Gamma(q-1)} = O\left(n^{q-2}\right). \tag{6.149}$$

Remark 6.7. *Note that*

$$N(3;n) = \frac{(2n+1)\,\Gamma(n+1)}{\Gamma(n+1)\Gamma(2)} = 2n+1, \tag{6.150}$$

$$N(2;n) = \frac{2n\,\Gamma(n)}{\Gamma(n+1)} = 2.$$

Each homogeneous polynomial of degree n can be uniquely decomposed in the form

$$H_n(q;x) = \sum_{j=0}^{\lfloor \frac{n}{2} \rfloor} |x|^{2j} K_{n-2j}(q;x), \quad x \in \mathbb{R}^q, \tag{6.151}$$

$K_{n-2j}(q;\cdot) \in \mathrm{Harm}_{n-2j}(\mathbb{R}^q)$, where $\lfloor \frac{n}{2} \rfloor$ is the largest integer which is less or equal than $\frac{n}{2}$.

We are now interested in explicitly giving the projection $\mathrm{Proj}_{\mathrm{Harm}_n(\mathbb{R}^q)} H_n(q;\cdot)$ of a given homogeneous polynomial $H_n(q;\cdot)$ to $\mathrm{Harm}_n(\mathbb{R}^q)$. We restrict ourselves to dimensions $q \geq 3$. Induction states that

$$(y \cdot \nabla_x)^n \frac{1}{|x|^{q-2}} = (-1)^n \frac{1}{|x|^{q+2n-2}} \frac{1}{\Gamma\left(\frac{q-2}{2}\right)} \tag{6.152}$$

$$\times \left(\sum_{m=0}^{\lfloor \frac{n}{2} \rfloor} \frac{(-1)^m \Gamma\left(\frac{q-2}{2}+n-m\right)}{\Gamma(m+1)} 2^{n-2m} |x|^{2m} \Delta^m \right) (y \cdot x)^n$$

is valid for every $y \in \mathbb{R}^q$. Now, as we have seen in (6.140), every $H_n(q;\cdot) \in \mathrm{Hom}_n(\mathbb{R}^q)$ can be represented for (suitable) $x_1, \ldots, x_{M(q;n)}$ as linear combination in terms of the functions $x \mapsto (x_i \cdot x)^n$, $x \in \mathbb{R}^q$, $i = 1, \ldots, M(q,n)$.

Thus, we are led to the following conclusion.

Lemma 6.9. *For all $H_n \in \mathrm{Hom}_n(\mathbb{R}^q)$, $q \geq 3$, and all $x \in \mathbb{R}^q$, $|x| \neq 0$, we have*

$$
H_n(q; \nabla_x) \frac{1}{|x|^{q-2}}
$$

$$
= (-1)^n \frac{1}{|x|^{q+2n-2}} \frac{2^n}{\Gamma\left(\frac{q-2}{2}\right)} \tag{6.153}
$$

$$
\times \left(\sum_{m=0}^{\lfloor \frac{n}{2} \rfloor} \frac{(-1)^m \Gamma\left(\frac{q-2}{2} + n - m\right)}{\Gamma(m+1)} 2^{-2m} |x|^{2m} \Delta^m \right) H_n(q; x).
$$

Using the decomposition $H_n(q; x) = K_n(q; x) + |x|^2 H_{n-2}(q; x)$, $x \in \mathbb{R}^q$, with $K_n(q; \cdot) \in \mathrm{Harm}_n(\mathbb{R}^q)$, $H_{n-2}(q; \cdot) \in \mathrm{Hom}_{n-2}(\mathbb{R}^q)$, we find for $|x| \neq 0$

$$
\left(H_n(q; \nabla_x) \right) \frac{1}{|x|^{q-2}} = K_n(q; \nabla_x) \frac{1}{|x|^{q-2}} + H_{n-2}(q; \nabla_x) \Delta_x \frac{1}{|x|^{q-2}}
$$

$$
= K_n(q; \nabla_x) \frac{1}{|x|^{q-2}} \tag{6.154}
$$

$$
= (-1)^n |x|^{2-2n-q} \frac{2^n \Gamma\left(\frac{q-2+2n}{2}\right)}{\Gamma\left(\frac{q-2}{2}\right)} K_n(q; x).
$$

This shows us that, for $H_n(q; \cdot) \in \mathrm{Hom}_n(\mathbb{R}^q)$,

$$
\mathrm{Proj}_{\mathrm{Harm}_n(\mathbb{R}^q)}(H_n(q; \cdot)) \tag{6.155}
$$

$$
= \sum_{m=0}^{\lfloor \frac{n}{2} \rfloor} (-1)^m \frac{2^{-2m} \Gamma\left(\frac{q-2}{2} + n - m\right)}{\Gamma\left(\frac{q-2}{2} + n\right) \Gamma(m+1)} |x|^{2m} \Delta^m H_n(q; x).
$$

Observing

$$
\Delta_x^k \frac{(x \cdot y)^n}{\Gamma(n+1)} = \frac{1}{\Gamma(n - 2k + 1)} |y|^{2k} (x \cdot y)^{n-2k}, \quad k \geq 0, \tag{6.156}
$$

we get for $|x| \neq 0$, $|y| \neq 0$

$$
\mathrm{Proj}_{\mathrm{Harm}_n(\mathbb{R}^q)} \left(\frac{(x \cdot y)^n}{\Gamma(n+1)} \right) \tag{6.157}
$$

$$
= \sum_{m=0}^{\lfloor \frac{n}{2} \rfloor} \frac{(-1)^m 2^{-2m} \Gamma\left(\frac{q-2}{2} + n - m\right)}{\Gamma\left(\frac{q}{2} + n - 1\right) \Gamma(m+1) \Gamma(n - 2m + 1)} |x|^{2m} |y|^{2m} (x \cdot y)^{n-2m}.
$$

Suppose that $\{K_{n,j}(q; \cdot)\}_{j=1,\ldots,N(q;n)}$ is an orthonormal system in $\mathrm{Hom}_n(\mathbb{R}^q)$ with respect to $(\cdot, \cdot)_{\mathrm{Hom}_n(\mathbb{R}^q)}$. Let $\{L_{n,j}(q; \cdot)\}_{j=1,\ldots,M(q;n)-N(q;n)}$ be an orthonormal system in $\mathrm{Harm}_n^{\perp}(\mathbb{R}^q)$. Then the union of both systems forms an orthonormal system in $\text{Harm}_n(\mathbb{R}^q)$. Therefore we are able to deduce that

$$\text{Proj}_{\text{Harm}_n(\mathbb{R}^q)}\left(\frac{(x \cdot y)^n}{\Gamma(n+1)}\right) \tag{6.158}$$

$$= \text{Proj}_{\text{Harm}_n(\mathbb{R}^q)}\left(\sum_{j=1}^{N(q;n)} K_{n,j}(q;x)\overline{K_{n,j}(q;y)}\right)$$

$$+ \text{Proj}_{\text{Harm}_n(\mathbb{R}^q)}\left(\sum_{j=1}^{M(q;n)-N(q;n)} L_{n,j}(q;x)\overline{L_{n,j}(q;y)}\right)$$

$$= \sum_{j=1}^{N(q;n)} K_{n,j}(q;x)K_{n,j}(q;y).$$

By comparison of (6.157) and (6.158) we obtain the *addition theorem of homogeneous harmonic polynomials in* $\mathbb{R}^q$.

Theorem 6.10. *Let* $\{K_{n,j}(q;\cdot)\}_{j=1,\dots,N(q;n)}$ *be an orthonormal system in the space* $\text{Harm}_n(\mathbb{R}^q)$ *with respect to inner product* $(\cdot,\cdot)_{\text{Hom}_n(\mathbb{R}^q)}$. *Then, for all* $x, y \in \mathbb{R}^q$, $q \geq 3$, $x = |x|\xi$, $y = |y|\eta$; $\xi, \eta \in \mathbb{S}^{q-1}$, *we have*

$$\sum_{j=1}^{N(q;n)} K_{n,j}(q;x)\overline{K_{n,j}(q;y)} = \frac{\Gamma\left(\frac{q}{2}\right)|x|^n|y|^n}{\Gamma\left(\frac{q}{2}+n\right)2^n} N(q;n)P_n(q,\xi\cdot\eta), \tag{6.159}$$

where we have used the abbreviation

$$P_n(q;t) = \frac{\Gamma(n+1)\Gamma(q-2)}{\Gamma(n+q-2)\Gamma\left(\frac{q-2}{2}\right)}\sum_{m=0}^{\lfloor\frac{n}{2}\rfloor} \frac{(-1)^m\Gamma\left(\frac{q-2}{2}+n-m\right)}{\Gamma(m+1)\Gamma(n-2m+1)}(2t)^{n-2m}, \tag{6.160}$$

$t \in [-1,1]$.

Next we discuss the important problem of how, for any pair of elements $H_n(q;\cdot) \in \text{Harm}_n(\mathbb{R}^q)$, $K_n(q;\cdot) \in \text{Harm}_n(\mathbb{R}^q)$, the inner product $(\cdot,\cdot)_{\text{Hom}_n(\mathbb{R}^q)}$ is related to the (usual) inner product $(\cdot,\cdot)_{\text{L}^2(\mathbb{S}^{q-1})}$.

Theorem 6.11. *For* $H_m(q;\cdot) \in \text{Harm}_m(\mathbb{R}^q)$, $K_n(q;\cdot) \in \text{Harm}_n(\mathbb{R}^q)$

$$\left(H_m(q;\cdot),K_n(q;\cdot)\right)_{\text{L}^2(\mathbb{S}^{q-1})} = \frac{\delta_{mn}}{\mu_n(q)}\left(H_n(q;\cdot),K_n(q;\cdot)\right)_{\text{Hom}_n(\mathbb{R}^q)}, \tag{6.161}$$

where $\mu_n(q)$ *is given by*

$$\mu_n(q) = \frac{2^n\Gamma\left(\frac{q}{2}+n\right)}{\|\mathbb{S}^{q-1}\|\Gamma\left(\frac{q}{2}\right)}. \tag{6.162}$$

Proof. By virtue of the third Green theorem (see Theorem 6.4) we find for all $x \in \mathbb{R}^q$, $|x| < 1$,

$$\overline{K_n(q;x)} \tag{6.163}$$
$$= \int_{\mathbb{S}^{q-1}} \left\{ F_q(|x-y|) \frac{\partial}{\partial \nu_y} \overline{K_n(q;y)} - \overline{K_n(q;y)} \frac{\partial}{\partial \nu_y} F_q(|x-y|) \right\} \, dS_{(q-1)}(y),$$

where $x \mapsto F_q(|x-y|)$, $x \in \mathbb{R}^q \setminus \{y\}$, is the fundamental solution to the Laplace operator in $\mathbb{R}^q$ at y and $\partial/\partial\nu$ denotes the derivative in the direction of the outer normal ν to $\mathbb{S}^{q-1}$. Interchanging differentiation and integration we obtain

$$\left(H_m(q; \nabla_x) \right) \overline{K_n(q;x)} \tag{6.164}$$
$$= \frac{1}{(q-2)\|\mathbb{S}^{q-1}\|} \int_{\mathbb{S}^{q-1}} \left\{ (H_m(q;\nabla_x))|x-y|^{2-q} \frac{\partial}{\partial \nu_y} \overline{K_n(q;y)} \right.$$
$$\left. - \overline{K_n(q;y)} \frac{\partial}{\partial \nu_y} (H_m(q;\nabla_x))|x-y|^{2-q} \right\} \, dS_{(q-1)}(y).$$

For $x \in \mathbb{R}^q \setminus \{y\}$ we have

$$(H_m(q;\nabla_x))|x-y|^{2-q} = (-1)^m |x-y|^{2-2m-q} \frac{2^m \Gamma\left(\frac{q-2}{2}+m\right)}{\Gamma\left(\frac{q-2}{2}\right)} H_m(q; x-y).$$
$$\tag{6.165}$$

Therefore, we get, because of the homogeneity of $H_m(q; \cdot) \in \mathrm{Hom}_m(\mathbb{R}^q)$,

$$\left(H_m(q;\nabla_x) \right) \overline{K_n(q;x)} \tag{6.166}$$
$$= \frac{2^m \Gamma\left(\frac{q-2}{2}+m\right)}{\|\mathbb{S}^{q-1}\| \Gamma\left(\frac{q-2}{2}\right)(q-2)}$$
$$\times \int_{\mathbb{S}^{q-1}} \left\{ H_m(q; y-x)|x-y|^{2-2m-q} \frac{\partial}{\partial \nu_y} \overline{K_n(q;y)} \right.$$
$$\left. - \overline{K_n(q;y)} \frac{\partial}{\partial \nu_y} H_m(q; y-x)|x-y|^{2-2m-q} \right\} \, dS_{(q-1)}(y).$$

Since

$$H_m(q;\nabla_x)\overline{K_n(q;x)}\Big|_{x=0} = H_m(q;\nabla_x)\overline{K_n(q;x)}$$
$$= \begin{cases} 0 & , \quad m \neq n \\ \left(H_m(q;\cdot), K_n(q;\cdot) \right)_{\mathrm{Hom}_n(\mathbb{R}^q)} & , \quad n = m, \end{cases}$$

this yields

$$\int_{\mathbb{S}^{q-1}} \left\{ H_m(q;y)|y|^{2-2m-q} \frac{\partial}{\partial \nu_y} \overline{K_n(q;y)} \right. \tag{6.167}$$

$$\left. - \overline{K_n(q;y)} \frac{\partial}{\partial \nu_y} H_m(q;y)|y|^{2-2m-q} \right\} dS_{(q-1)}(y)$$

$$= \begin{cases} 0 & , \quad m \neq n \\ \frac{\|\mathbb{S}^{q-1}\|\Gamma(\frac{q-2}{2})(q-2)}{2^n \Gamma(\frac{q-2}{2}+n)} (H_m, K_n)_{\mathrm{Hom}_n(\mathbb{R}^q)} & , \quad m = n. \end{cases}$$

Since the normal derivatives on $\mathbb{S}^{q-1}$ are radial derivatives we find

$$\int_{\mathbb{S}^{q-1}} \left\{ H_m(q;y)|y|^{2-2m-q} \frac{\partial}{\partial \nu_y} \overline{K_n(q;y)} \right. \tag{6.168}$$

$$\left. - \overline{K_n(q;y)} \frac{\partial}{\partial \nu_y} H_m(q;y)|y|^{2-2m-q} \right\} dS_{(q-1)}(y)$$

$$= \int_{\mathbb{S}^{q-1}} \left\{ n H_m(q;\eta)\overline{K_n(q;\eta)} + (m+q-2)\overline{K_n(q;\eta)}H_m(q;\eta) \right\} dS_{(q-1)}(\eta)$$

$$= (n+m+q-2) \int_{\mathbb{S}^{q-1}} H_m(q;\eta)\overline{K_n(q;\eta)} dS_{(q-1)}(\eta)$$

$$= (n+m+q-2)(H_m, K_n)_{\mathrm{L}^2(\mathbb{S}^{q-1})}.$$

Thus, combining our identities we have the desired result. $\qquad\square$

Theorem 6.10 leads to the following reformulation of the *addition theorem for homogeneous harmonic polynomials*.

Theorem 6.12. *Let* $K_{n,j}(q;\cdot) \in \mathrm{Harm}_n(\mathbb{R}^q)$, $j = 1, \ldots, N(q;n)$, *be an* $\mathrm{L}^2(\mathbb{S}^{q-1})$-*orthonormal system, i.e.,*

$$\int_{\mathbb{S}^{q-1}} K_{n,j}(q;x)\, \overline{K_{r,s}(q;x)}\, dS_{(q-1)}(x) = \delta_{nr}\delta_{js}. \tag{6.169}$$

Then, for $x, y \in \mathbb{R}^q$, $x = |x|\xi$, $y = |y|\eta$, $\xi, \eta \in \mathbb{S}^{q-1}$, *we have*

$$\sum_{j=1}^{N(q;n)} K_{n,j}(q;x)\, \overline{K_{n,j}(q;y)} = \frac{N(q;n)}{\|\mathbb{S}^{q-1}\|} |x|^n |y|^n P_n(q, \xi \cdot \eta), \tag{6.170}$$

where $P_n(q;\cdot)$ *is defined by* (6.160).

In other words, to any orthonormal system $\{K_{n,j}(q;\cdot)\}_{j=1,\ldots,N(q;n)}$ with respect to $(\cdot, \cdot)_{\mathrm{Hom}_n(\mathbb{R}^q)}$ there corresponds the $\mathrm{L}^2(\mathbb{S}^{q-1})$-orthonormal system $\left\{ \sqrt{\mu_n(q)} K_{n,j}(q;\cdot) \right\}_{j=1,\ldots,N(q;n)}$, and vice versa.

Legendre Polynomials

The function $P_n(q;\cdot) : t \mapsto P_n(q;t)$, $t \in [-1, +1]$, $n = 0, 1, \ldots$, occuring in the addition theorem for homogeneous harmonic polynomials of degree n in q-dimension (see (6.160))

$$P_n(q;t) = \frac{\Gamma(n+1)\Gamma(q-2)}{\Gamma(n+q-2)\Gamma\left(\frac{q-2}{2}\right)} \sum_{m=0}^{\lfloor \frac{n}{2} \rfloor} \frac{(-1)^m \Gamma\left(\frac{q-2}{2} + n - m\right)}{\Gamma(m+1)\Gamma(n-2m+1)} (2t)^{n-2m}$$

(6.171)

is called the *Legendre polynomial of degree n and dimension q.*

It is easily seen that $P_n(q;\cdot)$ satisfies the following properties:

(i) $P_n(q;\cdot)$ is a polynomial of degree n, i.e.,

$$P_n(q;t) = \alpha_n(q)\, t^n + \ldots ,$$

(6.172)

where

$$\alpha_n(q) = \frac{2^n \Gamma(n+\frac{q}{2})}{n!\Gamma(\frac{q}{2})N(q;n))} = \frac{n!}{2^n} \frac{\Gamma(\frac{q-1}{2})}{\Gamma(\frac{2n+q-1}{2})} \binom{2n+q-3}{n}.$$

(6.173)

(ii) $P_n(q;1) = 1$,

(iii) $\int_{-1}^{+1} P_n(q;t)P_m(q;t)(1-t^2)^{\frac{q-3}{2}}\, dt = 0, \quad n \neq m.$

$P_n(q;\cdot)$ is uniquely determined by the properties *(i)*, *(ii)*, *(iii)*.

It is not hard to see that

$$\int_{-1}^{+1} t^k P_n(q;t)(1-t^2)^{\frac{q-3}{2}}\, dt = \begin{cases} 0 & , \quad k-n < 0 \\ 0 & , \quad k-n > 0, \text{ odd} \\ \frac{n!}{2^n}\binom{k}{n}\frac{\Gamma\left(\frac{q-1}{2}\right)\Gamma\left(\frac{k-n+1}{2}\right)}{\Gamma\left(\frac{k+n+q}{2}\right)} & , \quad k-n > 0, \text{ even.} \end{cases}$$

(6.174)

Moreover, we have

$$\int_{-1}^{+1} P_n^2(q;t)(1-t^2)^{\frac{q-3}{2}}\, dt = \frac{\sqrt{\pi}\Gamma\left(\frac{q-1}{2}\right)}{\Gamma\left(\frac{q}{2}\right)} \frac{1}{N(q;n)} = \frac{\|\mathbb{S}^{q-1}\|}{\|\mathbb{S}^{q-2}\|N(q;n)}.$$

(6.175)

Applying the Cauchy–Schwarz inequality to the addition theorem, we obtain for $x, y \in \mathbb{R}^q$

$$\frac{N(q;n)}{\|\mathbb{S}^{q-1}\|}|x|^n|y|^n\,|P_n(q;\xi\cdot\eta)| = \left| \sum_{j=1}^{N(q;n)} K_{n,j}(q;x)K_{n,j}(q;y) \right|$$

$$\leq \sqrt{\sum_{j=1}^{N(q;n)} |K_{n,j}(q;x)|^2} \sqrt{\sum_{j=1}^{N(q;n)} |K_{n,j}(q;y)|^2}$$

$$= \frac{N(q;n)}{\|\mathbb{S}^{(q-1)}\|}|x|^n|y|^n P_n(q;1).$$

(6.176)

Consequently, we have

$$|P_n(q;t)| \le P_n(q;1) = 1 \tag{6.177}$$

for all $t \in [-1,+1]$.

By elementary calculations we find

$$\left((1-t^2)\left(\frac{d}{dt}\right)^2 - (q-1)t\frac{d}{dt} + n(n+q-2) \right) P_n(q;t) = 0. \tag{6.178}$$

More concretely, $P_n(q;\cdot)$ is the only twice continuously differentiable eigenfunction of the "Legendre operator" L

$$L_t = (1-t^2)\left(\frac{d}{dt}\right)^2 - (q-1)t\frac{d}{dt} \tag{6.179}$$

on the (one-dimensional) interval $[-1,+1]$, corresponding to the eigenvalues $L^\wedge(n) (= L_{[-1,1]}^\wedge(n)) = n(n+q-2), n = 0,1,\dots$, that is bounded on $[-1,+1]$ with $P_n(q;1) = 1$.

The Legendre polynomial can be expressed by the *Rodrigues formula*

$$P_n(q;t) = (-1)^n \frac{\Gamma\left(\frac{q-1}{2}\right)}{2^n \Gamma\left(\frac{q-1}{2}+n\right)} (1-t^2)^{\frac{3-q}{2}} \left(\frac{d}{dt}\right)^n (1-t^2)^{n+\frac{q-3}{2}}. \tag{6.180}$$

An easy calculation gives

$$
\begin{aligned}
P_0(q;t) &= 1, & (6.181)\\
P_1(q;t) &= t, & (6.182)\\
P_2(q;t) &= \frac{1}{q-1}(qt^2 - 1), & (6.183)\\
P_3(q;t) &= \frac{1}{q-1}\left((q+2)t^2 - 3\right)t. & (6.184)
\end{aligned}
$$

Legendre polynomials satisfy recursion formulas. We mention only the following relations:

(i)

$$(q-1)P_n'(q;t) = n(n+q-2)P_{n-1}(q+2;t), \tag{6.185}$$

(ii)

$$(n+q-2)P_{n+1}(q;t) - (2n+q-2)tP_n(q;t) + nP_{n-1}(q;t) = 0, \tag{6.186}$$

(iii)

$$(2n+q-2)(1-t^2)P_n(q+2;t) = (q-1)\left(P_{n-1}(q;t) - P_{n+1}(q;t)\right), \tag{6.187}$$

(iv)

$$(q-1)(2n+q-2)P_n(q;t) \tag{6.188}$$
$$= (n+q-1)(n+q-2)P_n(q+2;t) - n(n+1)P_{n-2}(q+2;t),$$

(v)

$$(1-t^2)P_n'(q;t) = -n\left(tP_n(q;t) - P_{n-1}(q;t)\right), \tag{6.189}$$

(vi)

$$(1-t^2)P_n'(q;t) = -(n+q-2)\left(P_{n+1}(q;t) - tP_n(q;t)\right). \tag{6.190}$$

(vii)

$$(1-t^2)P_n'(q;t) = -(q-3)\left(P_{n+1}(q-2;t) - tP_n(q;t)\right). \tag{6.191}$$

The polynomial $P_n(q;\cdot)$ has exactly n different zeros in $(-1,+1)$.

It is not difficult to verify that

$$\left(\frac{d}{dt}\right)^l P_n(q;t) = \frac{2^l\Gamma\left(l+\frac{q}{2}\right)}{\Gamma\left(\frac{q}{2}\right)} \frac{N(q+2l;n-l)}{N(q;n)} P_{n-l}(q+2l;t). \tag{6.192}$$

In connection with the Rodrigues formula (6.180) we get by partial integration the so–called *Rodrigues rule*

Lemma 6.10. *Let F be n times continuously differentiable on $[-1,+1]$. Then we have*

$$\int_{-1}^{+1} F(t)P_n(q;t)(1-t^2)^{\frac{q-3}{2}}\,dt \tag{6.193}$$
$$= \frac{\Gamma\left(\frac{q-1}{2}\right)}{2^n\Gamma\left(\frac{q-1}{2}+n\right)} \int_{-1}^{+1}(1-t^2)^{n+\frac{q-3}{2}}F^{(n)}(t)\,dt.$$

For $q \geq 2$ we have with $M = \min(m,n), \sigma = m+n, \tau = |m-n|$

$$\int_{-1}^{1} P_n(q,t)P_m(q,t)P_{|m-n|}(q,t)\left(1-t^2\right)^{\frac{q-3}{2}}\,dt \tag{6.194}$$
$$= \int_{-1}^{1} P_{\frac{\sigma+\tau}{2}}(q,t)P_{\frac{\sigma-\tau}{2}}(q,t)P_\tau(q,t)\left(1-t^2\right)^{\frac{q-3}{2}}\,dt$$
$$= \frac{\|\mathbb{S}^{q-1}\|}{\|\mathbb{S}^{q-2}\|} \frac{\Gamma\left(\tau+\frac{q}{2}\right)\Gamma\left(\frac{\sigma-\tau+q}{2}\right)\left(\frac{\sigma+\tau}{2}\right)!}{N(q,\tau)N\left(q,\frac{\sigma-\tau}{2}\right)\Gamma\left(\frac{q}{2}\right)\Gamma\left(\frac{\sigma+\tau+q}{2}\right)\tau!\left(\frac{\sigma-\tau}{2}\right)!}.$$

and

$$\int_{-1}^{1} P_n(q,t)P_m(q,t)P_{m+n}(q,t)\left(1-t^2\right)^{\frac{q-3}{2}}\,dt \tag{6.195}$$

$$= \int_{-1}^{1} P_{\frac{\sigma+\tau}{2}}(q,t)P_{\frac{\sigma-\tau}{2}}(q,t)P_{\sigma}(q,t)\left(1-t^2\right)^{\frac{q-3}{2}}\,dt$$

$$= \frac{\|\mathbb{S}^{q-1}\|}{\|\mathbb{S}^{q-2}\|}\,\frac{\Gamma\left(\frac{\sigma+\tau+q}{2}\right)\Gamma\left(\frac{\sigma-\tau+q}{2}\right)\sigma!}{N\left(q,\frac{\sigma+\tau}{2}\right)N\left(q,\frac{\sigma-\tau}{2}\right)\Gamma\left(\frac{q}{2}\right)\Gamma\left(\sigma+\frac{q}{2}\right)\left(\frac{\sigma+\tau}{2}\right)!\left(\frac{\sigma-\tau}{2}\right)!}\,.$$

Moreover, for $q \geq 2$ and $t \in [-1,1]$ we have

$$P_{\frac{\sigma+\tau}{2}}(q,t)P_{\frac{\sigma-\tau}{2}}(q,t) \;=\; \sum_{k=0}^{M}\alpha_k P_{\sigma-2k}(q,t), \tag{6.196}$$

where

$$\alpha_0 \;=\; \frac{N(q,\sigma)\Gamma\left(\frac{\sigma+\tau+q}{2}\right)\Gamma\left(\frac{\sigma-\tau+q}{2}\right)\sigma!}{N\left(q,\frac{\sigma+\tau}{2}\right)N\left(q,\frac{\sigma-\tau}{2}\right)\Gamma\left(\frac{q}{2}\right)\Gamma\left(\sigma+\frac{q}{2}\right)\left(\frac{\sigma+\tau}{2}\right)!\left(\frac{\sigma-\tau}{2}\right)!} \tag{6.197}$$

and

$$\alpha_M \;=\; \frac{\Gamma\left(\tau+\frac{q}{2}\right)\Gamma\left(\frac{\sigma-\tau+q}{2}\right)\left(\frac{\sigma+\tau}{2}\right)!}{N\left(q,\frac{\sigma-\tau}{2}\right)\Gamma\left(\frac{q}{2}\right)\Gamma\left(\frac{\sigma+\tau+q}{2}\right)\tau!\left(\frac{\sigma-\tau}{2}\right)!}\,. \tag{6.198}$$

The power series

$$\phi(r) = \sum_{n=0}^{\infty} c_n(q)P_n(q;t)r^n, \quad t \in [-1,+1], \tag{6.199}$$

with

$$c_n(q) = \frac{\Gamma(n+q-2)}{\Gamma(n+1)\Gamma(q-2)}, \quad q \geq 3, \tag{6.200}$$

is absolutely and uniformly convergent for all r with $|r| \leq r_0, r_0 \in [0,1)$. By termwise differentiation we find

$$\phi'(r) = \sum_{n=1}^{\infty} c_n(q)P_n(q;t)nr^{n-1}. \tag{6.201}$$

With $c_{n+1}(q) = \frac{n+q-2}{n+1}c_n(q)$ we then obtain

$$\phi'(r) \;=\; \sum_{n=0}^{\infty} c_{n+1}(q)(n+1)P_{n+1}(q;t)r^n \tag{6.202}$$

$$= \sum_{n=0}^{\infty} c_n(q)(n+q-2)P_{n+1}(q;t)r^n.$$

Adding

$$r\phi(r) + r^2\phi'(t) = \sum_{n=1}^{\infty} c_{n-1}(q)P_{n-1}(q;t)nr^n \tag{6.203}$$

and

$$(q-3)r\phi(r) = \sum_{n=1}^{\infty} c_{n-1}(q)(q-3)P_{n-1}(q;t)r^n \tag{6.204}$$

we get

$$r(q-2)\phi(r) + r^2\phi'(r) = \sum_{n=1}^{\infty} c_n(q)P_{n-1}(q;t)nr^n. \tag{6.205}$$

Altogether, by virtue of the recurrence relation (6.186), we obtain

$$\phi'(r)(1+r^2-2tr) + (q-2)(r-t)\phi(r) \tag{6.206}$$
$$= \sum_{n=0}^{\infty} c_n(q)r^n\left((n+q-2)P_{n+1}(q;t) + nP_{n-1}(q;t) - (2n+q-2)tP_n(q;t)\right)$$
$$= 0.$$

The differential equation

$$(1+r^2-2tr)\phi'(r) = -(q-2)(r-t)\phi(r) \tag{6.207}$$

under the initial condition $\phi(0) = 1$ is uniquely solvable. Since it is not hard to verify that

$$r \mapsto \phi(r) = (1+r^2-2rt)^{\frac{2-q}{2}} \tag{6.208}$$

solves the initial value problem, we are led to the following generating series expansion of $P_n(q;t)$

Lemma 6.11. *For $t \in [-1,1]$ and $|r| < 1$, $q \geq 3$*

$$\sum_{n=0}^{\infty} \frac{\Gamma(n+q-2)}{\Gamma(n+1)\Gamma(q-2)}P_n(q;t)r^n = \frac{1}{(1+r^2-2rt)^{\frac{q-2}{2}}}. \tag{6.209}$$

Remark 6.8. *The two-dimensional analogue of (6.209) reads under obvious assumptions*

$$\frac{1}{2}\ln\left(1-2rt+r^2\right) = -\sum_{n=1}^{\infty}\frac{1}{n}P_n(2;t)\,r^n. \tag{6.210}$$

As a special case we obtain with $x = |x|\xi$, $y = |y|\eta$, $\xi, \eta \in \mathbb{S}^{q-1}$, $q \geq 3$,

$$\frac{1}{|x-y|^{q-2}} = \frac{1}{|y|\left(1+\left(\frac{|x|}{|y|}\right)^2 - 2\frac{|x|}{|y|}\xi\cdot\eta\right)^{\frac{q-2}{2}}}$$
$$= \frac{1}{|y|}\sum_{n=0}^{\infty}\frac{\Gamma(n+q-2)}{\Gamma(n+1)\Gamma(q-2)}P_n(q;\xi\cdot\eta)\left(\frac{|x|}{|y|}\right)^n \tag{6.211}$$

provided that $|x| < |y|$. Moreover, we know that

$$|x - y|^{2-q} = \sum_{n=0}^{\infty} \frac{(-1)^n}{n!} |x|^n (\xi \cdot \nabla_y)^n |y|^{2-q}. \tag{6.212}$$

By comparison we therefore obtain the so–called *Maxwell representation*

$$(\xi \cdot \nabla_y)^n |y|^{2-q} = (-1)^n \frac{\Gamma(n + q - 2)}{\Gamma(q - 2)} \frac{P_n(q; \xi \cdot \eta)}{|y|^{n+q-2}}. \tag{6.213}$$

As $y \mapsto |y|^{2-q}, |y| \neq 0$, is (apart from a multiplicative constant) the fundamental solution of the Laplace equation in q dimensions, Maxwell's representation tells us that the Legendre polynomials may be obtained by repeated differentiations of the fundamental solution in the direction of ξ. The potential on the right side may be regarded as the potential of a pole of order n with the axis ξ at the origin.

Remark 6.9. *The Legendre polynomials $P_n(q; \cdot)$, $q \geq 3$, as defined in our approach by (6.160) coincide (apart from a multiplicative constant) with Gegenbauer polynomials. More precisely, we have (see, e.g., W. Magnus et al. [1966])*

$$\Gamma(q - 2 + n)P_n(q; t) = \Gamma(q - 2)\Gamma(n + 1)C_n^{\left(\frac{q-2}{2}\right)}(t), \quad t \in [-1, 1], \tag{6.214}$$

where

$$C_n^{(\alpha)}(t) = \frac{1}{\Gamma(\alpha)} \sum_{k=0}^{\lfloor \frac{n}{2} \rfloor} (-1)^k \frac{\Gamma(\alpha + n - k)}{k! \, (n - 2k)!} (2t)^{n-k}, \quad t \in [-1, 1]. \tag{6.215}$$

Spherical Harmonics

Let H_n be a homogeneous harmonic polynomial of degree n in $\mathbb{R}^q$, i.e., $H_n \in \text{Harm}_n(\mathbb{R}^q)$. Its decomposition into radial and angular parts is straightforward:

$$H_n(x) = r^n \, Y_n(\xi), \quad x = r\xi, \; r = |x|, \xi \in \mathbb{S}^{q-1}. \tag{6.216}$$

The restriction $Y_n = H_n|\mathbb{S}^{q-1}$ is called a *spherical harmonic of degree n and dimension q*. The set of all spherical harmonics of degree n and dimension q, i.e., the set of all restrictions $Y_n = H_n|\mathbb{S}^{q-1}$ with $H_n \in \text{Harm}_n(\mathbb{R}^q)$, is denoted by $\text{Harm}_n(\mathbb{S}^{q-1})$.

Remark 6.10. *In what follows we simply write Harm_n instead of $\text{Harm}_n(\mathbb{R}^q)$ (or $\text{Harm}_n(\mathbb{S}^{q-1})$) if no confusion is likely to arise. Our purpose is to list the essential properties of functions in Harm_n, i.e., spherical harmonics of degree n and dimension q. For more details on spherical harmonics the reader is referred to C. Müller [1952], W. Freeden [1979], R. Reuter [1982], W. Freeden et al. [1998], C. Müller [1966, 1998], W. Freeden, M. Schreiner [2009]. The last book also contains a vectorial as well as tensorial approach in Euclidean space $\mathbb{R}^3$.*

We already know that the linear space Harm_n is of dimension

$$N(q;n) = \dim(\text{Harm}_n) = \begin{cases} \frac{(2n+q-2)\Gamma(n+q-2)}{\Gamma(n+1)\Gamma(q-1)} &, \quad n \geq 1 \\ 1 &, \quad n = 0. \end{cases} \quad (6.217)$$

From Theorem 6.10 we know that the spherical harmonics of different degrees are orthogonal in the sense of the $L^2(\mathbb{S}^{q-1})$-inner product

$$\left(Y_n(q;\cdot), Y_m(q;\cdot)\right)_{L^2(\mathbb{S}^{q-1})} = \int_{\mathbb{S}^{q-1}} Y_n(q;\xi) \, \overline{Y_m(q;\xi)} \, dS_{(q-1)}(\xi) = 0 \quad (6.218)$$

if $n \neq m$.

From the addition theorem of homogeneous harmonic polynomials (Theorem 3.11) it is an easy task to obtain the *addition theorem for spherical harmonics*.

Theorem 6.13. *Let* $\{Y_{n,j}(q;\cdot)\}_{j=1,\ldots,N(q;n)}$ *be an* $L^2(\mathbb{S}^{q-1})$*-orthonormal system in* Harm_n. *Then, for any pair* $(\xi,\eta) \in \mathbb{S}^{q-1} \times \mathbb{S}^{q-1}$, *we have*

$$\sum_{j=1}^{N(q;n)} Y_{n,j}(q;\xi) \, \overline{Y_{n,j}(q;\eta)} = \frac{N(q;n)}{\|\mathbb{S}^{q-1}\|} P_n(q;\xi \cdot \eta). \quad (6.219)$$

An immediate consequence is that

$$Y_n(q;\xi) = \frac{N(q;n)}{\|\mathbb{S}^{q-1}\|} \int_{\mathbb{S}^{q-1}} P_n(q;\xi \cdot \eta) Y_n(q;\eta) \, dS_{(q-1)}(\eta) \quad (6.220)$$

holds for all $\xi \in \Omega$ and all $Y_n(q;\cdot) \in \text{Harm}_n$; i.e., $K_{\text{Harm}_n}(q;\cdot)$ defined by

$$K_{\text{Harm}_n}(q;\xi \cdot \eta) = \frac{N(q;n)}{\|\mathbb{S}^{q-1}\|} P_n(q;\xi \cdot \eta), \quad \xi,\eta \in \mathbb{S}^{q-1}, \quad (6.221)$$

is the reproducing kernel of the $N(q;n)$-dimensional space Harm_n.

Remark 6.11. *Within our notation the addition theorem for the circle* $\mathbb{S}^1$ *is based on "circular harmonics"* $H_{n,j}(2;\cdot) : \mathbb{R}^2 \to \mathbb{R}$, $j = 1,2$, *given by*

$$\begin{aligned} H_{n,1}(2;x_{(2)}) = H_{n,1}(2;x_1,x_2) &= \frac{1}{\sqrt{\pi}} \Im\left((x_2 + ix_1)^n\right) \quad (6.222) \\ &= \frac{1}{\sqrt{\pi}} |x|^n \sin\left(n\left(\frac{\pi}{2} - \varphi\right)\right) \\ &= \frac{(-1)^{n+1}}{\sqrt{\pi}} |x|^n \cos(n\varphi), \\ H_{n,2}(2;x_{(2)}) = H_{n,2}(2;x_1,x_2) &= \frac{1}{\sqrt{\pi}} \Re\left((x_2 + ix_1)^n\right) \quad (6.223) \\ &= \frac{1}{\sqrt{\pi}} |x|^n \cos\left(n\left(\frac{\pi}{2} - \varphi\right)\right) \\ &= \frac{(-1)^{n+1}}{\sqrt{\pi}} |x|^n \sin(n\varphi), \end{aligned}$$

$x_{(2)} \in \mathbb{R}^2$, $x_{(2)} = (x_1, x_2)^T$, $x_1 = r \cos \varphi$, $x_2 = r \sin \varphi$, $r = |x_{(2)}| = \sqrt{x_1^2 + x_2^2}$, $0 \le \varphi < 2\pi$. *Obviously, we have*

$$\int_{\substack{|x_{(2)}|=1 \\ x_{(2)} \in \mathbb{R}^2}} H_{n,j}\left(2; x_{(2)}\right) \overline{H_{k,l}\left(2; x_{(2)}\right)} \, dS_{(1)}(x_{(2)}) = \delta_{nk}\delta_{jl}. \tag{6.224}$$

In other words, we are able to introduce an $L^2(\mathbb{S}^1)$*–orthonormal system by*

$$Y_{n,1}(2; \xi_{(2)}) \quad = \quad \frac{1}{\sqrt{\pi}} \sin\left(n\left(\frac{\pi}{2} - \varphi\right)\right), \quad n = 0, 1, \ldots, \tag{6.225}$$

$$Y_{n,2}(2; \xi_{(2)}) \quad = \quad \frac{1}{\sqrt{\pi}} \cos\left(n\left(\frac{\pi}{2} - \varphi\right)\right), \quad n = 1, 2, \ldots, \tag{6.226}$$

such that

$$\int_{\xi_{(2)} \in \mathbb{S}^1} Y_{n,j}\left(2; \xi_{(2)}\right) \overline{Y_{k,l}\left(2; \xi_{(2)}\right)} \, dS_{(1)}(\xi_{(2)}) = \delta_{nk}\delta_{jl}. \tag{6.227}$$

Moreover, for $x_{(2)}, y_{(2)} \in \mathbb{R}^2$ *with* $x_{(2)} = |x_{(2)}| \, \xi_{(2)}$, $y_{(2)} = |y_{(2)}| \, \eta_{(2)}$ *and* $x_1 = |x_{(2)}| \cos \varphi$, $x_2 = |x_{(2)}| \sin \varphi$, $y_1 = |y_{(2)}| \cos \psi$, $y_2 = |y_{(2)}| \sin \psi$, $0 \le \varphi, \psi < 2\pi$, *we have*

$$\sum_{j=1}^{2} H_{n,j}\left(2; x_{(2)}\right) \overline{H_{n,j}\left(2; y_{(2)}\right)} \tag{6.228}$$

$$= \frac{|x_{(2)}|^n |y_{(2)}|^n}{\pi} \cos\left(n\left(\frac{\pi}{2} - \varphi\right)\right) \cos\left(n\left(\frac{\pi}{2} - \psi\right)\right)$$

$$+ \frac{|x_{(2)}|^n |y_{(2)}|^n}{\pi} \sin\left(n\left(\frac{\pi}{2} - \varphi\right)\right) \sin\left(n\left(\frac{\pi}{2} - \psi\right)\right)$$

$$= \frac{|x_{(2)}|^n |y_{(2)}|^n}{\pi} \cos(n(\varphi - \psi)),$$

such that

$$\sum_{j=1}^{2} Y_{n,j}\left(2; \xi_{(2)}\right) \overline{Y_{n,j}\left(2; \eta_{(2)}\right)} \tag{6.229}$$

$$= \frac{1}{\pi} \cos\left(n\left(\frac{\pi}{2} - \varphi\right)\right) \cos\left(n\left(\frac{\pi}{2} - \psi\right)\right)$$

$$+ \frac{1}{\pi} \sin\left(n(\frac{\pi}{2} - \varphi)\right) \sin\left(n\left(\frac{\pi}{2} - \psi\right)\right)$$

$$= \frac{1}{\pi} \cos(n(\varphi - \psi)).$$

It should be noted that

$$\xi_{(2)} \cdot \eta_{(2)} \quad = \quad \cos(\varphi - \psi), \tag{6.230}$$

such that

$$\sum_{j=1}^{2} Y_{n,j}\left(2;\xi_{(2)}\right)\overline{Y_{n,j}\left(2;\eta_{(2)}\right)} = \frac{1}{\pi}\cos\left(n\arccos\left(\xi_{(2)}\cdot\eta_{(2)}\right)\right). \tag{6.231}$$

The two-dimensional counterpart $P_n(2;\cdot)$ of the Legendre polynomial is known as the Chebyshev function $P_n(2;\cdot)$ given by

$$P_n(2;t) = \frac{1}{2}\left(\left(t+i\sqrt{1-t^2}\right)^n + \left(t-i\sqrt{1-t^2}\right)^n\right) \tag{6.232}$$

$$= \cos(n\arccos(t)) = \frac{n}{2}\sum_{s=0}^{\lfloor\frac{n}{2}\rfloor}(-1)^s\frac{\Gamma(n-s)}{s!\,(n-2s)!}(2t)^{n-2s}$$

$$= \left(-\frac{1}{2}\right)^n\frac{\Gamma\left(\frac{1}{2}\right)}{\Gamma\left(n+\frac{1}{2}\right)}(1-t^2)^{\frac{1}{2}}\left(\frac{d}{dt}\right)^n(1-t^2)^{n-\frac{1}{2}},$$

$t=\xi_{(2)}\cdot\eta_{(2)},\ t\in[-1,1]$.

Finally we obtain the two-dimensional analogue of the addition theorem

$$\sum_{j=1}^{2}H_{n,j}\left(2;x_{(2)}\right)\overline{H_{n,j}\left(2;y_{(2)}\right)} = \frac{|x_{(2)}|^n|y_{(2)}|^n}{\pi}P_n\left(2;\xi_{(2)}\cdot\eta_{(2)}\right), \tag{6.233}$$

i.e.,

$$\sum_{j=1}^{2}Y_{n,j}\left(2;x_{(2)}\right)\overline{Y_{n,j}\left(2;y_{(2)}\right)} = \frac{1}{\pi}P_n\left(2;\xi_{(2)}\cdot\eta_{(2)}\right). \tag{6.234}$$

Next we are interested in defining properties of the Legendre polynomial in $\mathbb{R}^q$. To this end we first show that the only function in Harm_n that is invariant under orthogonal transformations having one point fixed can be characterized by the Legendre polynomial.

Lemma 6.12. *Let $H_n(q;\cdot)$ be a homogeneous harmonic polynomial of degree n satisfying the following properties:*

(i) $H_n(q;\mathbf{t}x) = H_n(q;x)$ for all orthogonal transformations $\mathbf{t}$ satisfying $\det(\mathbf{t})=1$ and $\mathbf{t}\epsilon^q = \epsilon^q$,

(ii) $H_n(q;\epsilon^q) = 1$.

Then $H_n(q;\cdot)$ is uniquely determined, and we have with $x=r\xi$, $\xi\in\mathbb{S}^{q-1}$, $\xi = t\epsilon^q + \sqrt{1-t^2}\,\xi_{(q-1)},\ \xi_{(q-1)}\in\mathbb{S}^{q-2}$,

$$H_n(q;\xi) = r^n P_n(q;\xi\cdot\epsilon^q) = r^n P_n(q,t), \tag{6.235}$$

where

$$P_n(q;t) = n!\,\Gamma\left(\frac{q-1}{2}\right)\sum_{l=0}^{\lfloor\frac{n}{2}\rfloor}\left(-\frac{1}{4}\right)^l\frac{(1-t^2)^l\,t^{n-2l}}{l!\,(n-2l)!\,\Gamma\left(l+\frac{q-1}{2}\right)}. \tag{6.236}$$

Proof. Each homogeneous, harmonic $H_n(q; \cdot)$ can be written in the form

$$H_n(q; x_1, \ldots, x_q) = \sum_{k=0}^{n} x_q^k A_{n-k}(x_1, \ldots, x_{q-1}), \qquad (6.237)$$

where A_{n-k} are elements of $\mathrm{Hom}_{n-k}(\mathbb{R}^{q-1})$. The Laplace operator can be split as follows

$$\Delta_{(q)} = \Delta_{(q-1)} + \left(\frac{\partial}{\partial x_q} \right)^2. \qquad (6.238)$$

Hence, we obtain

$$0 = \Delta_{(q)} H_n(q; x_1, \ldots, x_q) = \sum_{k=0}^{n-2} x_q^k \Delta_{(q-1)} A_{n-k}(x_1, \ldots, x_{q-1}) \qquad (6.239)$$

$$+ \sum_{k=2}^{n} k(k-1) x_q^{k-2} A_{n-k}(x_1, \ldots, x_{q-1}).$$

For $k = 0, \ldots, n-2$ this yields

$$\Delta_{(q-1)} A_{n-k}(x_1, \ldots, x_{q-1}) = -(k+2)(k+1) A_{n-k-2}(x_1, \ldots, x_{q-1}). \quad (6.240)$$

That means that only A_n and A_{n-1} can be chosen arbitrarily. Under this choice, $A_j, j = 0, \ldots, n-2$, are uniquely determined. Now, Condition *(i)* implies that the polynomials $A_{n-k}, k = 0, \ldots, n$, depend only on $x_{(q-1)}^2 = x_1^2 + \ldots + x_{q-1}^2$. We, thus, find a constant C_l such that

$$A_{n-k}(x_{(q-1)}) = \begin{cases} C_l (x_{(q-1)})^{2l} & , \quad n-k = 2l \\ 0 & , \quad n-k = 2l+1. \end{cases} \qquad (6.241)$$

This leads to

$$H_n(q; x_{(q)}) = \sum_{l=0}^{\lfloor \frac{n}{2} \rfloor} C_l (x_{(q-1)})^{2l} x_q^{n-2l}. \qquad (6.242)$$

In connection with

$$\Delta_{(q-1)} (x_{(q-1)})^{2l} = 2l(2l+q-3) \left(x_{(q-1)} \right)^{2l-2} \qquad (6.243)$$

we then get the recursion relation

$$(2l+2)(2l+q-1) C_{l+1} + (n-2l)(n-1-2l) C_l = 0, \qquad (6.244)$$

$l = 0, \ldots, \lfloor \frac{n}{2} \rfloor - 1$. For $x = \epsilon^q$, Condition *(ii)* tells us that $x_{(q-1)} = 0$ and $x_q = 1$. Hence, $C_0 = 1$. Altogether, $H_n(q; \cdot)$ is uniquely determined. From the recursion relation together with $C_0 = 1$ we find

$$C_l = (-1)^l \frac{n! \, \Gamma \left(\frac{q-1}{2} \right)}{l! \, (n-2l)! \, \Gamma \left(\frac{q-1}{2} + l \right)}. \qquad (6.245)$$

Consequently, we obtain

$$H_n(q; x) = n! \, \Gamma\left(\frac{q-1}{2}\right) \sum_{l=0}^{\lfloor \frac{n}{2} \rfloor} \left(-\frac{1}{4}\right)^l \frac{(x_{(q-1)})^{2l} \, x_q^{n-2l}}{l! \, (n-2l)! \, \Gamma\left(\frac{q-1}{2}+l\right)}. \qquad (6.246)$$

Because of $x_{(q)} = x_q \epsilon^q + x_{(q-1)} = r\xi_{(q)} = r(t\epsilon^q) + \sqrt{1-t^2} \, \xi_{(q-1)}$ we finally get $|x_{(q-1)}| = r\sqrt{1-t^2}$. This shows Lemma 6.12. $\qquad \square$

We now come to the proof of the *Funk–Hecke formula* (cf. H. Funk [1916], E. Hecke [1918] for early versions, C. Müller [1952, 1966]).

Theorem 6.14. *Suppose that G is of class $\mathrm{L}^1([-1,1])$. Then, for all $(\xi, \eta) \in \mathbb{S}^{q-1} \times \mathbb{S}^{q-1}$ and all $n \in \mathbb{N}_0$,*

$$\int_{\mathbb{S}^{q-1}} G(\xi \cdot \zeta) P_n(q; \zeta \cdot \eta) \, dS_{(q-1)}(\zeta) = G^\wedge(n) \, P_n(q; \xi \cdot \eta), \qquad (6.247)$$

where

$$G^\wedge(n) = G^\wedge_{\mathbb{S}^{q-1}}(n) = \|\mathbb{S}^{q-2}\| \int_{-1}^{+1} G(s) P_n(q; s)(1-s^2)^{\frac{q-3}{2}} \, ds. \qquad (6.248)$$

Proof. For $n \in \mathbb{N}_0$, $\xi, \eta \in \mathbb{S}^{q-1}$, let $G_n(q; \cdot, \cdot)$ be given by

$$G_n(q; \xi, \eta) = \int_{\mathbb{S}^{q-1}} G(\xi \cdot \zeta) P_n(q; \zeta \cdot \eta) \, dS_{(q-1)}(\zeta). \qquad (6.249)$$

Then, with any orthogonal matrix $\mathbf{t}$ in $\mathbb{R}^q$ with $\det(\mathbf{t}) = 1$, we get

$$\begin{aligned} G_n(q; \mathbf{t}\xi, \mathbf{t}\eta) &= \int_{\mathbb{S}^{q-1}} G(\mathbf{t}\xi \cdot \zeta) P_n(q; \zeta \cdot \mathbf{t}\eta) \, dS_{(q-1)}(\zeta) \qquad (6.250) \\ &= \int_{\mathbb{S}^{q-1}} G(\xi \cdot \mathbf{t}^T \zeta) P_n(q; \mathbf{t}^T \zeta \cdot \eta) \, dS_{(q-1)}(\zeta), \end{aligned}$$

where $\mathbf{t}^T$ is the transpose of $\mathbf{t}$. We know that $dS_{(q-1)}(\mathbf{t}^T \zeta) = dS_{(q-1)}(\zeta)$ such that

$$G_n(q; \mathbf{t}\xi, \mathbf{t}\eta) = \int_{\mathbb{S}^{q-1}} G(\xi \cdot \mathbf{t}^T \zeta) P_n(q; \mathbf{t}^T \zeta \cdot \eta) \, dS_{(q-1)}(\mathbf{t}^T \zeta). \qquad (6.251)$$

On the one hand this shows us that

$$G_n(q; \mathbf{t}\xi, \mathbf{t}\eta) = G_n(q; \xi, \eta). \qquad (6.252)$$

On the other hand, with ξ fixed, $G_n(q; \xi, \cdot)$ is a member of Harm_n which is invariant under orthogonal transformations that leave ξ invariant. From Lemma 6.12 we are therefore able to deduce that there exists a constant $G_q^\wedge(n)$ such that

$$G_n(q; \xi, \eta) = G_q^\wedge(n) P_n(q; \xi \cdot \eta).$$

In order to determine $G_q^\wedge(n)$ we let $\xi = \eta$. Together with

$$\zeta = s\epsilon^q + \sqrt{1-s^2}\, \zeta_{(q-1)}, \quad \zeta_{(q-1)} \in \mathbb{S}^{q-2}, \tag{6.253}$$

and (cf. (3.42))

$$dS_{(q-1)}(\zeta) = (1-s^2)^{\frac{q-3}{2}} dS_{(q-2)}(\zeta_{(q-1)})\, ds \tag{6.254}$$

this yields

$$
\begin{aligned}
G^\wedge(n) \;&=\; G^\wedge(n) \underbrace{P_n(q;1)}_{=1} \\[2mm]
&=\; \int_{\mathbb{S}^{q-2}} dS_{(q-2)} \int_{-1}^{+1} G(s) P_n(q;s)(1-s^2)^{\frac{q-3}{2}}\, ds \\[2mm]
&=\; \|\mathbb{S}^{q-2}\| \int_{-1}^{+1} G(s) P_n(q;s)(1-s^2)^{\frac{q-3}{2}}\, ds.
\end{aligned}
\tag{6.255}
$$

This is the required result. $\qquad\qquad\qquad\qquad\qquad\qquad\square$

From Theorem 6.14 we find by multiplication with $Y_n(q;\eta)$ and integration with regard to η the following *variant of the Funk–Hecke formula.*

Theorem 6.15. *Let G be of class $L^1([-1,+1])$. Then, for $Y_n(q;\cdot) \in \mathrm{Harm}_n$,*

$$\int_{\mathbb{S}^{q-1}} G(\xi \cdot \eta) Y_n(q;\eta)\, dS_{(q-1)}(\eta) = G^\wedge(n)\, Y_n(q;\xi) \tag{6.256}$$

with

$$G^\wedge(n) = \|\mathbb{S}^{q-2}\| \int_{-1}^{+1} G(s) P_n(q;s)(1-s^2)^{\frac{q-3}{2}}\, ds. \tag{6.257}$$

Next we are interested in the Laplace representation of spherical harmonics. To this end we start from the integral

$$L_n(q;x) = \frac{1}{\|\mathbb{S}^{q-2}\|} \int_{\mathbb{S}^{q-2}} \left(x \cdot \epsilon^q + ix_{(q-1)} \cdot \zeta_{(q-1)}\right)^n dS_{(q-2)}(\zeta_{(q-1)}), \tag{6.258}$$

where $x \in \mathbb{R}^q$ is of the form $x = (x \cdot \epsilon^q)\epsilon^q + x_{(q-1)}$, $x_{(q-1)} \in \mathbb{R}^{q-1}$. $L_n(q;\cdot)$ is a homogeneous harmonic polynomial in $\mathbb{R}^q$ of degree n. In addition, for all orthogonal transformations $\mathbf{t}$ with $\det(\mathbf{t}) = 1$ and $\mathbf{t}\epsilon^q = \epsilon^q$, we have $L_n(q;\mathbf{t}x) = L_n(q;x)$ and $L_n(q;\epsilon^q) = 1$. Lemma 6.12 then tells us that

$$L_n(q;x) = r^n L_n(q;\xi) = r^n P_n(q;t), \tag{6.259}$$

where, as usual, $x = r\xi$, $\xi \in \mathbb{S}^{q-1}$ $\xi = t\epsilon^q + \sqrt{1-t^2}\xi_{(q-1)}$, $\xi_{(q-2)} \in \mathbb{S}^{q-2}$. Hence, we have

$$P_n(q;t) = \frac{1}{\|\mathbb{S}^{q-2}\|} \int_{\mathbb{S}^{q-2}} \left(t + i\sqrt{1-t^2}\xi_{(q-1)} \cdot \zeta_{(q-1)}\right)^n dS_{(q-2)}(\zeta_{(q-1)}). \tag{6.260}$$

Theorem 6.16. *(Laplace's Integral Representation) For $q \geq 3$, $n \geq 0$, and $t \in [-1,1]$*

$$P_n(q;t) = \frac{\|\mathbb{S}^{q-3}\|}{\|\mathbb{S}^{q-2}\|} \int_{-1}^{1} \left(t + is\sqrt{1-t^2}\right)^n (1-s^2)^{\frac{q-4}{2}} \, ds. \tag{6.261}$$

For $s, t \in [-1,1]$,

$$\left| t + is\sqrt{1-t^2} \right|^2 = 1 - (1-s^2)(1-t^2) \leq 1 \tag{6.262}$$

we get from Theorem 6.16 and $q \geq 3$

$$|P_n(q;t)| \leq \frac{\|\mathbb{S}^{q-3}\|}{\|\mathbb{S}^{q-2}\|} \int_{-1}^{+1} (1-s^2)^{\frac{q-4}{2}} \, ds = 1. \tag{6.263}$$

Utilizing again (6.262) and Theorem 6.16, for $q \geq 3$, $t \in (-1,1)$, we have with $1 - x \leq e^{-x}$, $x \geq 0$

$$\begin{aligned}
|P_n(q;t)| &\leq \frac{\|\mathbb{S}^{q-3}\|}{\|\mathbb{S}^{q-2}\|} \int_{-1}^{+1} \left(1 - (1-s^2)(1-t^2)\right)^{\frac{n}{2}} (1-s^2)^{\frac{q-4}{2}} \, ds \\
&\leq 2 \frac{\|\mathbb{S}^{q-3}\|}{\|\mathbb{S}^{q-2}\|} \int_{0}^{1} e^{-\frac{n}{2}(1-t^2)(1-s^2)} (1-s^2)^{\frac{q-4}{2}} \, ds.
\end{aligned} \tag{6.264}$$

Substituting $s = 1 - u$ we are able to derive, with $u \leq 1 - s^2 \leq 2u$, the estimate

$$\begin{aligned}
|P_n(q;t)| &\leq 2^{\frac{q-2}{2}} \frac{\|\mathbb{S}^{q-3}\|}{\|\mathbb{S}^{q-2}\|} \int_{0}^{1} e^{-\frac{n}{2}u(1-t^2)} u^{\frac{q-4}{2}} \, du \tag{6.265} \\
&< 2^{\frac{q-2}{2}} \frac{\|\mathbb{S}^{q-3}\|}{\|\mathbb{S}^{q-2}\|} \int_{0}^{\infty} e^{-\frac{n}{2}u(1-t^2)} u^{\frac{q-4}{2}} \, du \\
&= 2^{q-2} \frac{\|\mathbb{S}^{q-3}\|}{\|\mathbb{S}^{q-2}\|} \frac{\Gamma(\frac{q-2}{2})}{(n(1-t^2))^{\frac{q-2}{2}}} \\
&= \frac{\Gamma(\frac{q-1}{2})}{\sqrt{\pi}} \left(\frac{4}{n(1-t^2)}\right)^{\frac{q-2}{2}}.
\end{aligned}$$

This shows the following estimates for the Legendre polynomial.

Lemma 6.13. *For $q \geq 3$ the Legendre polynomials $P_n(q; \cdot)$ satisfy the inequalities*

$$|P_n(q;t)| \leq 1, \quad n \geq 0, \; t \in [-1,1], \tag{6.266}$$

$$|P_n(q;t)| \leq \frac{\Gamma(\frac{q-1}{2})}{\sqrt{\pi}} \left(\frac{4}{n(1-t^2)}\right)^{\frac{q-2}{2}}, \quad n \geq 1, \; t \in (-1,1). \tag{6.267}$$

The estimates remain valid for $q = 2$.

From Theorem 6.16 we additionally obtain

$$\frac{d}{dt}P_n(q;t) \tag{6.268}$$

$$= n\frac{\|\mathbb{S}^{q-3}\|}{\|\mathbb{S}^{q-2}\|} \int_{-1}^{1} \left(1 - is\frac{t}{\sqrt{1-t^2}}\right)\left(t + is\sqrt{1-t^2}\right)^{n-1}(1-s^2)^{\frac{q-4}{2}}\,ds.$$

This yields

$$\left|\frac{d}{dt}P_n(q;t)\right| \le \frac{\|\mathbb{S}^{q-3}\|}{\|\mathbb{S}^{q-2}\|}\frac{n}{\sqrt{1-t^2}}\int_{-1}^{1}\left|t + is\sqrt{1-t^2}\right|^{n-1}(1-s^2)^{\frac{q-4}{2}}\,ds. \tag{6.269}$$

Observing

$$\left|t + is\sqrt{1-t^2}\right| \le 1 \tag{6.270}$$

we get

$$\left|\frac{d}{dt}P_n(q;t)\right| \le \frac{n}{\sqrt{1-t^2}}, \quad n \ge 1, \quad t \in (-1,1). \tag{6.271}$$

Again, this estimate remains valid for $q = 2$.

Next we want to show the closure and completeness of the spherical harmonics in $L^2(\mathbb{S}^{q-1})$. To this end we consider the so–called *Bernstein kernel of degree n* (cf. C. Müller [1998], W. Freeden, M. Gutting [2008] for the vectorial and tensorial context in $\mathbb{R}^3$)

$$B_n(\xi \cdot \eta) = \left(\frac{1 + \xi \cdot \eta}{2}\right)^n, \quad \xi, \eta \in \mathbb{S}^{q-1}, \quad n \in \mathbb{N}_0. \tag{6.272}$$

As auxiliary material we need the following facts.

Lemma 6.14. *For $q \ge 2$, the following relations hold uniformly:*

(i) For all $\xi \in \mathbb{S}^{q-1}$,

$$\left(\frac{1}{4\pi}\right)^{\frac{q-1}{2}}\frac{\Gamma(n+q-1)}{\Gamma(n+\frac{q-1}{2})}\int_{\mathbb{S}^{q-1}}\left(\frac{1+\xi\cdot\eta}{2}\right)^n dS_{(q-1)}(\eta) = 1. \tag{6.273}$$

(ii) For $\xi, \eta \in \mathbb{S}^{q-1}$ with $\xi \cdot \eta < 1$,

$$\lim_{n\to\infty}\left(\frac{1}{4\pi}\right)^{q-1}\frac{\Gamma(n+q-1)}{\Gamma(n+\frac{q-1}{2})}\left(\frac{1+\xi\cdot\eta}{2}\right)^n = 0. \tag{6.274}$$

Proof. For $q \ge 2$, we have

$$\int_{\mathbb{S}^{q-1}}\left(\frac{1+\xi\cdot\eta}{2}\right)^n dS_{(q-1)}(\eta) = \|\mathbb{S}^{q-2}\|\int_{-1}^{+1}\left(\frac{1+t}{2}\right)^n(1-t^2)^{\frac{q-3}{2}}\,dt. \tag{6.275}$$

Substituting $t = 2u - 1$ we find

$$\|\mathbb{S}^{q-2}\| \int_{-1}^{+1} \left(\frac{1+t}{2} \right)^n (1 - t^2)^{\frac{q-3}{2}} \, dt \tag{6.276}$$

$$= 2^{q-2} \|\mathbb{S}^{q-2}\| \int_0^1 u^{n+\frac{q-3}{2}} (1 - u)^{\frac{q-3}{2}} \, du$$

$$= \frac{2^{q-1} \pi^{\frac{q-1}{2}}}{\Gamma(\frac{q-1}{2})} \frac{\Gamma\left(n + \frac{q-1}{2}\right) \Gamma\left(\frac{q-1}{2}\right)}{\Gamma(n+q-1)}.$$

This guarantees *(i)*. The relation *(ii)* follows from the Stirling formula, which tells us that

$$\lim_{n \to \infty} \frac{\frac{\Gamma(n+q-1)}{\Gamma\left(n+\frac{q-1}{2}\right)}}{n^{\frac{q-1}{2}}} = 1. \tag{6.277}$$

Consequently, for $\xi, \eta \in \mathbb{S}^{q-1}$ with $\xi \cdot \eta \leq 1 - \delta < 1$,

$$0 \leq \left(\frac{1}{4\pi} \right)^{\frac{q-1}{2}} \frac{\Gamma(n+q-1)}{\Gamma\left(n + \frac{q-1}{2}\right)} \left(\frac{1 + \xi \cdot \eta}{2} \right)^n \tag{6.278}$$

$$\leq \left(\frac{1}{4\pi} \right)^{\frac{q-1}{2}} \frac{\Gamma(n+q-1)}{\Gamma\left(n + \frac{q-1}{2}\right)} \left(1 - \frac{\delta}{2} \right)^n.$$

Thus, relation *(ii)* is verified. □

Now we are able to prove the following approximate identity on $C^{(0)}(\mathbb{S}^{q-1})$.

Theorem 6.17. *Let F be of class $C^{(0)}(\mathbb{S}^{q-1})$. Then*

$$\lim_{n \to \infty} \sup_{\xi \in \mathbb{S}^{q-1}} \left| \left(\frac{1}{4\pi} \right)^{\frac{q-1}{2}} \frac{\Gamma(n+q-1)}{\Gamma\left(n + \frac{q-1}{2}\right)} \int_{\mathbb{S}^{q-1}} \left(\frac{1 + \xi \cdot \eta}{2} \right)^n F(\eta) \, dS_{(q-1)}(\eta) - F(\xi) \right|$$

$$= 0. \tag{6.279}$$

Proof. We use the results of Lemma 6.14. From *(i)* we obtain

$$\left(\frac{1}{4\pi} \right)^{\frac{q-1}{2}} \frac{\Gamma(n+q-1)}{\Gamma\left(n + \frac{q-1}{2}\right)} \int_{\mathbb{S}^{q-1}} \left(\frac{1 + \xi \cdot \eta}{2} \right)^n F(\eta) \, dS_{(q-1)}(\eta) \tag{6.280}$$

$$= F(\xi)$$

$$+ \left(\frac{1}{4\pi} \right)^{\frac{q-1}{2}} \frac{\Gamma(n+q-1)}{\Gamma\left(n + \frac{q-1}{2}\right)} \int_{\mathbb{S}^{q-1}} \left(\frac{1 + \xi \cdot \eta}{2} \right)^n (F(\eta) - F(\xi)) \, dS_{(q-1)}(\eta).$$

We split $\mathbb{S}^{q-1}$ into two parts depending on a parameter $\delta \in (0, 1)$

$$\int_{\mathbb{S}^{q-1}} \cdots = \int_{\substack{\xi \cdot \eta \geq 1 - \delta \\ \eta \in \mathbb{S}^{q-1}}} \cdots + \int_{\substack{\xi \cdot \eta < 1 - \delta \\ \eta \in \mathbb{S}^{q-1}}} \cdots. \tag{6.281}$$

Case 1: $\xi \cdot \eta \geq 1 - \delta$, $\eta \in \mathbb{S}^{q-1}$. The function F is uniformly continuous on $\mathbb{S}^{q-1}$. Therefore, using the modulus of continuity $\mu(\delta)$, $\delta \in (0,1)$, with $\lim_{\delta \to 0} \mu(\delta) = 0$ we have

$$|F(\xi) - F(\eta)| \leq \mu(\delta) \tag{6.282}$$

provided that $\eta \in \mathbb{S}^{q-1}$ with $1 - \delta \leq \xi \cdot \eta \leq 1$. This shows us that

$$\left(\frac{1}{4\pi}\right)^{\frac{q-1}{2}} \frac{\Gamma(n+q-1)}{\Gamma\left(n+\frac{q-1}{2}\right)} \int_{\substack{\xi \cdot \eta \geq 1 - \delta \\ \eta \in \mathbb{S}^{q-1}}} \left(\frac{1+\xi \cdot \eta}{2}\right)^n |F(\eta) - F(\xi)| dS_{(q-1)}(\eta)$$

$$\leq \left(\frac{1}{4\pi}\right)^{\frac{q-1}{2}} \frac{\Gamma(n+q-1)}{\Gamma\left(n+\frac{q-1}{2}\right)} \mu(\delta) \int_{\substack{\xi \cdot \eta \geq 1 - \delta \\ \eta \in \mathbb{S}^{q-1}}} \left(\frac{1+\xi \cdot \eta}{2}\right)^n dS_{(q-1)}(\eta)$$

$$\leq \mu(\delta). \tag{6.283}$$

Hence, the integral tends to zero as $\delta \to 0$.

Case 2: $\xi \cdot \eta < 1 - \delta$, $\eta \in \mathbb{S}^{q-1}$. Now we have

$$\left(\frac{1}{4\pi}\right)^{\frac{q-1}{2}} \frac{\Gamma(n+q-1)}{\Gamma\left(n+\frac{q-1}{2}\right)} \tag{6.284}$$

$$\times \left| \int_{\substack{\xi \cdot \eta < 1 - \delta \\ \eta \in \mathbb{S}^{q-1}}} \left(\frac{1+\xi \cdot \eta}{2}\right)^n (F(\eta) - F(\xi)) dS_{(q-1)}(\eta) \right|$$

$$\leq 2 \left(\frac{1}{4\pi}\right)^{\frac{q-1}{2}} \frac{\Gamma(n+q-1)}{\Gamma\left(n+\frac{q-1}{2}\right)} \|F\|_{C^{(0)}(\mathbb{S}^{q-1})} \|\mathbb{S}^{q-1}\| \left(1 - \frac{\delta}{2}\right)^n.$$

For every $\delta \in (0,1)$ we have

$$\lim_{n \to \infty} \sup_{\xi \in \mathbb{S}^{q-1}} \left| \left(\frac{1}{4\pi}\right)^{q-1} \frac{\Gamma(n+q-1)}{\Gamma\left(n+\frac{q-1}{2}\right)} \int_{\mathbb{S}^{q-1}} \left(\frac{1+\xi \cdot \eta}{2}\right)^n F(\eta) \, dS_{(q-1)}(\eta) - F(\xi) \right|$$

$$\leq \mu(\delta) \tag{6.285}$$

such that $\lim_{\substack{\delta \to 0 \\ \delta > 0}} \mu(\delta) = 0$ verifies Theorem 6.17. $\square$

Next it is not difficult to see that

$$\left(\frac{1}{4\pi}\right)^{\frac{q-1}{2}} \frac{\Gamma(n+q-1)}{\Gamma\left(n+\frac{q-1}{2}\right)} \int_{-1}^{+1} \left(\frac{1+t}{2}\right)^n P_k(q;t)(1-t^2)^{\frac{q-3}{2}} \, dt = \frac{\beta_n^k(q)}{\|\mathbb{S}^{q-2}\|}, \tag{6.286}$$

where we have used the abbrevitation

$$\beta_n^k(q) = \frac{n!}{(n-k)!} \frac{(n+q-2)!}{(n+k+q-2)!}. \tag{6.287}$$

Note that

$$\beta_n^k(q) < \beta_{n+1}^k(q) \tag{6.288}$$

and

$$\lim_{n\to\infty} \beta_n^k(q) = 1. \tag{6.289}$$

Hence,

$$\left(\frac{1}{4\pi}\right)^{\frac{q-1}{2}} \frac{\Gamma(n+q-1)}{\Gamma\left(n+\frac{q-1}{2}\right)} \left(\frac{1+\xi\cdot\eta}{2}\right)^n = \sum_{k=0}^{n} \beta_n^k(q) \frac{N(q;k)}{\|\mathbb{S}^{q-1}\|} P_k(q;\xi\cdot\eta). \tag{6.290}$$

Summarizing our results we finally obtain in connection with the addition theorem for spherical harmonics

Theorem 6.18. *For all* $F \in C^{(0)}(\mathbb{S}^{q-1})$

$$\lim_{n\to\infty} \sup_{\xi\in\mathbb{S}^{q-1}} \left| F(\xi) - \sum_{k=0}^{n} \sum_{j=1}^{N(q;k)} \beta_n^k(q) \int_{\mathbb{S}^{q-1}} F(\eta)\overline{Y_{k,j}(q;\eta)}\, dS_{(q-1)}(\eta) Y_{k,j}(q;\xi) \right|$$

$$= 0. \tag{6.291}$$

Theorem 6.18 enables us to prove the closure of the system of spherical harmonics.

Corollary 6.4. *The system* $\{Y_{n,j}(q;\cdot)\}_{\substack{n=0,1,\ldots \\ j=1,\ldots,N(q;n)}}$ *is closed in* $C^{(0)}(\mathbb{S}^{q-1})$; *that is, for any given* $\varepsilon > 0$ *and each* $F \in C^{(0)}(\mathbb{S}^{q-1})$ *there exists a linear combination*

$$\sum_{k=0}^{n} \sum_{j=1}^{N(q;k)} d_{k,j} Y_{k,j}(q;\cdot) \tag{6.292}$$

such that

$$\left\| F - \sum_{k=0}^{n} \sum_{j=1}^{N(q;k)} d_{k,j} Y_{k,j}(q;\cdot) \right\|_{C^{(0)}(\mathbb{S}^{q-1})} \leq \varepsilon. \tag{6.293}$$

Proof. Given $F \in C^{(0)}(\mathbb{S}^{q-1})$. Then, for any given $\varepsilon > 0$, there exists an integer $n(= n(\varepsilon))$ such that

$$\sup_{\xi\in\mathbb{S}^{q-1}} \left| F(\xi) - \sum_{k=0}^{n} \sum_{j=1}^{N(q;k)} \underbrace{\beta_n^k(q) \int_{\mathbb{S}^{q-1}} F(\eta)\overline{Y_{k,j}(q;\eta)}\, dS_{(q-1)}(\eta)}_{=d_{k,j}} Y_{k,j}(q;\xi) \right| \leq \varepsilon.$$

$$\tag{6.294}$$

This proves Corollary 6.4. □

Next we are interested in closure and completeness in the Hilbert space $L^2(\mathbb{S}^{q-1})$. First we show

Lemma 6.15. *The system* $\{Y_{n,j}(q;\cdot)\}_{\substack{n=0,1,\ldots \\ j=1,\ldots,N(q;n)}}$ *is closed in* $C^{(0)}(\mathbb{S}^{q-1})$ *with respect to* $\|\cdot\|_{L^2(\mathbb{S}^{q-1})}$.

Proof. Indeed, Lemma 6.15 immediately follows from Corollary 6.4 by use of the norm estimate for $F \in C^{(0)}(\mathbb{S}^{q-1})$

$$\|F\|_{L^2(\mathbb{S}^{q-1})} \leq \sqrt{\|\mathbb{S}^{q-1}\|} \, \|F\|_{C^{(0)}(\mathbb{S}^{q-1})}. \tag{6.295}$$

$\square$

Finally we arrive at the following result.

Theorem 6.19. *The system* $\{Y_{n,j}(q;\cdot)\}_{\substack{n=0,1,\ldots \\ j=1,\ldots,N(q,n)}}$ *is closed in the space* $L^2(\mathbb{S}^{q-1})$ *with respect to* $\|\cdot\|_{L^2(\mathbb{S}^{q-1})}$.

Proof. $C^{(0)}(\mathbb{S}^{q-1})$ is dense in $L^2(\mathbb{S}^{q-1})$, that is, for every $\varepsilon > 0$ and every $F \in L^2(\mathbb{S}^{q-1})$, there exists a function $G \in C^{(0)}(\mathbb{S}^{q-1})$ with $\|F-G\|_{L^2(\mathbb{S}^{q-1})} \leq \varepsilon$. The function $G \in C^{(0)}(\mathbb{S}^{q-1})$ admits an arbitrarily close approximation by finite linear combinations of spherical harmonics. Therefore, the proof of the closure is clear. $\square$

From constructive approximation (see, e.g., P.J. Davis [1963]), we know the equivalence of closure and completeness within the Hilbert space $L^2(\mathbb{S}^{q-1})$.

Corollary 6.5. *The closure in* $L^2(\mathbb{S}^{q-1})$ *is equivalent to each of the following properties:*

(i) The orthogonal expansion of $F \in L^2(\mathbb{S}^{q-1})$ *converges in the* $L^2(\mathbb{S}^{q-1})$*-norm to* F, *i.e.,*

$$0 = \lim_{n\to\infty} \left\| F - \sum_{k=0}^{n} \sum_{j=1}^{N(q;k)} \int_{\mathbb{S}^{q-1}} F(\eta)\overline{Y_{k,j}(q;\eta)} \, dS_{(q-1)}(\eta) \, Y_{k,j}(q;\cdot) \right\|_{L^2(\mathbb{S}^{q-1})}$$

(ii) Parseval's identity holds. That is, for any $F \in L^2(\mathbb{S}^{q-1})$,

$$\|F\|_{L^2(\mathbb{S}^{q-1})}^2 = \sum_{k=0}^{\infty} \sum_{j=1}^{N(q;k)} \left(\int_{\mathbb{S}^{q-1}} F(\eta)\overline{Y_{k,j}(q;\eta)} \, dS_{(q-1)}(\eta) \right)^2. \tag{6.296}$$

(iii) The extended Parseval's identity holds. That is, for any $F, G \in L^2(\mathbb{S}^{q-1})$

$$(F,G)_{L^2(\mathbb{S}^{q-1})} = \sum_{k=0}^{\infty} \sum_{j=1}^{N(q;k)} \left(\int_{\mathbb{S}^{q-1}} F(\eta)\overline{Y_{k,j}(q;\eta)} \, dS_{(q-1)}(\eta) \right)$$

$$\times \left(\int_{\mathbb{S}^{q-1}} G(\eta)\overline{Y_{k,j}(q;\eta)} \, dS_{(q-1)}(\eta) \right). \tag{6.297}$$

(iv) There is no strictly larger orthonormal system containing the orthonormal system $\{Y_{n,j}(q;\cdot)\}_{\substack{n=0,1,\ldots \\ j=1,\ldots,N(q;n)}}$.

(v) The system $\{Y_{n,j}(q;\cdot)\}_{\substack{n=0,1,\ldots \\ j=1,\ldots,N(q;n)}}$ has the completeness property. That is, $F \in L^2(\mathbb{S}^{q-1})$ and $\int_{\mathbb{S}^{q-1}} F(\eta) Y_{k,j}(q;\eta)\, dS_{(q-1)}(\eta) = 0$ for all k,j implies $F = 0$.

Associated Legendre Polynomials

By virtue of the Funk–Hecke formula (i.e., Theorem 6.15) any spherical harmonic $Y_n(q;\cdot) \in \mathrm{Harm}_n(\mathbb{S}^{q-1})$ given by

$$Y_n(q;\xi) \tag{6.298}$$
$$= C_{n,l}(q) \int_{\mathbb{S}^{q-2}} \left(\xi \cdot \epsilon^q + i\, \xi \cdot \eta_{(q-1)}\right)^n Y_l(q-1;\eta_{(q-1)})\, dS_{(q-2)}(\eta_{(q-1)})$$

can be represented in the form

$$Y_n(q;\xi) = P_{n,l}(q;t)\, Y_l(q-1;\xi_{(q-1)}), \tag{6.299}$$

$(\xi = t\epsilon^q + \sqrt{1-t^2}\,\xi_{(q-1)})$, where $P_{n,l}(q;\cdot)$ is given by

$$P_{n,l}(q;t) \tag{6.300}$$
$$= C_{n,l}(q)\, \|\mathbb{S}^{q-3}\| \int_{-1}^{1} \left(t + i\sqrt{1-t^2}s\right)^n P_l(q-1;s)(1-s^2)^{\frac{q-4}{2}}\, ds.$$

From the Rodrigues formula (6.180) we are able to deduce that

$$P_{n,l}(q;t) \tag{6.301}$$
$$= C_{n,l}(q)\, \frac{i^l l!}{2^l} \binom{n}{l} \frac{2\pi^{\frac{q-2}{2}}}{\Gamma(l+\frac{q-2}{2})}$$
$$\times (1-t^2)^{\frac{l}{2}} \int_{-1}^{1} \left(t + i\sqrt{1-t^2}s\right)^{n-l} (1-s^2)^{l+\frac{q-4}{2}}\, ds.$$

From the Laplace representation of $P_{n-l}(q+2l,\cdot)$ given by

$$P_{n-l}(q+2l,t) = \frac{\|\mathbb{S}^{q+2l-3}\|}{\|\mathbb{S}^{q+2l-2}\|} \int_{-1}^{1} \left(t + i\sqrt{1-t^2}s\right)^{n-l} (1-s^2)^{l+\frac{q-4}{2}}\, ds \tag{6.302}$$

we get

$$P_{n,l}(q;t) = C_{n,l}(q)\, \frac{i^l l!}{2^l} \binom{n}{l} \frac{2\pi^{\frac{q-1}{2}}}{\Gamma\left(l+\frac{q-1}{2}\right)}(1-t^2)^{\frac{l}{2}} P_{n-l}(q+2l;t). \tag{6.303}$$

By use of

$$P_{n-l}(q+2l;t) = \frac{N(q;n)}{N(q+2l;n-l)}\frac{\Gamma(\frac{q}{2})}{\Gamma(l+\frac{q}{2})} 2^{-l} P_n^{(l)}(q;t) \tag{6.304}$$

we find

$$P_{n,l}(q;t) = C_{n,l}\, i^l \,\frac{2\pi^{\frac{q-1}{2}}}{\Gamma(\frac{q-1}{2})}(1-t^2)^{\frac{l}{2}} P_n^{(l)}(q;t). \tag{6.305}$$

Until now, the coefficients $C_{n,l}(q)$ are arbitrary. Next, they will be determined in such a way that

$$\int_{-1}^{1} \left(P_{n,l}(q;t)\right)^2 (1-t^2)^{\frac{q-3}{2}}\, dt = 1. \tag{6.306}$$

Using

$$B_{n,l}(q) = C_{n,l}(q)\frac{i^l l!}{2^l}\binom{n}{l}\frac{2\pi^{\frac{q-1}{2}}}{\Gamma(l+\frac{q-1}{2})} \tag{6.307}$$

we have

$$P_{n,l}(q;t) = B_{n,l}(q)(1-t^2)^{\frac{l}{2}} P_{n-l}(q+2l;t) \tag{6.308}$$

such that

$$\int_{-1}^{1} \left(P_{n,l}(q;t)\right)^2 (1-t^2)^{\frac{q-3}{2}}\, dt \tag{6.309}$$

$$= B_{n,l}^2(q) \int_{-1}^{1} \left(P_{n,l}(q+2l;t)\right)^2 (1-t^2)^{\frac{q+2l-3}{2}}\, dt$$

$$= B_{n,l}^2(q)\frac{\|\mathbb{S}^{q+2l-1}\|}{\|\mathbb{S}^{q+2l-2}\|}\frac{1}{N(q+2l;n-l)}$$

$$= B_{n,l}^2(q)\frac{\pi^{\frac{1}{2}}\Gamma\left(l+\frac{q-1}{2}\right)}{\Gamma\left(l+\frac{q}{2}\right)N(q+2l;n-l)}.$$

In other words, the coefficients $C_{n,l}(q)$ satisfying (6.306) are related to $B_{n,l}(q)$ given by

$$B_{n,l}(q) = \sqrt{\frac{\|\mathbb{S}^{q+2l-2}\|}{\|\mathbb{S}^{q+2l-1}\|}\, N(q+2l;n-l)} \tag{6.310}$$

as follows:

$$C_{n,l}(q) = B_{n,l}(q)\frac{2^l}{i^l\, l!\,\binom{n}{l}}\frac{\Gamma\left(l+\frac{q-1}{2}\right)}{2\pi^{\frac{q-1}{2}}}. \tag{6.311}$$

Explicitly written out we therefore have

$$P_{n,l}(q;t) = \sqrt{\frac{\|\mathbb{S}^{q+2l-2}\|}{\|\mathbb{S}^{q+2l-1}\|}}\, N(q+2l;n-l)\,(1-t^2)^{\frac{l}{2}} P_{n-l}(q+2l;t). \tag{6.312}$$

Hence, under this normalization, we finally obtain

$$\int_{-1}^{1} P_{n,l}(q;t)\, P_{m,l}(q;t)\,(1-t^2)^{\frac{q-3}{2}}\, dt = \delta_{nm}. \tag{6.313}$$

Definition 6.4. *The function $P_{n,l}(q; \cdot)$ is called (normalized) associated Leg-endre polynomial of degree n and order l of dimension q.*
The system

$$\left\{ P_{n,l}(q; t) \, Y_{l,j}(q-1; \xi_{(q-1)}) \right\}_{\substack{l=0,\ldots n \\ j=1,\ldots,N(q-1;l)}} \tag{6.314}$$

forms an $L^2(\mathbb{S}^{q-1})$-orthonormal system in $\mathrm{Harm}_n(\mathbb{S}^{q-1})$ provided that the set $\left\{ Y_{l,j}(q-1; \cdot) \right\}_{\substack{l=0,\ldots,n \\ j=1,\ldots,N(q-1,l)}}$ forms an $L^2(\mathbb{S}^{q-2})$-orthonormal system.

Setting

$$\xi = t\epsilon^q + \sqrt{1-t^2}\,\xi_{(q-1)}, \quad \eta = s\epsilon^q + \sqrt{1-s^2}\,\eta_{(q-1)} \tag{6.315}$$

we find

$$\xi \cdot \eta = ts + \sqrt{1-t^2}\sqrt{1-s^2}\,\xi_{(q-1)} \cdot \eta_{(q-1)}. \tag{6.316}$$

The addition theorem of spherical harmonics therefore gives us

$$\frac{1}{\|\mathbb{S}^{q-2}\|} \sum_{l=0}^{n} N(q-1,l) P_{n,l}(q;t) P_{n,l}(q;s) \, P_l\left(q-1;\xi_{(q-1)} \cdot \eta_{(q-1)}\right)$$

$$= \frac{N(q;n)}{\|\mathbb{S}^{q-1}\|} P_n\left(q; ts + \sqrt{1-t^2}\sqrt{1-s^2}\,\xi_{(q-1)} \cdot \eta_{(q-1)}\right). \tag{6.317}$$

By integration of (6.317) with respect to $\xi_{(q-1)} \in \mathbb{S}^{q-1}$ we finally obtain

$$\int_{-1}^{1} P_n\left(q; ts + \sqrt{1-t^2}\sqrt{1-s^2}\,u\right) (1-u^2)^{\frac{q-4}{2}}\, du = P_n(q;t)\, P_n(q;s). \tag{6.318}$$

Pointwise Expansion Theorem

We start our considerations with a special power series that shows the Legen-dre polynomials as generating coefficients.

Lemma 6.16. *For $q \geq 3$, $|r| < 1$, and $t \in [-1,1]$ we have*

$$\sum_{n=0}^{\infty} N(q,n) r^n P_n(q;t) = \frac{1 - r^2}{(1 + r^2 - 2rt)^{\frac{q}{2}}}. \tag{6.319}$$

Proof. Let ϕ be given by (6.199) and (6.200). Then, we are able to express

$$\frac{1}{q-2}\left(2r\phi'(r) + (q-2)\phi(r)\right) \tag{6.320}$$

as series

$$\sum_{n=0}^{\infty} c_n(q) \frac{(2n+q-2)}{q-2}\, r^n\, P_n(q;t). \tag{6.321}$$

Furthermore, we find

$$\frac{1}{q-2}\left[\frac{2r(q-2)(t-r)}{(1+r^2-2rt)^{\frac{q}{2}}}+\frac{(q-2)(1+r^2-2rt)}{(1+r^2-2rt)^{\frac{q}{2}}}\right]=\frac{1-r^2}{(1+r^2-2rt)^{\frac{q}{2}}}.$$

(6.322)

Observing the fact that

$$N(q,n)=\frac{(2n+q-2)\Gamma(n+q-2)}{\Gamma(n+1)(q-2)\Gamma(q-2)}=\frac{2n+q-2}{q-2}c_n(q)$$

(6.323)

we obtain the desired result of Lemma 6.16.   □

From Lemma 6.16 it follows that

$$\int_{-1}^{1}\frac{(1-r^2)(1-t^2)^{\frac{q-3}{2}}}{(1+r^2-2rt)^{\frac{q}{2}}}\,dt=\int_{-1}^{1}(1-t^2)^{\frac{q-3}{2}}\,dt=\frac{\|\mathbb{S}^{q-1}\|}{\|\mathbb{S}^{q-2}\|}$$

(6.324)

for all $r\in\mathbb{R}$ with $0\le r<1$. This leads to

Lemma 6.17. *Let F be of class* $C^{(0)}([-1,1])$. *Then*

$$\lim_{\substack{r\to 1 \\ r<1}}\int_{-1}^{1}\frac{(1-r^2)F(t)(1-t^2)^{\frac{q-3}{2}}}{(1+r^2-2rt)^{\frac{q}{2}}}\,dt=F(1)\frac{\|\mathbb{S}^{q-1}\|}{\|\mathbb{S}^{q-2}\|}.$$

(6.325)

Proof. We have to verify that

$$\lim_{\substack{r\to 1 \\ r<1}}\int_{-1}^{1}\frac{(1-r^2)(F(t)-F(1))}{(1+r^2-2rt)^{\frac{q}{2}}}(1-t^2)^{\frac{q-3}{2}}\,dt=0.$$

(6.326)

In accordance with our assumption there exists a positive function M with $\lim_{s\to 0}M(s)=0$, such that

$$\sup_{1-s\le t\le 1}|F(t)-F(1)|=M(s).$$

(6.327)

Moreover, there exists a positive constant C such that

$$\sup_{t\in[-1,1]}|F(t)-F(1)|=C.$$

(6.328)

Now we have for $-1\le t\le 1-s$ and $r>0$

$$1+r^2-2rt=(1-r)^2+2r(1-t)\ge 2rs.$$

(6.329)

In addition, we see that for $-1\le t\le 1-s$ and $\frac{1}{2}<r<1$

$$\frac{1-r^2}{(1+r^2-2rt)^{\frac{q}{2}}}\le\frac{1-r^2}{(2rs)^{\frac{q}{2}}}$$

(6.330)

$$=\frac{(1+r)(1-r)}{(2r)^{\frac{q}{2}}s^{\frac{q}{2}}}$$

$$\le 2\frac{(1-r)}{s^{\frac{q}{2}}}.$$

We divide the interval $[-1, 1]$ into $[-1, 1-s]$ and $(1-s, 1]$. Then, we obtain

$$\int_{-1}^{1-s} \frac{(1-r^2)(F(t) - F(1))}{(1+r^2-2rt)^{\frac{q}{2}}} (1-t^2)^{\frac{q-3}{2}}\, dt = O\left(\frac{\|\mathbb{S}^{q-1}\|}{\|\mathbb{S}^{q-1}\|} s\right) \quad (6.331)$$

and

$$\int_{1-s}^{1} \frac{(1-r^2)(F(t) - F(1))}{(1+r^2-2rt)^{\frac{q}{2}}} (1-t^2)^{\frac{q-3}{2}}\, dt \quad (6.332)$$

$$= O\left(M(s) \int_{-1}^{1} \frac{(1-r^2)(1-t^2)^{\frac{q-3}{2}}}{(1+r^2-2rt)^{\frac{q}{2}}}\, dt\right).$$

With $s^{\frac{q}{2}} = (1-r)^{\frac{1}{2}}$ Lemma 6.17 follows from the last estimate. $\qquad\Box$

Next we are now able to formulate the *Poisson integral formula*.

Theorem 6.20. *Suppose that G is of class $C^{(0)}(\mathbb{S}^{q-1})$. Then we have*

$$\lim_{\substack{r \to 1 \\ r<1}} \frac{1}{\|\mathbb{S}^{q-1}\|} \int_{\mathbb{S}^{q-1}} \frac{(1-r^2)G(\eta)}{(1+r^2-2r\,\xi\cdot\eta)^{\frac{q}{2}}}\, dS_{(q-1)}(\eta) = G(\xi) \quad (6.333)$$

uniformly with respect to $\xi \in \mathbb{S}^{q-1}$.

Proof. Since $\mathbb{S}^{q-1}$ is compact, the continuity of G implies the existence of a positive function M, such that

$$|G(\xi) - G(\eta)| \le M(\tau) \quad (6.334)$$

holds for $1 - \tau \le \xi\cdot\eta \le 1$. For $\xi = \epsilon^q$ we let

$$F(t) = \int_{\mathbb{S}^{q-2}} G\left(t\epsilon^q + \sqrt{1-t^2}\eta_{(q-1)}\right) dS_{(q-2)}(\eta_{(q-1)})\,. \quad (6.335)$$

Then we have

$$F(1) = \|\mathbb{S}^{q-2}\| G(\epsilon^q). \quad (6.336)$$

From (6.334) it follows for $1 - \tau \le t \le 1$ that

$$|F(1) - F(t)| \le \|\mathbb{S}^{q-2}\| M(\tau). \quad (6.337)$$

The integral in (6.333) can be written in the form

$$\frac{1}{\|\mathbb{S}^{q-1}\|} \int_{-1}^{1} \frac{(1-r^2)F(t)(1-t^2)^{\frac{q-3}{2}}}{(1-r^2-2rt)^{\frac{q}{2}}}\, dt. \quad (6.338)$$

In connection with Lemma 6.17 this yields for $\xi = \epsilon^q$

$$\lim_{\substack{r \to 1 \\ r<1}} \frac{1}{\|\mathbb{S}^{q-1}\|} \int_{\Omega_q} \frac{(1-r^2)G(\eta)}{(1+r^2-2r\xi\cdot\eta)^{\frac{q}{2}}}\, dS_{(q-1)}(\eta)$$

$$= \lim_{\substack{r \to 1 \\ r<1}} \frac{1}{\|\mathbb{S}^{q-1}\|} \int_{-1}^{1} \frac{(1-r^2)F(t)(1-t^2)^{\frac{q-3}{2}}}{(1+r^2-2rt)^{\frac{q}{2}}}\, dt$$

$$= \frac{1}{\|\mathbb{S}^{q-1}\|} \frac{\|\mathbb{S}^{q-1}\|}{\|\mathbb{S}^{q-2}\|} F(1) = G(\epsilon^q). \quad (6.339)$$

Since every point on $\mathbb{S}^{q-1}$ can be transformed into ϵ^q, the argumentation is valid for all $\xi \in \mathbb{S}^{q-1}$. From (6.337), it follows that the limit relation holds uniformly. $\square$

Combining Theorem 6.20 and Lemma 6.16 we obtain

Theorem 6.21. *(Expansion Theorem) Suppose that G is of class $C^{(0)}(\mathbb{S}^{q-1})$. Then*

$$G(\xi) = \lim_{\substack{r \to 1 \\ r < 1}} \sum_{n=0}^{\infty} r^n \sum_{j=1}^{N(q,n)} \int_{\mathbb{S}^{q-1}} G(\eta)\overline{Y_{n,j}(q;\eta)} \, dS_{(q-1)}(\eta) \, Y_{nj}(q;\xi) \qquad (6.340)$$

holds uniformly with respect to $\xi \in \mathbb{S}^{q-1}$.

As a result of our considerations we are able to solve the Dirichlet problem corresponding to continuous boundary values on the unit sphere.

Theorem 6.22. *(Dirichlet's Problem) Let F be of class $C^{(0)}(\mathbb{S}^{q-1})$. Then the series*

$$\sum_{n=0}^{\infty} r^n \sum_{j=1}^{N(q,n)} \int_{\mathbb{S}^{q-1}} F(\eta)\overline{Y_{n,j}(q;\eta)} \, dS_{(q-1)}(\eta) \, Y_{nj}(q;\xi) \qquad (6.341)$$

converges for all $r \leq r_0 < 1$ absolutely and uniformly. The function $U : \overline{\mathbb{B}_1^q} \to \mathbb{R}$ given by

$$U(x) = U(r\xi) = \sum_{n=0}^{\infty} r^n \sum_{j=1}^{N(q,n)} \int_{\mathbb{S}^{q-1}} F(\eta)\overline{Y_{n,j}(q;\eta)} \, dS_{(q-1)}(\eta) \, Y_{nj}(q;\xi)$$

$$(6.342)$$

represents the uniquely determined solution of the Dirichlet problem $U \in C^{(0)}\left(\overline{\mathbb{B}_1^q}\right) \cap C^{(2)}(\mathbb{B}_1^q)$ with $\Delta U = 0$ in $\mathbb{B}_1^q$ satisfying the boundary condition

$$\lim_{\substack{r \to 1 \\ r < 1}} U(r\xi) = F(\xi), \quad \xi \in \mathbb{S}^{q-1}, \qquad (6.343)$$

uniformly with respect to $\xi \in \mathbb{S}^{q-1}$.

Next we are concerned with the pointwise representation of a function by its convergent orthogonal (Fourier) spherical harmonic expansions. Roughly speaking, the result is that the continuity of a function together with the convergence of its Fourier series expansion at a point on the unit sphere $\mathbb{S}^{q-1}$ assures the equality of the functional value and the value of its Fourier expansion at the point under consideration (cf. C. Müller [1998]).

Theorem 6.23. *(Pointwise Expansion Theorem) Suppose F is continuous in* the point $\xi_0 \in \mathbb{S}^{q-1}$ and uniformly bounded on $\mathbb{S}^{q-1}$. Assume that the sequence $\{S_n(\xi)\}_{n=0,1,\ldots}$ with

$$S_n(\xi) = \sum_{k=0}^{n} \int_{\mathbb{S}^{q-1}} F(\eta)\overline{Y_{k,j}(q;\eta)} \, dS_{(q-1)}(\eta) \, Y_{k,j}(q;\xi) \qquad (6.344)$$

converges for $\xi \in \mathbb{S}^{q-1}$. Then

$$F(\xi_0) = \lim_{n \to \infty} S_n(\xi_0) \qquad (6.345)$$

$$= \sum_{k=0}^{\infty} \sum_{j=1}^{N(q;n)} \int_{\mathbb{S}^{q-1}} F(\eta) \, \overline{Y_{k,j}(q;\eta)} \, dS_{(q-1)}(\eta) \, Y_k(q;\xi_0).$$

Proof. From Theorem 6.21 it is known that

$$\lim_{\substack{r \to 1 \\ r<1}} \sum_{k=0}^{\infty} \sum_{j=1}^{N(q,k)} r^k \int_{\mathbb{S}^{q-1}} F(\eta) \, \overline{Y_{k,j}(q;\eta)} \, dS_{(q-1)}(\eta) \, Y_{k,j}(q;\xi_0) = F(\xi_0). \quad (6.346)$$

We rewrite the series expansion as

$$\sum_{k=1}^{\infty} r^k \left(S_k(\xi_0) - S_{k-1}(\xi_0) \right) + S_k(\xi_0). \qquad (6.347)$$

Of course, we are able to use the knowledge of the limits

$$\lim_{\substack{r \to 1 \\ r<1}} (1-r) \sum_{k=0}^{\infty} r^k S_k(\xi_0) = F(\xi_0), \qquad (6.348)$$

and

$$\lim_{k \to \infty} S_k(\xi_0) = S_\infty(\xi_0). \qquad (6.349)$$

For every $\varepsilon > 0$ we have a number $K_0(\varepsilon)$ such that

$$S_\infty(\xi_0) - \varepsilon \le S_k(\xi_0) \le S_\infty(\xi_0) + \varepsilon \qquad (6.350)$$

holds true for all $k > K_0(\varepsilon)$. This leads to the estimate

$$\frac{r^{k_0}}{1-r}(S_\infty(\xi_0) - \varepsilon) \le \sum_{k=k_0}^{\infty} r^k S_k(\xi_0) \le \frac{r^{k_0}}{1-r}(S_\infty(\xi_0) + \varepsilon), \qquad (6.351)$$

such that the limit $r \to 1$ gives

$$S_\infty(\xi_0) - \varepsilon \le F(\xi_0) \le S_\infty(\xi_0) + \varepsilon. \qquad (6.352)$$

This proves the assertion of Theorem 6.23. $\qquad \square$

Remark 6.12. *An analogous result is valid for the dimension $q = 2$.*

Asymptotic Relations for the Spherical Harmonic Coefficients

Now we discuss the pointwise representation theorem by means of spherical harmonics under the assumption of sufficient differentiability ("smoothness") imposed on the function. If F is assumed to be continuously differentiable on $\mathbb{S}^{q-1}$, then the first Green surface theorem shows for all $\xi \in \mathbb{S}^{q-1}$ and $Y_n(q; \cdot) \in \mathrm{Harm}_n$ of the form

$$Y_n(q; \xi) = n(n+q-2) \int_{\mathbb{S}^{q-1}} F(\eta) P_n(q; \xi \cdot \eta) \, dS_{(q-1)}(\eta) \qquad (6.353)$$

that

$$|Y_n(q; \xi)| = \left| \int_{\mathbb{S}^{q-1}} \nabla_\eta^* P_n(q; \xi \cdot \eta) \cdot \nabla_\eta^* F(\eta) \, dS_{(q-1)}(\eta) \right|. \qquad (6.354)$$

The Cauchy–Schwarz inequality yields the estimate

$$
\begin{aligned}
|Y_n(q; \xi)| \;\leq\; & \left(\int_{\mathbb{S}^{q-1}} \left| \nabla_\eta^* P_n(q; \xi \cdot \eta) \right|^2 \, dS_{(q-1)}(\eta) \right)^{\frac{1}{2}} \\
& \times \left(\int_{\mathbb{S}^{q-1}} \left| \nabla_\eta^* F(\eta) \right|^2 \, dS_{(q-1)}(\eta) \right)^{\frac{1}{2}}.
\end{aligned}
\qquad (6.355)
$$

Now, the second Green surface theorem gives

$$
\begin{aligned}
& \int_{\mathbb{S}^{q-1}} \left| \nabla_\eta^* P_n(q; \xi \cdot \eta) \right|^2 \, dS_{(q-1)}(\eta) \qquad &&(6.356) \\
&= -\int_{\mathbb{S}^{q-1}} P_n(q; \xi \cdot \eta) \, \Delta_\eta^* P_n(q; \xi \cdot \eta) \, dS_{(q-1)}(\eta) \\
&= n(n+q-2) \int_{\mathbb{S}^{q-1}} (P_n(q; \xi \cdot \eta))^2 \, dS_{(q-1)}(\eta) \\
&= \frac{\|\mathbb{S}^{q-1}\|}{N(q; n)} \, n(n+q-2).
\end{aligned}
$$

From (6.355), in connection with (6.356), we therefore obtain for $n > 0$

$$
\begin{aligned}
\|Y_n(q; \cdot)\|_{C^{(0)}(\mathbb{S}^{q-1})} &= \sup_{\xi \in \mathbb{S}^{q-1}} |Y_n(q; \xi)| \\
&\leq \left(\frac{N(q; n)}{n(n+q-2)\|\mathbb{S}^{q-1}\|} \right)^{\frac{1}{2}} \left(\int_{\mathbb{S}^{q-1}} \left| \nabla_\eta^* F(\eta) \right|^2 \, dS_{(q-1)}(\eta) \right)^{\frac{1}{2}}.
\end{aligned}
$$

From the theory of spherical harmonics (Section 6.4) we know that $N(q; n) = O\left(n^{q-2}\right)$ for $n \to \infty$ such that

$$\|Y_n(q; \cdot)\|_{C^{(0)}(\mathbb{S}^{q-1})} = O\left(n^{\frac{q-4}{2}}\right). \qquad (6.357)$$

Assuming that F is twice continuously differentiable on $\mathbb{S}^{q-1}$ we find for $n > 0$

$$\frac{N(q;n)}{\|\mathbb{S}^{q-1}\|} \int_{\mathbb{S}^{q-1}} F(\eta) P_n(q;\xi\cdot\eta)\, dS_{(q-1)}(\eta) \tag{6.358}$$

$$= \frac{1}{n(n+q-2)} \frac{N(q;n)}{\|\mathbb{S}^{q-1}\|} \int_{\mathbb{S}^{q-1}} (\Delta_\eta^* F(\eta))\, (\Delta_\eta^* P_n(q;\xi\cdot\eta))\, dS_{(q-1)}(\eta)$$

such that $Y_n(q;\cdot)$ of the form (6.353) satisfies

$$\|Y_n(q;\cdot)\|_{C^{(0)}(\mathbb{S}^{q-1})} \tag{6.359}$$

$$\leq \frac{1}{n(n+q-2)} \left(\frac{N(q;n)}{\|\mathbb{S}^{q-1}\|}\right)^{\frac{1}{2}} \left(\int_{\mathbb{S}^{q-1}} |\Delta_\eta^* F(\eta)|^2\, dS_{(q-1)}(\eta)\right)^{\frac{1}{2}}.$$

This leads to the relation

$$\|Y_n(q;\cdot)\|_{C^{(0)}(\mathbb{S}^{q-1})} = O\left(n^{\frac{q-6}{2}}\right) \tag{6.360}$$

for $n \to \infty$. Continuing in this way we finally arrive at

Lemma 6.18. *If F is of class $C^{(k)}(\mathbb{S}^{q-1})$, then*

$$\left|\frac{N(q;n)}{\|\mathbb{S}^{q-1}\|} \int_{\mathbb{S}^{q-1}} F(\eta)\, P_n(q;\xi\cdot\eta)\, dS_{(q-1)}(\eta)\right| = O\left(n^{\frac{q-2k-2}{2}}\right), \quad n \to \infty. \tag{6.361}$$

If $k > \frac{q}{2}$, then we are able to deduce from Lemma 6.18 that the series

$$\sum_{n=0}^{\infty} \frac{N(q;n)}{\|\mathbb{S}^{q-1}\|} \int_{\mathbb{S}^{q-1}} F(\eta)\, P_n(q;\xi\cdot\eta)\, dS_{(q-1)}(\eta) \tag{6.362}$$

is absolutely and uniformly convergent. The Pointwise Expansion Theorem of the theory of spherical harmonics (Theorem 6.23), in connection with the addition theorem, therefore shows that

$$F(\xi) = \sum_{n=0}^{\infty} \frac{N(q;n)}{\|\mathbb{S}^{q-1}\|} \int_{\mathbb{S}^{q-1}} F(\eta)\, P_n(q;\xi\cdot\eta)\, dS_{(q-1)}(\eta) \tag{6.363}$$

$$= \sum_{n=0}^{\infty} \sum_{j=1}^{N(q;n)} \int_{\mathbb{S}^{q-1}} F(\eta)\, \overline{Y_{n,j}(q;\eta)}\, dS_{(q-1)}(\eta)\, Y_{n,j}(q;\xi)$$

holds true for all $\xi \in \mathbb{S}^{q-1}$ whenever $\{Y_{n,j}(q;\cdot)\}_{\substack{n=0,1,\dots \\ j=1,\dots,N(q;n)}}$ is an $L^2(\mathbb{S}^{q-1})$-orthonormal system of spherical harmonics.

6.5   Integral Theorems for the Helmholtz–Beltrami Operator

Next we consider the *eigenvalue problem*

$$(\Delta^* + \lambda)^m Y = 0, \quad Y \in C^{(2m)}(\mathbb{S}^{q-1}), \ m \in \mathbb{N}. \tag{6.364}$$

It is known that every $K_n(q; \cdot) \in \mathrm{Harm}_n(\mathbb{R}^q)$ can be written in the form $K_n(q; x) = r^n Y_n(q; \xi)$, $x = r\xi$, $\xi \in \mathbb{S}^{q-1}$. In addition, an easy calculation shows

$$r^{1-q} \frac{d}{dr}\left(r^{q-1}\frac{d}{dr}r^n\right) = n(n+q-2)r^{n-2}. \tag{6.365}$$

Thus, observing the representation (2.33) of the Laplace operator we obtain

$$\Delta_x K_n(q; x) = 0 = r^{n-2}\left(n(n+q-2)\,Y_n(q; \xi)\ +\ \Delta^*_\xi Y_n(q; \xi)\right), \tag{6.366}$$

i.e.,

$$(\Delta^* + (\Delta^*)^\wedge(n))\,Y_n(q; \cdot) = 0 \tag{6.367}$$

for all $Y_n \in \mathrm{Harm}_n(\mathbb{S}^{q-1})$, where

$$(\Delta^*)^\wedge(n) = n(n+q-2), \ n = 0, 1, \ldots\ . \tag{6.368}$$

It is noteworthy that the sequence $\{(\Delta^*)^\wedge(n)\}_{n\in\mathbb{N}_0}$ is monotonically increasing and its only "accumulation point" is at infinity.

Sphere Function for the Helmholtz–Beltrami Operator

The differential equation (6.367) motivates the introduction of the $\mathbb{S}^{q-1}$-sphere function, $q \geq 3$, for the *Helmholtz–Beltrami operator* $\Delta^* + \lambda$, $\lambda \in \mathbb{R}$ (cf. W. Freeden [1979], W. Freeden, R. Reuter [1982]). Note that the case $q = 2$ leads to the classical Fourier theory (as presented in Section 4.1). Therefore, we always assume $q \geq 3$, here. Moreover, for the case $\lambda = 0$, the reader is referred to Definition 6.2.

Definition 6.5. *Let λ be a real number. A function $G(\Delta^* + \lambda; \cdot) : (\xi, \eta) \mapsto G(\Delta^* + \lambda; \xi \cdot \eta)$, $-1 \leq \xi \cdot \eta < 1$, is called the Green function for the Helmholtz–Beltrami operator $\Delta^* + \lambda$ on the unit sphere $\mathbb{S}^{q-1}$ (briefly called, $\mathbb{S}^{q-1}$-sphere function for $\Delta^* + \lambda$) if it satisfies the following properties:*

(i) For each fixed $\xi \in \mathbb{S}^{q-1}, \eta \mapsto G(\Delta^* + \lambda; \xi \cdot \eta)$ is twice continuously differentiable with respect to $\eta \in \mathbb{S}^{q-1}$, $1 - \xi \cdot \eta \neq 0$, with

$$(\Delta^* + \lambda)G(\Delta^* + \lambda; \xi \cdot \eta) = -\sum_{\substack{(\Delta^*+\lambda)^\wedge(n)=0 \\ n\in\mathbb{N}_0}}\ \sum_{j=1}^{N(q;n)} Y_{n,j}(q; \xi)\overline{Y_{n,j}(q; \eta)}. \tag{6.369}$$

(ii) In the neighborhood of the point $\xi \in \mathbb{S}^{q-1}$ the estimates

$$
\begin{aligned}
G(\Delta^* + \lambda; \xi \cdot \eta) - \tfrac{1}{\|\mathbb{S}^2\|} \ln(1 - \xi \cdot \eta) &= O(1) \\
\nabla_\eta^* G(\Delta^* + \lambda; \xi \cdot \eta) - \tfrac{1}{\|\mathbb{S}^2\|} \nabla_\eta^* \ln(1 - \xi \cdot \eta) &= O(1)
\end{aligned}
\qquad (q = 3) \qquad (6.370)
$$

and

$$
G(\Delta^* + \lambda; \xi \cdot \eta) + \frac{1}{q-3} \frac{2^{\frac{3-q}{2}}}{\|\mathbb{S}^{q-2}\|} (1 - \xi \cdot \eta)^{\frac{3-q}{2}} \tag{6.371}
$$
$$
= \begin{cases} O\left(\ln(1 - \xi \cdot \eta)\right), & q = 5 \\ O\left((1 - \xi \cdot \eta)^{\frac{5-q}{2}}\right) & q = 4, q \geq 6 \end{cases}
$$

$$
\nabla_\eta^* G(\Delta^* + \lambda; \xi \cdot \eta) + \frac{1}{q-3} \frac{2^{\frac{3-q}{2}}}{\|\mathbb{S}^{q-2}\|} \nabla_\eta^*(1 - \xi \cdot \eta)^{\frac{3-q}{2}} \tag{6.372}
$$
$$
= \begin{cases} O\left(\nabla_\eta^* \ln(1 - \xi \cdot \eta)\right), & q = 5 \\ O\left(\nabla_\eta^*(1 - \xi \cdot \eta)^{\frac{5-q}{2}}\right) & q = 4, q \geq 6 \end{cases}
$$

are valid.

(iii) For all orthogonal transformations **t**

$$
G(\Delta^* + \lambda; \xi \cdot \eta) = G(\Delta^* + \lambda, \mathbf{t}\xi \cdot \mathbf{t}\eta). \tag{6.373}
$$

(iv) For all n with $(\Delta^* + \lambda)^\wedge(n) = 0$ and $j = 1, \ldots, N(q; n)$,

$$
\int_{\mathbb{S}^{q-1}} G(\Delta^* + \lambda; \xi \cdot \eta) \, \overline{Y_{n,j}(q; \eta)} \, dS_{(q-1)}(\eta) = 0. \tag{6.374}
$$

Remark 6.13. *It should be noted that, for all $Y_n \in \mathrm{Harm}_n$,*

$$
(\Delta^* + \lambda + (\Delta^* + \lambda)^\wedge(n)) \, Y_n = 0 \tag{6.375}
$$

such that

$$
(\Delta^* + \lambda)^\wedge(n) = -(\lambda - n(n + q - 2) = (\Delta^*)^\wedge(n) - \lambda. \tag{6.376}
$$

Therefore, for an operator $\Delta^* + \lambda$, the symbol $\sum_{(\Delta^*+\lambda)^\wedge(n)=0}$ means that the sum is to be taken over all non-negative integers $n \in \mathbb{N}_0$ for which $(\Delta^*)^\wedge(n) = n(n+q-2) = \lambda$. In the case of $(\Delta^* + \lambda)^\wedge(n) \neq 0$ for all non-negative integers $n \in \mathbb{N}_0$, the sum is assumed to be zero.

By the defining properties stated in Definition 6.5 the $\mathbb{S}^{q-1}$-sphere function for $\Delta^* + \lambda$, $\lambda \in \mathbb{R}$, is uniquely determined (cf. W. Freeden [1979, 1980b]).

The spherical harmonics $Y_n(q; \cdot) \in \mathrm{Harm}_n$ are eigenvalues in the sense of the integral equation

$$-\Delta^{*\wedge}(n) \int_{\mathbb{S}^{q-1}} G(\Delta^*; \xi \cdot \zeta) \, \overline{Y_n(q; \zeta)} \, dS_{(q-1)}(\zeta) = (1 - \delta_{0n}) \, Y_n(q; \xi). \quad (6.377)$$

Following Hilbert's approach to the theory of Green's functions (see D. Hilbert [1912]) we are able to prove the existence of $G(\Delta^* + \lambda; \xi \cdot \eta)$ by first giving an explicit representation of the $\mathbb{S}^{q-1}$-sphere function for Δ^*, i.e., $\lambda = 0$ (cf. W. Freeden [1979], R. Reuter [1982]).

In order to guarantee the existence of $G(\Delta^* + \lambda; \xi \cdot \eta)$ for the case of a non-eigenvalue λ, i.e., $(\Delta^* + \lambda)^\wedge(n) \neq 0$ for all non-negative integers n, we consider the integral equation

$$G(\Delta^* + \lambda; \xi \cdot \eta) = G(\Delta^*; \xi \cdot \eta) \quad (6.378)$$
$$- \lambda \int_{\mathbb{S}^{q-1}} G(\Delta^* + \lambda; \xi \cdot \zeta) G(\Delta^*; \eta \cdot \zeta) \, dS_{(q-1)}(\zeta) \; - \; \frac{1}{\lambda \|\mathbb{S}^{q-1}\|},$$

which establishes the close relation between the $\mathbb{S}^{q-1}$-sphere function and the resolvent of the kernel $G(\Delta^*; \cdot)$. The value λ is an eigenvalue of the kernel $G(\Delta^*; \cdot)$ if and only if λ is an eigenvalue with respect to the Beltrami operator Δ^*. Thus the existence of $G(\Delta^*; \cdot)$ follows from standard arguments of the Fredholm theory of integral equations. In the eigenvalue case of $(\Delta + \lambda)^\wedge(n) = 0$ for an integer $n > 0$ we discuss the integral equation

$$G(\Delta^* + \lambda; \xi \cdot \eta) = G(\Delta^*; \xi \cdot \eta) \quad (6.379)$$
$$- \lambda \int_{\mathbb{S}^{q-1}} G(\Delta^* + \lambda; \xi \cdot \zeta) G(\Delta^*; \eta \cdot \zeta) \, dS_{(q-1)}(\zeta)$$
$$- \frac{1}{\lambda \|\mathbb{S}^{q-1}\|} + \frac{1}{-(\Delta^*)^\wedge(n)} \sum_{j=1}^{N(q;n)} Y_{n,j}(q; \xi) \overline{Y_{n,j}(q, \eta)}.$$

Hence it is not difficult to see that in the case $(\Delta + \lambda)^\wedge(n) = 0$ for an integer $n > 0$

$$\int_{\mathbb{S}^{q-1}} G(\Delta^* + \lambda; \xi \cdot \eta) \, \overline{Y_n(q; \eta)} \, dS_{(q-1)}(\eta) \quad (6.380)$$
$$= \int_{\mathbb{S}^{q-1}} \left(-\frac{1}{\lambda \|\mathbb{S}^{q-1}\|} + \frac{1}{-(\Delta^*)^\wedge(n)} \sum_{j=1}^{N(q;n)} \overline{Y_{n,j}(q; \xi)} Y_{n,j}(q; \eta) \right) \overline{Y_n(q; \eta)} \, dS_{(q-1)}(\eta)$$

for all spherical harmonics $Y_n(q; \cdot)$ of degree $n > 0$. Therefore, the integral equation (6.379) has a solution which is uniquely determined by the conditions

$$\int_{\mathbb{S}^{q-1}} G(\Delta^* + \lambda; \xi \cdot \eta) \, \overline{Y_{n,j}(q; \eta)} \, dS_{(q-1)}(\eta) = 0 \quad (6.381)$$

for $j = 1, \ldots, N(q; n)$.

Finally the analogue to the *Poisson differential equation* in potential theory should be mentioned briefly.

Lemma 6.19. *Let F be a bounded function on $\mathbb{S}^{q-1}$ satisfying a Lipschitz-condition in the neighborhood of the point $\xi \in \mathbb{S}^{q-1}$. Then*

$$U(\xi) = \int_{\mathbb{S}^{q-1}} G(\Delta^* + \lambda; \; \xi \cdot \zeta) \, F(\zeta) \, dS_{(q-1)}(\zeta) \tag{6.382}$$

is twice continuously differentiable at $\xi \in \mathbb{S}^{q-1}$ with

$$(\Delta_\xi^* + \lambda) \, U(\xi) = F(\xi) \tag{6.383}$$

$$- \sum_{\substack{(\Delta^* + \lambda)^\wedge(n) = 0 \\ n \in \mathbb{N}_0}} \sum_{j=1}^{N(q;n)} Y_{n,j}(q; \xi) \int_{\mathbb{S}^{q-1}} F(\zeta) \, \overline{Y_{n,j}(q; \zeta)} \, dS_{(q-1)}(\zeta).$$

Indeed, the proof can be given by quite similar conclusions as known from potential theory. The details are omitted here.

Integral Formulas for the Helmholtz–Beltrami Operator

Now, our purpose is to formulate a counterpart to the Third Surface Green Theorem on the unit sphere $\mathbb{S}^{q-1}$ for the Beltrami operator Δ^* (see W. Freeden [1979, 1981], R. Reuter [1982]) and to derive some extensions to the operator $\Delta^* + \lambda$, $\lambda \in \mathbb{R}$. Suppose that F is a twice continuously differentiable function on $\mathbb{S}^{q-1}$, i.e., $F \in C^{(2)}(\mathbb{S}^{q-1})$. Then, for each sufficiently small $\varepsilon > 0$ and for each number $\lambda \in \mathbb{R}$, the Second Green Surface Theorem gives

$$\int_{\substack{\sqrt{1-\xi\cdot\eta} \geq \varepsilon \\ \eta \in \mathbb{S}^{q-1}}} \left\{ G(\Delta^* + \lambda; \xi \cdot \eta)(\Delta_\eta^* + \lambda) F(\eta) \right. \tag{6.384}$$

$$\left. - F(\eta)(\Delta_\eta^* + \lambda) G(\Delta^* + \lambda; \xi \cdot \eta) \right\} \, dS_{(q-1)}(\eta)$$

$$= \int_{\substack{\sqrt{1-\xi\cdot\eta} = \varepsilon \\ \eta \in \mathbb{S}^{q-1}}} \left\{ G(\Delta^* + \lambda; \xi \cdot \eta) \frac{\partial}{\partial \nu_\eta} F(\eta) \right.$$

$$\left. - F(\eta) \, \frac{\partial}{\partial \nu_\eta} G(\Delta^* + \lambda; \xi \cdot \eta) \right\} \, dS_{(q-2)}(\eta),$$

where $dS_{(q-2)}$ denotes the surface element in $\mathbb{R}^{q-1}$, while ν is the (unit) vector normal to $\{\eta \in \mathbb{S}^{q-1} \mid \sqrt{1-\xi\cdot\eta} = \varepsilon\}$ and tangential on $\mathbb{S}^{q-1}$ and directed into the exterior of $\{\eta \in \mathbb{S}^{q-1} \mid \sqrt{1-\xi\cdot\eta} \geq \varepsilon\}$. Inserting the differential equations of the $\mathbb{S}^{q-1}$-sphere function we obtain

$$\int_{\substack{\sqrt{1-\xi\cdot\eta} \geq \varepsilon \\ \eta \in \mathbb{S}^{q-1}}} F(\eta) \, \left((\Delta_\eta^* + \lambda) G(\Delta^* + \lambda; \xi \cdot \eta)\right) \, dS_{(q-1)}(\eta) \tag{6.385}$$

$$= - \sum_{\substack{(\Delta^* + \lambda)^\wedge(n) = 0 \\ n \in \mathbb{N}_0}} \sum_{j=1}^{N(q;n)} \int_{\substack{\sqrt{1-\xi\cdot\eta} \geq \varepsilon \\ \eta \in \mathbb{S}^{q-1}}} F(\eta) \, \overline{Y_{n,j}(q; \eta)} \, dS_{(q-1)}(\eta) \, Y_{n,j}(q; \xi).$$

Observing the characteristic singularity of the $\mathbb{S}^{q-1}$-sphere function we are able to prove by analogous conclusions as known in potential theory

$$\int_{\substack{\sqrt{1-\xi\cdot\eta}=\varepsilon \\ \eta\in\mathbb{S}^{q-1}}} G(\Delta^* + \lambda; \xi\cdot\eta)\frac{\partial}{\partial\nu_\eta}F(\eta)\, dS_{(q-2)}(\eta) \;=\; o(1), \quad \varepsilon\to 0, \qquad (6.386)$$

and

$$\int_{\substack{\sqrt{1-\xi\cdot\eta}=\varepsilon \\ \eta\in\mathbb{S}^{q-1}}} F(\eta)\frac{\partial}{\partial\nu_\eta}G(\Delta^* + \lambda; \xi\cdot\eta)\, dS_{(q-2)}(\eta) = -F(\xi) + o(1), \;\; \varepsilon\to 0.$$

$$(6.387)$$

Summarizing our results we therefore get the following analogue to the third Green theorem for the unit sphere $\mathbb{S}^{q-1}$.

Theorem 6.24. *(Integral Formula for the Operator $\Delta^* + \lambda$). If $\lambda \in \mathbb{R}$, $\xi \in \mathbb{S}^{q-1}$, and $F \in C^{(2)}(\mathbb{S}^{q-1})$, then*

$$F(\xi) \;=\; \sum_{\substack{(\Delta^*+\lambda)^\wedge(n)=0 \\ n\in\mathbb{N}_0}} \sum_{j=1}^{N(q;n)} \int_{\mathbb{S}^{q-1}} F(\eta)\,\overline{Y_{n,j}(q;\eta)}\, dS_{(q-1)}(\eta)\, Y_{n,j}(q;\xi)$$

$$+ \int_{\mathbb{S}^{q-1}} G(\Delta^* + \lambda; \xi\cdot\eta)\big((\Delta^*_\eta + \lambda)F(\eta)\big)\, dS_{(q-1)}(\eta). \quad (6.388)$$

The integral formula (Theorem 6.24) may serve as the point of departure for purposes of numerical integration on the sphere (see, e.g., W. Freeden [1979, 1980b, 1981]). It provides the explicit knowledge of a remainder term involving the Helmholtz–Beltrami operator $\Delta^* + \lambda$.

Finally it should be mentioned that Theorem 6.24 can be formulated for iterated operators $(\Delta^* + \lambda)^m$, $\lambda \in \mathbb{R}$. For that purpose we introduce

Definition 6.6. *Let $G\big((\Delta^* + \lambda)^m; \xi\cdot\eta\big)$, $m = 1, 2, \ldots$, be defined by the convolution*

$$G\big((\Delta^* + \lambda)^m; \xi\cdot\eta\big) \tag{6.389}$$

$$= \int_{\mathbb{S}^{q-1}} G\big((\Delta^* + \lambda)^{m-1}; \xi\cdot\zeta\big)G(\Delta^* + \lambda; \zeta\cdot\eta)\, dS_{(q-1)}(\zeta),$$

$$m = 2, 3, \ldots$$

$$G\big((\Delta^* + \lambda)^m; \xi\cdot\eta\big) = G(\Delta^* + \lambda; \xi\cdot\eta), \qquad m = 1. \quad (6.390)$$

Then, $G\big((\Delta^* + \lambda)^m; \cdot, \cdot\big)$ is called the $\mathbb{S}^{q-1}$-sphere function for $(\Delta^* + \lambda)^m$, $\lambda \in \mathbb{R}$.

In analogy to techniques known in potential theory it can be proved that

$$G\big((\Delta^* + \lambda)^m, \xi\cdot\eta\big) = \begin{cases} O\left((1 - \xi\cdot\eta)^{m-\frac{q-1}{2}}\ln(1 - \xi\cdot\eta)\right), \\ \hspace{3cm} 2m + 2 \leq q - 1, \quad q \text{ odd} \\ O\left((1 - \xi\cdot\eta)^{m-\frac{q-1}{2}}\right), \hspace{2cm} \text{otherwise.} \end{cases}$$

Hence, if $m > \frac{q-1}{2}$, $G\big((\Delta^* + \lambda)^m, \xi \cdot \big)$ is continuous on the whole sphere $\mathbb{S}^{q-1}$. Furthermore, for $m > \frac{q-1}{2}$, the bilinear expansion

$$\sum_{\substack{(\Delta^*+\lambda)^\wedge(n)\neq 0 \\ n\in\mathbb{N}_0}} \frac{1}{-((\Delta^*+\lambda)^m)^\wedge(n)} \sum_{j=1}^{N(q;n)} Y_{n,j}(q;\xi)\, \overline{Y_{n,j}(q;\eta)} \tag{6.391}$$

is absolutely and uniformly convergent both in ξ and η and uniformly in ξ and η together, where $((\Delta^* + \lambda)^m)^\wedge(n) = -(\lambda - n(n + q - 2))^m$. Therefore, in connection with the addition theorem for spherical harmonics, we have the following result.

Lemma 6.20. *If $m > \frac{q-1}{2}$, then*

$$G\big((\Delta^* + \lambda)^m; \xi \cdot \eta\big) = \sum_{\substack{(\Delta^*+\lambda)^\wedge(n)\neq 0 \\ n\in\mathbb{N}_0}} \frac{1}{-((\Delta^* + \lambda)^m)^\wedge(n)} \frac{N(q;n)}{\|\mathbb{S}^{q-1}\|} P_n(q;\xi \cdot \eta).$$

$$\tag{6.392}$$

Observing the differential equation

$$(\Delta^* + \lambda)\, G\big((\Delta^* + \lambda)^m; \xi \cdot \eta\big) = G\big((\Delta^* + \lambda)^{m-1}, \xi \cdot \eta\big), \tag{6.393}$$

$-1 \leq \xi \cdot \eta < 1$, we obtain by successive integration by parts the following extension of Theorem 6.24.

Theorem 6.25. *(Integral Formula for the Operator $(\Delta^* + \lambda)^m$) Let $\xi \in \mathbb{S}^{q-1}$ and $F \in C^{(2m)}(\mathbb{S}^{q-1})$. Then*

$$
\begin{aligned}
F(\xi) \;=\;& \sum_{\substack{(\Delta^*+\lambda)^\wedge(n)=0 \\ n\in\mathbb{N}_0}} \sum_{j=1}^{N(q;n)} \int_{\mathbb{S}^{q-1}} F(\eta)\, \overline{Y_{n,j}(q;\eta)}\, dS_{(q-1)}(\eta)\, Y_{n,j}(q;\xi) \\
&+ \int_{\mathbb{S}^{q-1}} G\big((\Delta^* + \lambda)^m; \xi \cdot \eta\big)\, \big((\Delta^*_\eta + \lambda)^m F(\eta)\big)\, dS_{(q-1)}(\eta).
\end{aligned}
$$

$$\tag{6.394}$$

An immediate consequence of Theorem 6.25 is the following corollary.

Corollary 6.6. *Under the assumptions of Theorem 6.25 we have*

$$
\begin{aligned}
F(\xi) \;=\;& \sum_{\substack{(\Delta^*+\lambda)^\wedge(n)=0 \\ n\in\mathbb{N}_0}} \frac{N(q;n)}{\|\mathbb{S}^{q-1}\|} \int_{\mathbb{S}^{q-1}} F(\eta)\, P_n(q;\xi \cdot \eta)\, dS_{(q-1)}(\eta) \\
&+ \int_{\mathbb{S}^{q-1}} (\Delta^*_\eta + \lambda)^m G\big((\Delta^* + \lambda)^{2m}; \xi \cdot \eta\big) \big((\Delta^*_\eta + \lambda)^m F(\eta)\big)\, dS_{(q-1)}(\eta).
\end{aligned}
$$

Helmholtz–Beltrami Differential Equation

Theorem 6.25 will be used now to discuss the differential equation

$$(\Delta^* + \lambda)^m V = W, \quad V \in C^{(2m)}(\mathbb{S}^{q-1}). \tag{6.395}$$

From the Extended Second Green Theorem (6.67) it is obvious that

$$\int_{\mathbb{S}^{q-1}} ((\Delta^* + \lambda)^m V(\eta))\ Y(\eta)\ dS_{(q-1)}(\eta) \tag{6.396}$$

$$= \int_{\mathbb{S}^{q-1}} V(\eta)\ ((\Delta^* + \lambda)^m Y(\eta))\ dS_{(q-2)}(\eta)$$

$$= 0$$

holds for all elements Y belonging to the null space (kernel) of the operator $\Delta^* + \lambda$. Clearly, any function Y of the null space of $\Delta^* + \lambda$ can be added to V without changing the differential equation. However, if we require that V is orthogonal to the null space of $\Delta^* + \lambda$, then the differential equation is uniquely solvable.

Theorem 6.26. *Let W be a function of class $C^{(0)}(\mathbb{S}^{q-1})$ orthogonal to the null space of the operator $\Delta^* + \lambda$. Then the function V given by*

$$V(\xi) = \int_{\mathbb{S}^{q-1}} G((\Delta^* + \lambda)^m; \xi \cdot \eta)\ W(\eta)\ dS_{(q-1)}(\eta) \tag{6.397}$$

represents the only $(2m)$-times continuously differentiable solution of the differential equation

$$(\Delta^* + \lambda)^m V = W \tag{6.398}$$

on the sphere $\mathbb{S}^{q-1}$, which is orthogonal to the null space of $\Delta^* + \lambda$.

Spherical Harmonics as Eigenfunctions

The integral formula (Theorem 6.24) enables us to justify that the spherical harmonics are the only everywhere on the unit sphere $\mathbb{S}^{q-1}$ twice continuously differentiable eigenfunctions of the Beltrami differential operator Δ^*.

Lemma 6.21. *Let K be of class $C^{(2)}(\mathbb{S}^{q-1})$ satisfying*

$$(\Delta^*_\xi + \lambda)K(\xi) = 0, \quad \xi \in \mathbb{S}^{q-1}. \tag{6.399}$$

(i) If $\lambda \notin \operatorname{Spect}_{\Delta^*}(\mathbb{S}^{q-1})$, i.e., $\lambda \neq (\Delta^*)^\wedge(n)$, for all $n = 0, 1, \ldots$, then $K = 0$.

(ii) If $\lambda \in \operatorname{Spect}_{\Delta^*}(\mathbb{S}^{q-1})$, i.e., $\lambda = (\Delta^*)^\wedge(n)$, $n \in \mathbb{N}_0$, then K is a member of class $\operatorname{Harm}_n(\mathbb{S}^{q-1})$.

Summarizing our results about harmonics we are therefore led to the following conclusions:

- The functions $x \mapsto H_n(q;x) = r^n Y_n(q;\xi)$, $x \in \mathbb{R}^q$, are polynomials in Cartesian coordinates which satisfy the Laplace equation $\Delta H_n(q;\cdot) = 0$ in $\mathbb{R}^q$ and are homogeneous of degree n. Conversely, every homogeneous harmonic polynomial of degree n when restricted to the unit sphere $\mathbb{S}^{q-1}$ is a spherical harmonic of degree n and dimension q.

- The Legendre polynomial $P_n(q;\cdot)$ is the only everywhere on the interval $[-1,1]$ twice continuously differentiable eigenfunction of the Legendre (differential) equation

$$\left(\underbrace{(1-t)^2 \left(\frac{d}{dt}\right)^2 - (q-1)t\frac{d}{dt}}_{=L_t} + L^{\wedge}(n) \right) P_n(q;t) = 0, \ t \in [-1,1],$$

(6.400)

$n = 0, 1, \ldots$, which in $t = 1$ satisfy $P_n(q;1) = 1$ (note that $L^{\wedge}(n) = n(n+q-2)$, $n = 0,1,\ldots$).

- The spherical harmonics $Y_n(q;\cdot)$ of degree n and dimension q are the everywhere on the unit sphere $\mathbb{S}^{q-1}$ twice continuously differentiable eigenfunctions of the Beltrami (differential) equation

$$(\Delta^* + (\Delta^*)^{\wedge}(n)) \, Y_n(q;\xi) = 0 \tag{6.401}$$

corresponding to the eigenvalues

$$(\Delta^*)^{\wedge}(n) \, (= (\Delta^*)^{\wedge}_{\mathbb{S}^{q-1}}(n)) = n(n+q-2), \tag{6.402}$$

$n = 0, 1, \ldots$.

Lattice Point Generated Spherical Equidistribution

Next we briefly explain the essential features of equidistribution. Our purpose is not to derive an extended theory of the equidistribution on the sphere (for more details the reader is referred to E. Hlawka [1981, 1984], who based his considerations on the integral formulas of the thesis W. Freeden [1979]). In this approach we restrict ourselves to the problem of generating an equidistributed point system, that is obtained by projection of lattice points of $\mathbb{Z}^q \backslash \{0\}$ to $\mathbb{S}^{q-1}$, $q \geq 3$: to be more accurate, for $n \in \mathbb{N}$, let us consider the integer solutions $g = (n_1, \ldots, n_q)^T \in \mathbb{Z}^q$ with $n_i = g \cdot \epsilon^i$, $i = 1, \ldots, q$, of the equation $|g|^2 = n_1^2 + \cdots + n_q^2 = n$, i.e., the number $r_q(n)$ of representations of $n \in \mathbb{N}$ as sum of q squares. It is already known from the Fermat–Euler Theorem that $r_q(n) > 0$ provided that $q \geq 4$. For the dimension $q = 3$, the situation is more difficult (cf. Section 5.3). Anyway, for all dimensions $q \geq 3$, there exists a sequence $\{n_j\}_{j=1,2,\ldots} \subset \mathbb{N}$ such that $N_j = r_q(n_j) > 0$ and $\lim_{j \to \infty} n_j = \infty$.

Keeping this sequence $\{n_j\}_{j=1,2,\ldots} \subset \mathbb{N}$ in mind we let $X_{N_j} = \left\{x_1^{N_j}, \ldots, x_{N_j}^{N_j}\right\}$ be the set of points of $\mathbb{Z}^q$ on the sphere $\mathbb{S}_{\sqrt{n_j}}^{q-1}$ around the origin 0 with radius $\sqrt{n_j}$, i.e.,

$$X_{N_j} = \mathbb{Z}^q \cap \mathbb{S}_{\sqrt{n_j}}^{q-1}. \tag{6.403}$$

Furthermore, we understand by $H_{N_j} \subset \mathbb{S}^{q-1}$ the set

$$H_{N_j} = \left\{\eta_1^{N_j}, \ldots, \eta_{N_j}^{N_j}\right\}, \tag{6.404}$$

where

$$\eta_i^{N_j} = \frac{1}{\sqrt{n_j}} x_i^{N_j}, \tag{6.405}$$

$j = 1, \ldots, N_j$. In other words, H_{N_j} is the projection of the point set X_{N_j} to the unit sphere $\mathbb{S}^{q-1} \subset \mathbb{R}^q$. From the Third Green Surface Theorem on $\mathbb{S}^{q-1}$ (cf. Theorem 6.24) and the property $1 = (N_j)^{-1} \sum_{\eta \in H_{N_j}} 1$, it follows that

$$\frac{1}{\|\mathbb{S}^{q-1}\|} \int_{\mathbb{S}^{q-1}} F(\eta) \, dS_{(q-1)}(\eta) = \frac{1}{N_j} \sum_{\eta \in H_{N_j}} F(\eta) \tag{6.406}$$

$$- \frac{1}{N_j} \sum_{\eta \in H_{N_j}} \int_{\mathbb{S}^{q-1}} G((\Delta^*)^m; \eta \cdot \zeta) \, (\Delta_\zeta^*)^m F(\zeta) \, dS_{(q-1)}(\zeta)$$

holds true for $j = 1, 2, \ldots$, provided that F is a function of class $C^{(2m)}(\mathbb{S}^{q-1})$, $m \in \mathbb{N}$. The identity (6.406) can be understood as a *spherical counterpart of the Hlawka–Koksma formula* (for the classical approach in Euclidean spaces $\mathbb{R}^q$ see, e.g., L. Kuipers, H. Niederreiter [1974], I.H. Sloan, S. Joe [1994] and the references therein) formulated for the "projected" lattice point set (6.404) to $\mathbb{S}^{q-1}$.

We are interested in the difference $R_{N_j}(F)$ of the integral mean

$$I(F) = \frac{1}{\|\mathbb{S}^{q-1}\|} \int_{\mathbb{S}^{q-1}} F(\zeta) \, dS_{(q-1)}(\zeta) \tag{6.407}$$

and the "approximating sum"

$$L_{N_j}(F) = \frac{1}{N_j} \sum_{\eta \in H_{N_j}} F(\eta); \tag{6.408}$$

i.e., the *"remainder term"* $R_{N_j}(F)$ is given by

$$R_{N_j}(F) = I(F) - L_{N_j}(F) \tag{6.409}$$

within the reference space $C^{(2m)}(\mathbb{S}^{q-1})$ (note that $R_{N_j}(1) = 0$, more generally, $R_{N_j}(Y_{0,1}(q;\cdot)) = 0$ for $Y_{0,1}(q;\cdot) \in \text{Harm}_0(\mathbb{S}^{q-1})$). In this notational framework, the *Hlawka–Koksma formula* (6.406) can be rewritten as follows

Lemma 6.22. *For $F \in C^{(2m)}(\mathbb{S}^{q-1})$, $m \in \mathbb{N}$,*

$$R_{N_j}(F) = \int_{\mathbb{S}^{q-1}} R_{N_j} G((\Delta^*)^m; \cdot \zeta) \; (\Delta_\zeta^*)^m F(\zeta) \; dS_{(q-1)}(\zeta). \qquad (6.410)$$

The error term (6.409) can be estimated from above in different norms:

For $m > \frac{q-1}{4}$ we obtain from the Cauchy–Schwarz inequality

$$|R_{N_j}(F)| \le \left(D_{\mathrm{L}^{(2)}(\mathbb{S}^{q-1})}(H_{N_j})\right)^{\frac{1}{2}} \left(\int_{\mathbb{S}^{q-1}} |(\Delta_\zeta^*)^m F(\zeta)|^2 \; dS_{(q-1)}(\zeta)\right)^{\frac{1}{2}},$$
$$(6.411)$$

where the expression $D_{\mathrm{L}^{(2)}(\mathbb{S}^{q-1})}(H_{N_j})$ given by

$$D_{\mathrm{L}^{(2)}(\mathbb{S}^{q-1})}(H_{N_j}) \qquad (6.412)$$

$$= \quad L_{N_j} L_{N_j} G((\Delta^*)^{2m}; \cdot)$$

$$= \quad \left(\frac{1}{N_j}\right)^2 \sum_{\eta \in H_{N_j}} \sum_{\delta \in H_{N_j}} G((\Delta^*)^{2m}; \eta \cdot \delta)$$

$$= \quad \left(\frac{1}{N_j}\right)^2 \sum_{\eta \in H_{N_j}} \sum_{\delta \in H_{N_j}} \int_{\mathbb{S}^{q-1}} G((\Delta^*)^m; \eta \cdot \zeta) \; G((\Delta^*)^m; \delta \cdot \zeta) \; dS_{(q-1)}(\zeta)$$

is called the $\mathrm{L}^{(2)}\left(\mathbb{S}^{q-1}\right)$-*discrepancy of* H_{N_j} (*of order* m, $m > \frac{q-1}{4}$).

Remark 6.14. *For the particularly important case* $q = 3$, *the* $\mathbb{S}^2$-*sphere function* $G((\Delta^*)^2; \cdot)$ *is explicitly available in terms of elementary functions (cf. W. Freeden, M. Schreiner [2009]), viz.*

$$G((\Delta^*)^2; \xi \cdot \eta) = \sum_{n=1}^{\infty} \frac{2n+1}{4\pi} \frac{1}{(-n(n+1))^2} \; P_n(\xi \cdot \eta) \qquad (6.413)$$

$$= \begin{cases} \frac{1}{4\pi} & , \quad 1 - \xi \cdot \eta = 0 \\[2mm] \begin{aligned} &-\frac{1}{4\pi} \ln(1 - \xi \cdot \eta) \ln(1 + \xi \cdot \eta) \\ &+\frac{\ln 2}{4\pi} \ln\left(1 - (\xi \cdot \eta)^2\right) - \frac{1}{4\pi}\mathcal{L}_2\left(\frac{1-\xi \cdot \eta}{2}\right) \\ &+\frac{1}{4\pi}\left(1 - (\ln 2)^2\right) \end{aligned} & , \quad 1 \pm \xi \cdot \eta \ne 0 \\[2mm] \frac{1}{4\pi} - \frac{\pi}{24} & , \quad 1 + \xi \cdot \eta = 0. \end{cases}$$

where the "dilogarithm" is given by

$$\mathcal{L}_2\left(\frac{1-\xi \cdot \eta}{2}\right) = \sum_{k=1}^{\infty} \left(\frac{1-\xi \cdot \eta}{2}\right)^k \frac{1}{k^2}. \qquad (6.414)$$

For $m > \frac{q-1}{2}$, the function $\eta \mapsto G((\Delta^*)^m; \xi \cdot \eta)$, $\eta \in \mathbb{S}^{q-1}$, is continuous, and its bilinear expansion

$$G((\Delta^*)^m; \xi \cdot \eta) = \sum_{n=1}^{\infty} \frac{N(q;n)}{\|\mathbb{S}^{q-1}\|} \frac{1}{(-n(n+q-2))^m} P_n(q; \xi \cdot \eta) \qquad (6.415)$$

is absolutely and uniformly convergent on $\mathbb{S}^{q-1}$.

For $m > \frac{q-1}{2}$ we are able to formulate an estimate of the error term (6.409) in the $C^{(0)}$-norm (see, e.g., E. Hlawka [1981]) as follows.

Lemma 6.23. *Assume that F is of class $C^{(2m)}(\mathbb{S}^{q-1})$, $m > \frac{q-1}{2}$. Then*

$$\left| R_{N_j}(F) \right| \leq D_{C^{(0)}(\mathbb{S}^{q-1})}(H_{N_j}) \int_{\mathbb{S}^{q-1}} \left| (\Delta_\zeta^*)^m F(\zeta) \right| \, dS_{(q-1)}(\zeta). \qquad (6.416)$$

The expression

$$D_{C^{(0)}(\mathbb{S}^{q-1})}(H_{N_j}) = \sup_{\zeta \in \mathbb{S}^{q-1}} \left| \frac{1}{N_j} \sum_{\eta \in H_{N_j}} G((\Delta^*)^m; \eta \cdot \zeta) \right| \qquad (6.417)$$

is called the $C^{(0)}(\mathbb{S}^{q-1})$-*discrepancy of* H_{N_j} (of order m, $m > \frac{q-1}{2}$).

Obviously, the sequence $\{H_{N_j}\}_{j=1,2,\ldots}$ is equidistributed in the sense that $\lim_{j \to \infty} R_{N_j}(F) = 0$, if the discrepancy of H_{N_j} tends to 0.

More concretely, in the case of the $C^{(0)}(\mathbb{S}^{q-1})$-discrepancy of H_{N_j}, we are confronted with the following situation: for $q \geq 4$, C. Pommerenke [1959], A.V. Malyshev [1962] proved that

$$D_{C^{(0)}(\mathbb{S}^{q-1})}(H_{N_j}) \leq \frac{C_m(q)}{n_j^{\frac{q-1}{4}}}, \qquad m > q-1 \qquad (6.418)$$

and

$$D_{C^{(0)}(\mathbb{S}^{q-1})}(H_{N_j}) \leq \frac{C_m(q)}{n_j^{\frac{1}{4}(2m-\frac{q-1}{4})}}, \qquad \frac{q-1}{2} < m \leq q-1. \qquad (6.419)$$

The case $q = 3$ is known from R. F. Arenstorf, D. Johnson [1979]. They verified (under certain circumstances not specified here)

$$D_{C^{(0)}(\mathbb{S}^2)}(H_{N_j}) \leq \frac{C_m(2)}{\ln(\ln(n_j))}, \qquad m > 1. \qquad (6.420)$$

All proofs are rather technical.

Remark 6.15. *An overview on the activities in the theory of equidistribution by projection of lattice points of $\mathbb{Z}^q \backslash \{0\}$ from certain spheres $\mathbb{S}^{q-1}_{\sqrt{n_j}}$ to the (unit) sphere $\mathbb{S}^{q-1}$ is due to E. Hlawka [1981]. The note by E. Hlawka [1984] also delivers economical and efficient numerical perspectives for the application to partial differential equations, e.g. the wave equation and related equations, in Euclidean spaces $\mathbb{R}^q$.*

Next we adopt a famous criterion due to H. Weyl [1916] (see also W. Freeden et al. [1998] for the three-dimensional case); i.e., we consider the sequence $\{L_{N_j} Y_{n,r}(q; \cdot)\}_{\substack{n=1,2,\ldots \\ r=1,\ldots,N(q;n)}}$ of "*Weyl-sums*"

$$L_{N_j}(Y_{n,r}(q; \cdot)) \quad = \quad \frac{1}{N_j} \sum_{\zeta \in H_{N_j}} Y_{n,r}(q; \zeta), \qquad (6.421)$$

where $Y_{n,r}(q; \cdot) \in \mathrm{Harm}_n(\mathbb{S}^{q-1})$, $r = 1, \ldots, N(q,n)$.

Theorem 6.27. *The following statements are equivalent:*

(i) $\displaystyle \lim_{j \to \infty} \frac{1}{N_j} \sum_{\zeta \in H_{N_j}} Y_{n,r}(q; \zeta) = 0$

$\quad$ *for $n = 1, 2, \ldots$ and $r = 1, \ldots, N(q; n)$.*

(ii) $\displaystyle \lim_{j \to \infty} \frac{1}{N_j} \sum_{\zeta \in H_{N_j}} G((\Delta^*)^m; \xi \cdot \zeta) = 0$

$\quad$ *for all $\xi \in \mathbb{S}^{q-1}$, $m > \frac{q-1}{2}$.*

(iii) $\displaystyle \frac{1}{\|\mathbb{S}^{q-1}\|} \int_{\mathbb{S}^{q-1}} F(\zeta) \, dS_{(q-1)}(\zeta) = \lim_{j \to \infty} \frac{1}{N_j} \sum_{\zeta \in H_{N_j}} F(\zeta)$

$\quad$ *for all $F \in \mathrm{C}^{(2m)}(\mathbb{S}^{q-1})$, $m > \frac{q-1}{2}$.*

Proof. Assume that *(i)* is true. Then, for $m > \frac{q-1}{2}$, the Hlawka–Koksma formula (6.406) yields

$$\lim_{j \to \infty} \int_{\mathbb{S}^{q-1}} \frac{1}{N_j} \sum_{\zeta \in H_{N_j}} G((\Delta^*)^m; \xi \cdot \zeta) \, Y_{n,r}(q; \xi) \, dS_{(q-1)}(\xi) = 0 \qquad (6.422)$$

for $n = 1, 2, \ldots$ and $r = 1, \ldots, N(q; n)$. Hence, *(ii)* is valid. From Lemma 6.23 it follows that $R_{N_j}(F) \to 0$ as $j \to \infty$, i.e., *(iii)* is true. This finally implies $R_{N_j}(Y_{n,r}(q; \cdot)) = 0$ for $n = 1, 2, \ldots, r = 1, \ldots, N(q; n)$, i.e., *(i)* follows from *(iii)*. $\qquad \square$

Theorem 6.27 leads us to the impression that equidistribution can be formulated in all reference spaces for which the spherical harmonics show the property of closure. In our context we only prove the result within $\mathrm{C}^{(0)}(\mathbb{S}^{q-1})$.

Lemma 6.24. *The sequence* $\{H_{N_j}\}_{j=1,2,\ldots}$ *satisfies*

$$\frac{1}{\|\mathbb{S}^{q-1}\|} \int_{\mathbb{S}^{q-1}} F(\zeta)\, dS_{(q-1)}(\zeta) = \lim_{j \to \infty} \frac{1}{N_j} \sum_{\zeta \in H_{N_j}} F(\zeta) \qquad (6.423)$$

for all $F \in C^{(0)}(\mathbb{S}^{q-1})$.

Proof. From Theorem 6.27 we already know that (6.423) is valid for all functions of class $C^{(2m)}(\mathbb{S}^{q-1})$, $m > \frac{q-1}{2}$. Take now a continuous function F on $\mathbb{S}^{q-1}$. For $\varepsilon > 0$ arbitrary, the closure of the system of spherical harmonics of dimension q then tells us that there exists a function $P \in C^{(2m)}(\mathbb{S}^{q-1})$ such that

$$\sup_{\zeta \in \mathbb{S}^{q-1}} |F(\zeta) - P(\zeta)| \le \frac{\varepsilon}{3}. \qquad (6.424)$$

Therefore, it follows from the triangular inequality, that

$$\left| \frac{1}{\|\mathbb{S}^{q-1}\|} \int_{\mathbb{S}^{q-1}} F(\zeta)\, dS_{(q-1)}(\zeta) - \frac{1}{N_j} \sum_{\zeta \in H_{N_j}} F(\zeta) \right| \qquad (6.425)$$

$$\le \frac{1}{\|\mathbb{S}^{q-1}\|} \int_{\mathbb{S}^{q-1}} \left| F(\zeta) - P(\zeta) \right| dS_{(q-1)}(\zeta)$$

$$+ \left| \frac{1}{\|\mathbb{S}^{q-1}\|} \int_{\mathbb{S}^{q-1}} P(\zeta)\, dS_{(q-1)}(\zeta) - \frac{1}{N_j} \sum_{\zeta \in H_{N_j}} P(\zeta) \right|$$

$$+ \frac{1}{N_j} \sum_{\zeta \in H_{N_j}} |P(\zeta) - F(\zeta)|.$$

Hence, we have for all $F \in C^{(0)}(\mathbb{S}^{q-1})$

$$\left| \frac{1}{\|\mathbb{S}^{q-1}\|} \int_{\mathbb{S}^{q-1}} F(\zeta)\, dS_{(q-1)}(\zeta) - \frac{1}{N_j} \sum_{\zeta \in H_{N_j}} F(\zeta) \right| \le \frac{2}{3}\varepsilon + \left| R_{N_j}(P) \right|.$$

For j sufficiently large, we have $|R_{N_j}(P)| \le \frac{\varepsilon}{3}$, as required. $\qquad\square$

6.6 Radial and Angular Decomposition of Metaharmonics

This section deals with the decomposition of certain metaharmonic functions into radial and angular parts. In particular, we are interested in the (radial) cylinder functions of dimension q under the particular aspect of their applicability in our lattice point theory. Special emphasis is laid on asymptotic expansions of entire (integral) solutions of the Helmholtz equation.

Bessel Functions

By a simple coordinate transformation the equation $(\Delta+\lambda)\,U = 0$, $\lambda \in \mathbb{R}\backslash\{0\}$, can be reduced to $(\Delta+1)U = 0$ or $(\Delta-1)U = 0$, respectively. The best known solutions of these Helmholtz equations are $x \mapsto e^{ix\cdot\eta}$ or $x \mapsto e^{x\cdot\eta}$, $x \in \mathbb{R}^q$, $\eta \in \mathbb{S}^{q-1}$, respectively. Since the Helmholtz operators $\Delta\pm1$ are linear, more general solutions can be obtained by superposition. This is the basic idea of (regular) Bessel and Hankel functions (for more details the reader is referred to, e.g., G.N. Watson [1944], G. Herglotz and C. Müller [1952], A. Sommerfeld [1966], C. Müller [1998], and the references therein).

The point of departure for our work is the entire solution $U_n : \mathbb{R}^q \to \mathbb{C}$ of the Helmholtz equation $\Delta U_n + U_n = 0$ in $\mathbb{R}^q$ of the form

$$U_n(x) = \frac{i^{-n}}{\|\mathbb{S}^{q-1}\|} \int_{\mathbb{S}^{q-1}} e^{ix\cdot\eta}\, Y_n(q;\eta)\, dS_{(q-1)}(\eta), \qquad (6.426)$$

where $Y_n(q;\cdot)$ is a member of Harm_n. In terms of (standard) polar coordinates $x = r\xi, r = |x|, \xi \in \mathbb{S}^{q-1}$, the Funk–Hecke formula (cf. Theorem 6.14) yields the decomposition

$$U_n(r\xi) = i^{-n}\frac{\|\mathbb{S}^{q-2}\|}{\|\mathbb{S}^{q-1}\|} \int_{-1}^{1} e^{irt} P_n(q;t)(1-t^2)^{\frac{q-3}{2}}\, dt\; Y_n(q;\xi). \qquad (6.427)$$

In other words, a separation of the variables into a radial and an angular part is achieved in the form

$$U_n(x) = J_n(q;r)\, Y_n(q;\xi), \quad x = r\xi, \; \xi \in \mathbb{S}^{q-1}. \qquad (6.428)$$

Definition 6.7. *The function $J_n(q;\cdot)$ given by*

$$J_n(q;r) = i^{-n}\frac{\|\mathbb{S}^{q-2}\|}{\|\mathbb{S}^{q-1}\|} \int_{-1}^{1} e^{irt} P_n(q;t)(1-t^2)^{\frac{q-3}{2}}\, dt, \quad r \geq 0, \qquad (6.429)$$

is called the Bessel function of order n and dimension q (more accurately, regular Bessel function of order n and dimension q).

Applying the Rodrigues rule (i.e., Lemma 6.10) we get

$$J_n(q;r) = \frac{\Gamma\left(\frac{q}{2}\right)\left(\frac{r}{2}\right)^n}{\Gamma\left(n+\frac{q-1}{2}\right)\Gamma\left(\frac{1}{2}\right)} \int_{-1}^{1} e^{irt}(1-t^2)^{n+\frac{q-3}{2}}\, dt \qquad (6.430)$$

such that

$$J_n(q;r) = \frac{\Gamma\left(\frac{q}{2}\right)\left(\frac{r}{2}\right)^n}{\Gamma\left(n+\frac{q-1}{2}\right)\Gamma\left(\frac{1}{2}\right)} \int_{-1}^{1} (1-t^2)^{n+\frac{q-3}{2}}\cos(rt)\, dt. \qquad (6.431)$$

In connection with the well known results of the theory of the Gamma function we have

$$\left| \int_{-1}^{1} (1-t^2)^{n+\frac{q-3}{2}} \cos(rt) \, dt \right| \leq 2 \int_{0}^{1} (1-t^2)^{n+\frac{q-3}{2}} \, dt = \frac{\Gamma\left(\frac{1}{2}\right) \Gamma\left(n+\frac{q-1}{2}\right)}{\Gamma\left(n+\frac{q}{2}\right)}.$$

$$(6.432)$$

Lemma 6.25. *For fixed* $r \in [0, \infty)$ *and* $n \to \infty$,

$$|J_n(q;r)| \leq \frac{\Gamma\left(\frac{q}{2}\right)}{\Gamma\left(n+\frac{q}{2}\right)} \left(\frac{r}{2}\right)^n.$$

$$(6.433)$$

The Bessel functions can be seen as orthogonal (i.e., Fourier) coefficients of an orthogonal series expansion in terms of Legendre polynomials.

Lemma 6.26. *For* $x \in \mathbb{R}^q$, $x = r\xi$, $r = |x|$, $\xi \in \mathbb{S}^{q-1}$, *and* $\eta \in \mathbb{S}^{q-1}$ *we have*

$$e^{ix\cdot\eta} = e^{ir\xi\cdot\eta} = \sum_{n=0}^{\infty} i^n N(q;n) J_n(q;r) P_n(q;\xi \cdot \eta),$$

$$(6.434)$$

where, for each $r \in [0, \infty)$, *the series converges absolutely and uniformly with respect to* $\xi \cdot \eta$.

Proof. In connection with the Funk–Hecke formula of the theory of spherical harmonics we obtain the following decomposition into radial and angular components

$$\frac{N(q;n)}{\|\mathbb{S}^{q-1}\|} \int_{\mathbb{S}^{q-1}} e^{ir\xi\cdot\zeta} P_n(q;\eta \cdot \zeta) \, dS_{(q-1)}(\zeta) = i^n \, N(q;n) \, J_n(q;r) \, P_n(q;\xi \cdot \eta).$$

$$(6.435)$$

This explains the identity (6.434) of Lemma 6.26.   ☐

The series representation (Lemma 6.26) has an immediate consequence known as the *addition theorem of Bessel functions*. To this end we consider, for $x, y \in \mathbb{R}^q$, the integral

$$J_0(q; |x-y|) = \frac{1}{\|\mathbb{S}^{q-1}\|} \int_{\mathbb{S}^{q-1}} e^{ix\cdot\zeta} e^{-iy\cdot\zeta} \, dS_{(q-1)}(\zeta).$$

$$(6.436)$$

The Parseval identity for the (orthogonal system of) Legendre polynomials shows that

$$\sum_{n=0}^{\infty} (N(q;n))^2 J_n(q; |x|) J_n(q; |y|) \int_{\mathbb{S}^{q-1}} P_n(q;\xi \cdot \zeta) P_n(q;\eta \cdot \zeta) \, dS_{(q-1)}(\zeta)$$

$$= \|\mathbb{S}^{q-1}\| \sum_{n=0}^{\infty} N(q;n) J_n(q; |x|) J_n(q; |y|) P_n(q;\xi \cdot \eta).$$

$$(6.437)$$

This gives

Lemma 6.27. *(Addition Theorem of Bessel Functions) For $x, y \in \mathbb{R}^q$ with $x = |x|\xi, y = |y|\eta, \xi, \eta \in \mathbb{S}^{q-1}$, we have*

$$J_0(q; |x - y|) = \sum_{n=0}^{\infty} N(q; n) J_n(q; |x|) J_n(q; |y|) P_n(q; \xi \cdot \eta). \tag{6.438}$$

Observing the well known inequality for the cosine function

$$\sum_{k=0}^{2n+1} \frac{(-1)^k}{(2k)!} t^{2k} \le \cos(t) \le \sum_{k=0}^{2n} \frac{(-1)^k}{(2k)!} t^{2k}, \quad t \in \mathbb{R}, \tag{6.439}$$

and the identity known from the theory of the Gamma function

$$\int_{-1}^{+1} t^{2k} (1 - t^2)^{n + \frac{q-3}{2}} \, dt = \frac{\Gamma\left(k + \frac{1}{2}\right) \Gamma\left(n + \frac{q-1}{2}\right)}{\Gamma\left(n + k + \frac{q}{2}\right)} \tag{6.440}$$

we obtain via (6.431) the inequalities

$$J_n(q; r) \le \Gamma\left(\frac{q}{2}\right) \left(\frac{r}{2}\right)^n \sum_{k=0}^{2n} \frac{(-1)^k}{k!} \frac{\left(\frac{r}{2}\right)^{2k}}{\Gamma\left(n + k + \frac{q}{2}\right)} \tag{6.441}$$

and

$$J_n(q; r) \ge \Gamma\left(\frac{q}{2}\right) \left(\frac{r}{2}\right)^n \sum_{k=0}^{2n+1} \frac{(-1)^k}{k!} \frac{\left(\frac{r}{2}\right)^{2k}}{\Gamma\left(n + k + \frac{q}{2}\right)} \tag{6.442}$$

(note that we have used in both cases the formula $\sqrt{\pi}(2k)! = 2^{2k} \Gamma\left(k + \frac{1}{2}\right) k!$ known for the Gamma function).

This leads to the following series representation of the Bessel function.

Lemma 6.28. *For $r \in [0, \infty)$, the Bessel function $J_n(q; \cdot)$ permits the representation by the power series*

$$J_n(q; r) = \Gamma\left(\frac{q}{2}\right) \left(\frac{r}{2}\right)^n \sum_{k=0}^{\infty} \frac{(-1)^k}{k!} \frac{\left(\frac{r}{2}\right)^{2k}}{\Gamma\left(n + k + \frac{q}{2}\right)}. \tag{6.443}$$

Furthermore, an error estimate between the Bessel function and its truncated power series is given by

$$\left| J_n(q; r) - \Gamma\left(\frac{q}{2}\right) \left(\frac{r}{2}\right)^n \sum_{k=0}^{m} \frac{(-1)^k}{k!} \frac{\left(\frac{r}{2}\right)^{2k}}{\Gamma\left(n + k + \frac{q}{2}\right)} \right| \tag{6.444}$$

$$\le \Gamma\left(\frac{q}{2}\right) \left(\frac{r}{2}\right)^n \frac{1}{(m+1)!} \frac{\left(\frac{r}{2}\right)^{2m+2}}{\Gamma\left(n + m + 1 + \frac{q}{2}\right)}.$$

Since the spherical harmonics of degree n and dimension q are the eigenfunctions of the Beltrami operator corresponding to the eigenvalues

$$(\Delta^*)^\wedge(n) = n(n + q - 2), \quad n = 0, 1, \ldots, \tag{6.445}$$

i.e., $(\Delta^* + (\Delta^*)^\wedge(n))Y_n(q;\cdot) = 0$, $Y_n(q;\cdot) \in \mathrm{Harm}_n$, we get from the Helmholtz equation

$$\left(\underbrace{\frac{1}{r^{q-1}}\frac{\partial}{\partial r}r^{q-1}\frac{\partial}{\partial r} + \frac{1}{r^2}\Delta^*_{(q)}}_{=\Delta_{(q)}} + 1\right) J_n(q;r)Y_n(q;\xi) = 0 \tag{6.446}$$

the following differential equation for the Bessel functions

$$J''_n(q;r) + \frac{q-1}{r}J'_n(q;r) + \left(1 - \frac{n(n+q-2)}{r^2}\right)J_n(q;r) = 0. \tag{6.447}$$

In connection with the recursion relation for the Legendre polynomial

$$(n+q-2)P_{n+1}(q;t) = (2n+q-2)tP_n(q;t) - nP_{n-1}(q;t) \tag{6.448}$$

we are able to deduce the recursion relation

$$(n+q-2)J_{n+1}(q;r) + (2n+q-2)J'_n(q;r) = nJ_{n-1}(q;r). \tag{6.449}$$

Even more, we are able to verify the following recursion relations.

Lemma 6.29. *The following recurrence relations for Bessel functions hold true:*

(i)

$$J_{n-1}(q;r) + J_{n+1}(q;r) = \frac{2n+q-2}{r}J_n(q;r), \tag{6.450}$$

(ii)

$$J_{n-1}(q;r) - J_{n+1}(q;r) = 2J'_n(q;r) + \frac{q-2}{r}J_n(q;r), \tag{6.451}$$

(iii)

$$J_{n+1}(q;r) = -r^n\frac{d}{dr}\left(r^{-n}J_n(q;r)\right), \tag{6.452}$$

(iv)

$$J_{n-1}(q;r) = r^{2-q-n}\frac{d}{dr}\left(r^{n+q-2}J_n(q;r)\right). \tag{6.453}$$

For later use in our number theoretical context of Hardy–Landau summation (cf. Section 10.5) we mention the following integral representation, which is an immediate consequence of (6.436) and (6.453).

Lemma 6.30. *For $R > 0$, $w \in \mathbb{R}^q$, we have*

$$\int_{\substack{|x| \leq R \\ x \in \mathbb{R}^q}} e^{-2\pi i w \cdot x} \, dV_{(q)}(x) = \int_0^R r^{q-1} \int_{\mathbb{S}^{q-1}} e^{-2\pi i r w \cdot \xi} \, dS_{(q-1)}(\xi) \, dr$$

$$= \|\mathbb{S}^{q-1}\| \int_0^R r^{q-1} J_0(q; 2\pi r |w|) \, dr$$

$$= \|\mathbb{S}^{q-1}\| R^q \frac{J_1(q; 2\pi |w| R)}{2\pi |w| R}. \tag{6.454}$$

Let $H_n(q; \cdot)$ be a homogeneous harmonic polynomial of the form $H_n(q; x) = r^n Y_n(q; \xi)$, $x \in \mathbb{R}^q$, $x = r\xi$, $r = |x|$, $x \in \mathbb{S}^{q-1}$. Then it is not hard to see that

$$H_n(q; \nabla_x) \, e^{ix \cdot \eta} = i^n e^{ix \cdot \eta} \, H_n(q; \eta), \quad \eta \in \mathbb{S}^{q-1}. \tag{6.455}$$

In connection with

$$J_0(q; |x|) = \frac{1}{\|\mathbb{S}^{q-1}\|} \int_{\mathbb{S}^{q-1}} e^{ix \cdot \eta} \, dS_{(q-1)}(\eta) \tag{6.456}$$

this yields

$$H_n(q; \nabla_x) J_0(q; |x|) = \frac{i^n}{\|\mathbb{S}^{q-1}\|} \int_{\mathbb{S}^{q-1}} e^{ix \cdot \eta} Y_n(q; \eta) \, dS_{(q-1)}(\eta) \tag{6.457}$$

such that

$$H_n(q; \nabla_x) J_0(q; |x|) = (-1)^n J_n(q; r) H_n(q; \xi). \tag{6.458}$$

In other words, Bessel functions of any order may be obtained by differentiation of the Bessel function $J_0(q; \cdot)$ of order 0.

After these preliminaries we are prepared to prove that $J_n(q; r)$ is asymptotically equal to the term

$$\frac{\Gamma\left(\frac{q}{2}\right) \left(\frac{r}{2}\right)^n}{\Gamma\left(n + \frac{q}{2}\right)}. \tag{6.459}$$

Lemma 6.31. *For fixed $r > 0$,*

$$\lim_{n \to \infty} \frac{J_n(q; r)}{\frac{\Gamma\left(\frac{q}{2}\right)\left(\frac{r}{2}\right)^n}{\Gamma\left(n + \frac{q}{2}\right)}} = 1. \tag{6.460}$$

Proof. Using $|e^{irt} - 1| \leq r|t|$ we find

$$\left| \int_{-1}^1 e^{irt} (1 - t^2)^{n + \frac{q-3}{2}} \, dt - \int_{-1}^1 (1 - t^2)^{n + \frac{q-3}{2}} \, dt \right| \tag{6.461}$$

$$\leq r \int_{-1}^1 |t| (1 - t^2)^{n + \frac{q-3}{2}} \, dt = \frac{r}{n + \frac{q-1}{2}}.$$

We already know (cf. (6.432)) that

$$\int_{-1}^{1} (1-t^2)^{n+\frac{q-3}{2}} \, dt = \frac{\sqrt{\pi}\,\Gamma\left(n+\frac{q-1}{2}\right)}{\Gamma\left(n+\frac{q}{2}\right)}. \tag{6.462}$$

Moreover, from Stirling's formula of the theory of the Gamma function, we get

$$\lim_{n\to\infty} \frac{\frac{\sqrt{\pi}\Gamma\left(n+\frac{q-1}{2}\right)}{\Gamma\left(n+\frac{q}{2}\right)}}{\sqrt{\frac{\pi}{n}}} = 1. \tag{6.463}$$

In connection with (6.461) this leads to the following result

$$\lim_{n\to\infty} \frac{\int_{-1}^{1} e^{irt}(1-t^2)^{n+\frac{q-3}{2}} \, dt}{\frac{\sqrt{\pi}\Gamma\left(n+\frac{q-1}{2}\right)}{\Gamma\left(n+\frac{q}{2}\right)}} = 1. \tag{6.464}$$

After a simple manipulation this shows Lemma 6.31.   □

Lemma 6.31 allows us to formulate the following expansion theorem.

Lemma 6.32. *Let* $\{Y_n(q;\cdot)\}_{n=0,1,\dots}$ *be a sequence of spherical harmonics of dimension q such that*

$$\sum_{n=0}^{\infty} J_n(q;R)Y_n(q;\xi), \quad \xi \in \mathbb{S}^{q-1}, \tag{6.465}$$

is convergent for $R > 0$. Then

$$\sum_{n=0}^{\infty} n^\alpha J_n(q;r)Y_n(q;\xi), \quad \xi \in \mathbb{S}^{q-1}, \tag{6.466}$$

is convergent for all $r \in [0, R)$ and $\alpha \geq 0$, where the series (6.466) is absolutely and uniformly convergent for all $\overline{\mathbb{B}_r^q} \subset \mathbb{B}_R^q$.

Proof. According to (6.465) we have

$$\lim_{n\to\infty} |J_n(q;R)Y_n(q;\xi)| = 0, \tag{6.467}$$

i.e., for sufficiently large n

$$|Y_n(q;\xi)| \leq \frac{1}{|J_n(q;R)|}. \tag{6.468}$$

From the limit relation (Lemma 6.31) we obtain for $n \to \infty$

$$|Y_n(q;\xi)| = O\left(2^{n+\frac{q-2}{2}} n^{\frac{q-2}{2}} \Gamma\left(n+\frac{q}{2}\right) R^{-n}\right). \tag{6.469}$$

Hence, there exist constants A and $N(= N(R))$ such that

$$|n^\alpha J_n(q;r)Y_n(q;\xi)| \leq A\, n^\alpha \left(\frac{r}{R}\right)^n \tag{6.470}$$

for all $n \geq N$ and $r < R$. This proves Lemma 6.32.   □

From the power series of the Bessel functions we are immediately able to verify the following asymptotic relation.

Lemma 6.33. *For fixed n we have*

$$J_n(q;r) = \Gamma\left(\frac{q}{2}\right)\left(\frac{r}{2}\right)^n + O(r^{n+2}), \quad r \to 0. \tag{6.471}$$

The relevant properties of Bessel functions for our later work can be summarized as follows:

Lemma 6.34. *The Bessel function $J_n(q; \cdot)$ satisfies the following relations:*

(i) For $r > 0$, $\xi, \eta \in \mathbb{S}^{q-1}$,

$$e^{r\xi \cdot \eta} = \sum_{n=0}^{\infty} N(q; n) J_n(q; r) P_n(q, \xi \cdot \eta),$$

(ii) For $x = |x|\xi$, $y = |y|\eta$

$$J_0(q; |x - y|) = \sum_{n=0}^{\infty} N(q; n) J_n(q; |x|) J_n(q; |y|) P_n(q; \xi \cdot \eta),$$

(iii) $J_n(q; \cdot)$ *is a solution of the differential equation*

$$J_n''(q;r) + \frac{q-1}{r} J_n'(q;r) + \left(1 - \frac{n(n+q-2)}{r^2}\right) J_n(q;r) = 0,$$

(iv) $J_n(q; \cdot)$ *satisfies the recursion relation*

$$(n + q - 2)J_{n+1}(q;r) + (2n + q - 2)J_n'(q;r) - nJ_{n-1}(q;r) = 0,$$

(v)

$$J_n(q;r) = (-1)^n \left(\frac{1}{r}\frac{d}{dr}\right)^n J_n(q;r),$$

$$J_{n-1}(q;r) = r^{2-q-n}\frac{d}{dr}\left(r^{n+q-2} J_n(q;r)\right),$$

(vi)

$$\lim_{n \to \infty} \frac{J_n(q;r)}{\frac{\Gamma(\frac{q}{2})(\frac{r}{2})^n}{\Gamma(n+\frac{q}{2})}} = 1.$$

(vii) For homogeneous harmonic polynomials $H_n(q; \cdot)$ of degree n

$$H_n(q; \nabla_x)J_0(q; |x|) = (-1)^n J_n(q; |x|)H_n(q; \xi),$$

(viii) For n fixed and $r \to 0$,

$$J_n(q;r) = \Gamma\left(\frac{q}{2}\right)\left(\frac{r}{2}\right)^n + O(r^{n+2}).$$

The traditional notation, which is mostly used in the literature, is based on the observation that the Bessel functions of integral order n and dimension q are expressible as Bessel functions of order $n + \frac{q-2}{2}$ and dimension 2. We discuss this aspect in more detail: letting, as usual (see G.N. Watson [1944]),

$$J_\nu(r) = J_\nu(2; r) = \left(\frac{r}{2}\right)^\nu \sum_{k=0}^\infty \frac{(-1)^k}{k!} \frac{\left(\frac{r}{2}\right)^{2k}}{\Gamma(\nu + k + 1)}, \quad \nu \geq 0, \tag{6.472}$$

and observing

$$J_n(q; r) = \left(\frac{r}{2}\right)^n \Gamma\left(\frac{q}{2}\right) \sum_{k=0}^\infty \frac{(-1)^k}{k!} \frac{\left(\frac{r}{2}\right)^{2k}}{\Gamma(n + k + \frac{q}{2})}, \quad n \in \mathbb{N}_0, \tag{6.473}$$

we are led by comparison to the relation

$$J_n(q; r) = \Gamma\left(\frac{q}{2}\right) \left(\frac{r}{2}\right)^{\frac{2-q}{2}} J_{n+\frac{q-2}{2}}(r), \quad n \in \mathbb{N}_0. \tag{6.474}$$

In particular, we have

$$J_0(3; r) = \sqrt{\frac{\pi}{2}} \frac{J_{\frac{1}{2}}(r)}{\sqrt{r}}. \tag{6.475}$$

Furthermore we see that $J_0(3; \cdot)$ is the sinc-function

$$
\begin{aligned}
J_0(3; r) &= \frac{1}{2}\sqrt{\pi} \sum_{k=0}^\infty \frac{(-1)^k r^{2k}}{2^{2k} \Gamma(k+1) \Gamma\left(k + \frac{3}{2}\right)} \\
&= \sum_{k=0}^\infty \frac{(-1)^k r^{2k}}{(2k+1)!} \\
&= \frac{\sin(r)}{r}.
\end{aligned}
\tag{6.476}
$$

This yields

$$J_0(3; r) = \frac{\sin(r)}{r} = \operatorname{sinc}(r). \tag{6.477}$$

Finally it should be remarked that the relations (6.472), (6.473), and (6.474) also allow us to understand $J_\nu(q; \cdot)$ for non-negative real values ν by

$$J_\nu(q; r) = \Gamma\left(\frac{q}{2}\right) \left(\frac{r}{2}\right)^{\frac{2-q}{2}} J_{\nu+\frac{q-2}{2}}(r). \tag{6.478}$$

Modified Bessel Functions

Solutions of the equation $\Delta U - U = 0$ can be obtained in an analogous way. For that purpose we start in the standard polar coordinates $x = r\xi, r = |x|, \xi \in \mathbb{S}^{q-1}$, from the radial and angular separation

$$\frac{1}{\|\mathbb{S}^{q-1}\|} \int_{\mathbb{S}^{q-1}} e^{x \cdot \eta} Y_n(q; \eta) \, dS_{(q-1)}(\eta) = I_n(q; r) Y_n(q; \xi). \tag{6.479}$$

Definition 6.8. $I_n(q; \cdot)$ *given by*

$$I_n(q; r) = \frac{\Gamma\left(\frac{q}{2}\right)\left(\frac{r}{2}\right)^n}{\sqrt{\pi}\,\Gamma\left(n + \frac{q-1}{2}\right)} \int_{-1}^{1} e^{rt}(1 - t^2)^{n + \frac{q-3}{2}}\, dt \tag{6.480}$$

is called a modified Bessel function of order n and dimension q.

By comparison we find the relation $I_n(q; r) = (-1)^n J_n(q; ir)$. Hence, the properties of the modified Bessel function of order n and dimension q can be collected in the following lemma.

Lemma 6.35. *The modified Bessel function* $I_n(q; \cdot)$ *satisfies the following relations:*

(i) For r > 0, $\xi, \eta \in \mathbb{S}^{q-1}$,

$$e^{r\xi \cdot \eta} = \sum_{n=0}^{\infty} N(q; n) I_n(q; r) P_n(q, \xi \cdot \eta),$$

(ii) For $x = |x|\xi$, $y = |y|\eta$

$$I_0(q; |x - y|) = \sum_{n=0}^{\infty} N(q; n) I_n(q; |x|) I_n(q; |y|) P_n(q; \xi \cdot \eta),$$

(iii) $I_n(q; \cdot)$ is a solution of the differential equation

$$I_n''(q; r) + \frac{q-1}{r} I_n'(q; r) - \left(1 + \frac{n(n + q - 2)}{r^2}\right) I_n(q; r) = 0,$$

(iv) $I_n(q; \cdot)$ satisfies the recursion relation

$$(n + q - 2) I_{n+1}(q; r) + (2n + q - 2) I_n'(q; r) - n I_{n-1}(q; r) = 0,$$

(v)

$$I_n(q; r) = r^n \left(\frac{1}{r}\frac{d}{dr}\right)^n I_0(q; r),$$

(vi)

$$\lim_{n \to \infty} \frac{I_n(q; r)}{\frac{\Gamma\left(\frac{q}{2}\right)\left(\frac{r}{2}\right)^n}{\Gamma\left(n + \frac{q}{2}\right)}} = 1.$$

202 — Metaharmonic Lattice Point Theory

(vii) For homogeneous harmonic polynomials $H_n(q;\cdot)$ of degree n

$$H_n(q;\nabla_x)I_0(q;|x|) = I_n(q;|x|)H_n(q;\xi),$$

(viii) For n fixed and $r \to 0$,

$$I_n(q;r) = \Gamma\left(\frac{q}{2}\right)\left(\frac{r}{2}\right)^n + O(r^{n+2}).$$

In particular, we have

$$I_0(3;r) = \frac{\sinh(r)}{r} \tag{6.481}$$

and

$$I_n(3;r) = r^n\left(\frac{1}{r}\frac{d}{dr}\right)^n\frac{\sinh(r)}{r}. \tag{6.482}$$

Hankel Functions

The Bessel functions are regular at 0. We now discuss a pair of functions which turn out to be regular at infinity.

Definition 6.9. *For $r > 0$, the functions $H_n^{(1)}(q;\cdot)$, $H_n^{(2)}(q;\cdot)$ defined by*

$$H_n^{(1)}(q;r) = -2i^{-n}\frac{\|\mathbb{S}^{q-2}\|}{\|\mathbb{S}^{q-1}\|}\int_{1+0i}^{1+\infty i} e^{irt}P_n(q;t)(1-t^2)^{\frac{q-3}{2}}dt, \tag{6.483}$$

$$H_n^{(2)}(q;r) = 2i^{-n}\frac{\|\mathbb{S}^{q-2}\|}{\|\mathbb{S}^{q-1}\|}\int_{-1+0i}^{-1+\infty i} e^{irt}P_n(q;t)(1-t^2)^{\frac{q-3}{2}}dt \tag{6.484}$$

are called Hankel functions of the first and second kind of order n and dimension q, respectively.

It is well known that the two paths of integration in this definition may be deformed to the curves as depicted in Figure 6.4. For example, we get for the path of $H_n^{(2)}(q;\cdot)$

$$\frac{1}{2}i^n\frac{\|\mathbb{S}^{q-1}\|}{\|\mathbb{S}^{q-2}\|}H_n^{(2)}(q;r) = \int_{-1}^0 e^{irt}P_n(q;t)(1-t^2)^{\frac{q-3}{2}}dt \tag{6.485}$$

$$+ \int_0^\infty e^{-rs}P_n(q;is)(1+s^2)^{\frac{q-3}{2}}ds$$

$$= (-1)^n\int_0^1 e^{-irt}P_n(q;t)(1-t^2)^{\frac{q-3}{2}}dt$$

$$+ i^{n+1}\int_0^\infty e^{-rs}P_n(q;is)i^{-n}(1+s^2)^{\frac{q-3}{2}}ds.$$

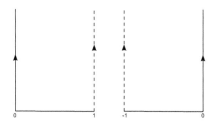

FIGURE 6.4
The two integration paths of the Hankel functions $H_n^{(1)}(q;\cdot)$ (left) and $H_n^{(2)}(q;\cdot)$ (right), respectively.

Remark 6.16. *The representation of the Legendre polynomial as a power series shows us that $i^{-n}P_n(q;is)$ is a polynomial of degree n in the variable s possessing exclusively real coefficients (cf. (6.524))*

$$i^{-n}P_n(q;is) = \frac{2^{n-1}\Gamma\left(n+\frac{q-2}{2}\right)\Gamma(q-1)}{\Gamma\left(\frac{q}{2}\right)\Gamma(n+q-2)}s^n + O\left((1+s)^{n-2}\right), \qquad (6.486)$$

$s \in [0,\infty)$, $n+q \geq 3$.

From Definition 6.9 it follows that

$$\frac{\|\mathbb{S}^{q-1}\|}{\|\mathbb{S}^{q-2}\|}H_n^{(2)}(q;r) = 2i^n \int_0^1 e^{-irt}P_n(q;t)(1-t^2)^{\frac{q-3}{2}}\,dt \qquad (6.487)$$

$$+ 2i\int_0^\infty e^{-rs}P_n(q;is)i^{-n}(1+s^2)^{\frac{q-3}{2}}\,ds$$

$$= \frac{\|\mathbb{S}^{q-1}\|}{\|\mathbb{S}^{q-2}\|}\overline{H_n^{(1)}(q;r)}.$$

The Hankel functions have certain characteristic properties for $r \to \infty$ and fixed n. With the substitution $t = 1+is$, $s \in [0,\infty)$, we get

$$H_n^{(1)}(q;r) = E_n(q;r)\int_0^\infty e^{-rs}P_n(q;1+is)(s(s+2i))^{\frac{q-3}{2}}\,ds, \qquad (6.488)$$

where

$$E_n(q;r) = 2\frac{\|\mathbb{S}^{q-2}\|}{\|\mathbb{S}^{q-1}\|}e^{i(r-n\frac{\pi}{2}-\frac{\pi}{4}(q-1))}. \qquad (6.489)$$

Note that $t = 1+is$ yields

$$(1-t^2)^{\frac{q-3}{2}} = (-is(2+is))^{\frac{q-3}{2}} = e^{-\frac{\pi}{4}i(q-3)}(2s)^{\frac{q-3}{2}} + \ldots, \qquad (6.490)$$

such that by standard arguments of complex analysis (according to our definition of the root function)

$$(1-t^2)^{\frac{q-3}{2}} = e^{-\frac{\pi}{4}i(q-3)}(2s)^{\frac{q-3}{2}} + \ldots.$$

With the abbreviation

$$F(s) = (2 + is)^{\frac{q-3}{2}} P_n(q; 1 + is)$$
(6.491)

we have

$$H_n^{(1)}(q; r) = E_n(q; r) \int_0^\infty e^{-rs} F(s) s^{\frac{q-3}{2}} \, ds.$$
(6.492)

It can be deduced from well known results for the Legendre polynomial (see C. Müller [1998]) that

$$F(s) = 2^{\frac{q-3}{2}} \left(1 + \frac{(2n + q - 1)(2n + q - 3)}{q - 1} \frac{is}{4} + \dots \right).$$
(6.493)

In connection with Lemma 3.18 we therefore obtain the following result.

Lemma 6.36. *For n fixed and $r \to \infty$ we have*

$$
\begin{aligned}
H_n^{(1)}(q; r) &= \frac{2}{\|S^{q-1}\|} \left(\frac{2\pi}{r} \right)^{\frac{q-1}{2}} e^{i(r - n\frac{\pi}{2} - (q-1)\frac{\pi}{4})} + O(r^{-\frac{q+1}{2}}) \\
&= \frac{\Gamma(\frac{q}{2})}{\sqrt{\pi}} \left(\frac{2}{r} \right)^{\frac{q-1}{2}} i^{-n - \frac{q-1}{2}} e^{ir} + O\left(r^{-\frac{q+1}{2}} \right), \\
H_n^{(2)}(q; r) &= \frac{\Gamma(\frac{q}{2})}{\sqrt{\pi}} \left(\frac{2}{r} \right)^{\frac{q-1}{2}} i^{n + \frac{q-1}{2}} e^{-ir} + O\left(r^{-\frac{q+1}{2}} \right).
\end{aligned}
$$

Observing the identity

$$J_n(q; r) = \frac{1}{2} \left(H_n^{(1)}(q; r) + H_n^{(2)}(q; r) \right)$$
(6.494)

we obtain from Lemma 6.36

Lemma 6.37. *For n fixed and $r \to \infty$,*

$$J_n(q; r) = \frac{\Gamma(\frac{q}{2})}{\sqrt{\pi}} \left(\frac{2}{r} \right)^{\frac{q-1}{2}} \cos\left(n\frac{\pi}{2} + (q - 1)\frac{\pi}{4} - r \right) + O\left(r^{-\frac{q+1}{2}} \right).$$

Based on techniques due to G.N. Watson [1944] (see also W. Magnus et al. [1949]) the O-term in Lemma 6.37 can be written out in more detail.

Lemma 6.38. *For $n \in \mathbb{N}_0$, $m \in \mathbb{N}$ fixed and $r \to \infty$*

$$
\begin{aligned}
J_n(q; r) = {} & \frac{\Gamma(\frac{q}{2})}{2^{\frac{3-q}{2}} \sqrt{\pi}} r^{\frac{1-q}{2}} \\
& \times \left(e^{i(r - \frac{\pi}{2}(n + \frac{q-2}{2}) - \frac{\pi}{4})} \sum_{l=0}^{m-1} \frac{(n + \frac{q-2}{2}, l)}{(-2ir)^l} \right. \\
& \left. + e^{-i(r - \frac{\pi}{2}(n + \frac{q-2}{2}) - \frac{\pi}{4})} \sum_{l=0}^{m-1} \frac{(n + \frac{q-2}{2}, l)}{(2ir)^l} \right) \\
& + O\left(r^{-(m + \frac{q-1}{2})} \right),
\end{aligned}
$$
(6.495)

where

$$\left(n + \frac{q-2}{2}, 0\right) = 1 \tag{6.496}$$

and

$$\left(n + \frac{q-2}{2}, l\right) = \frac{1}{l!}\left(n(n+q-2) + \frac{(q-3)(q-1)}{4}\right) \tag{6.497}$$

$$\cdots\cdot\left(n(n+q-2) + \frac{(q+2l-3)(q-2l-1)}{4}\right).$$

Next we are interested in an asymptotic relation for $n \to \infty$ and fixed r.

Lemma 6.39. *For $n \to \infty$ and fixed $r > 0$*

$$\lim_{n\to\infty} \frac{H_n^{(1)}(q;r)}{\frac{i}{\pi}\Gamma\left(\frac{q}{2}\right)\Gamma\left(n + \frac{q}{2} - 1\right)\left(\frac{2}{r}\right)^{n+q-2}} = 1. \tag{6.498}$$

Proof. The Rodrigues rule yields

$$H_n^{(1)}(q;r) = -2\frac{\Gamma\left(\frac{q}{2}\right)}{\sqrt{\pi}\,\Gamma\left(n + \frac{q-1}{2}\right)}\left(\frac{r}{2}\right)^n \int_{1+0i}^{1+\infty i} e^{irt}(1-t^2)^{n+\frac{q-3}{2}}\,dt. \tag{6.499}$$

Obviously,

$$\int_{1+0i}^{1+\infty i} e^{irt}(1-t^2)^{n+\frac{q-3}{2}}\,dt \tag{6.500}$$

$$= \int_1^0 e^{irt}(1-t^2)^{n+\frac{q-3}{2}}\,dt + \int_{0i}^{\infty i} e^{irt}(1-t^2)^{n+\frac{q-3}{2}}\,dt.$$

The first integral is uniformly bounded and does not give any contribution to the assertion. The second integral can be written as follows

$$\int_{0i}^{\infty i} e^{irt}(1-t^2)^{n+\frac{q-3}{2}}\,dt = i\int_0^\infty e^{-rs}(1+s^2)^{n+\frac{q-3}{2}}\,ds. \tag{6.501}$$

In order to prove Lemma 6.39 we observe that

$$\lim_{n\to\infty} \frac{\int_0^\infty e^{-rs}(1+s^2)^{n+\frac{q-3}{2}}\,ds}{\int_0^\infty e^{-rs}s^{2n+q-3}\,ds} = 1. \tag{6.502}$$

Moreover, we have

$$\int_0^\infty e^{-rs}s^{2n+q-3}\,ds = \frac{\Gamma(2n+q-2)}{r^{2n+q-2}}. \tag{6.503}$$

Now two estimates come into play

$$(1+s)^{2k} = (1+2s+s^2)^k > (1+s^2)^k, \quad s > 0, k > 0, \tag{6.504}$$

and

$$(1+s^2)^k - s^{2k} \leq k(1+s^2)^{k-1} \leq k(1+s)^{2k-2} \tag{6.505}$$

(note that the estimate (6.505) follows from the mean value theorem of one-dimensional analysis). This leads to the estimate

$$\int_0^\infty e^{-rs} \left((1+s^2)^{n+\frac{q-3}{2}} - s^{2n+q-3} \right) ds \tag{6.506}$$

$$\leq \left(n + \frac{q-3}{2} \right) \int_0^\infty e^{-rs}(1+s)^{2n+q-5} \, ds$$

$$\leq e^r \left(n + \frac{q-3}{2} \right) \int_0^\infty e^{-ru} u^{2n+q-5} \, du.$$

The last integral can be calculated explicitly.

$$e^r \left(n + \frac{q-3}{2} \right) \int_0^\infty e^{-ru} u^{2n+q-5} \, du \tag{6.507}$$

$$= e^r \left(n + \frac{q-3}{2} \right) \frac{\Gamma(2n+q-4)}{r^{2n+q-4}}$$

$$= \frac{e^r}{2r^{2n+q-4}} \frac{\Gamma(2n+q-2)}{2n+q-4}.$$

Consequently, we are able to see that

$$\left| \int_0^\infty e^{-rs}(1+s^2)^{n+\frac{q-3}{2}} \, ds - \frac{\Gamma(2n+q-2)}{r^{2n+q-2}} \right| \tag{6.508}$$

$$\leq \frac{e^r}{r^{2n+q-4}} \frac{\Gamma(2n+q-2)}{2n+q-4}.$$

In connection with (6.500) and (6.501) we therefore obtain for $n \to \infty$

$$\lim_{n\to\infty} \frac{\int_{0i}^{\infty i} e^{irt}(1-t^2)^{n+\frac{q-3}{2}} \, dt}{i \int_0^\infty e^{-rs}(1+s^2)^{n+\frac{q-3}{2}} \, ds} \tag{6.509}$$

$$= \lim_{n\to\infty} \frac{\int_{0i}^{\infty i} e^{irt}(1-t^2)^{n+\frac{q-3}{2}} \, dt}{i \frac{\Gamma(2n+q-2)}{r^{2n+q-2}}} = 1.$$

Lemma 6.39 then follows with the aid of the formula

$$\sqrt{\pi}\Gamma(2n+q-2) = 2^{2n+q-3}\Gamma\left(n + \frac{q-2}{2} \right)\Gamma\left(n + \frac{q-1}{2} \right) \tag{6.510}$$

known from the theory of the Gamma function. $\qquad \square$

Kelvin Functions

Next we come to the Neumann function which together with the Bessel function implies the Hankel functions.

Definition 6.10. *For $r > 0$, the Neumann function $N_n(q; \cdot)$ of order n and dimension q is defined by*

$$
\begin{aligned}
N(q; r) = \; & 2 \frac{\|\mathbb{S}^{q-2}\|}{\|\mathbb{S}^{q-1}\|} \int_0^1 \sin\left(rt - n\frac{\pi}{2}\right) P_n(q; t)(1 - t^2)^{\frac{q-3}{2}} \, dt \\
& - 2i^{-n} \frac{\|\mathbb{S}^{q-2}\|}{\|\mathbb{S}^{q-1}\|} \int_0^\infty e^{-rt} P_n(q; it)(1 - t^2)^{\frac{q-3}{2}} \, dt. \quad (6.511)
\end{aligned}
$$

As already announced, Hankel functions can be obtained by combination of Bessel and Neumann functions. More explicitly, from (6.485) and (6.487) we get the following identities.

Lemma 6.40. *For $r > 0$,*

$$
\begin{aligned}
H_n^{(1)}(q; r) &= J_n(q; r) + iN(q; r), \\
H_n^{(2)}(q; r) &= J_n(q; r) - iN(q; r).
\end{aligned}
$$

Let $C_n(q; \cdot)$ stand for any of the so-called "*cylinder functions*" $J_n(q; \cdot)$, $N_n(q; \cdot)$, $H_n^{(1)}(q; \cdot)$, and $H_n^{(2)}(q; \cdot)$. Then, for $r > 0$, the following recursion relation holds true

$$
C_{n-1}(q; r) + C_{n+1}(q; r) = \frac{2n + q - 2}{r} C_n(q; r). \quad (6.512)
$$

Furthermore, we have

$$
(2n + q - 2)C_n'(q; r) = nC_{n-1}(q; r) - (n + q - 2)C_{n+1}(q; r). \quad (6.513)
$$

These identities immediately follow from the recursion relation for the Legendre polynomial.

Definition 6.11. *For $r > 0$, the function $K_n(q; \cdot)$ given by*

$$
K_n(q; r) = \int_1^\infty e^{-rt} P_n(q; t)(t^2 - 1)^{\frac{q-3}{2}} \, dt \quad (6.514)
$$

is called the Kelvin function (or modified Hankel function) of order n and dimension q.

It can be shown that

$$
H_n^{(1)}(q; ir) = 2 \frac{\|\mathbb{S}^{q-2}\|}{\|\mathbb{S}^{q-1}\|} i^{1-q-n} K_n(q; r), \quad (6.515)
$$

$$
H_n^{(2)}(q; -ir) = 2 \frac{\|\mathbb{S}^{q-2}\|}{\|\mathbb{S}^{q-1}\|} i^{n+q-1} K_n(q; r). \quad (6.516)
$$

Moreover, we find for $r > 0$ (see (6.512) and (6.513), respectively)

$$K_{n-1}(q;r) - K_{n+1}(q;r) = -\frac{2n+q-2}{r}K_n(q;r) \tag{6.517}$$

and

$$nK_{n-1}(q;r) + (n+q-2)K_{n+1}(q;r) = (2n+q-2)K_n'(q;r), \tag{6.518}$$

which again follows from the recursion relation for the Legendre polynomial.

Keeping r fixed we obtain by similar techniques as used for the proof of Lemma 6.39.

Lemma 6.41. *For $n \to \infty$ and fixed $r > 0$ we have*

$$\lim_{n\to\infty} \frac{K_n(q;r)}{\frac{1}{2\pi}\Gamma(\frac{q-1}{2})\Gamma\left(n+\frac{q}{2}-1\right)\left(\frac{2}{r}\right)^{n+q-2}} = 1. \tag{6.519}$$

Proof. The Rodrigues rule gives the integral representation

$$K_n(q;r) = \frac{\Gamma\left(\frac{q-1}{2}\right)r^n}{\sqrt{\pi}\Gamma\left(n+\frac{q-1}{2}\right)}\int_1^\infty e^{-rt}(t^2-1)^{n+\frac{q-3}{2}}\,dt. \tag{6.520}$$

Now we have

$$\lim_{n\to\infty} \frac{\int_1^\infty e^{-rt}(t^2-1)^{n+\frac{q-3}{2}}\,dt}{\int_0^\infty e^{-rt}t^{2n+q-3}\,dt} = 1, \tag{6.521}$$

where the integral in the denominator can be calculated explicitly

$$\int_0^\infty e^{-rt}t^{2n+q-3}\,dt = \frac{\Gamma(2n+q-2)}{r^{2n+q-2}}. \tag{6.522}$$

From the estimate

$$0 \le t^{2k} - (t^2-1)^k \le kt^{2k-2}, \quad t \ge 1, \tag{6.523}$$

we get

$$\begin{aligned}
0 &\ge \int_1^\infty e^{-rt}(t^2-1)^{n+\frac{q-3}{2}}\,dt - \int_1^\infty e^{-rt}t^{2n+q-3}\,dt \\
&\ge -\left(n+\frac{q-3}{2}\right)\int_0^\infty e^{-rt}t^{2n-q-5}\,dt
\end{aligned} \tag{6.524}$$

such that

$$\begin{aligned}
0 &\ge \int_1^\infty e^{-rt}(t^2-1)^{n+\frac{q-3}{2}}\,dt - \int_0^\infty e^{-rt}t^{2n+q-3}\,dt \tag{6.525} \\
&\ge -\left(n+\frac{q-3}{2}\right)\frac{\Gamma(2n+q-4)}{r^{2n+q-4}} = -\frac{1}{2}\frac{\Gamma(2n+q-4)}{2n+q-4}\left(\frac{1}{r}\right)^{2n+q-4}.
\end{aligned}$$

This is the desired result. $\square$

The Hankel functions as well as the Kelvin functions are not defined at the origin, but they have characteristic singularities there.

Lemma 6.42. *For* $r \to 0$ *and* $n + q - 3 \geq 0$ *we have*

$$H_n^{(1)}(q;r) = \frac{i}{\pi} \Gamma \left(n + \frac{q-2}{2} \right) \Gamma \left(\frac{q}{2} \right) \left(\frac{2}{r} \right)^{n+q-2} \tag{6.526}$$
$$+ O \left(r^{-n-q+4} \right),$$

$$K_n(q;r) = \frac{\Gamma(q-1)}{2^{q-1}\Gamma\left(\frac{q}{2}\right)} \Gamma \left(n + \frac{q-2}{2} \right) \left(\frac{2}{r} \right)^{n+q-2} \tag{6.527}$$
$$+ O \left(r^{-n-q+4} \right).$$

Proof. First we deal with the asymptotic relation (6.526). For that purpose we observe that

$$P_n(q;t) = \frac{\Gamma(q-1)}{\Gamma\left(\frac{q}{2}\right)} \frac{2^{n-1}n!}{(n+q-3)!} \sum_{l=0}^{\lfloor \frac{n}{2} \rfloor} \left(-\frac{1}{4} \right)^l \frac{\Gamma\left(n - l + \frac{q-2}{2} \right)}{l!(n-2l)!} t^{n-2l}$$
$$= a_n^0(q)t^n - + \cdots, \tag{6.528}$$

where (cf. (6.486))

$$a_n^0(q) = \frac{2^{n-1}\Gamma\left(n + \frac{q-2}{2} \right)\Gamma(q-1)}{\Gamma(n+q-2)\Gamma\left(\frac{q}{2}\right)}. \tag{6.529}$$

Therefore,

$$i^{-n}P_n(q;it) = a_n^0(q)t^n + O\left((1+t)^{n-2} \right), \tag{6.530}$$

such that

$$-2i^{-n}\frac{\|\mathbb{S}^{q-2}\|}{\|\mathbb{S}^{q-1}\|} \int_0^\infty e^{-rt} P_n(q;it)(1+t^2)^{\frac{q-3}{2}} dt \tag{6.531}$$
$$= -2^{n+q-2}\frac{\Gamma\left(n + \frac{q-2}{2} \right)\Gamma\left(\frac{q}{2}\right)}{\pi r^{n+q-2}} + O\left(r^{-n-q+4} \right).$$

We now discuss the asymptotic relation (6.527). For $n + q - 3 \geq 0$ we obtain after some manipulations

$$K_n(q;r) = e^{-r} \int_0^\infty e^{-rs} P_n(q;1+s)(s(1+s))^{n+q-5} ds \tag{6.532}$$
$$= a_n^0(q) \int_0^\infty e^{-rs} s^{n+q-3} ds + O\left(\int_0^\infty e^{-rs}(s+1)^{n+q-5} ds \right).$$

Replacing the integral $\int_0^\infty \ldots$ in the second term by $\int_{-1}^\infty \ldots$ we find for all n with $n + q - 3 \geq 0$

$$K_n(q;r) = \frac{\Gamma(q-1)}{2^{q-1}\Gamma\left(\frac{q}{2}\right)} \Gamma \left(n + \frac{q-2}{2} \right) \left(\frac{2}{r} \right)^{n+q-2} + O\left(r^{-n-q+4} \right). \tag{6.533}$$

This is the desired result. □

The identity

$$
K_0(2; r) = \int_1^\infty e^{-rt}(t^2 - 1)^{-\frac{1}{2}}\, dt = \int_r^\infty e^{-u}(u^2 - r^2)^{-\frac{1}{2}}\, du
$$

$$
= -e^{-r}\ln(r) + \int_r^\infty e^{-u}\ln\left(u + \sqrt{u^2 + r^2}\right)\, du \qquad (6.534)
$$

shows us that $K_0(2, r)$ has a logarithmic singularity. In fact, for $n = 0$, $q = 2$, we have

$$
K_0(2; r) = -\ln(r) + O(1). \qquad (6.535)
$$

In the same way we obtain

$$
H_0^{(1)}(2; r) = \frac{2i}{\pi}\ln(r) + O(1). \qquad (6.536)
$$

Remark 6.17. *The function $x \mapsto K_0(q; |x|)$, $x \in \mathbb{R}^q\backslash\{0\}$, shows the same singularity behavior at the origin as the fundamental solution $x \mapsto F_q(|x|)$, $x \in \mathbb{R}^q\backslash\{0\}$, for the Laplace operator in $\mathbb{R}^q$. This fact is of particular significance for the characterization of the singularity behavior of the multi-dimensional lattice function (see Definition 8.1) in lattice points.*

For $r \to \infty$ and $n \geq 0$ fixed we obtain

$$
K_n(q; r) = \int_1^\infty e^{-rt} P_n(q; t)(t^2 - 1)^{\frac{q-3}{2}}\, dt \qquad (6.537)
$$

$$
\overset{t=1+s}{=} e^{-r}\int_0^\infty e^{-rs} P_n(q; 1 + s)((2 + s)s)^{\frac{q-2}{2}}\, ds
$$

$$
= e^{-r}\left(2^{\frac{q-3}{2}}\frac{\Gamma\left(\frac{q-1}{2}\right)}{r^{\frac{q-1}{2}}} + O(r^{-\frac{q+1}{2}})\right).
$$

This leads to the following asymptotic relations.

Lemma 6.43. *For $r \to \infty$ and $n \geq 0$ fixed,*

$$
K_n(q; r) = \frac{e^{-r}}{2}\Gamma\left(\frac{q-1}{2}\right)\left(\left(\frac{2}{r}\right)^{\frac{q-1}{2}} + O\left(r^{-\frac{q+1}{2}}\right)\right). \qquad (6.538)
$$

Moreover, for $r \to \infty$, we have

$$
K_n'(q; r) - K_n(q; r) = O\left(e^{-r} r^{-\frac{q+1}{2}}\right). \qquad (6.539)
$$

The properties of the Kelvin function relevant for our purposes in the analytic theory of numbers can be summarized as follows.

Lemma 6.44. *The Kelvin function $K_n(q; \cdot)$ satisfies the following relations:*

(i) For $x = |x|\xi$, $y = |y|\eta$,

$$K_0(q; |x - y|) = \sum_{n=0}^{\infty} N(q; n) I_n(q; |x|) K_n(q; |y|) P_n(q; \xi \cdot \eta),$$

(ii) $K_n(q; \cdot)$ is a solution of the differential equation

$$K_n''(q; r) - \frac{q-1}{r} K_n'(q; r) - \left(1 + \frac{n(n+q-2)}{r^2}\right) K_n(q; r) = 0,$$

(iii) $K_n(q; \cdot)$ satisfies the recursion relations

$$(n + q - 2) K_{n+1}(q; r) + (2n + q - 2) K_n'(q; r) + n K_{n-1}(q; r) = 0,$$

$$K_{n-1}(q; r) - K_{n+1}(q; r) = -\frac{2n + q - 2}{r} K_n(q; r),$$

(iv)

$$(-1)^m r^{-n-m} K_{n+m}(q; r) = \left(\frac{1}{r}\frac{d}{dr}\right)^m \left(K_n(q; r) \, r^{-n}\right),$$

$$(-1)^m r^{n-m+q-2} K_{n-m}(q; r) = \left(\frac{1}{r}\frac{d}{dr}\right)^m \left(r^{n+q-2} K_n(q; r)\right),$$

(v)

$$\lim_{n \to \infty} \frac{2\pi \, K_n(q; r)}{\Gamma\left(\frac{q-1}{2}\right) \left(\frac{2}{r}\right)^{n+q-2} \Gamma\left(n + \frac{q}{2} - 1\right)} = 1,$$

(vi) For $n + q - 3 > 0$ fixed and $r \to 0$,

$$K_n(q; r) = \frac{\Gamma(q-1)}{2^{q-1}\Gamma\left(\frac{q}{2}\right)} \Gamma\left(n + \frac{q-2}{2}\right) \left(\frac{2}{r}\right)^{n+q-2} + O(r^{-n-q+4})$$

and, for $n = 0$, $q = 2$ and $r \to 0$,

$$K_0(q; r) = -\ln(r) + O(1).$$

(vii) For $r \to \infty$ and $n \geq 0$ fixed

$$K_n(q; r) = \frac{e^{-r}}{2} \Gamma\left(\frac{q-1}{2}\right) \left(\left(\frac{2}{r}\right)^{\frac{q-1}{2}} + O\left(r^{-\frac{q+1}{2}}\right)\right).$$

In particular,

$$K_0(3; r) = \frac{e^{-r}}{r} \tag{6.540}$$

and

$$K_n(3;r) = r^n \left(\frac{1}{r}\frac{d}{dr}\right)^n \frac{e^{-r}}{r}. \tag{6.541}$$

It is interesting to relate the Kelvin and Hankel functions of dimension q to their counterparts of dimension 2. For real values ν we have (see G.N. Watson [1944], W. Magnus et al. [1966])

$$K_\nu(r) = K_\nu(2;r) = \frac{\sqrt{\pi}}{\Gamma\left(\nu + \frac{1}{2}\right)} \left(\frac{r}{2}\right)^\nu \int_1^\infty e^{-rt}(t^2 - 1)^{\nu - \frac{1}{2}}\, dt. \tag{6.542}$$

By comparison we find

$$K_n(q;r) = \frac{\Gamma\left(\frac{q-1}{2}\right)}{\sqrt{\pi}} \left(\frac{r}{2}\right)^{\frac{2-q}{2}} K_{n+\frac{q-2}{2}}(r). \tag{6.543}$$

Similarly,

$$H_n^{(1)}(q;r) = \Gamma\left(\frac{q}{2}\right)\left(\frac{r}{2}\right)^{\frac{2-q}{2}} H_{n+\frac{q-2}{2}}^{(1)}(r), \tag{6.544}$$

where

$$H_\nu^{(1)}(r) = H_\nu^{(1)}(2;r) = \frac{-2}{\sqrt{\pi}\Gamma\left(\nu + \frac{1}{2}\right)} \left(\frac{r}{2}\right)^\nu \int_{1+0i}^{1+\infty i} e^{irt}(1 - t^2)^{\nu - \frac{1}{2}}\, dt. \tag{6.545}$$

Expansion Theorems

The solutions of the Helmholtz equation can be subdivided into three classes depending on their domain of definition:

- solutions for the "inner space" of a fixed ball $\mathbb{B}_R^q$, $R > 0$,

- solutions for the "outer space" of a fixed ball $\mathbb{B}_R^q$, $R > 0$,

- "entire" solutions valid for the whole Euclidean space $\mathbb{R}^q$.

From the literature about the Helmholtz equation (see, e.g., H. Niemeyer [1962]) we borrow the following expansion theorems for inner and/or outer space solutions involving Bessel and Hankel functions.

Theorem 6.28. *(Expansion for the Inner Space of a Ball) Let U be a function of class $\mathrm{C}^{(2)}\left(\mathbb{B}_R^q\right)$ such that $\Delta U + U = 0$ in $\mathbb{B}_R^q$. Then there exists a sequence $\{Y_n(q;\cdot)\}_{n=0,1,\dots}$ of spherical harmonics such that*

$$U(x) = \sum_{n=0}^\infty J_n(q;r)Y_n(q;\xi), \tag{6.546}$$

$x = r\xi$, $\xi \in \mathbb{S}^{q-1}$, *where the series is absolutely and uniformly convergent for all $x \in \mathbb{B}_r^q$ with $r = |x| < R$.*

Conversely, if (6.546) holds uniformly for all $x \in \overline{\mathbb{B}_r^q}$ with $r = |x| < R$, then U is of class $\mathrm{C}^{(2)}\left(\mathbb{B}_R^q\right)$ satisfying $\Delta U + U = 0$ in $\mathbb{B}_R^q$.

Proof. The functions $U_{n,j}, j = 1, \ldots, N(q; n)$, given by

$$U_{n,j}(r) = \int_{\mathbb{S}^{q-1}} U(r\xi)\overline{Y_{n,j}(q; \xi)} \, dS_{(q-1)}(\xi) \tag{6.547}$$

are twice continuously differentiable for all $r \in [0, R]$. Applying Green's formula we obtain

$$
\begin{aligned}
r^{q-1}U'_{n,j}(r) &= \int_{\mathbb{S}^{q-1}_r} \frac{\partial}{\partial \nu} U(x)\overline{Y_{n,j}\left(q; \frac{x}{|x|}\right)} \, dS_{(q-1)}(x) \tag{6.548} \\
&= \int_{\mathbb{B}^q_r} \left\{ \overline{Y_{n,j}\left(q; \frac{x}{|x|}\right)} (\Delta U(x)) - U(x)\overline{\Delta Y_{n,j}\left(q; \frac{x}{|x|}\right)} \right\} \, dV_{(q)}(x)
\end{aligned}
$$

such that

$$r^{q-1}U'_{n,j}(r) = -\int_0^r \left\{ s^{q-1}U_{n,j}(s) - n(n+q-2)s^{q-2}U_{n,j}(s) \right\} \, ds. \tag{6.549}$$

By differentiation with respect to r we are able to show that

$$U''_{n,j}(r) + \frac{q-1}{r}U'_{n,j}(r) + \left(1 - \frac{n(n+q-2)}{r^2}\right)U_{n,j}(r) = 0. \tag{6.550}$$

Apart from a multiplicative constant the only solution of (6.550) that is bounded for $r \to 0$ is the Bessel function $J_n(q; \cdot)$. Hence there exists a constant $C_{n,j}$ such that

$$U_{n,j}(r) = C_{n,j}J_n(q; r) \tag{6.551}$$

for all $r \in [0, R)$. We let

$$Y_n(q; \xi) = \sum_{j=1}^{N(q;n)} C_{n,j}Y_{n,j}(q; \xi), \quad \xi \in \mathbb{S}^{q-1}, \tag{6.552}$$

and

$$C_n^2 = \int_{\mathbb{S}^{q-1}} |Y_n(q; \xi)|^2 \, dS_{(q-1)}(\xi). \tag{6.553}$$

The $L^2(\mathbb{S}^{q-1})$-orthonormality of the spherical harmonics then yields

$$C_n^2 = \sum_{j=1}^{N(q;n)} |C_{n,j}|^2. \tag{6.554}$$

For $r \in [0, R)$ it therefore follows from Lemma 6.32 that the series

$$\Phi(r; \xi) = \sum_{n=0}^{\infty} J_n(q; r)Y_n(q; \xi) \tag{6.555}$$

is convergent for all $r \in [0, R)$. More precisely, the series (6.555) is absolutely and uniformly convergent on every ball $\overline{\mathbb{B}}_r^q$, $r \in [0, R)$. In addition

$$\int_{\mathbb{S}^{q-1}} (U(r\xi) - \Phi(r; \xi)) \, \overline{Y_{n,j}(q; \xi)} \, dS_{(q-1)}(\xi) = 0 \tag{6.556}$$

for all n, j. The completeness of the system of spherical harmonics, therefore, implies $U(r\xi) = \Phi(r; \xi)$, $r \in [0, R)$, which shows the first part of Theorem 6.28.

The second part follows from the fact that all partial sums of U satisfy the Helmholtz equation for every $\overline{\mathbb{B}}_r^q \subset \mathbb{B}_R^q$, $r < R$, and converge uniformly to U. Under these assumptions (see, e.g., H. Niemeyer [1962]) U is of class $C^{(2)}\left(\overline{\mathbb{B}}_r^q\right)$ for every $r \in [0, R)$ and, in addition, U satisfies the differential equation $\Delta U + U = 0$ in $\mathbb{B}_R^q$. $\qquad \square$

Theorem 6.29. *(Expansion for the Outer Space of a Ball) Let U be of class* $C^{(2)}\left(\mathbb{R}^q \backslash \overline{\mathbb{B}}_R^q\right)$ *such that $\Delta U + U = 0$ in $\mathbb{R}^q \backslash \overline{\mathbb{B}}_R^q$. Then there exist sequences* $\{Y_n^{(i)}(q; \cdot)\}_{n=0,1,\dots}$, $i = 1, 2$, *of spherical harmonics such that*

$$U(x) = \sum_{n=0}^{\infty} \left\{ H_n^{(1)}(q; r) Y_n^{(1)}(q; \xi) + H_n^{(2)}(q; r) Y_n^{(2)}(q; \xi) \right\}, \tag{6.557}$$

$x = r\xi$, $\xi \in \mathbb{S}^{q-1}$, *where the series (6.557) is absolutely and uniformly convergent for all $x \in \overline{\mathbb{B}}_r^q$ with $r = |x| > R$.*

Conversely, if (6.557) holds uniformly for all $x \in \mathbb{R}^q \backslash \overline{\mathbb{B}}_r^q$ with $r = |x| > R$, then U is of class $C^{(2)}\left(\mathbb{R}^q \backslash \overline{\mathbb{B}}_R^q\right)$ satisfying $\Delta U + U = 0$ in $\mathbb{R}^q \backslash \overline{\mathbb{B}}_R^q$.

Proof. The functions $U_{n,j}$, $j = 1, \dots, N(q; n)$, given by

$$U_{n,j}(r) = \int_{\mathbb{S}^{q-1}} U(r\xi) \, Y_{n,j}(q; \xi) \, dS_{(q-1)}(\xi) \tag{6.558}$$

are defined for all $\mathbb{R}^q \backslash \overline{\mathbb{B}}_r^q$, $r > R$, and satisfy the Bessel differential equations

$$U_{n,j}''(r) + \frac{q-1}{r} U_{n,j}'(r) + \left(1 - \frac{n(n+q-2)}{r^2}\right) U_{n,j}(r) = 0. \tag{6.559}$$

Hence, there are coefficients $C_{n,j}^{(i)}$, $i = 1, 2$, such that for all $r > R$

$$U_{n,j}(r) = C_{n,j}^{(1)} H_n^{(1)}(q; r) + C_{n,j}^{(2)} H_n^{(2)}(q; r). \tag{6.560}$$

It follows that

$$\int_{\mathbb{S}^{q-1}} |U(r\xi)|^2 \, dS_{(q-1)}(\xi) \tag{6.561}$$

$$= \sum_{n=0}^{\infty} \sum_{j=1}^{N(q;n)} \left| C_{n,j}^{(1)} H_n^{(1)}(q; r) + C_{n,j}^{(2)} H_n^{(2)}(q; r) \right|^2. \tag{6.562}$$

Therefore, we are able to verify that the series

$$\sum_{n=0}^{\infty} \sum_{j=1}^{N(q;n)} \left(C_{n,j}^{(1)} H_n^{(1)}(q;r) + C_{n,j}^{(2)} H_n^{(2)}(q;r) \right) Y_{n,j}(q;\xi) \qquad (6.563)$$

is absolutely and uniformly convergent on $\mathbb{R}^3 \backslash \overline{\mathbb{B}_r^q}$ for all $r > R$. Letting

$$Y_n^{(i)}(q;\xi) = \sum_{j=1}^{N(q;n)} C_{n,j}^{(i)} Y_{n,j}(q;\xi), \quad i = 1, 2, \qquad (6.564)$$

we finally get

$$U(x) = \sum_{n=0}^{\infty} \left\{ H_n^{(1)}(q;r) Y_n^{(1)}(q;\xi) + H_n^{(2)}(q;r) Y_n(q;\xi) \right\}, \qquad (6.565)$$

as required for the first part of Theorem 6.29. The second part follows by analogous arguments as described in the proof of Theorem 6.28. □

Evidently, for entire solutions both theorems (i.e., Theorem 6.28 as well as Theorem 6.29) hold true.

6.7 Tools Involving Helmholtz Operators

As is well known, any finite sum of the form

$$U^{(N)}(x) = (2\pi)^{\frac{1-q}{2}} \left\| \mathbb{S}^{q-1} \right\| \sum_{n=0}^{N} i^n J_n(q;r) Y_n(q;\xi) \qquad (6.566)$$

with $x = r\xi$, $r = |x|$, $\xi \in \mathbb{S}^{q-1}$, $Y_n(q; \cdot) \in \text{Harm}_n(\mathbb{S}^{q-1})$, and $N \in \mathbb{N}_0$, satisfies the Helmholtz wave equation

$$\Delta U^{(N)}(x) + U^{(N)}(x) = 0, \quad x \in \mathbb{R}^q; \qquad (6.567)$$

i.e., $U^{(N)}$ is an entire solution of the Helmholtz equation (6.567) in $\mathbb{R}^q$.

Remembering the asymptotic expansion of the Bessel function (Lemma 6.37)

$$J_n(q;r) = \frac{\Gamma\left(\frac{q}{2}\right)}{\sqrt{\pi}} \left(\frac{2}{r}\right)^{\frac{q-1}{2}} \cos\left(n\frac{\pi}{2} + (q-1)\frac{\pi}{4} - r\right) + O\left(r^{-\frac{q+1}{2}}\right) \qquad (6.568)$$

we get

$$r^{\frac{q-1}{2}} U^{(N)}(r\xi) = \sum_{n=0}^{N} i^n \left(e^{i\left(n\frac{\pi}{2} + (q-1)\frac{\pi}{4} - r\right)} + e^{-i\left(n\frac{\pi}{2} + (q-1)\frac{\pi}{4} - r\right)} \right) Y_n(q;\xi)$$
$$+ \; o(1), \qquad (6.569)$$

i.e.,

$$r^{\frac{q-1}{2}}U^{(N)}(r\xi) = e^{i\left(r-(q-1)\frac{\pi}{4}\right)}F^{(N)}(q;\xi) + e^{-i\left(r-(q-1)\frac{\pi}{4}\right)}F^{(N)}(q;-\xi) + o(1),$$
(6.570)

where

$$F^{(N)}(q;\xi) = \sum_{n=0}^{N} Y_n(q;\xi), \quad F^{(N)}(q;-\xi) = \sum_{n=0}^{N}(-1)^n Y_n(q;\xi). \quad (6.571)$$

Using the estimate (Lemma 6.38)

$$
J_n(q;r) = \frac{\Gamma\left(\frac{q}{2}\right)}{\sqrt{\pi}2^{\frac{3-q}{2}}}r^{\frac{1-q}{2}}
$$
(6.572)
$$
\times \left(e^{i\left(r-\frac{\pi}{2}\left(n+\frac{q-2}{2}\right)-\frac{\pi}{4}\right)} \sum_{l=0}^{m-1} \frac{\left(n+\frac{q-2}{2},l\right)}{(-2ir)^l} \right.
$$
$$
+ e^{-i\left(r-\frac{\pi}{2}\left(n+\frac{q-2}{2}\right)-\frac{\pi}{4}\right)} \sum_{l=0}^{m-1} \frac{\left(n+\frac{q-2}{2},l\right)}{(-2ir)^l} \right)
$$
$$
+ O\left(\frac{1}{r^{m+\frac{q-1}{2}}}\right), \quad m \geq 1,
$$

with (see (6.496) and (6.497))

$$\left(n+\frac{q-2}{2}, 0\right) = 1 \qquad (6.573)$$

and

$$
\left(n+\frac{q-2}{2}, l\right) = \frac{1}{l!}\left(n(n+q-2) + \frac{(q-1)(q-3)}{4}\right) \qquad (6.574)
$$
$$
\cdots \cdot \left(n(n+q-2) + \frac{(q+2l-3)(q-2l-1)}{4}\right)
$$

we are led to introduce the "operators"

$$
O(q;0) = 1, \qquad (6.575)
$$
$$
O(q;l) = \frac{1}{l!}\left(\frac{(q-1)(q-3)}{4} - \Delta^*\right) \qquad (6.576)
$$
$$
\cdots \cdot \left(\frac{(q+2l-3)(q-2l-1)}{4} - \Delta^*\right).
$$

In terms of (6.575) and (6.576) we obtain the asymptotic relation

$$
r^{\frac{q-1}{2}} U^{(N)}(r\xi) = (2\pi)^{\frac{1-q}{2}} \left\| \mathbb{S}^{q-1} \right\| \sum_{n=0}^{N} i^n J_n(q;r) Y_n(q;\xi) \tag{6.577}
$$

$$
= i^{\frac{1-q}{2}} e^{ir} \sum_{l=0}^{m-1} \left(\frac{1}{-2\pi ir} \right)^l O(q;l) F^{(N)}(q;\xi)
$$

$$
+ i^{\frac{q-1}{2}} e^{-ir} \sum_{l=0}^{m-1} \left(\frac{1}{2\pi ir} \right)^l O(q;l) F^{(N)}(q;-\xi)
$$

$$
+ o\left(r^{1-m} \right), \quad r \to \infty.
$$

Asymptotic expansions of the type (6.570) and (6.577), respectively, are characteristic for entire solutions of the Helmholtz wave equation.

Remark 6.18. *Unfortunately, when the attempt is made to formulate an asymptotic expansion of type (6.577) for $N \to \infty$, we are confronted with serious problems of the convergence for the right side of (6.577). A way out can be found in certain mean values as proposed by G. Herglotz and C. Müller (see C. Müller [1952] and, in addition, W. Magnus et al. [1949], P. Hartmann [1959], P. Hartmann, C. Wilcox [1961] and many others). However, this concept is not of interest for our purposes in the lattice point theory.*

An Asymptotic Expansion for an Entire Integral Solution

Our number theoretical applications require an asymptotic expansion of the integral

$$
U(r\xi) = \left(\frac{r}{2\pi} \right)^{\frac{q-1}{2}} \int_{\mathbb{S}^{q-1}} F(\eta) e^{ir\xi \cdot \eta} \, dS_{(q-1)}(\eta), \tag{6.578}
$$

where certain smoothness imposed on the integrand F defined on $\mathbb{S}^{q-1}$ is mandatory to circumvent problems of convergence. Clearly, the integral U as defined by (6.578) satisfies the equation $\Delta U(x) + U(x) = 0$, $x \in \mathbb{R}^q$, i.e., U is an entire solution. In connection with Lemma 6.38 these observations motivate us to claim an asymptotic relation of the type

$$
\left(\frac{r}{2\pi} \right)^{\frac{q-1}{2}} \int_{\mathbb{S}^{q-1}} e^{ir\xi \cdot \eta} F(\eta) \, dS_{(q-1)}(\eta) \tag{6.579}
$$

$$
= i^{\frac{1-q}{2}} e^{ir} \sum_{l=0}^{m-1} \left(\frac{1}{-2ir} \right)^l O(q;l) F(\xi) + i^{\frac{q-1}{2}} e^{-ir} \sum_{l=0}^{m-1} \left(\frac{1}{2ir} \right)^l O(q;l) F(-\xi)
$$

$$
+ o\left(r^{1-m} \right), \quad r \to \infty.
$$

In order to verify (6.579) under suitable smoothness imposed on the function F we must have a closer look at the integral (6.578). From Lemma 6.26 we know that

$$e^{ir\xi\cdot\eta} = \sum_{n=0}^{\infty} i^n N(q;n) J_n(q;r) P_n(q;\xi\cdot\eta), \tag{6.580}$$

where $P_n(q;\cdot)$ is the Legendre polynomial of degree n and dimension q (note that the series (6.580) is absolutely and uniformly convergent). Inserting (6.580) into (6.578) we find

$$U(r\xi) = \left(\frac{r}{2\pi}\right)^{\frac{q-1}{2}} \sum_{n=0}^{\infty} i^n N(q;n) J_n(q;r) \int_{\mathbb{S}^{q-1}} F(\eta) P_n(q;\xi\cdot\eta)\, dS_{(q-1)}(\eta)\,. \tag{6.581}$$

Furthermore, in accordance with (6.429), we have

$$J_n(q;r) = i^{-n} \frac{\|\mathbb{S}^{q-2}\|}{\|\mathbb{S}^{q-1}\|} \int_{-1}^{1} e^{irt} P_n(q;t)(1-t^2)^{\frac{q-3}{2}}\, dt. \tag{6.582}$$

In addition, we let

$$Y_n(q;\xi) = \frac{N(q;n)}{\|\mathbb{S}^{q-1}\|} \int_{\mathbb{S}^{q-1}} F(\eta) P_n(q;\xi\cdot\eta)\, dS_{(q-1)}(\eta)\,. \tag{6.583}$$

Then it follows that

$$U(r\xi) = \|\mathbb{S}^{q-2}\| \left(\frac{r}{2\pi}\right)^{\frac{q-1}{2}} \sum_{n=0}^{\infty} Y_n(q;\xi) \int_{-1}^{1} e^{irt} P_n(q;t)(1-t^2)^{\frac{q-3}{2}}\, dt\,. \tag{6.584}$$

Applying the Rodrigues formula (6.180) we find

$$U(r\xi) = \|\mathbb{S}^{q-2}\| \left(\frac{r}{2\pi}\right)^{\frac{q-1}{2}} \Gamma\left(\frac{q-1}{2}\right) \tag{6.585}$$

$$\times \sum_{n=0}^{\infty} \frac{(-\frac{1}{2})^n}{\Gamma\left(n+\frac{q-1}{2}\right)} Y_n(q;\xi) \int_{-1}^{1} e^{irt} \left(\frac{d}{dt}\right)^n (1-t^2)^{n+\frac{q-3}{2}}\, dt.$$

Collecting our results we therefore obtain

Lemma 6.45. *Let F be of class $C^{(0)}(\mathbb{S}^{q-1})$. Then*

$$U(r\xi) = \left(\frac{r}{2\pi}\right)^{\frac{q-1}{2}} \int_{\mathbb{S}^{q-1}} F(\eta)\, e^{ir\xi\cdot\eta}\, dS_{q-1}(\eta) \tag{6.586}$$

can be represented in the form

$$U(r\xi) = \frac{\|\mathbb{S}^{q-1}\|}{\sqrt{\pi}} \Gamma\left(\frac{q}{2}\right) \left(\frac{r}{2\pi}\right)^{\frac{q-1}{2}} \sum_{n=0}^{\infty} \frac{(-\frac{1}{2})^n}{\Gamma\left(n+\frac{q-1}{2}\right)} A_n(q;r) Y_n(q;\xi), \tag{6.587}$$

where

$$A_n(q;r) = \int_{-1}^{1} e^{irt} \left(\frac{d}{dt}\right)^n (1-t^2)^{n+\frac{q-3}{2}}\, dt. \tag{6.588}$$

In what follows, the integral $A_n(q;r)$ will be evaluated separately for even dimensions $q = 2p+2$ and odd dimensions $q = 2p+3$ $(p \geq 0)$. In doing so we are led back to tools of the one-dimensional Fourier theory listed in Subsection 3.2. Moreover, we have to remember our results on the convergence of the series $\sum_{n=0}^{\infty} Y_n(q;\xi)$, where $Y_n(q;\xi)$ is defined by (6.583) (see Subsection 6.1). In fact, assuming that F is of class $C^{(k)}(\mathbb{S}^{q-1})$ with $k \geq \frac{q}{2}+1$ we are able to use Lemma 6.45 in order to develop an asymptotic expansion of type (6.577) for the particular integral representation U given by (6.578).

From Lemma 6.45 we are able to derive the following asymptotic relation which plays an important role in our later work concerned with lattice point sums.

Theorem 6.30. *Let F be of class $C^{(k)}(\mathbb{S}^{q-1})$, $k \geq \frac{q}{2}+1$. Then, uniformly with respect to all $\xi \in \mathbb{S}^{q-1}$, we have for $r \to \infty$*

$$\left(\frac{r}{2\pi}\right)^{\frac{q-1}{2}} \int_{\mathbb{S}^{q-1}} F(\eta) e^{ir\xi\cdot\eta} \, dS_{(q-1)}(\eta) \tag{6.589}$$

$$= i^{\frac{1-q}{2}} e^{ir} F(\xi) + i^{\frac{q-1}{2}} e^{-ir} F(-\xi) + o(1).$$

Proof. The point of departure is the series expansion (6.587).

For $q = 2p+2$ $(p \geq 0)$, elementary calculations yield

$$\int_{-1}^{1} e^{irt} \left(\frac{d}{dt}\right)^n (1-t^2)^{n+\frac{q-3}{2}} \, dt \tag{6.590}$$

$$= \left(-\frac{1}{ir}\right)^p (-2)^{n+p} \frac{\Gamma(n+\frac{2p-1}{2})}{\sqrt{\pi}} \int_{-1}^{1} e^{irt} P_{n+p}(2;t)(1-t^2)^{-\frac{1}{2}} \, dt$$

such that

$$U(r\xi) = \frac{2^{\frac{q-3}{2}}}{\sqrt{\pi}} \frac{\|\mathbb{S}^{q-1}\|}{(2\pi)^{\frac{q-1}{2}}} i^{\frac{2-q}{2}} \Gamma\left(\frac{q}{2}\right) \sqrt{\frac{2r}{\pi}} \int_{-1}^{1} e^{irt} \Phi(t)(1-t^2)^{-\frac{1}{2}} \, dt, \tag{6.591}$$

where

$$\Phi(t) = \sum_{n=0}^{\infty} Y_n(q;\xi) P_{n+p}(2;t). \tag{6.592}$$

The series $t \mapsto \Phi(t)$, $t \in [-1,1]$, is absolutely and uniformly convergent; hence, it is continuous in $[-1,1]$, continuously differentiable in $(-1,1)$, and absolutely integrable on the interval $[-1,1]$. From the one-dimensional Fourier theory (i.e., Lemma 3.16) we are therefore able to borrow the following limit relation

$$\sqrt{\frac{2r}{\pi}} \int_{-1}^{1} e^{irt} \Phi(t)(1-t^2)^{-\frac{1}{2}} \, dt \tag{6.593}$$

$$= \Phi(1) e^{ir} i^{-\frac{1}{2}} + \Phi(-1) e^{-ir} i^{\frac{1}{2}} + o(1)$$

for $r \to \infty$, where

$$\Phi(1) = \sum_{n=0}^{\infty} Y_n(q; \xi) \tag{6.594}$$

$$= \sum_{n=0}^{\infty} \frac{N(q; n)}{\|\mathbb{S}^{q-1}\|} \int_{\mathbb{S}^{q-1}} F(\eta) P_n(q; \xi \cdot \eta) \, dS_{(q-1)}(\eta)$$

$$= F(\xi)$$

and

$$\Phi(-1) = (-1)^p \sum_{n=0}^{\infty} Y_n(q; -\xi) = (-1)^p F(-\xi). \tag{6.595}$$

Thus, with $q = 2p + 2$ $(p \geq 0)$, we obtain for $r \to \infty$

$$U(r\xi) = \left(\frac{r}{2\pi} \right)^{\frac{q-1}{2}} \int_{\mathbb{S}^{q-1}} F(\eta) e^{ir\xi \cdot \eta} \, dS_{(q-1)}(\eta) \tag{6.596}$$

$$= \frac{\sqrt{\pi} \, r^{\frac{q-1}{2}}}{\|\mathbb{S}^{q-1}\| \, \Gamma\left(\frac{q}{2} \right) 2^{\frac{q-3}{2}}} \int_{\mathbb{S}^{q-1}} F(\eta) e^{ir\xi \cdot \eta} \, dS_{(q-1)}(\eta)$$

$$= e^{i\left(r - \frac{\pi}{4}(q-1)\right)} F(\xi) + e^{-i\left(r - \frac{\pi}{4}(q-1)\right)} F(-\xi) + o(1).$$

For $q = 2p + 3$ $(p \geq 0)$ we again insert the Rodrigues formula (6.180) and find by partial integration

$$U(r\xi) = \frac{\|\mathbb{S}^{2p+2}\|}{\sqrt{\pi}} \Gamma\left(\frac{2p+3}{2} \right) \frac{2^p}{(2\pi)^{p+1}} i^{-p} \, r \int_{-1}^{1} e^{irt} \Psi(t) \, dt, \tag{6.597}$$

where

$$\Psi(t) = \sum_{n=0}^{\infty} Y_n(q; \xi) P_{n+p}(3; t). \tag{6.598}$$

It follows after simple manipulations that

$$U(r\xi) = \frac{\|\mathbb{S}^{2p+2}\|}{\sqrt{\pi}} \Gamma\left(\frac{2p+3}{2} \right) \frac{2^p}{(2\pi)^{p+1}} i^{-\frac{2p-1}{2}} \left(e^{irt} \Psi(t) \right) \Big|_{t=-1}^{t=1} \tag{6.599}$$

$$- \frac{\|\mathbb{S}^{2p+2}\|}{\sqrt{\pi}} \Gamma\left(\frac{2p+3}{2} \right) \frac{2^p}{(2\pi)^{p+1}} i^{-p} \int_{-1}^{1} e^{irt} \Psi'(t) \, dt.$$

The one-dimensional Fourier theory (i.e., Lemma 3.15) tells us that

$$\lim_{r \to \infty} \int_{-1}^{1} e^{irt} \Psi'(t) \, dt = 0 \tag{6.600}$$

(note that $\Psi'(t)$ defines a continuous function on $(-1, +1)$ and an absolutely integrable function on the interval $[-1, +1]$). In addition, we have

$$\Psi(1) = F(\xi), \quad \Psi(-1) = (-1)^p F(-\xi). \tag{6.601}$$

Thus, with $q = 2p + 3$ $(p \geq 0)$, we find

$$U(r\xi)$$
$$= \frac{\|\mathbb{S}^{q-1}\|}{\sqrt{\pi}} \Gamma\left(\frac{q}{2}\right) \frac{2^{\frac{q-3}{2}}}{(2\pi)^{\frac{q-1}{2}}} \left(e^{i(r - \frac{\pi}{4}(q-1))} F(\xi) + e^{-i(r - \frac{\pi}{4}(q-1))} F(-\xi)\right)$$
$$+ o(1), \qquad r \to \infty. \tag{6.602}$$

Summarizing our results for odd as well as even dimensions we arrive at the desired result of Theorem 6.30 for all dimensions $q \geq 2$. $\qquad\square$

Canonical Extensions

In an analogous way as in the theory of Bessel functions, Theorem 6.30 admits a canonical extension.

Corollary 6.7. *Let F be of class $C^{(k)}(\mathbb{S}^{q-1})$ with $k \geq \frac{q}{2} + 2m - 1, m \in \mathbb{N}$. Then*

$$\left(\frac{r}{2\pi}\right)^{\frac{q-1}{2}} \int_{\mathbb{S}^{q-1}} e^{ir\xi \cdot \eta} F(\eta) \, dS_{(q-1)}(\eta) \tag{6.603}$$

$$= i^{\frac{1-q}{2}} e^{ir} \sum_{l=0}^{m-1} \left(\frac{1}{-2ir}\right)^l O(q;l) F(\xi) + i^{\frac{q-1}{2}} e^{-ir} \sum_{l=0}^{m-1} \left(\frac{1}{2ir}\right)^l O(q;l) F(-\xi)$$

$$+ o\left(r^{1-m}\right),$$

where

$$O(q;0) = 1, \tag{6.604}$$

$$O(q;l) = \frac{1}{l!} \prod_{j=1}^{l} \left(\frac{(q + 2j - 3)(q - 2j - 1)}{4} - \Delta^*\right), \ l \in \mathbb{N}, \tag{6.605}$$

and, as always, Δ^* is the Laplace–Beltrami operator on the unit sphere $\mathbb{S}^{q-1}$.

Obviously, for the constant function $F = 1$, Corollary 6.7 leads back to Lemma 6.38.

7

Preparatory Tools of Fourier Analysis

CONTENTS

7.1   Periodical Polynomials and Fourier Expansions 224
      Periodical Polynomials 224
      Eigenspectrum of the Laplace Operator 225
      Absolute Convergent Fourier Series 226
7.2   Classical Fourier Transform 227
      Definition 227
      Convolutions 228
      Important Operations 228
7.3   Poisson Summation and Periodization 229
      First Periodization 230
      Second Periodization 230
      Poisson Summation Formula 230
7.4   Gauß–Weierstraß and Abel–Poisson Transforms 232
      Exponential Integrals Involving Bessel Functions 233
      Gauß–Weierstraß Transform 236
      Abel–Poisson Transform 241
7.5   Hankel Transform and Discontinuous Integrals 243
      Hankel Transform 243
      Discontinuous Integrals 245

This chapter presents basic material on the classical Fourier theory in higher dimensions. It starts with multi-dimensional orthonormal periodical polynomials and their role in Fourier expansions (in Section 7.1). The Fourier transform is discussed in the standard reference space of (Lebesgue) absolutely integrable functions over the Euclidean space $\mathbb{R}^q$ (in Section 7.2). Our main interest is the relation between functions being absolutely integrable over $\mathbb{R}^q$ as well as periodical with respect to the unit lattice $\mathbb{Z}^q \subset \mathbb{R}^q$. In consequence, we are immediately led to the process of "periodization" as "bridging tool", i.e., the Poisson summation formula in $\mathbb{R}^q$. The results are formulated for dimensions $q \geq 2$ (in Section 7.3). For our lattice point summation (as intended, e.g., in Chapters 10 and 14), however, it must be emphasized that our technique of realizing the process of periodization in $\mathbb{R}^q$, $q \geq 2$, is different from those developed in the classical literature, e.g., by E.M. Stein, G. Weiss [1971]. In fact, these authors verify the Poisson summation formula under the strong assumption of the absolute convergence of all occurring sums. The essential calamity in lattice point theory, however, is the convergence behavior of the Fourier series expansion of the lattice function in dimensions $q \geq 2$. This fact is a striking difference to the one-dimensional theory. Moreover, in contrast to the one-dimensional case, another serious difficulty is the Fourier inversion formula. As a powerful remedy, the inversion formula of Fourier integrals can be understood in the terminology of certain means (such as Gauß–Weierstraß or Abel–Poisson transforms in Section 7.4). Furthermore it turns out that the integral transform for discontinuous functions possessing a "potato-like" regular region as a local support is critical for Fourier inversion. Nevertheless, at least in the case of spherical geometry, the Hankel transform provides a way out for handling alternating, not absolutely convergent series expansions in terms of Bessel functions.

7.1   Periodical Polynomials and Fourier Expansions

First the standard Λ-periodical polynomials (orthonormal in $\mathrm{L}_\Lambda^2(\mathbb{R}^q)$-sense) are listed. Equivalent conditions for the closure and completeness are formulated within the space $\mathrm{L}_\Lambda^2(\mathbb{R}^q)$ of square-integrable Λ-periodical functions in $\mathbb{R}^q$.

Periodical Polynomials

Let Λ be a lattice in $\mathbb{R}^q$. The functions Φ_h, $h \in \Lambda^{-1}$, defined by

$$\Phi_h(x) = \frac{1}{\sqrt{\|\mathcal{F}\|}}\, e(h \cdot x) = \frac{1}{\sqrt{\|\mathcal{F}\|}}\, e^{2\pi i h \cdot x}, \quad x \in \mathbb{R}^q, \tag{7.1}$$

are Λ-*periodical*, i.e.,

$$\Phi_h(x+g) = \Phi_h(x) \tag{7.2}$$

for all $x \in \mathbb{R}^q$ and all $g \in \Lambda$.

Remark 7.1. *There exists a natural identification of $\mathbb{R}^q/\mathbb{Z}^q$ with the q-torus*

$$\left\{ \left(e^{2\pi i x_1}, \ldots, e^{2\pi i x_q}\right)^T \in \mathbb{C}^q \;\middle|\; (x_1, \ldots, x_q)^T \in \mathbb{R}^q \right\}. \tag{7.3}$$

This identification is given by the mapping

$$(x_1, \ldots, x_q)^T \mapsto \left(e^{2\pi i x_1}, \ldots, e^{2\pi i x_q}\right)^T. \tag{7.4}$$

From (7.4) we obtain the standard identification of $\mathbb{Z}^q$-periodical functions on $\mathbb{R}^q$ with functions on the q-torus.

The space of all $F \in \mathrm{C}^{(m)}(\mathbb{R}^q)$ that are Λ-periodical is denoted by $\mathrm{C}_\Lambda^{(m)}(\mathbb{R}^q)$, $0 \le m \le \infty$. $\mathrm{L}_\Lambda^p(\mathbb{R}^q)$, $1 \le p < \infty$, is the space of all $F : \mathbb{R}^q \to \mathbb{C}$ that are Λ-periodical and are Lebesgue-measurable on $\mathcal{F}$ with

$$\|F\|_{\mathrm{L}_\Lambda^p(\mathbb{R}^q)} = \left(\int_{\mathcal{F}} |F(x)|^p \, dV(x) \right)^{\frac{1}{p}} < \infty. \tag{7.5}$$

Clearly, $C_\Lambda^{(0)}(\mathbb{R}^q) \subset L_\Lambda^p(\mathbb{R}^q)$.

As is well known, $L_\Lambda^2(\mathbb{R}^q)$ is the completion of $C_\Lambda^{(0)}(\mathbb{R}^q)$ with respect to the norm $\|\cdot\|_{L_\Lambda^2(\mathbb{R}^q)}$:

$$L_\Lambda^2(\mathbb{R}^q) = \overline{C_\Lambda^{(0)}(\mathbb{R}^q)}^{\|\cdot\|_{L_\Lambda^2(\mathbb{R}^q)}}. \tag{7.6}$$

Eigenspectrum of the Laplace Operator

An easy calculation shows

$$\int_\mathcal{F} \Phi_h(x)\overline{\Phi_{h'}(x)} \, dV(x) = \delta_{hh'} = \begin{cases} 1 & , \quad h = h' \\ 0 & , \quad h \neq h'. \end{cases} \tag{7.7}$$

In other words, the system $\{\Phi_h\}_{h\in\Lambda^{-1}}$ of multi-dimensional "periodic polynomials" is orthonormal with respect to the $L_\Lambda^2(\mathbb{R}^q)$-inner product:

$$(\Phi_h, \Phi_{h'})_{L_\Lambda^2(\mathbb{R}^q)} = \int_\mathcal{F} \Phi_h(x)\overline{\Phi_{h'}(x)} \, dV(x) = \delta_{hh'}. \tag{7.8}$$

An elementary calculation yields

$$(\Delta_x + \Delta^\wedge(h)) \, \Phi_h(x) = 0, \quad \Delta^\wedge(h) = 4\pi^2 h^2, \quad h \in \Lambda^{-1}. \tag{7.9}$$

We shall say that λ is an *eigenvalue of the lattice* Λ with respect to the operator Δ if there is a non-trivial solution U of the differential equation $(\Delta + \lambda) \, U = 0$ satisfying the "boundary condition of periodicity" $U(x + g) = U(x)$ for all $g \in \Lambda$. The function U is then called an *eigenfunction of the lattice* Λ with respect to the eigenvalue λ and the operator Δ.

In analogy to the one-dimensional case we are able to see that the functions Φ_h are the only eigenfunctions. Furthermore, the scalars

$$\Delta^\wedge(h) = 4\pi^2 h^2, \; h \in \Lambda^{-1}, \tag{7.10}$$

are the only eigenvalues of Δ with respect to the lattice Λ (note that we simply write $\Delta^\wedge(h)$ instead of $\Delta_\Lambda^\wedge(h)$ if no confusion is likely to arise). The set of all eigenvalues $\Delta^\wedge(h)$ with respect to Δ is the *spectrum* $\mathrm{Spect}_\Delta(\Lambda)$:

$$\mathrm{Spect}_\Delta(\Lambda) = \{\Delta^\wedge(h)| \; \Delta^\wedge(h) = 4\pi^2 h^2, h \in \Lambda^{-1}\}. \tag{7.11}$$

Clearly, an analogue of Theorem 4.1 also holds true for the multi-dimensional case; i.e., the system $\{\Phi_h\}_{h\in\Lambda^{-1}}$ is closed and complete in the pre-Hilbert space $(C_\Lambda^{(0)}(\mathbb{R}^q); \|\cdot\|_{L_\Lambda^2(\mathbb{R}^q)})$ as well as in the Hilbert space $(L_\Lambda^2(\mathbb{R}^q); \|\cdot\|_{L_\Lambda^2(\mathbb{R}^q)})$. In this respect, the fundamental result in Fourier analysis is that each $F \in L_\Lambda^2(\mathbb{R}^q)$ can be represented by its (orthogonal) Fourier series in the sense

$$\lim_{N\to\infty} \left\| F - \sum_{\substack{|h|\leq N \\ h\in\Lambda^{-1}}} F_\Lambda^\wedge(h)\Phi_h \right\|_{L_\Lambda^2(\mathbb{R}^q)} = 0, \tag{7.12}$$

where the Fourier coefficients $F_\Lambda^\wedge(h)$ of F are given by

$$F_\Lambda^\wedge(h) = \int_{\mathcal{F}} F(x)\overline{\Phi_h(x)}\, dV(x), \quad h \in \Lambda^{-1}. \tag{7.13}$$

The Parseval identity then tells us that, for each $F \in L_\Lambda^2(\mathbb{R}^q)$,

$$\int_{\mathcal{F}} |F(x)|^2\, dV(x) = \sum_{h \in \Lambda^{-1}} |F_\Lambda^\wedge(h)|^2. \tag{7.14}$$

Absolute Convergent Fourier Series

A useful corollary of (7.12) is that any function $F \in L_\Lambda^1(\mathbb{R}^q)$ with

$$\sum_{h \in \Lambda^{-1}} |F_\Lambda^\wedge(h)| < \infty \tag{7.15}$$

can be modified on a set of measure zero so that it is in $C_\Lambda^{(0)}(\mathbb{R}^q)$ and equals the Fourier series $\sum_{h \in \Lambda^{-1}} F_\Lambda^\wedge(h)\Phi_h$ for all $x \in \mathcal{F}$ (see, e.g., E.M. Stein, G. Weiss [1971]).

Suppose that F is of class $C_\Lambda^{(k)}(\mathbb{R}^q)$ with $k > \frac{q}{2}$. Then

$$\int_{\mathcal{F}} (\nabla^\alpha F)(x)\, \overline{\Phi_h(x)}\, dV(x) = (-2\pi i h)^\alpha F_\Lambda^\wedge(h), \tag{7.16}$$

whenever $F \in C_\Lambda^{(k)}(\mathbb{R}^q)$ and $\alpha = (\alpha_1,\ldots,\alpha_q)^T$ is chosen such that $[\alpha] = \alpha_1 + \ldots + \alpha_q \leq k$. Since $\nabla^\alpha F$ is continuous in $\mathcal{F}$ it must belong to $L_\Lambda^2(\mathbb{R}^q)$. In other words, for all $n \leq k$,

$$\sum_{[\alpha]=n} \left(\sum_{h \in \Lambda^{-1}} |F_\Lambda^\wedge(h)|^2 ((2\pi h)^\alpha)^2 \right) < \infty. \tag{7.17}$$

Moreover, there exits a constant C (dependent on k, q) such that

$$C|h|^{2k} \leq \sum_{[\alpha]=k} ((2\pi h)^\alpha)^2. \tag{7.18}$$

From the Cauchy–Schwarz inequality we therefore obtain for all $N > 0$

$$
\sum_{\substack{0<|h|\leq N \\ h\in\Lambda^{-1}}} |F_\Lambda^\wedge(h)| \leq \sum_{\substack{0<|h|\leq N \\ h\in\Lambda^{-1}}} |F_\Lambda^\wedge(h)| \left(\sum_{[\alpha]=k} ((2\pi h)^\alpha)^2 \right)^{1/2} C^{-1/2} |h|^{-k}
$$

$$
\leq \left(\sum_{\substack{0<|h|\leq N \\ h\in\Lambda^{-1}}} |F_\Lambda^\wedge(h)|^2 \sum_{[\alpha]=k} ((2\pi h)^\alpha)^2 \right)^{1/2} \tag{7.19}
$$

$$
\times \left(\sum_{\substack{0<|h|\leq N \\ h\in\Lambda^{-1}}} |h|^{-2k} \right)^{1/2} C^{-1/2}.
$$

If $k > \frac{q}{2}$, the sum $\sum_{0<|h|\leq N} |h|^{-2k}$ is finite for $N \to \infty$; hence, the last expression must be finite. This leads to the following statement relating the smoothness of a function and the absolute convergence of its Fourier series.

Theorem 7.1. *If F is of class $C_\Lambda^{(k)}(\mathbb{R}^q)$, where k is an integer with $k > \frac{q}{2}$, then $\sum_{h\in\Lambda^{-1}} |F_\Lambda^\wedge(h)| < \infty$, where $F_\Lambda^\wedge(h)$ are the Fourier coefficients of F.*

7.2   Classical Fourier Transform

This section presents basic results on the multi-dimensional Fourier transform on Euclidean spaces, however, restricted to the standard reference space $L^1(\mathbb{R}^q)$ of (Lebesgue) absolutely integrable functions in the Euclidean space $\mathbb{R}^q$. The Fourier inversion formula is explained in its classical framework that a function as well as its Fourier transform are of class $L^1(\mathbb{R}^q)$.

Definition

In a standard way we deal with various spaces of functions defined on $\mathbb{R}^q$. $L^p(\mathbb{R}^q)$, $1 \leq p < \infty$, are the spaces of all measurable functions $F : \mathbb{R}^q \to \mathbb{C}$ such that

$$
\|F\|_{L^p(\mathbb{R}^q)} = \left(\int_{\mathbb{R}^q} |F(x)|^p \, dV(x) \right)^{1/p} < \infty. \tag{7.20}
$$

The number $\|F\|_{L^p(\mathbb{R}^q)}$ is called the L^p-norm of F. The *Fourier transform* of $F : \mathbb{R}^q \to \mathbb{C}$ at $x \in \mathbb{R}^q$ is defined by

$$
\mathcal{F}_{\mathbb{R}^q}(F)(x) = F_{\mathbb{R}^q}^\wedge(x) = \int_{\mathbb{R}^q} F(y) \, e^{-2\pi i x \cdot y} \, dV(y). \tag{7.21}
$$

If $F \in L^1(\mathbb{R}^q)$, then it is well known (for more details see, e.g., E.M. Stein, G. Weiss [1971]) that the function $\mathcal{F}_{\mathbb{R}^q}(F)$ is uniformly continuous. Moreover, the Riemann–Lebesgue theorem of Fourier analysis tells us that $F_{\mathbb{R}^q}^{\wedge}(x) \to 0$ as $|x| \to \infty$.

Convolutions

$L^1(\mathbb{R}^q)$ is endowed with a *convolution* defined in the following way: If F, G belong to $L^1(\mathbb{R}^q)$, their convolution $H = F * G$ is the function whose value at $x \in \mathbb{R}^q$ is given by

$$H(x) = \int_{\mathbb{R}^q} F(x - y)G(y) \, dV(y). \tag{7.22}$$

Clearly, $H \in L^1(\mathbb{R}^q)$ and $\|H\|_{L^1(\mathbb{R}^q)} \leq \|F\|_{L^1(\mathbb{R}^q)} \|G\|_{L^1(\mathbb{R}^q)}$. The operation is commutative and associative. Even more, if $F \in L^p(\mathbb{R}^q)$, $1 \leq p < \infty$, and $G \in L^1(\mathbb{R}^q)$, then $H = F * G$ belongs to $L^p(\mathbb{R}^q)$, and we have

$$\|H\|_{L^p(\mathbb{R}^q)} \leq \|F\|_{L^p(\mathbb{R}^q)} \|G\|_{L^1(\mathbb{R}^q)}. \tag{7.23}$$

An essential feature is that the Fourier transform of the convolution of two functions is the (pointwise) product of their Fourier transforms. More concretely, if F and G belong to $L^1(\mathbb{R}^q)$, then

$$\mathcal{F}_{\mathbb{R}^q}(F * G) = (F * G)_{\mathbb{R}^q}^{\wedge} = F_{\mathbb{R}^q}^{\wedge} G_{\mathbb{R}^q}^{\wedge} = \mathcal{F}_{\mathbb{R}^q}(F)\mathcal{F}_{\mathbb{R}^q}(G). \tag{7.24}$$

Important Operations

Many other important operations have simple relations with the Fourier transform:

If τ_d denotes the *translation* by $d \in \mathbb{R}^q$ (by this we mean the operator mapping the function $F \in L^1(\mathbb{R}^q)$ into the function $\tau_d F = F(\cdot - d)$), then

$$(\tau_d F)_{\mathbb{R}^q}^{\wedge}(x) = e^{-2\pi i d \cdot x} F_{\mathbb{R}^q}^{\wedge}(x), \quad x \in \mathbb{R}^q \tag{7.25}$$

and

$$(e^{2\pi i d \cdot x} F)_{\mathbb{R}^q}^{\wedge}(x) = (\tau_d F_{\mathbb{R}^q}^{\wedge})(x), \quad x \in \mathbb{R}^q. \tag{7.26}$$

If $\delta_\tau, \tau > 0$, denotes the *dilation* (by this we mean the operator mapping $F \in L^1(\mathbb{R}^q)$ into the function $\delta_\tau F = F(\tau \cdot)$), then

$$\tau^q (\delta_\tau F)_{\mathbb{R}^q}^{\wedge}(x) = F_{\mathbb{R}^q}^{\wedge}(\tau^{-1} x), \quad x \in \mathbb{R}^q. \tag{7.27}$$

If $x \mapsto P(x), x \in \mathbb{R}^q$, is a polynomial in q variables $x = (x_1, \ldots, x_q)^T$ and $P(\nabla)$

is the associated differential operator (i.e., replace $x_i = x \cdot \epsilon^i$ by $\frac{\partial}{\partial x_i} = \nabla \cdot \epsilon^i$, $i \in \{1, \dots, q\}$), then

$$P(\nabla)F_{\mathbb{R}^q}^\wedge(x) = (P(-2\pi i \cdot)F)_{\mathbb{R}^q}^\wedge(x), \quad x \in \mathbb{R}^q, \tag{7.28}$$

and

$$(P(\nabla)F)_{\mathbb{R}^q}^\wedge(x) = P(-2\pi i x)F_{\mathbb{R}^q}^\wedge(x), \quad x \in \mathbb{R}^q. \tag{7.29}$$

Applying Fubini's theorem we immediately have the identity

$$\int_{\mathbb{R}^q} F_{\mathbb{R}^q}^\wedge(x)G(x)\, dV(x) = \int_{\mathbb{R}^q} F(x)G_{\mathbb{R}^q}^\wedge(x)\, dV(x) \tag{7.30}$$

provided that F, G belong to $\mathrm{L}^1(\mathbb{R}^q)$. Moreover, if $F \in \mathrm{L}^1(\mathbb{R}^q)$, $G \in \mathrm{L}^p(\mathbb{R}^q)$, $1 \le p \le 2$, then $H = F * G \in \mathrm{L}^p(\mathbb{R}^q)$ satisfies $H_{\mathbb{R}^q}^\wedge(x) = F_{\mathbb{R}^q}^\wedge(x)G_{\mathbb{R}^q}^\wedge(x)$ (for almost every x). Finally, we mention the (Fourier) *inversion formula*

$$F(x) = \int_{\mathbb{R}^q} F_{\mathbb{R}^q}^\wedge(y)e^{2\pi i x \cdot y}\, dV(y) \tag{7.31}$$

(for almost every x) in its standard framework that $F, F_{\mathbb{R}^q}^\wedge$ belong to $\mathrm{L}^1(\mathbb{R}^q)$. In operator notation the inversion formula briefly reads as follows

$$\mathcal{F}_{\mathbb{R}^q}\left(\mathcal{F}_{\mathbb{R}^q}(F)\right)(-x) = F(x). \tag{7.32}$$

Remark 7.2. *The assumption that $F, F_{\mathbb{R}^q}^\wedge$ belong to $\mathrm{L}^1(\mathbb{R}^q)$ is usually not satisfied in the lattice point theory as discussed in this work. For example, for two-dimensional weighted Hardy–Landau identities (as, e.g., worked out in Chapter 13) we need a pointwise approach to the Fourier inversion formula that is applicable for alternating functions that do not belong to the class $\mathrm{L}^1(\mathbb{R}^2)$.*

7.3 Poisson Summation and Periodization

As already announced, the particular emphasis of this section is the relation between functions being absolutely integrable over $\mathbb{R}^q$ as well as periodical with respect to the lattice $\mathbb{Z}^q \subset \mathbb{R}^q$. This leads to a first understanding of *"periodization"*.

Suppose we are given an (appropriate) function F in the Euclidean space $\mathbb{R}^q$. The question now is what is its periodical analogue with respect to the lattice $\mathbb{Z}^q$? Our purpose is to show that there are at least two viable ways of obtaining a periodization of the function under consideration which fortunately turn out to be equivalent under suitable circumstances.

First Periodization

The *"first periodization"* with $\mathbb{Z}^q$ is straightforward:

$$x \mapsto \sum_{g \in \mathbb{Z}^q} F(x + g). \tag{7.33}$$

Since this (formal) sum is extended over all points of the lattice $\mathbb{Z}^q$ it is obviously $\mathbb{Z}^q$-periodical.

Second Periodization

The second concept of periodization is based on the discretization of the "inversion formula"

$$F(x) = \int_{\mathbb{R}^q} F^\wedge_{\mathbb{R}^q}(y) \, e^{2\pi i x \cdot y} \, dV(y), \tag{7.34}$$

where

$$F^\wedge_{\mathbb{R}^q}(y) = \int_{\mathbb{R}^q} F(x) \, e^{-2\pi i y \cdot x} \, dV(x). \tag{7.35}$$

Hence, the *"second periodization"* amounts to the discretization of (7.34), i.e., the "Poisson summation" within the framework of the lattice $\mathbb{Z}^q$ as follows

$$x \mapsto \sum_{h \in \mathbb{Z}^q} F^\wedge_{\mathbb{R}^q}(h) e^{2\pi i h \cdot x}. \tag{7.36}$$

Poisson Summation Formula

The *Poisson summation formula in Euclidean space* $\mathbb{R}^q$ tells us that the two approaches to a $\mathbb{Z}^q$-periodical analogue of F, i.e., the two periodizations (7.33) and (7.36), are identical. Mathematically this conclusion is expressible precisely in many ways.

First we recapitulate a standard manifestation (see, e.g., E.M. Stein, G. Weiss [1971]) formulated by the following theorem.

Theorem 7.2. *Suppose that F is of class* $\mathrm{L}^1(\mathbb{R}^q)$. *Then the function*

$$x \mapsto \sum_{g \in \mathbb{Z}^q} F(x + g) \tag{7.37}$$

converges in the norm $\| \cdot \|_{\mathrm{L}^1_{\mathbb{Z}^q}(\mathbb{R}^q)}$. *The resulting function (i.e., the first periodization) in* $\mathrm{L}^1_{\mathbb{Z}^q}(\mathbb{R}^q)$ *has the Fourier expansion (i.e., the second periodization)*

$$\sum_{h \in \mathbb{Z}^q} \int_{\mathbb{R}^q} F(y) \, e^{-2\pi i h \cdot y} \, dV(y) \, e^{2\pi i h \cdot x}. \tag{7.38}$$

In other words, $\{F^\wedge_{\mathbb{R}^q}(h)\}_{h \in \mathbb{Z}^q}$ *are the Fourier coefficients of the* L^1-*function (7.37) defined by the series* $\sum_{g \in \mathbb{Z}^q} F(x + g)$.

Proof. As usual, denote by $\mathcal{F} - \{g\}$ the translate of $\mathcal{F}$ by the lattice point $g \in \mathbb{Z}^q$. It is not difficult to verify that

$$\int_{\mathcal{F}} \left| \sum_{g \in \mathbb{Z}^q} F(x+g) \right| dV(x) \leq \sum_{g \in \mathbb{Z}^q} \int_{\mathcal{F}} |F(x+g)| \, dV(x) \qquad (7.39)$$

$$= \sum_{g \in \mathbb{Z}^q} \int_{\mathcal{F} - \{g\}} |F(x)| \, dV(x).$$

Since the fundamental cell $\mathcal{F}$ of the lattice $\mathbb{Z}^q$ defines translates $\mathcal{F} - \{g\}, g \in \mathbb{Z}^q$, that are mutually disjoint, their union is $\mathbb{R}^q$. This implies the identity

$$\sum_{g \in \mathbb{Z}^q} \int_{\mathcal{F} - \{g\}} |F(x)| \, dV(x) = \int_{\mathbb{R}^q} |F(x)| \, dV(x). \qquad (7.40)$$

Thus, according to our assumption imposed on F, the series $\sum_{g \in \mathbb{Z}^q} F(x+g)$ converges in the norm $\| \cdot \|_{L^1_{\mathbb{Z}^q}(\mathbb{R}^q)}$. The Fourier coefficients of the series are calculable as follows

$$\int_{\mathcal{F}} \sum_{g \in \mathbb{Z}^q} F(x+g) \, \overline{e^{2\pi i h \cdot x}} \, dV(x) = \sum_{g \in \mathbb{Z}^q} \int_{\mathcal{F} - \{g\}} F(x) \, e^{-2\pi i h \cdot x} \, dV(x)$$

$$= \int_{\mathbb{R}^q} F(x) \, e^{-2\pi i h \cdot x} \, dV(x) \qquad (7.41)$$

$$= F_{\mathbb{R}^q}^{\wedge}(h).$$

Note that the L^1-convergence allows the interchange of summation and integration. Hence, $\sum_{g \in \mathbb{Z}^q} F(x+g)$ has the Fourier series $\sum_{h \in \mathbb{Z}^q} F_{\mathbb{R}^q}^{\wedge}(h) e^{2\pi i h \cdot x}$, as indicated by Theorem 7.2. $\qquad \square$

The following corollary is a first attempt to develop the Poisson summation formula for the standard lattice $\mathbb{Z}^q$ in Euclidean space $\mathbb{R}^q$, however, under conditions implying the absolute convergence of all occurring series involved in the process of periodization. A consequence is the continuity of the function F as well as its Fourier transform $F_{\mathbb{R}^q}^{\wedge}$.

Theorem 7.3. *Suppose that the functions F and $F_{\mathbb{R}^q}^{\wedge}$ satisfy the conditions*

$$F(x) = O\left(\frac{1}{|x|^{q+\varepsilon}}\right), \quad \varepsilon > 0, \ |x| \to \infty, \qquad (7.42)$$

$$F_{\mathbb{R}^q}^{\wedge}(x) = O\left(\frac{1}{|x|^{q+\varepsilon}}\right), \quad \varepsilon > 0, \ |x| \to \infty \qquad (7.43)$$

with

$$F(x) = \int_{\mathbb{R}^q} F_{\mathbb{R}^q}^{\wedge}(y) e^{2\pi i x \cdot y} \, dV(y), \quad x \in \mathbb{R}^q. \qquad (7.44)$$

Then, F and $F_{\mathbb{R}^q}^{\wedge}$ are continuous in $\mathbb{R}^q$ such that

$$\sum_{g\in\mathbb{Z}^q} F(x+g) = \sum_{h\in\mathbb{Z}^q} \int_{\mathbb{R}^q} F(y)\,e^{-2\pi i h\cdot y}\,dV(y)\ e^{2\pi i h\cdot x}. \qquad (7.45)$$

In particular,

$$\sum_{g\in\mathbb{Z}^q} F(g) = \sum_{h\in\mathbb{Z}^q} \int_{\mathbb{R}^q} F(y)e^{-2\pi i h\cdot y}\,dV(y). \qquad (7.46)$$

The four series in the last two identities converge absolutely.

Proof. According to the assumptions of Theorem 7.3, $\Sigma_{g\in\mathbb{Z}^q} F(x+g)$ can be modified on a set of measure zero so as to be equal to the continuous sum $\Sigma_{h\in\mathbb{Z}^q} F_{\mathbb{R}^q}^{\wedge}(h)e^{2\pi i h\cdot x}$ everywhere. By comparison with $\Sigma_{g\in\mathbb{Z}^q}(1+|g|)^{-(q+\varepsilon)}$, $\varepsilon > 0$, we see that $\Sigma_{g\in\mathbb{Z}^q} F(x+g)$ is a uniformly convergent series whose terms constitute continuous functions. Thus, as its limit, the sum is continuous on $\mathbb{R}^q$; hence, the Poisson summation formula (7.44) holds true for all $x \in \mathbb{R}^q$. $\square$

Remark 7.3. *In Chapter 10, weaker (sufficient) conditions for the validity of the Poisson summation formula in Euclidean spaces $\mathbb{R}^q$ can be formulated such that the absolute convergence is not implied for the series on the left side of (7.46) (as stated in Theorem 7.3); hence, the summation process becomes important for purposes of the convergence. Indeed, in our (meta)harmonic approach we are canonically led to the spherical summation, i.e., partial sums of lattice points inside spheres, which provide significant results in analytic theory of numbers.*

7.4 Gauß–Weierstraß and Abel–Poisson Transforms

Next our aim is to discuss some weighted variants of the (inverse) Fourier transform, namely the Gauß and Weierstraß integral transform as well as the Poisson and Abel integral transform in their particular interrelation to the classical Fourier transform. In the usual $L^1(\mathbb{R}^q)$-nomenclature, the theory of these integral transforms is well understood (see, e.g., E.M. Stein, G. Weiss [1971]). In our approach, however, we are essentially inspired by C. Müller [1998] to transfer the theory of Fourier transforms (defined as spherical principal values) to so–called spherically continuous functions. In doing so we get structures and settings that are adequate in inversion procedures of lattice point theory.

Exponential Integrals Involving Bessel Functions

Our point of departure is the integral

$$\int_0^\infty e^{-r} J_0(q; ar) r^{q-1} \, dr, \quad q \geq 2. \tag{7.47}$$

Clearly, (7.47) exists for all $a \in \mathbb{R}$. We already know from the theory of Bessel functions that

$$J_0(q; ar) = \frac{\|\mathbb{S}^{q-2}\|}{\|\mathbb{S}^{q-1}\|} \int_{-1}^1 e^{iart}(1 - t^2)^{\frac{q-3}{2}} \, dt. \tag{7.48}$$

Observing (7.48) and integrating (7.47) in reversed order, we obtain

$$\int_0^\infty e^{-r} J_0(q; ar) r^{q-1} dr = \frac{\|\mathbb{S}^{q-2}\|}{\|\mathbb{S}^{q-2}\|} \Gamma(q) \int_{-1}^1 \frac{(1 - t^2)^{\frac{q-3}{2}}}{(1 - iat)^q} \, dt. \tag{7.49}$$

The last integral defines a function, which is holomorphic in the strip of all $a \in \mathbb{C}$ with $-1 < \Im(a) < 1$. Furthermore, for all $a \in \mathbb{C}$ with $|a| < 1$, we have

$$\frac{\Gamma(q)}{(1 - iat)^q} = \sum_{k=0}^\infty \frac{(iat)^k}{k!} \Gamma(k + q). \tag{7.50}$$

The odd powers of t do not contribute to (7.49), and we get

$$\Gamma(q) \int_{-1}^1 \frac{(1 - t^2)^{\frac{q-3}{2}}}{(1 - iat)^q} \, dt \tag{7.51}$$

$$= \sum_{k=0}^\infty \frac{(-a^2)^k}{(2k)!} \Gamma(2k + q) \int_{-1}^1 t^{2k}(1 - t^2)^{\frac{q-3}{2}} \, dt$$

$$= \sum_{k=0}^\infty (-a^2)^k \frac{\Gamma(2k + q)\Gamma(k + \frac{1}{2})\Gamma(\frac{q-1}{2})}{\Gamma(2k + 1)\Gamma\left(k + \frac{q}{2}\right)}.$$

The coefficient of the term $(-a^2)^k$ equals

$$\frac{2^{2k+q-1}\Gamma\left(k + \frac{q+1}{2}\right)\Gamma\left(\frac{q-1}{2}\right)}{2^{2k}\Gamma(k + 1)} = \frac{2^{q-1}\Gamma(k + \frac{q+1}{2})\Gamma\left(\frac{q-1}{2}\right)}{k!}. \tag{7.52}$$

Thus, we have for all $a \in \mathbb{C}$ with $|a| < 1$

$$\Gamma(q) \int_{-1}^1 \frac{(1 - t^2)^{\frac{q-3}{2}}}{(1 - iat)^q} \, dt = \frac{2^{q-1}\Gamma(\frac{q-1}{2})\Gamma\left(\frac{q+1}{2}\right)}{(1 + a^2)^{\frac{q+1}{2}}}. \tag{7.53}$$

Consequently, in connection with (7.49) and by observation of the well known explicit values of $\|\mathbb{S}^{q-1}\|, \|\mathbb{S}^{q-2}\|$, we find after a simple calculation

$$\int_0^\infty e^{-r} J_0(q; ar) r^{q-1} \, dr = \frac{\Gamma(q)}{(1 + a^2)^{\frac{q+1}{2}}}. \tag{7.54}$$

In the strip of all $a \in \mathbb{C}$ with $-1 < \Im(a) < 1$ both sides of the identity (7.54) are holomorphic such that the identity (7.54) is also valid for all $a \in \mathbb{R}$.

By a simple substitution the identity (7.54) leads to the following result.

Lemma 7.1. *For $t \geq 0$, $a \in \mathbb{R}$, we have*

$$\|S^{q-1}\| \int_0^\infty e^{-2\pi tr} J_0(q; 2\pi ar) r^{q-1} \, dr = \frac{2}{\|S^q\|} \frac{t}{(a^2 + t^2)^{\frac{q+1}{2}}}. \tag{7.55}$$

The next integral takes a particular role in the theory of Theta functions (see Section 12.1).

Lemma 7.2. *For $\rho > 0$, $n \in \mathbb{N}_0$, and $\sigma \in \mathbb{C}$ with $\Re(\sigma) > 0$ we have*

$$\int_0^\infty e^{-\pi \sigma r^2} J_n(q; 2\pi r\rho) \, r^{n+q-1} \, dr = \tfrac{1}{2} \Gamma \left(\tfrac{q}{2} \right) \frac{\rho^n}{\pi^{\frac{q}{2}} \sigma^{n+\frac{q}{2}}} e^{-\frac{\pi \rho^2}{\sigma}}.$$

Proof. Using the power series expansion of the Bessel function (see Lemma 6.28) we get

$$\int_0^\infty e^{-\pi \sigma r^2} J_n(q; 2\pi r\rho) r^{n+q-1} \, dr \tag{7.56}$$

$$= \lim_{T \to \infty} \frac{\Gamma \left(\tfrac{q}{2} \right)}{2^n} \int_0^T e^{-\pi \sigma r^2} r^{n+q-1} \left(\sum_{k=0}^\infty \frac{(-1)^k (2\pi r\rho)^{n+2k}}{2^{2k} \Gamma(k+1) \Gamma(n+k+\frac{q}{2})} \right) dr.$$

Because of $|e^{-\pi \sigma r^2}| < 1$ the series on the right of (7.56) is uniformly convergent, and the members of the series are continuous. Thus we are allowed to write

$$\int_0^\infty e^{-\pi \sigma r^2} J_n(q; 2\pi r\rho) \, r^{n+q-1} \, dr \tag{7.57}$$

$$= \lim_{T \to \infty} \frac{\Gamma \left(\tfrac{q}{2} \right)}{2^n} \sum_{k=0}^\infty \frac{(-1)^k (2\pi \rho)^{n+2k}}{2^{2k} \Gamma(k+1) \Gamma \left(n+k+\frac{q}{2} \right)} \int_0^T e^{-\pi \sigma r^2} \, r^{2n+2k+q-1} dr.$$

The series on the right side of (7.57) converges uniformly with respect to T. In fact, we have for sufficiently large positive N', N'' with ρ, σ, and n fixed

$$\left| \sum_{k=N'}^{N''} \frac{(-1)^k (2\pi \rho)^{n+2k}}{2^{2k} \Gamma(k+1) \Gamma(n+k+\frac{q}{2})} \int_0^T e^{-\pi \sigma r^2} r^{2n+2k+q-1} \, dr \right| \tag{7.58}$$

$$\leq \sum_{k=N'}^{N''} \frac{(2\pi \rho)^{n+2k}}{2^{2k} \Gamma(k+1) \Gamma(n+k+\frac{q}{2})} \int_0^\infty e^{-\pi r^2 \Re(\sigma)} r^{2n+2k+q-1} \, dr$$

$$\leq \frac{1}{2\pi \Re(\sigma)} \sum_{k=N'}^{N''} \frac{(2\pi \rho)^{n+2k}}{2^{2k} \Gamma(k+1) \Gamma(n+k+\frac{q}{2})} \int_0^\infty e^{-r} \left(\frac{r}{\pi \Re(\sigma)} \right)^{n+k+\frac{q}{2}-1} dr,$$

where

$$\int_0^\infty e^{-u} u^{n+k+\frac{q}{2}-1} \, du = \Gamma\left(n+k+\frac{q}{2}\right). \tag{7.59}$$

Thus, the sum and the limit on the right side of (7.57) may be interchanged

$$\int_0^\infty e^{-\pi\sigma r^2} J_n(q; 2\pi r\rho) \, r^{n+q-1} \, dr \tag{7.60}$$

$$= \frac{\Gamma(\frac{q}{2})}{2^n} \sum_{k=0}^\infty \frac{(-1)^k (2\pi\rho)^{n+2k}}{2^{2k}\Gamma(k+1)\Gamma\left(n+k+\frac{q}{2}\right)} \int_0^\infty e^{-\pi\sigma r^2} r^{2n+2k+q-1} \, dr.$$

Together with (7.59) this implies

$$\int_0^\infty e^{-\pi\sigma r^2} J_n(q; 2\pi r\rho) \, r^{n+q-1} \, dr \tag{7.61}$$

$$= \frac{1}{2}\Gamma\left(\frac{q}{2}\right) \left(\frac{1}{\pi\sigma}\right)^{\frac{q}{2}} \left(\frac{\rho}{\sigma}\right)^n \sum_{k=0}^\infty \frac{(-1)^k}{\Gamma(k+1)} \left(\frac{\pi\rho^2}{\sigma}\right)^k.$$

Summing up the last exponential series we arrive at the identity

$$\int_0^\infty e^{-\pi\sigma r^2} J_n(q; 2\pi r\rho) r^{n+q-1} \, dr = \frac{1}{2}\Gamma\left(\frac{q}{2}\right) \frac{\rho^n}{\pi^{\frac{q}{2}}\sigma^{n+\frac{q}{2}}} e^{-\frac{\pi\rho^2}{\sigma}}. \tag{7.62}$$

This is the desired result of Lemma 7.2. $\qquad\square$

In particular with $x, y \in \mathbb{R}^q$ and $\rho = |x-y|$ we get the identity (cf. Lemma 7.2)

$$\|\mathbb{S}^{q-1}\| \int_0^\infty e^{-\pi\sigma r^2} J_0(q; 2\pi|x-y|r) r^{q-1} \, dr = \left(\frac{1}{t}\right)^{\frac{q}{2}} e^{-\frac{\pi}{\sigma}(x-y)^2}. \tag{7.63}$$

A different technique is used to verify the following integral estimate.

Lemma 7.3. *If $q \geq 3$, $t > 0$, $a > 0$, and $\tau > 0$, then*

$$\left| \int_0^\infty e^{-2\pi\tau r} J_0(q; 2\pi ar) \, dr \right| \leq \frac{1}{2} \frac{\|\mathbb{S}^{q-2}\|}{\|\mathbb{S}^{q-1}\|} \frac{1}{a}. \tag{7.64}$$

Furthermore,

$$\int_0^\infty J_0(q; 2\pi ar) \, dr = \frac{1}{2} \frac{\|\mathbb{S}^{q-2}\|}{\|\mathbb{S}^{q-1}\|} \frac{1}{a}. \tag{7.65}$$

Proof. We start from

$$J_0(q; ar) = \frac{\|\mathbb{S}^{q-2}\|}{\|\mathbb{S}^{q-1}\|} \int_{-1}^1 e^{iart}(1-t^2)^{\frac{q-3}{2}} \, dt \tag{7.66}$$

such that

$$\int_0^\infty e^{-r} J_0(q; ar) \, dr = \frac{\|\mathbb{S}^{q-2}\|}{\|\mathbb{S}^{q-1}\|} \int_{-1}^1 \frac{(1-t^2)^{\frac{q-3}{2}}}{1 - iat} \, dt \tag{7.67}$$

$$= 2 \frac{\|\mathbb{S}^{q-2}\|}{\|\mathbb{S}^{q-1}\|} \int_0^1 \frac{(1-t^2)^{\frac{q-3}{2}}}{1 + a^2 t^2} \, dt.$$

The integral on the right side of (7.67) can be calculated as follows

$$\int_0^\infty e^{-2\pi\tau r} J_0(q; 2\pi ar) \, dr = \frac{\|\mathbb{S}^{q-2}\|}{\|\mathbb{S}^{q-1}\|} \frac{1}{\pi} \int_0^1 \frac{\tau (1-t^2)^{\frac{q-3}{2}}}{\tau^2 + a^2 t^2} \, dt \tag{7.68}$$

$$= \frac{\|\mathbb{S}^{q-2}\|}{\|\mathbb{S}^{q-1}\|} \frac{1}{\pi a} \int_0^a \frac{\tau}{\tau^2 + r^2} \left(1 - \frac{r^2}{a^2}\right)^{\frac{q-3}{2}} dr$$

$$\leq \frac{\|\mathbb{S}^{q-2}\|}{\|\mathbb{S}^{q-1}\|} \frac{1}{\pi a} \int_0^\infty \frac{\tau}{\tau^2 + r^2} \, dr$$

$$= \frac{1}{2} \frac{\|\mathbb{S}^{q-2}\|}{\|\mathbb{S}^{q-1}\|} \frac{1}{a}.$$

In consequence,

$$\lim_{\tau \to 0} \int_0^\infty e^{-2\pi\tau r} J_0(q; 2\pi ar) \, dr = \frac{1}{2a} \frac{\|\mathbb{S}^{q-2}\|}{\|\mathbb{S}^{q-1}\|}, \tag{7.69}$$

which proves the second assertion, because the last integral exists.   □

There are many more interesting integrals involving Bessel functions that can be evaluated in a similar way. Nevertheless, the important cases to be needed for our Fourier inversion techniques are covered by our examples.

Gauß–Weierstraß Transform

We start from the *"Gaussian function"* $G : \mathbb{R}^q \to \mathbb{C}$ given by

$$G(x) = e^{2\pi ix \cdot z} e^{-\pi\tau x^2}, \quad z \in \mathbb{R}^q, \ \tau > 0. \tag{7.70}$$

An elementary calculation yields

$$G_{\mathbb{R}^q}^\wedge(x) = \int_{\mathbb{R}^q} G(y) e^{-2\pi ix \cdot y} \, dV_{(q)}(y) \tag{7.71}$$

$$= \int_{\mathbb{R}^q} e^{2\pi iy \cdot z} e^{-\pi\tau y^2} e^{-2\pi ix \cdot y} \, dV_{(q)}(y)$$

$$= \int_{\mathbb{R}^q} e^{-\pi\tau y^2} e^{2\pi i(z-x)\cdot y} \, dV_{(q)}(y).$$

With $y = \rho\eta$ we find

$$G_{\mathbb{R}^q}^\wedge(x) = \int_0^\infty \rho^{q-1} e^{-\pi\tau\rho^2} \left(\int_{\mathbb{S}^{q-1}} e^{2\pi i\rho(z-x)\cdot\eta} \, dS_{(q-1)}(\eta) \right) d\rho. \tag{7.72}$$

Remembering the definition of the Bessel function of order 0, we find

$$
\int_{\mathbb{S}^{q-1}} e^{2\pi i \rho((z-x)\cdot\eta)} \, dS_{(q-1)}(\eta) = \|\mathbb{S}^{q-1}\| \int_{-1}^{1} e^{2\pi i \rho |z-x|s} (1-s^2)^{\frac{q-3}{2}} \, ds
$$
$$
= \|\mathbb{S}^{q-1}\| \, J_0(q; 2\pi|z-x|\rho). \tag{7.73}
$$

For $\tau > 0$, $a \in \mathbb{R}$, we know from Lemma 7.2 that

$$
\|\mathbb{S}^{q-1}\| \int_0^\infty e^{-\pi\tau r^2} J_0(q; 2\pi a r) r^{q-1} \, dr = \tau^{-\frac{q}{2}} e^{-\frac{\pi}{\tau} a^2}. \tag{7.74}
$$

In other words, with $x, z \in \mathbb{R}^q$ and $a = |x-z|$, we have

$$
\|\mathbb{S}^{q-1}\| \int_0^\infty e^{-\pi\tau r^2} J_0(q; 2\pi|x-z|r) \, r^{q-1} \, dr = \tau^{-\frac{q}{2}} e^{-\frac{\pi}{\tau}(x-z)^2}. \tag{7.75}
$$

Altogether this shows us that

$$
G_{\mathbb{R}^q}^{\wedge}(x) = \left(\frac{1}{\tau}\right)^{\frac{q}{2}} e^{-\frac{\pi}{\tau}(x-z)^2}. \tag{7.76}
$$

Next we consider the *"Weierstraß function"* $W : \mathbb{R}^q \to \mathbb{C}$ given by

$$
W(x) = \left(\frac{1}{\tau}\right)^{\frac{q}{2}} e^{-\frac{\pi}{\tau}(z-x)^2}, \quad z \in \mathbb{R}^q, \ \tau > 0. \tag{7.77}
$$

Now we have

$$
W_{\mathbb{R}^q}^{\wedge}(-x) = \left(\frac{1}{\tau}\right)^{\frac{q}{2}} \int_{\mathbb{R}^q} e^{-\frac{\pi}{\tau}(z-y)^2} e^{2\pi i x \cdot y} \, dV_{(q)}(y) \tag{7.78}
$$
$$
= \left(\frac{1}{\tau}\right)^{\frac{q}{2}} e^{2\pi i x \cdot z} \int_{\mathbb{R}^q} e^{-\frac{\pi}{\tau}(y-z)^2} e^{2\pi i x \cdot (y-z)} \, dV_{(q)}(y).
$$

By use of polar coordinates $y = z + \rho\eta$ the integral can be transformed to

$$
\int_0^\infty \rho^{q-1} e^{-\frac{\pi}{\tau}\rho^2} \left(\int_{\mathbb{S}^{q-1}} e^{2\pi i \rho(x\cdot\eta)} \, dS_{(q-1)}(\eta) \right) d\rho \tag{7.79}
$$
$$
= \|\mathbb{S}^{q-1}\| \int_0^\infty \rho^{q-1} e^{-\frac{\pi}{\tau}\rho^2} J_0(q; 2\pi|x|\rho) \, d\rho.
$$

The last integral is known from Lemma 7.2. For $x \in \mathbb{R}^q$ we have

$$
W_{\mathbb{R}}^{\wedge}(-x) = e^{2\pi i x \cdot z} e^{-\pi\tau x^2} = G(x). \tag{7.80}
$$

In particular, we have
$$
W_{\mathbb{R}}^{\wedge}(0) = 1 = G(0). \tag{7.81}
$$

Summarizing our results we therefore obtain the following relations.

Lemma 7.4. *For all $x \in \mathbb{R}^q$ we have*

$$W_{\mathbb{R}^q}^\wedge(-x) \;\; = \;\; G(x) = \int_{\mathbb{R}^q} \underbrace{G_{\mathbb{R}^q}^\wedge(y)}_{=W(y)} e^{2\pi i x \cdot y} \, dV_{(q)}(y),$$

$$G_{\mathbb{R}^q}^\wedge(x) \;\; = \;\; W(x) = \int_{\mathbb{R}^q} \underbrace{W_{\mathbb{R}^q}^\wedge(y)}_{=G(-y)} e^{2\pi i x \cdot y} \, dV_{(q)}(y).$$

We are now in the position to introduce the so–called Gauß transform and Weierstraß transform as integrals involving the kernels G and W, respectively.

Definition 7.1. *For $\tau > 0$, the Gauß transform $\mathcal{G}_\tau(F)$ of $F : \mathbb{R}^q \to \mathbb{C}$ at $x \in \mathbb{R}^q$ is defined by*

$$\mathcal{G}_\tau(F)(x) = \left(\frac{1}{\tau}\right)^{q/2} \int_{\mathbb{R}^q} e^{-\frac{\pi}{\tau}(x-y)^2} F(y) \, dV_{(q)}(y), \tag{7.82}$$

while the Weierstraß transform $\mathcal{W}_\tau(F)$ of $F : \mathbb{R}^q \to \mathbb{C}$ at $x \in \mathbb{R}^q$ is defined by

$$\mathcal{W}_\tau(F)(x) = \int_{\mathbb{R}^q} e^{-\pi \tau y^2} e^{2\pi i x \cdot y} F(y) \, dV_{(q)}(y). \tag{7.83}$$

The relation between the Gauß transform and the Weierstraß transform is characterized by the following identities.

Lemma 7.5. *For $F \in \mathrm{L}^1(\mathbb{R}^q)$, $\tau > 0$, and $z \in \mathbb{R}^q$*

$$\mathcal{W}_\tau(\mathcal{F}_{\mathbb{R}^q}(F))(z) \;\; = \;\; \mathcal{F}_{\mathbb{R}^q}(\mathcal{W}_\tau(F))(z) = \mathcal{G}_\tau(F)(z),$$
$$\mathcal{G}_\tau(\mathcal{F}_{\mathbb{R}^q}(F))(z) \;\; = \;\; \mathcal{F}_{\mathbb{R}^q}(\mathcal{G}_\tau(F))(z) = \mathcal{W}_\tau(F)(-z).$$

Proof. $\mathcal{W}_\tau(\mathcal{F}_{\mathbb{R}^q}(F))(z)$ written out reads as follows:

$$\mathcal{W}_\tau(\mathcal{F}_{\mathbb{R}^q}(F))(z) = \int_{\mathbb{R}^q} e^{-\pi \tau y^2} \int_{\mathbb{R}^q} F(x) e^{-2\pi i x \cdot y} \, dV_{(q)}(x) \, e^{2\pi i z \cdot y} \, dV_{(q)}(y).$$
$$\tag{7.84}$$

The order of the integrations may be reversed such that

$$\mathcal{W}_\tau(\mathcal{F}_{\mathbb{R}^q}(F))(z) \;\; = \;\; \int_{\mathbb{R}^q} F(x) \left(\int_{\mathbb{R}^q} e^{-\pi \tau y^2} e^{2\pi i (z-x) \cdot y} \, dV_{(q)}(y) \right) dV_{(q)}(x)$$

$$= \;\; \left(\frac{1}{\tau}\right)^{q/2} \int_{\mathbb{R}^q} e^{-\frac{\pi}{\tau}(x-z)^2} F(x) \, dV_{(q)}(x)$$

$$= \;\; \mathcal{G}_\tau(F)(z). \tag{7.85}$$

$\mathcal{G}_\tau(\mathcal{F}_{\mathbb{R}^q}(F))(z)$ can be written out in the form

$$\mathcal{G}_\tau(\mathcal{F}_{\mathbb{R}^q}(F))(z) = \left(\frac{1}{\tau}\right)^{q/2} \int_{\mathbb{R}^q} e^{-\frac{\pi}{\tau}(z-y)^2} \int_{\mathbb{R}^q} F(x) e^{-2\pi i x \cdot y} \, dV_{(q)}(x) \, dV_{(q)}(y). \tag{7.86}$$

Interchanging the order of integration leads to

$$
\begin{aligned}
\mathcal{G}_\tau(\mathcal{F}_{\mathbb{R}^q}(F))(z) &= \int_{\mathbb{R}^q} F(x) \int_{\mathbb{R}^q} \left(\frac{1}{\tau}\right)^{q/2} e^{-\frac{\pi}{\tau}(z-y)^2} e^{-2\pi i x \cdot y} \, dV_{(q)}(y) \, dV_{(q)}(x) \\
&= \int_{\mathbb{R}^q} e^{-\pi \tau x^2} e^{-2\pi i z \cdot x} F(x) \, dV_{(q)}(x) \\
&= \mathcal{W}_\tau(F)(-z).
\end{aligned}
\tag{7.87}
$$

This is the desired result of Lemma 7.5. $\qquad\square$

Next our aim is to show that the "integral means" $\mathcal{W}_\tau(F)(x)$ and $\mathcal{G}_\tau(F)(x)$ tend to $F(x)$ as $\tau \to 0$, if suitable conditions are imposed on the position x as well as the function F under consideration.

We begin with the Weierstraß transform.

Lemma 7.6. *Suppose that F is continuous and uniformly bounded in $\mathbb{R}^q$ such that, for $z \in \mathbb{R}^q$, the "spherical principal value" of $\mathcal{F}_{\mathbb{R}^q}(F)(z)$; i.e., the limit*

$$
\lim_{N \to \infty} \int_{\substack{|x| \leq N \\ x \in \mathbb{R}^q}} F(x) \, e^{-2\pi i z \cdot x} \, dV_{(q)}(x)
\tag{7.88}
$$

exists. Then

$$
\lim_{\substack{\tau \to 0 \\ \tau > 0}} \mathcal{W}_\tau(F)(z) = \lim_{N \to \infty} \int_{\substack{|x| \leq N \\ x \in \mathbb{R}^q}} F(x) \, e^{-2\pi i x \cdot z} \, dV_{(q)}(x).
\tag{7.89}
$$

Proof. For $r > 0$, the functions

$$
r \mapsto \Phi(r; z) = \int_{\mathbb{S}^{q-1}} F(r\xi) e^{2\pi i r z \cdot \xi} \, dS_{(q-1)}(\xi),
\tag{7.90}
$$

$$
r \mapsto \Psi(r; z) = \int_r^\infty s^{q-1} \Phi(s; z) \, ds
\tag{7.91}
$$

are continuous and uniformly bounded such that

$$
\frac{\partial \Psi(r; z)}{\partial r} = - r^{q-1} \Phi(r; z)
\tag{7.92}
$$

and

$$
\lim_{r \to \infty} \Psi(r; z) = 0.
\tag{7.93}
$$

We consider the limit

$$
\lim_{\substack{\tau \to 0 \\ \tau > 0}} \mathcal{W}_\tau(F)(z) = \lim_{\substack{\tau \to 0 \\ \tau > 0}} \int_0^\infty e^{-\pi \tau r^2} \Phi(r; z) r^{q-1} \, dr.
\tag{7.94}
$$

From Lemma 3.18 we get

$$
\lim_{\substack{\tau \to 0 \\ \tau > 0}} \int_0^\infty e^{-\pi \tau r^2} \Phi(r; z) r^{q-1} \, dr = \int_0^\infty \Phi(r, z) r^{q-1} \, dr.
\tag{7.95}
$$

This is the assertion of Lemma 7.6. $\qquad\square$

Before we discuss the Gauß transform, we introduce the notion of "spherical continuity" at a point $z \in \mathbb{R}^q$.

Definition 7.2. *A function $F : \mathbb{R}^q \to \mathbb{C}$ is called spherically continuous at $z \in \mathbb{R}^q$ if the spherical mean*

$$\Phi(r; z) = \frac{1}{\|\mathbb{S}^{q-1}\|} \int_{\mathbb{S}^{q-1}} F(z + r\xi) \, dS_{(q-1)}(\xi) \tag{7.96}$$

exists for all values $r > 0$ and is continuous for $r \to 0$ with $\Phi(0; z) = F(z)$.

This property is of particular significance if F is a "discontinuous" function. We come back to this aspect later.

Lemma 7.7. *Suppose that $F \in L^1(\mathbb{R}^q)$ is spherically continuous at $z \in \mathbb{R}^q$. Then*

$$\lim_{\substack{\tau \to 0 \\ \tau > 0}} \mathcal{G}_\tau(F)(z) = \lim_{\substack{\tau \to 0 \\ \tau > 0}} \mathcal{W}_\tau(F^\wedge_{\mathbb{R}^q})(z) = F(z). \tag{7.97}$$

Proof. We start from

$$\mathcal{G}_\tau(F)(z) = \left(\frac{1}{\tau}\right)^{\frac{q}{2}} \int_{\mathbb{R}^q} e^{-\frac{\pi}{\tau}(z-x)^2} F(x) \, dV_{(q)}(x). \tag{7.98}$$

By introducing polar coordinates $x = z + r\xi$, $r > 0$, $\xi \in \mathbb{S}^{q-1}$, we obtain

$$\mathcal{G}_\tau(F)(z) = \left(\frac{1}{\tau}\right)^{\frac{q}{2}} \int_0^\infty e^{-\frac{\pi}{\tau}r^2} \Phi(r; z) r^{q-1} \, dr, \tag{7.99}$$

where

$$\Phi(r; z) = \int_{\mathbb{S}^{q-1}} F(z + r\xi) \, dS_{(q-1)}(\xi). \tag{7.100}$$

Observing the spherical continuity of $F \in L^1(\mathbb{R}^q)$ at $z \in \mathbb{R}^q$, we get

$$\Phi(0; z) = \|\mathbb{S}^{q-1}\| \, F(z). \tag{7.101}$$

By virtue of Lemma 3.20 we find from (7.101)

$$\lim_{\substack{\tau \to 0 \\ \tau > 0}} \left(\frac{1}{\tau}\right)^{\frac{q}{2}} \int_0^\infty e^{-\frac{\pi}{\tau}r^2} \Phi(r; z) \, r^{q-1} \, dr = \frac{\Phi(0; z)}{\|\mathbb{S}^{q-1}\|} = F(z), \tag{7.102}$$

as required. $\square$

It should be pointed out that Lemma 7.7 allows the interpretation as pointwise (Gauß) inversion formula.

Theorem 7.4. *Suppose that* $F \in L^1(\mathbb{R}^q)$ *is spherically continuous at* $z \in \mathbb{R}^q$. *Furthermore, assume that the "spherical principal value"*

$$\lim_{N \to \infty} \int_{\substack{|y| \le N \\ y \in \mathbb{R}^q}} e^{2\pi i z \cdot y} F_{\mathbb{R}^q}^{\wedge}(y) \, dV_{(q)}(y) \tag{7.103}$$

exists. Then we have

$$F(z) = \lim_{N \to \infty} \int_{\substack{|y| \le N \\ y \in \mathbb{R}^q}} e^{2\pi i z \cdot y} F_{\mathbb{R}^q}^{\wedge}(y) \, dV_{(q)}(y). \tag{7.104}$$

If the spherical principal value does not exist, the limit relation

$$F(z) = \lim_{\substack{\tau \to 0 \\ \tau > 0}} \int_{\mathbb{R}^q} e^{-\pi \tau y^2} e^{2\pi i z \cdot y} F_{\mathbb{R}^q}^{\wedge}(y) \, dV_{(q)}(y) \tag{7.105}$$

remains valid as a "mean representation" of F *at the point* z *of spherical continuity.*

Abel–Poisson Transform

For $\tau > 0$, $z \in \mathbb{R}^3$, we consider the *"Abel function"* A given by

$$A(x) = e^{-2\pi \tau |x|} e^{2\pi i z \cdot x}, \quad x \in \mathbb{R}^q. \tag{7.106}$$

For $\tau > 0$, $a \in \mathbb{R}$, it is well known from Lemma 7.1 that

$$\|\mathbb{S}^{q-1}\| \int_0^\infty e^{-2\pi \tau r} J_0(q; 2\pi a r) r^{q-1} \, dr = \frac{2}{\|\mathbb{S}^q\|} \frac{\tau}{(a^2 + \tau^2)^{\frac{q-1}{2}}}. \tag{7.107}$$

For $y \in \mathbb{R}^q$, this gives us

$$
\begin{aligned}
A_{\mathbb{R}^q}^{\wedge}(y) &= \int_{\mathbb{R}^q} e^{-2\pi \tau |x|} e^{2\pi i x \cdot z} e^{-2\pi i x \cdot y} dV_{(q)}(x) \tag{7.108} \\
&= \frac{2}{\|\mathbb{S}^q\|} \frac{\tau}{(\tau^2 + (y-z)^2)^{\frac{q+1}{2}}} = P(y),
\end{aligned}
$$

where P stands for the *"Poisson function"*. Furthermore, we know that

$$
\begin{aligned}
P_{\mathbb{R}^q}^{\wedge}(y) &= \frac{2}{\|\mathbb{S}^q\|} \int_{\mathbb{R}^q} \frac{\tau}{(\tau^2 + (x-z)^2)^{\frac{q+1}{2}}} e^{-2\pi i x \cdot y} dV_{(q)}(y) \tag{7.109} \\
&= e^{-2\pi \tau |y|} e^{-2\pi i y \cdot z} = A(-y).
\end{aligned}
$$

These identities motivate the following definitions.

Definition 7.3. *For* $\tau > 0$, *the Abel transform* $\mathcal{A}_\tau(F)$ *of* $F : \mathbb{R}^q \to \mathbb{C}$ *at* $x \in \mathbb{R}^q$ *is defined by*

$$\mathcal{A}_\tau(F)(x) = \int_{\mathbb{R}^q} e^{-2\pi \tau |y|} e^{2\pi i y \cdot x} F(y) \, dV_{(q)}(y), \tag{7.110}$$

while the Poisson transform $\mathcal{P}_\tau(F)$ of $F : \mathbb{R}^q \to \mathbb{C}$ at $x \in \mathbb{R}^q$ is defined by

$$\mathcal{P}_\tau(F)(x) = \frac{2}{\|\mathbb{S}^q\|} \int_{\mathbb{R}^q} \frac{\tau}{(\tau^2 + (x-y)^2)^{\frac{q+1}{2}}} F(y) \, dV_{(q)}(y). \tag{7.111}$$

By analogous arguments as those leading to Lemma 7.5, we find the following relations.

Lemma 7.8. *For $F \in \mathrm{L}^1(\mathbb{R}^q)$, $\tau > 0$, and $z \in \mathbb{R}^q$*

$$\begin{aligned}
\mathcal{A}_\tau(\mathcal{F}_{\mathbb{R}^q}(F))(z) &= \mathcal{F}_{\mathbb{R}^q}(\mathcal{A}_\tau(F))(z) = \mathcal{P}_\tau(F)(z), \\
\mathcal{P}_\tau(\mathcal{F}_{\mathbb{R}^q}(F))(z) &= \mathcal{F}_{\mathbb{R}^q}(\mathcal{P}_\tau(F))(z) = \mathcal{A}_\tau(F)(-z).
\end{aligned}$$

For the Poisson transform we are able to verify

Lemma 7.9. *Suppose that $F \in \mathrm{L}^1(\mathbb{R}^q)$ is spherically continuous at $z \in \mathbb{R}^q$. Then*

$$\lim_{\substack{\tau \to 0 \\ \tau > 0}} \mathcal{P}_\tau(F)(z) = \lim_{\substack{\tau \to 0 \\ \tau > 0}} \mathcal{A}_\tau(F_{\mathbb{R}^q}^\wedge)(z) = F(z). \tag{7.112}$$

Proof. We have

$$\mathcal{P}_\tau(z) = \frac{2}{\|\mathbb{S}^q\|} \int_{\mathbb{R}^q} \frac{\tau}{(\tau^2 + (x-z)^2)^{\frac{q+1}{2}}} F(x) \, dV_{(q)}(x). \tag{7.113}$$

Polar coordinates $x = z + r\xi$, $r > 0$, $\xi \in \mathbb{S}^{q-1}$, yield

$$\mathcal{P}_\tau(F)(z) = 2 \frac{\|\mathbb{S}^{q-1}\|}{\|\mathbb{S}^q\|} \int_0^\infty \frac{\tau}{(\tau^2 + r^2)^{\frac{q+1}{2}}} \Phi(r;z) r^{q-1} \, dr, \tag{7.114}$$

where

$$\Phi(r;z) = \frac{1}{\|\mathbb{S}^{q-1}\|} \int_{\mathbb{S}^{q-1}} F(z + r\xi) \, dS_{(q-1)}(\xi) \tag{7.115}$$

and

$$\Phi(0;z) = F(z). \tag{7.116}$$

The application of Lemma 3.21 then yields

$$\lim_{\substack{\tau \to 0 \\ \tau > 0}} \frac{2\|\mathbb{S}^{q-1}\|}{\|\mathbb{S}^q\|} \int_0^\infty \frac{\tau}{(\tau^2 + r^2)^{\frac{q+1}{2}}} \Phi(r;z) r^{q-1} \, dr = \Phi(0;z) = F(z). \tag{7.117}$$

This is the desired result.   □

It should be mentioned that the formulation of the pointwise (Fourier) inversion formula (Theorem 7.4) remains unchanged by this alternative limit relation (more precisely, by use of $\mathcal{A}_\tau$ instead of $\mathcal{W}_\tau$).

7.5 Hankel Transform and Discontinuous Integrals

The Fourier transform preserves the spherical symmetry in the sense that the orthogonal invariance of the original function in $\mathbb{R}^q$ remains unchanged for the transformed function. In this respect, it is worthwhile to mention that the orthogonal invariance reduces the Fourier transform to a one-dimensional integral transform. This feature applied to spherically discontinuous functions leads to the so–called Hankel transform. The Hankel transform turns out to be a helpful tool for evaluating discontinuous integrals occurring in alternating lattice point sums (understood in spherical convergence). Examples are multi-dimensional extensions of the Hardy–Landau identities.

Hankel Transform

We begin our considerations with a function $F_n : \mathbb{R}^q \to \mathbb{C}$, $n \in \mathbb{N}_0$, given in the separated radial and angular form

$$F_n(x) = \Phi_n(r)Y_n(q;\xi), \quad x = r\xi, r = |x|, \quad \xi \in \mathbb{S}^{q-1}, \tag{7.118}$$

where $Y_n(q;\cdot)$ is a spherical harmonic of degree n and dimension q and the radial function $r \mapsto \Phi(r)$, $r \geq 0$, is assumed to be continuous with

$$\int_0^\infty |\Phi_n(r)| \, r^{q-1} \, dr < \infty. \tag{7.119}$$

Under these assumptions the Fourier transform $(F_n)^\wedge_{\mathbb{R}^q}$ at $y, y = s\eta$, $s = |y|$, $\eta \in \mathbb{S}^{q-1}$, admits the representation

$$\mathcal{F}_{\mathbb{R}^q}(F_n)(y) = \int_{\mathbb{R}^q} F_n(x)e^{-2\pi i y \cdot x} \, dV_{(q)}(x) = i^n \Psi_n(s)Y_n(q;\eta), \tag{7.120}$$

where

$$\Psi_n(s) = \|\mathbb{S}^{q-1}\| \int_0^\infty r^{q-1}\Phi_n(r) \, J_n(q; 2\pi sr) \, dr. \tag{7.121}$$

Note that

$$\int_{\mathbb{S}^{q-1}} e^{-2\pi i r s(\xi \cdot \eta)} Y_n(q;\xi) \, dS_{(q-1)}(\xi) = i^n \|\mathbb{S}^{q-1}\| J_n(q; 2\pi sr)Y_n(q;\eta). \tag{7.122}$$

In other words, the Fourier transform

$$\mathcal{F}_{\mathbb{R}^q}(F_n)(y) = F^\wedge_{\mathbb{R}^q}(y) = i^n \Psi_n(s)Y_n(q;\eta) \tag{7.123}$$

also shows a splitting into a radial and an angular part (as the function F_n itself given by (7.118)).

In what follows we replace the continuity of Φ_n on $(0,\infty)$ by the so–called symmetrical continuity.

Definition 7.4. *Suppose that Φ_n is symmetrically continuous on $(0, \infty)$, i.e., for all $r > 0$*

$$\lim_{\substack{s \to 0 \\ s > 0}} \frac{1}{2} \left(\Phi_n(r+s) + \Phi_n(r-s) \right) = \Phi_n(r). \tag{7.124}$$

Then the integral transform Ψ_n as defined by (7.121) is called the Hankel transform of Φ_n of degree n, i.e.,

$$\mathcal{H}_n(\Phi_n)(\rho) = \Psi_n(\rho) = \|\mathbb{S}^{q-1}\| \int_0^\infty r^{q-1} \Phi_n(r) J_n(q; 2\pi\rho r) \, dr. \tag{7.125}$$

Our purpose is to verify the following inversion formula.

Lemma 7.10. *Suppose that Φ is symmetrically continuous on $(0, \infty)$. Moreover, assume that (7.119) holds true. If*

$$\mathcal{H}_n(\Phi_n)(s) = \Psi_n(s) = \|\mathbb{S}^{q-1}\| \int_0^\infty r^{q-1} \Phi_n(r) J_n(q; 2\pi s r) \, dr, \tag{7.126}$$

then

$$\mathcal{H}_n(\Psi_n)(r) = \Phi_n(r) = \|\mathbb{S}^{q-1}\| \int_0^\infty s^{q-1} \Psi_n(s) J_n(q; 2\pi s r) \, ds, \tag{7.127}$$

provided the last integral exists.

Proof. Under the assumptions (7.124) and (7.119) the Fourier inversion formula is valid for the function F_n defined by (7.118), i.e.,

$$F_n(x) = \int_{\mathbb{R}^q} (F_n)_{\mathbb{R}^q}^\wedge(y) \, e^{2\pi i x \cdot y} \, dV_{(q)}(y), \tag{7.128}$$

if the integral exists as a principal value. We find in connection with the homogeneity of the spherical harmonic $Y_n(q; \cdot) \in \mathrm{Harm}_n(\mathbb{S}^{q-1})$

$$\mathcal{F}_{\mathbb{R}^q}(\mathcal{F}_{\mathbb{R}^q}(F_n))(-x)$$
$$= (-1)^n \|\mathbb{S}^{q-1}\| \int_0^\infty s^{q-1} \Psi(s) J_n(q; 2\pi r s) \, ds \, Y_n(q; -\xi)$$
$$= \|\mathbb{S}^{q-1}\| \int_0^\infty s^{q-1} \Psi(s) J_n(q; 2\pi r s) \, ds \, Y_n(q; \xi). \tag{7.129}$$

This assures Lemma 7.10. $\square$

The spherical symmetry, in fact, reduces the complexity. The integral transform is of one-dimensional nature. In addition, it can be handled within the framework of the theory of Bessel functions.

Discontinuous Integrals

We discuss an example that is of particular significance in analytical theory of numbers: we assume F_n to be separated in a radial and angular part in the form

$$F_n(x) = \Phi_n(r)Y_n(q;\xi), \quad x = r\xi, \; \xi \in \mathbb{S}^{q-1}, \tag{7.130}$$

where $Y_n(q;\cdot)$ is a member of $\mathrm{Harm}_n(\mathbb{S}^{q-1})$ and $\Phi_n : r \mapsto \Phi_n(r)$ is given by

$$\Phi_n(r) = \begin{cases} r^n & , \quad 0 \le r < R \\ \frac{1}{2}r^n & , \quad r = R \\ 0 & , \quad r > R. \end{cases} \tag{7.131}$$

The function F_n clearly is of class $L^1(\mathbb{R}^q)$, and is discontinuous at $r = R$, but everywhere spherically continuous. As already known, the pointwise (Fourier) inversion formula holds true

$$F(x) = \int_{\mathbb{R}^q} F_{\mathbb{R}^q}^\wedge(y)e^{2\pi i x \cdot y} \, dV_{(q)}(y), \tag{7.132}$$

provided that the integral exists as a spherical principal value. The Hankel transform is given by

$$\mathcal{H}_n(\Phi_n)(s) = \Psi_n(s) = \|\mathbb{S}^{q-1}\| \int_0^R r^{n+q-1} J_n(q; 2\pi sr) \, dr. \tag{7.133}$$

As Φ_n is symmetrically continuous, Lemma 7.10 is applicable, and we find

$$\Phi_n(r) = \|\mathbb{S}^{q-1}\| \int_0^\infty s^{q-1} \Psi_n(s) J_n(q; 2\pi sr) \, ds, \tag{7.134}$$

if the integral exists (which must be investigated in more detail). In order to show the existence of (7.134) we remember (cf. (6.453))

$$\int_0^R r^{n+q-1} J_n(q; 2\pi sr) \, dr = \frac{R^{n+q-1}}{2\pi s} J_{n+1}(q; 2\pi sR). \tag{7.135}$$

Consequently, we find

$$\Psi_n(s) = \|\mathbb{S}^{q-1}\| \frac{R^{n+q-1}}{2\pi s} J_{n+1}(q; 2\pi sR). \tag{7.136}$$

Now, the existence of the integral (7.134) is a consequence of the asymptotic relation of the Bessel function

$$J_n(q; 2\pi sr) = \frac{2}{\|\mathbb{S}^{q-1}\|} \left(\frac{1}{Rs}\right)^{\frac{q-1}{2}} \cos\left(2\pi s - \left(n\frac{\pi}{2} + (q-1)\frac{\pi}{4}\right)\right) + O\left(s^{-\frac{q-1}{2}}\right), \tag{7.137}$$

which yields after a simple calculation

$$s^{q-1}\Psi_n(s)J_n(q;2\pi sr) \tag{7.138}$$

$$= \frac{R^{n+q-1}}{2\pi} s^{q-2} J_{n+1}(q;2\pi sR)J_n(q;2\pi sr)$$

$$= \frac{C}{s}\left(\sin\left(2\pi s(r+R) - 2\left(n\frac{\pi}{2} + (q-1)\frac{\pi}{4}\right) + \sin\left(2\pi s(r-R)\right)\right)\right)$$

$$+ O\left(\frac{1}{s^2}\right)$$

with a certain explicitly calculable constant C (whose value is not of importance in the discussion of the convergence). This shows that the integral (7.134) is convergent.

Finally, in connection with (7.134) and (7.136), we obtain from Lemma 7.10.

Lemma 7.11. *For $R > 0$,*

$$\int_0^\infty s^{q-2}J_{n+1}(q;2\pi sR)J_n(q;2\pi sr)\,ds \tag{7.139}$$

$$= \frac{2\pi}{\|\mathbb{S}^{q-1}\|}\frac{1}{R^{q-1}}\left(\frac{r}{R}\right)^n \begin{cases} 1 & , & 0 \le r < R \\ \frac{1}{2} & , & r = R \\ 0 & , & r > R. \end{cases}$$

The expression (7.139) is the so–called *Weber–Schlafheitlin discontinuous integral* (for more details on discontinuous integrals involving the theory of Bessel functions the reader is referred, e.g., to the monograph of G.N. Watson [1944]).

8

Lattice Function for the Iterated Helmholtz Operator

CONTENTS

8.1   Lattice Function for the Helmholtz Operator 248
      Defining Constituents 248
      Uniqueness 249
      Existence 251
8.2   Lattice Function for the Iterated Helmholtz Operator 255
      Definition 255
      Essential Properties 255
8.3   Lattice Function in Terms of Circular Harmonics 256
      Representation in Terms of Circle Harmonics 256
8.4   Lattice Function in Terms of Spherical Harmonics 265
      Representation in Terms of Spherical Harmonics 265

This chapter deals with multi-dimensional analogues of the Bernoulli polynomials related to iterated Helmholtz operators. As in the one-dimensional theory, the multi-dimensional "Bernoulli polynomials" are understood as Green functions for the Helmholtz operators $\Delta + \lambda$, $\lambda \in \mathbb{R}$, and the "boundary condition" of Λ-lattice periodicity.

The setup of this chapter is as follows: in Section 8.1 we introduce the Green function for a Helmholtz operator and the "boundary condition" of Λ-periodicity - briefly called Λ-lattice function $G\big((\Delta + \lambda); \cdot\big)$ - by its constituting properties, i.e., differential equation, characteristic singularity, boundary condition, and normalization. Section 8.1 also assures the uniqueness of this function. Contrary to the one-dimensional case, the existence of the Λ-lattice function $G\big((\Delta + \lambda); \cdot\big)$ cannot be guaranteed just by the representation in terms of its bilinear expansion. Instead we need an auxiliary mathematical tool. In this respect, it is canonical to show the existence of $G(\Delta + \lambda; \cdot)$ by the Fredholm theory of linear (weakly) singular integral equations (Section 8.1). The introduction of Λ-lattice functions for iterated Helmholtz operators $(\Delta + \lambda)^k$, $\lambda \in \mathbb{R}$, $k \in \mathbb{N}$, is given in Section 8.2. From the theory of linear (weakly) singular integral equations we are able to deduce that each iteration, i.e. each convolution over the fundamental cell, reduces the order of the singularity by two, such that $G\big((\Delta + \lambda)^m; \cdot\big)$, $m \geq \frac{q}{2}$, turns out to be a continuous function in $\mathbb{R}^q$. Section 8.3 is concerned with the orthogonal expansion of the two-dimensional Λ-lattice function $G(\Delta^m; \cdot)$ by means of "circular harmon-

247

ics". More concretely, standard Fourier theory together with classical means of complex analysis help us to express $G(\Delta; \cdot)$ in terms of the Theta function Θ and the (Weierstraß) $\mathcal{P}$-function. Finally, Section 8.4 gives the orthogonal expansion of the q-dimensional Λ-lattice function $G(\Delta^m; \cdot)$ by means of "spherical harmonics" in a (punctured) neighborhood of the origin.

8.1   Lattice Function for the Helmholtz Operator

Our considerations start with the definition of the Λ-lattice function by its constituting ingredients. In accordance with the one-dimensional case (cf. Section 4.1) and seen from the point of mathematical physics, the Λ-lattice function as introduced in Definition 8.1 is nothing else than the Green function for the Helmholtz operator $\Delta + \lambda$, $\lambda \in \mathbb{R}$, in Euclidean space $\mathbb{R}^q$ corresponding to the "boundary condition" of periodicity with regard to the lattice Λ.

Defining Constituents

Definition 8.1. $G(\Delta + \lambda; \cdot) : \mathbb{R}^q \setminus \Lambda \to \mathbb{R}$, $\lambda \in \mathbb{R}$ *fixed, is called the Λ-lattice function for the operator $\Delta + \lambda$ if it satisfies the following properties:*

(i) (Periodicity) For all $x \in \mathbb{R}^q \setminus \Lambda$ and $g \in \Lambda$

$$G(\Delta + \lambda; x + g) = G(\Delta + \lambda; x) \tag{8.1}$$

(ii) (Differential equation) $G(\Delta + \lambda; \cdot)$ is twice continuously differentiable for all $x \notin \Lambda$.

For $\lambda \notin \mathrm{Spect}_\Delta(\Lambda)$,

$$(\Delta + \lambda)G(\Delta + \lambda; x) = 0, \tag{8.2}$$

For $\lambda \in \mathrm{Spect}_\Delta(\Lambda)$,

$$(\Delta + \lambda)G(\Delta + \lambda; x) = -\frac{1}{\sqrt{\|\mathcal{F}\|}} \sum_{\substack{(\Delta+\lambda)^\wedge(h)=0 \\ h \in \Lambda^{-1}}} \Phi_h(x). \tag{8.3}$$

(iii) (Characteristic singularity) In the neighborhood of the origin

$$G(\Delta + \lambda; x) + F_q(|x|) = \begin{cases} O(1) & , \quad q = 2, \\ O\left(|x|^{3-q} \ln |x|\right) & , \quad q = 4, \\ O\left(|x|^{3-q}\right) & , \quad q \neq 2, 4 \end{cases} \tag{8.4}$$

and

$$\nabla_x G(\Delta + \lambda; x) + \nabla_x F_q(|x|) = \begin{cases} O(1) & , \quad q = 2, \\ O\left(|x|^{2-q} \ln |x|\right) & , \quad q = 4, \\ O\left(|x|^{2-q}\right) & , \quad q \neq 2, 4, \end{cases} \tag{8.5}$$

where $F_q : x \mapsto F_q(|x|), x \neq 0$, *is the fundamental solution in* $\mathbb{R}^q$ *with respect to the Laplace operator* Δ *(see (6.18)).*

(iv) (Normalization) For all $h \in \Lambda^{-1}$ *with* $(\Delta + \lambda)^\wedge(h) = 0$

$$\int_{\mathcal{F}} G(\Delta + \lambda; x) \overline{\Phi_h(x)} \, dV_{(q)}(x) = 0. \tag{8.6}$$

Remark 8.1. *If there is no confusion likely to arise, we write* $G(\Delta + \lambda; x)$ *instead of the lengthy expression* $G(\Delta_{(q)} + \lambda; \Lambda_{(q)}; x_{(q)})$.

Uniqueness

Our first purpose is to justify the uniqueness of the Λ-lattice function for the Helmholtz operator $\Delta + \lambda$, $\lambda \in \mathbb{R}$, by showing that the difference of two Λ-lattice functions as introduced by Definition 8.1 is the zero function.

Let Λ be an arbitrary lattice in $\mathbb{R}^q$. As usual, let us denote by Λ^{-1} its inverse lattice. By application of the Second Green Theorem we obtain for every (sufficiently small) $\varepsilon > 0$ and all lattice points $h \in \Lambda^{-1}$ satisfying $(\Delta + \lambda)^\wedge(h) \neq 0$ the integral identity

$$\int_{\substack{x \in \mathcal{F} \\ |x| \geq \varepsilon}} ((\Delta + \lambda) G(\Delta + \lambda; x)) \, \overline{\Phi_h(x)} \, dV_{(q)}(x) \tag{8.7}$$

$$- \int_{\substack{x \in \mathcal{F} \\ |x| \geq \varepsilon}} G(\Delta + \lambda; x) \left((\Delta + \lambda) \, \overline{\Phi_h(x)}\right) dV_{(q)}(x)$$

$$= \int_{\partial \mathcal{F}} \left(\frac{\partial G(\Delta + \lambda; x)}{\partial \nu}\right) \overline{\Phi_h(x)} \, dS_{(q-1)}(x)$$

$$- \int_{\partial \mathcal{F}} G(\Delta + \lambda; x) \left(\frac{\partial \overline{\Phi_h(x)}}{\partial \nu}\right) dS_{(q-1)}(x)$$

$$+ \int_{\substack{|x| = \varepsilon \\ x \in \mathcal{F}}} \left(\frac{\partial G(\Delta + \lambda; x)}{\partial \nu}\right) \overline{\Phi_h(x)} \, dS_{(q-1)}(x)$$

$$- \int_{\substack{|x| = \varepsilon \\ x \in \mathcal{F}}} G(\Delta + \lambda; x) \left(\frac{\partial \overline{\Phi_h(x)}}{\partial \nu}\right) dS_{(q-1)}(x),$$

where ν is the outward directed (unit) normal field on the boundary surfaces (as depicted in Figure 8.1 for the two-dimensional case). In other words, the normal is directed to the origin on the inner sphere $\mathbb{S}_\varepsilon^{q-1}$.

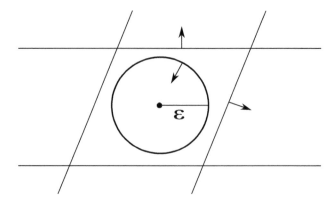

FIGURE 8.1
The geometric illustration for Equation (8.7) in Euclidean space $\mathbb{R}^2$.

Because of the Λ-periodicity of the Λ-lattice function (cf. Condition *(i)* of Definition 8.1) as well as the functions Φ_h, $h \in \Lambda^{-1}$, the integrals over the boundary $\partial\mathcal{F}$ of the fundamental cell $\mathcal{F} \subset \mathbb{R}^q$ vanish in the identity (8.7). Observing the differential equation (i.e., Condition *(ii)* for the Λ-lattice function) we get

$$(\Delta + \lambda)^\wedge(h) \int_{\substack{x \in \mathcal{F} \\ |x| \geq \varepsilon}} G(\Delta + \lambda; x)\overline{\Phi_h(x)} \, dV_{(q)}(x) \qquad (8.8)$$

$$= \int_{\substack{|x|=\varepsilon \\ x \in \mathcal{F}}} \left(\frac{\partial G(\Delta + \lambda; x)}{\partial \nu} \right) \overline{\Phi_h(x)} \, dS_{(q-1)}(x)$$

$$- \int_{\substack{|x|=\varepsilon \\ x \in \mathcal{F}}} G(\Delta + \lambda; x) \left(\frac{\partial \overline{\Phi_h(x)}}{\partial \nu} \right) dS_{(q-1)}(x).$$

Letting ε tend to 0 we obtain, in connection with Lemma 6.1, the identity

$$(\Delta + \lambda)^\wedge(h) \int_{\mathcal{F}} G(\Delta + \lambda; x) \, \overline{\Phi_h(x)} \, dV_{(q)}(x) = -\frac{1}{\sqrt{\|\mathcal{F}\|}}. \qquad (8.9)$$

Consequently, for all $h \in \Lambda^{-1}$ with $(\Delta + \lambda)^\wedge(h) \neq 0$, the Fourier coefficients of $G(\Delta + \lambda; \cdot)$ read as follows:

$$\int_{\mathcal{F}} G(\Delta + \lambda; x) \, \overline{\Phi_h(x)} dV_{(q)}(x) = -\frac{1}{\sqrt{\|\mathcal{F}\|}} \frac{1}{(\Delta + \lambda)^\wedge(h)}. \qquad (8.10)$$

In addition, the normalization condition *(iv)* of Definition 8.1 tells us that for all $h \in \Lambda^{-1}$ with $(\Delta + \lambda)^\wedge(h) = 0$, i.e, $\Delta^\wedge(h) = 4\pi^2 h^2 = \lambda$

$$\int_{\mathcal{F}} G(\Delta + \lambda; x) \, \overline{\Phi_h(x)} \, dV_{(q)}(x) = 0. \qquad (8.11)$$

Combining (8.9) and (8.11) we therefore find

Lemma 8.1. *For $h \in \Lambda^{-1}$ and $\lambda \in \mathbb{R}$,*

$$\int_{\mathcal{F}} G\left(\Delta + \lambda; x\right) \overline{\Phi_h(x)} \, dV_{(q)}(x) = \begin{cases} 0 & , \quad (\Delta + \lambda)^{\wedge}(h) = 0, \\ -\frac{1}{\sqrt{\|\mathcal{F}\|}} \frac{1}{(\Delta + \lambda)^{\wedge}(h)} & , \quad (\Delta + \lambda)^{\wedge}(h) \neq 0. \end{cases}$$

$$(8.12)$$

From Lemma 8.1 we are immediately able to verify the uniqueness of the Λ-lattice function $G\left(\Delta + \lambda; \cdot\right)$ by virtue of the completeness of the system $\{\Phi_h\}_{h \in \Lambda^{-1}}$ in the Hilbert space $\mathrm{L}^2_\Lambda(\mathbb{R}^q)$.

Theorem 8.1. *(Uniqueness) For each number $\lambda \in \mathbb{R}$ there exists one and only one Λ-lattice function $G\left(\Delta + \lambda; \cdot\right)$ satisfying the constituting conditions (i) – (iv) listed under Definition 8.1.*

Proof. From Lemma 8.1 it follows that the difference of two Λ-lattice functions for the Helmholtz operator $\Delta + \lambda$, $\lambda \in \mathbb{R}$, has only vanishing Fourier coefficients. Moreover, the difference is a Λ-periodical, continuous function on $\mathbb{R}^q$. Therefore, the completeness of the system $\{\Phi_h\}_{h \in \Lambda^{-1}}$ tells us that the difference vanishes everywhere in $\mathbb{R}^q$. $\qquad \square$

Existence

Unfortunately, for $q \geq 2$, the series expansion of the function $G(\Delta + \lambda; \cdot)$

$$G(\Delta + \lambda; x) \sim \frac{1}{\sqrt{\|\mathcal{F}\|}} \sum_{\substack{(\Delta + \lambda)^{\wedge}(h) \neq 0 \\ h \in \Lambda^{-1}}} \frac{\Phi_h(x)}{-(\Delta + \lambda)^{\wedge}(h)} \qquad (8.13)$$

does not show the pointwise convergence in $\mathbb{R}^q$ (as was the case for the dimension $q = 1$). Following Hilbert's approach to the theory of Green's functions (see, e.g., D. Hilbert [1912], R. Courant, D. Hilbert [1924]), however, we are able to guarantee the existence of $G(\Delta + \lambda; \cdot)$ by methods known from linear (weakly singular) integral equations. To do so we first discuss the explicit representation particularly for the Λ-lattice function $G(\Delta{-}1; \cdot)$. As a matter of fact, it turns out that $G(\Delta{-}1; \cdot)$ allows a series expansion in terms of the Kelvin function $K_0(q; \cdot)$ (see Definition 6.11). Motivated by this fact we are led to apply the Fredholm theory in order to guarantee the existence for all types of Λ-lattice functions $G(\Delta + \lambda; \cdot)$ by suitably investigating the difference function $G(\Delta + \lambda; \cdot) - G(\Delta - 1; \cdot)$. In other words, we start the proof of the existence with the following identity comparing the Green function for an operator $\Delta + \lambda$ with the Green function for the operator $\Delta + \lambda^*$, where λ^* is a non-eigenvalue, i.e., $\lambda^* \notin \mathrm{Spect}_\Delta(\Lambda)$.

Lemma 8.2. *Under the assumption that* $\lambda \neq \lambda^*$, $\lambda^* \notin \mathrm{Spect}_\Delta(\Lambda)$, *the identity*

$$G\left(\Delta + \lambda; x\right) \;=\; G(\Delta + \lambda^*; x) + \frac{1}{\sqrt{\|\mathcal{F}\|}} \sum_{\substack{(\Delta + \lambda)^\wedge(h)=0 \\ h \in \Lambda^{-1}}} \frac{1}{\lambda - \lambda^*} \Phi_h(x) \quad (8.14)$$

$$+ \,(\lambda^* - \lambda) \int_{\mathcal{F}} G(\Delta + \lambda^*; x - y) G\left(\Delta + \lambda; y\right)\, dV(y)$$

holds true for all $x \in \mathcal{F}\backslash\{0\}$, *where the finite sum on the right side occurs only if* λ *is an eigenvalue, i.e.,* $\lambda \in \mathrm{Spect}_\Delta(\Lambda)$.

Proof. Let x be a point in the interior of $\mathcal{F}$ different from the origin 0. Then, it follows from the Second Green Theorem that

$$\int_{\substack{y \in \mathcal{F} \\ |y| \geq \varepsilon \\ |x-y| \geq \varepsilon}} \left(\Delta_y G(\Delta + \lambda^*; x - y)\right) G\left(\Delta + \lambda; y\right)\, dV(y) \qquad (8.15)$$

$$- \int_{\substack{y \in \mathcal{F} \\ |y| \geq \varepsilon \\ |x-y| \geq \varepsilon}} G(\Delta + \lambda^*; x - y)\left(\Delta_y G\left(\Delta + \lambda; y\right)\right)\, dV(y)$$

$$= \left\{ \int_{\partial\mathcal{F}} + \int_{\substack{|x-y|=\varepsilon \\ y \in \mathcal{F}}} + \int_{\substack{|y|=\varepsilon \\ y \in \mathcal{F}}} \right\} \left(\frac{\partial}{\partial\nu_y} G(\Delta + \lambda^*; x - y)\right) G\left(\Delta + \lambda; y\right) dS(y)$$

$$- \left\{ \int_{\partial\mathcal{F}} + \int_{\substack{|x-y|=\varepsilon \\ y \in \mathcal{F}}} + \int_{\substack{|y|=\varepsilon \\ y \in \mathcal{F}}} \right\} G(\Delta + \lambda^*; x - y) \left(\frac{\partial}{\partial\nu_y} G\left(\Delta + \lambda; y\right)\right) dS(y)$$

for (sufficiently small) $\varepsilon > 0$, where ν is the outward drawn (unit) normal field. In (8.15) we observe the differential equation for the Λ-lattice function. Under the assumption of Lemma 8.2 this shows that

$$(\lambda - \lambda^*) \int_{\substack{y \in \mathcal{F} \\ |y| \geq \varepsilon \\ |x-y| \geq \varepsilon}} G(\Delta + \lambda^*; x - y)\, G\left(\Delta + \lambda; y\right)\, dV(y) \qquad (8.16)$$

$$+ \,\frac{1}{\sqrt{\|\mathcal{F}\|}} \sum_{\substack{(\Delta + \lambda)^\wedge(h)=0 \\ h \in \Lambda^{-1}}} \int_{\substack{|y| \geq \varepsilon \\ |x-y| \geq \varepsilon \\ y \in \mathcal{F}}} G(\Delta + \lambda^*; x - y)\, \Phi_h(y)\, dV(y)$$

$$= \left\{ \int_{\partial\mathcal{F}} + \int_{\substack{|x-y|=\varepsilon \\ y \in \mathcal{F}}} + \int_{\substack{|y|=\varepsilon \\ y \in \mathcal{F}}} \right\} \left(\frac{\partial}{\partial\nu_y} G(\Delta + \lambda^*; x - y)\right) G\left(\Delta + \lambda; y\right) dS(y)$$

$$- \left\{ \int_{\partial\mathcal{F}} + \int_{\substack{|x-y|=\varepsilon \\ y \in \mathcal{F}}} + \int_{\substack{|y|=\varepsilon \\ y \in \mathcal{F}}} \right\} G(\Delta + \lambda^*; x - y) \left(\frac{\partial}{\partial\nu_y} G\left(\Delta + \lambda; y\right)\right) dS(y).$$

Because of the Λ-periodicity of the integrands all surface integrals $\int_{\partial\mathcal{F}} \cdots$ vanish. Passing to the limit $\varepsilon \to 0$ we get the wanted result of Lemma 8.2 from well known techniques of potential theory (see, e.g., O.D. Kellogg [1929]). $\square$

As alluded above, our aim is to verify that the Λ-lattice function for the operator $\Delta - 1$, i.e., $\lambda^* = -1$, is expressible by a series expansion in terms of the Kelvin function $K_0(q; \cdot)$ of dimension q and order 0.

Lemma 8.3. *For $q \geq 2$ and $x \notin \Lambda$*

$$G(\Delta - 1; x) = -\frac{1}{\Gamma(q-1)\,\|\mathbb{S}^{q-1}\|} \sum_{g \in \Lambda} K_0\left(q; |x+g|\right). \qquad (8.17)$$

Proof. From the theory of the Kelvin function (see Section 6.6) we know that the expression $-\left(\Gamma(q-1)\|\mathbb{S}^{q-1}\|\right)^{-1} K_0(q; |x|)$ possesses the characteristic singularity of the Λ-lattice function for $\Delta - 1$. Moreover, it is clear that the series on the right side of (8.17) is Λ-periodical. In addition, we have

$$\Delta K_0(q; |x|) = K_0(q; |x|), \quad |x| \neq 0. \qquad (8.18)$$

For $|x| \neq 0$, the function $x \mapsto K_0(q; |x|)$ is analytic. Furthermore, from Lemma 6.43 it is clear that

$$\left|K_0(q; |x|)\right| = O\left(|x|^{\frac{1-q}{2}} e^{-|x|}\right) \qquad (8.19)$$

for all $x \in \mathbb{R}^q$ with $|x| > 1$. In consequence, the series on the right of (8.17) converges for all $x \notin \Lambda$. Hence, it shows all defining properties of the Λ-lattice function for the operator $\Delta - 1$ (cf. Definition 8.1), and the uniqueness theorem (Theorem 8.1) guarantees the equality of the right and left sides of (8.17). $\square$

Poisson's differential equation of potential theory (Theorem 6.4) admits the following transfer to the theory of the Λ-lattice function.

Lemma 8.4. *Assume that $F : y \mapsto F(y)$, $y \in \mathcal{F}$, is a bounded, Λ-periodical function that satisfies a Lipschitz-condition in the neighborhood of a point $x \in \mathcal{F}$. Then the function*

$$U(x) = \int_{\mathcal{F}} G\left(\Delta + \lambda; x - y\right) F(y)\, dV(y) \qquad (8.20)$$

is twice continuously differentiable in $x \in \mathcal{F}$ such that

$$(\Delta_x + \lambda)\, U(x) = F(x) - \sum_{\substack{(\Delta+\lambda)^\wedge(h)=0 \\ h \in \Lambda}} \int_{\mathcal{F}} F(y)\, \overline{\Phi_h(y)}\, dV(y)\, \Phi_h(x), \qquad (8.21)$$

where the sum on the right side of (8.21) occurs only if $\lambda \in \mathrm{Spect}_\Delta(\Lambda)$.

After these preparations about the uniqueness, the explicit representation of the special Λ-lattice function $G(\Delta - 1; \cdot)$, and the Poisson formula for the Λ-lattice functions, the Fredholm theory of (weakly singular) integral equations should come into play to guarantee the existence of all $G(\Delta + \lambda; \cdot)$ for all values $\lambda \in \mathbb{R}$. To this end, we consider the linear (Fredholm) integral equation provided by Lemma 8.2

$$H(\lambda; x) = G(\Delta - 1; x) + \frac{1}{\sqrt{\|\mathcal{F}\|}} \sum_{\substack{(\Delta+\lambda)^\wedge(h)=0 \\ h \in \Lambda-1}} \frac{1}{\lambda+1} \Phi_h(x) \qquad (8.22)$$

$$- (\lambda+1) \int_{\mathcal{F}} G(\Delta - 1; x - y) H(\lambda; y)\, dV(y).$$

The solvability of the (weakly) singular linear integral equation (8.22) can be handled in accordance with the well known Fredholm alternative.

Theorem 8.2. *(Existence of the Λ-lattice function $G((\Delta + \lambda); \cdot))$*
(ia) If λ is not an eigenvalue, i.e., $\lambda \notin \text{Spect}_\Delta(\Lambda)$, then the integral equation (8.22) possesses a unique solution $H(\lambda; \cdot)$.
(ib) If λ is an eigenvalue, i.e., $\lambda \in \text{Spect}_\Delta(\Lambda)$, then the integral equation (8.22) possesses a unique solution $H(\lambda; \cdot)$ under the condition that

$$\int_{\mathcal{F}} H(\lambda; x)\overline{\Phi_h(x)}\, dV(x) = 0 \tag{8.23}$$

holds for all $h \in \Lambda^{-1}$ with $(\Delta + \lambda)^\wedge(h) = 0$.

(ii) $H(\lambda; \cdot)$, as specified by (ia) or (ib), satisfies all defining conditions of the function $G(\Delta + \lambda; \cdot)$; hence, in light of the uniqueness theorem (Theorem 8.1),

$$H(\lambda; \cdot) = G(\Delta + \lambda; \cdot). \tag{8.24}$$

Proof. The parts *(ia)* and *(ib)* follow immediately from the well known Fredholm alternative.

Part *(ii)* is valid if $H(\lambda; \cdot)$ satisfies all defining properties of the Λ-lattice function. In fact, the Λ-periodicity and the characteristic singularity are clear from the construction. The differential equation follows from the Poisson equation (Lemma 8.4). In fact, it shows that

$$
\begin{aligned}
(\Delta_x - 1)H(\lambda; x) \;=\;& (\Delta_x - 1)G(\Delta - 1; x) \tag{8.25}\\
& - (1 + \lambda)(\Delta_x - 1)\int_{\mathcal{F}} G(\Delta - 1; x - y)H(\lambda; y)\, dV(y)\\
& + (\Delta_x - 1)\frac{1}{\sqrt{\|\mathcal{F}\|}} \sum_{\substack{(\Delta+\lambda)^\wedge(h)=0 \\ h \in \Lambda^{-1}}} \frac{1}{\lambda + 1}\Phi_h(x)
\end{aligned}
$$

is equivalent to

$$(\Delta_x - 1)H(\lambda; x) = -(1 + \lambda)H(\lambda; x) - \frac{1}{\sqrt{\|\mathcal{F}\|}} \sum_{\substack{(\Delta+\lambda)^\wedge(h)=0 \\ h \in \Lambda^{-1}}} \Phi_h(x). \tag{8.26}$$

Thus we obtain the identity

$$(\Delta_x + \lambda)H(\lambda; x) = -\frac{1}{\sqrt{\|\mathcal{F}\|}} \sum_{\substack{(\Delta+\lambda)^\wedge(h)=0 \\ h \in \Lambda^{-1}}} \Phi_h(x). \tag{8.27}$$

Consequently, the orthonormalization condition is satisfied in the case of $\lambda \in \text{Spect}_\Delta(\Lambda)$. By collecting all results this justifies the assertion of Theorem 8.2. $\qquad\square$

8.2 Lattice Function for the Iterated Helmholtz Operator

The Λ-lattice functions for operators $(\Delta + \lambda)^m$, $\lambda \in \mathbb{R}$, $m \in \mathbb{N}$, can be based on the Λ-lattice functions for operators $\Delta + \lambda$, $\lambda \in \mathbb{R}$, by forming convolution integrals in the usual way as known from mathematical physics.

Definition

The function $G\left((\Delta + \lambda)^m; \cdot\right)$, $m = 1, 2, \ldots$, $\lambda \in \mathbb{R}$, defined by

$$G\left((\Delta + \lambda)^1; x\right) = G(\Delta + \lambda; x), \tag{8.28}$$

$$G\left((\Delta + \lambda)^m; x\right) = \int_{\mathcal{F}} G\left((\Delta + \lambda)^{m-1}; z\right) G(\Delta + \lambda; x - z) \ dV(z),$$

$m = 2, 3, \ldots$, is called the Λ-*lattice function for the operator* $(\Delta + \lambda)^m$, $\lambda \in \mathbb{R}$.

Essential Properties

Obviously, for all $x \notin \Lambda$ and $g \in \Lambda$,

$$G\left((\Delta + \lambda)^m; x + g\right) = G\left((\Delta + \lambda)^m; x\right) \tag{8.29}$$

is satisfied; i.e., $G\left((\Delta + \lambda)^m; \cdot\right)$ is Λ-periodical. In analogy to well known techniques of potential theory it can be proved that

$$G\left((\Delta + \lambda)^m; x\right) = \begin{cases} O\left(|x|^{2m-q} \ln|x|\right) & , \quad q = 2m \\ O\left(|x|^{2m-q}\right), & q > 2m. \end{cases} \tag{8.30}$$

The differential equation

$$(\Delta + \lambda) G\left((\Delta + \lambda)^m; x\right) = G\left((\Delta + \lambda)^{m-1}; x\right), \quad x \notin \Lambda, \tag{8.31}$$

$m = 2, 3, \ldots$, represents a recursion relation relating the Λ-lattice function for the operator $(\Delta + \lambda)^m$ to the Λ-lattice function for the operator $(\Delta + \lambda)^{m-1}$. The series expansion of $G\left((\Delta + \lambda)^m; \cdot\right)$ in terms of eigenfunctions, which is equivalent to the (formal) Fourier (orthogonal) expansion, reads for iteration orders $m = 2, 3, \ldots$

$$\frac{1}{\sqrt{\|\mathcal{F}\|}} \sum_{\substack{(\Delta + \lambda)^\wedge(h) \neq 0 \\ h \in \Lambda^{-1}}} \frac{\Phi_h(x)}{-((\Delta + \lambda)^m)^\wedge(h)}, \tag{8.32}$$

where

$$((\Delta + \lambda)^m)^\wedge(h) = -(\lambda - (\Delta)^\wedge(h))^m = -(\lambda - 4\pi^2 h^2)^m. \tag{8.33}$$

For $m > \frac{q}{2}$, therefore, it follows that there is a constant $C > 0$ such that

$$
\left| \sum_{\substack{(\Delta+\lambda)^\wedge(h)\neq 0 \\ h\in\Lambda^{-1}}} \frac{\Phi_h(x)}{-((\Delta + \lambda)^m)^\wedge(h)} \right|
=
\left| \frac{1}{\sqrt{\|\mathcal{F}\|}} \sum_{\substack{4\pi^2 h^2 \neq \lambda \\ h\in\Lambda^{-1}}} \frac{e^{2\pi i h\cdot x}}{(\lambda - 4\pi^2 h^2)^m} \right|
$$

$$
\leq C \sum_{h\in\Lambda^{-1}} \frac{1}{(1 + h^2)^m} < \infty. \tag{8.34}
$$

So it is clear that the Fourier series converges absolutely and uniformly in $\mathbb{R}^q$, and $G\left((\Delta + \lambda)^m; \cdot\right)$ is continuous in $\mathbb{R}^q$ provided that $m > \frac{q}{2}$.

Altogether we are able to formulate

Lemma 8.5. *For $m > \frac{q}{2}$, the Λ-lattice function $G\left((\Delta + \lambda)^m; \cdot\right)$ is continuous in $\mathbb{R}^q$, and its bilinear series reads*

$$
G\left((\Delta + \lambda)^m; x - y\right) = \sum_{\substack{(\Delta+\lambda)^\wedge(h)\neq 0 \\ h\in\Lambda^{-1}}} \frac{\Phi_h(x)\overline{\Phi_h(y)}}{-((\Delta + \lambda)^m)^\wedge(h)} \quad x, y \in \mathbb{R}^q. \tag{8.35}
$$

8.3   Lattice Function in Terms of Circular Harmonics

In the next sections we present spherical harmonic expansions of the Λ-lattice function $G\left(\Delta; \cdot\right)$ for the Laplace operator (note that, because of the Λ-periodicity, it suffices to investigate the (punctured) neighborhood of the origin). For the convenience of the reader we start with the two-dimensional case, where the (Fourier) expansion in terms of "circular harmonics", i.e., trigonometric polynomials, can be handled especially within the framework of complex analysis as well as classical Fourier theory. This part, i.e., Section 8.3, is strongly influenced by the work of C. Müller [1954a]. Section 8.4 presents the investigation of the Λ-lattice function for the Laplace operator in the neighborhood of lattice points by means of a spherical harmonic expansion in dimensions $q \geq 3$.

Representation in Terms of Circle Harmonics

The two-dimensional Fourier expansion of the two-dimensional Λ-lattice function in terms of the basis system $\{\Phi_h\}_{h\in\Lambda^{-1}}$ reads as follows

$$
G\left(\Delta; \cdot\right) \sim \frac{1}{\sqrt{\|\mathcal{F}\|}} \sum_{\substack{|h|>0 \\ h\in\Lambda^{-1}}} \frac{1}{-\Delta^\wedge(h)} \Phi_h, \tag{8.36}
$$

where $\Delta^\wedge(h) = 4\pi^2 h^2, h \in \Lambda^{-1} \subset \mathbb{R}^2$.

Let $g_1, g_2 \in \mathbb{R}^2$ be the basis vectors of the lattice $\Lambda \subset \mathbb{R}^2$. Then, in connection with (8.36), it is not hard to see that

$$G(\Delta; x) + G\left(\Delta; x + \frac{1}{2}g_1\right) + G\left(\Delta; x + \frac{1}{2}g_2\right) + G\left(\Delta; x + \frac{1}{2}(g_1 + g_2)\right)$$

$$\sim \frac{1}{\|\mathcal{F}\|} \sum_{\substack{|h|>0 \\ h \in \Lambda^{-1}}} \left(1 + e^{2\pi i \frac{1}{2}g_1 \cdot h}\right)\left(1 + e^{2\pi i \frac{1}{2}g_2 \cdot h}\right) \frac{e^{2\pi i h \cdot x}}{-\Delta^\wedge(h)}$$

$$\sim \frac{1}{\sqrt{\|\mathcal{F}\|}} \sum_{\substack{|h|>0 \\ h \in \Lambda^{-1}}} \frac{\Phi_h(2x)}{-\Delta^\wedge(h)}, \tag{8.37}$$

holds for $x \in \mathbb{R}^2 \backslash \frac{1}{2}\Lambda$, since only lattice points $h \in \Lambda^{-1}$, $h = n_1 h_1 + n_2 h_2$, must be observed in the sum corresponding to even integers n_1, n_2. Therefore, by comparison of the Fourier coefficients we immediately obtain

Lemma 8.6. *If $g_1, g_2 \in \mathbb{R}^2$ are the basis vectors of the lattice $\Lambda \subset \mathbb{R}^2$, then*

$$G(\Delta; x) + G\left(\Delta; x + \frac{1}{2}g_1\right) + G\left(\Delta; x + \frac{1}{2}g_2\right) + G\left(\Delta; x + \frac{1}{2}(g_1 + g_2)\right)$$

$$= G(\Delta; 2x). \tag{8.38}$$

holds for all $x \in \mathbb{R}^2 \backslash \frac{1}{2}\Lambda$.

According to Condition *(ii)* of Definition 8.1, $G(\Delta; \cdot)$ satisfies in $\mathbb{R}^2 \setminus \Lambda$ the differential equation

$$\Delta G(\Delta; \cdot) = -\frac{1}{\|\mathcal{F}\|}. \tag{8.39}$$

In the neighborhood of the origin 0, the function $G(\Delta; \cdot)$ can be represented in the form

$$G(\Delta; x) = \frac{1}{2\pi} \ln |x| - \frac{1}{4\|\mathcal{F}\|}|x|^2 + H(x) \tag{8.40}$$

where H is a harmonic function. In terms of two-dimensional polar coordinates $x = r\xi$, $r = |x|$, $\xi \in \mathbb{S}^1$, we are allowed to express H by the following series expansion in terms of "circular harmonics"

$$H(x) = Y^\wedge(0, 1)Y_{0,1}(2; \xi) + \sum_{n=1}^{\infty} r^n \sum_{j=1}^{2} Y^\wedge(n, j)Y_{n,j}(2; \xi), \quad \xi \in \mathbb{S}^1, \tag{8.41}$$

where all coefficients $Y^\wedge(0, 1)$, $Y^\wedge(n, j)$, $n \in \mathbb{N}$, $j = 1, 2$, are dependent on the planar lattice Λ under consideration; hence, we have to write more precisely

$$H(x) = \underbrace{Y^\wedge(0, 1; \Lambda)Y_{0,1}(2; \xi)}_{= Y_0(\Lambda)} + \sum_{n=1}^{\infty} r^n \sum_{j=1}^{2} Y^\wedge(n, j; \Lambda)Y_{n,j}(2; \xi), \quad \xi \in \mathbb{S}^1.$$

$$\tag{8.42}$$

It should be noted that the property

$$G(\Delta; x) = G(\Delta; -x), \quad x \in \mathbb{R}^2 \backslash \Lambda, \tag{8.43}$$

implies

$$Y^\wedge(2n+1, 1; \Lambda) = Y^\wedge(2n+1, 2; \Lambda) = 0, \quad n = 0, 1, \dots. \tag{8.44}$$

Therefore we obtain from (8.45)

$$H(x) = Y_0(\Lambda) + \sum_{n=1}^\infty r^{2n} \sum_{j=1}^2 Y^\wedge(2n, j; \Lambda) Y_{2n,j}(2; \xi), \quad \xi \in \mathbb{S}^1. \tag{8.45}$$

From the series expansion (8.45) we therefore obtain, for $x \in \mathbb{R}^2 \setminus \frac{1}{2}\Lambda$, via Lemma 8.6,

$$
\begin{aligned}
G(\Delta; 2x) - G(\Delta; x) &= G\left(\Delta; x + \frac{1}{2}g_1\right) + G\left(\Delta; x + \frac{1}{2}g_2\right) \tag{8.46} \\
&\quad + G\left(\Delta; x + \frac{1}{2}(g_1 + g_2)\right) \\
&= \frac{1}{2\pi}\ln(2|x|) - \frac{1}{4\|\mathcal{F}\|}|2x|^2 + H(2x) \\
&\quad - \frac{1}{2\pi}\ln(|x|) + \frac{1}{4\|\mathcal{F}\|}|x|^2 - H(x).
\end{aligned}
$$

Hence, letting $x \to 0$ we find

$$\lim_{x \to 0} (G(\Delta; 2x) - G(\Delta; x)) = \frac{1}{2\pi}\ln(2). \tag{8.47}$$

In other words, we are led to

Lemma 8.7. *Let $g_1, g_2 \in \mathbb{R}^2$ be the basis vectors of the lattice $\Lambda \subset \mathbb{R}^2$. Then*

$$G\left(\Delta; \frac{1}{2}g_1\right) + G\left(\Delta; \frac{1}{2}g_2\right) + G\left(\Delta; \frac{1}{2}(g_1 + g_2)\right) = \frac{1}{2\pi}\ln(2). \tag{8.48}$$

Next, for $x \in \mathbb{R}^2$, we particularly set

$$x = w_1 g_1 + w_2 g_2, \tag{8.49}$$

$w_i \in \mathbb{R}$, $i = 1, 2$, where $g_1, g_2 \in \mathbb{R}^2$ are the generating vectors of the planar lattice Λ. Then it is easy to see that

$$
\begin{aligned}
h \cdot x &= (n_1 h_1 + n_2 h_2) \cdot (w_1 g_1 + w_2 g_2) \tag{8.50} \\
&= n_1 w_1 + n_2 w_2
\end{aligned}
$$

and

$$h \cdot h = (n_1 h_1 + n_2 h_2) \cdot (n_1 h_1 + n_2 h_2) \tag{8.51}$$
$$= n_1^2 \gamma^{11} + n_1 n_2 \gamma^{12} + n_2 n_1 \gamma^{21} + n_2^2 \gamma^{22}$$
$$= n_1^2 \gamma^{11} + 2 n_1 n_2 \gamma^{12} + n_2^2 \gamma^{22},$$

where $h_1, h_2 \in \mathbb{R}^2$ are the generating vectors of the inverse lattice Λ^{-1}. In terms of the variables w_1, w_2 the Λ-lattice function can be expressed in the form

$$\tilde{G}(w_1, w_2) = G(\Delta; w_1 g_1 + w_2 g_2). \tag{8.52}$$

The Fourier expansion in this terminology reads as follows

$$-\frac{1}{4\pi^2 \|\mathcal{F}\|} \sum_{\substack{n_1^2 + n_2^2 > 0 \\ n_1, n_2 \in \mathbb{Z}}} \frac{e^{2\pi i n_1 w_1} e^{2\pi i n_2 w_2}}{\gamma^{11} n_1^2 + 2\gamma^{12} n_1 n_2 + \gamma^{22} n_2^2}. \tag{8.53}$$

For $w_1 \in (0, 1)$ fixed, the one-dimensional function $\tilde{G}(w_1, \cdot)$ is analytic and $\mathbb{Z}$-periodical such that

$$\tilde{G}(w_1, w_2 + 1) = \tilde{G}(w_1, w_2) \tag{8.54}$$

for all w_2. Consequently, for $w_1 \in (0, 1)$, we have (as a function of w_2)

$$\tilde{G}(w_1, w_2) = \sum_{n \in \mathbb{Z}} A_n(w_1) e^{2\pi i n w_2} \tag{8.55}$$

with

$$A_n(w_1) = \int_0^1 \tilde{G}(w_1, s) e^{-2\pi i n s} \, ds. \tag{8.56}$$

For $n = 0$ we find by virtue of our results from the one-dimensional Λ-lattice function theory (see Theorem 4.2) together with (5.23)

$$A_0(w_1) = \frac{1}{\gamma^{11} \|\mathcal{F}\|} \sum_{\substack{n \neq 0 \\ n \in \mathbb{Z}}} \frac{e^{2\pi i w_1 n}}{-4\pi^2 n^2} = \frac{\|\mathcal{F}\|}{\gamma^{22}} \left(-\frac{(w_1)^2}{2} + \frac{w_1}{2} - \frac{1}{12} \right). \tag{8.57}$$

Integration within the framework of complex analysis by use of the calculus of the residuum shows the following statement.

Lemma 8.8. *For $w_1 \in (0, 1)$ fixed,*

$$G(\Delta; w_1 g_1 + w_2 g_2) = \sum_{n \in \mathbb{Z}} A_n(w_1) e^{2\pi i n w_2} \tag{8.58}$$

with $A_n(w_1)$, $n \neq 0$, given by

$$A_n(w_1) = -\frac{1}{4\pi n} \left(\frac{e^{2\pi i \tau n w_1}}{1 - e^{2\pi i \tau n}} + \frac{e^{2\pi i \bar{\tau} n (w_1 - 1)}}{1 - e^{-2\pi i \bar{\tau} n}} \right) \tag{8.59}$$
$$= \frac{1}{4\pi n} \left(\frac{e^{2\pi i \tau n (w_1 - 1)}}{1 - e^{-2\pi i \tau n}} + \frac{e^{2\pi i \bar{\tau} n w_1}}{1 - e^{2\pi i \bar{\tau} n}} \right),$$

where $\tau \in \mathbb{C}$ is defined by

$$\mathcal{J}(\tau) > 0 \tag{8.60}$$

and

$$\gamma^{11}\tau^2 + 2\gamma^{12}\tau + \gamma^{22} = 0. \tag{8.61}$$

Remark 8.2. *Note that*

$$\tau = -\frac{\gamma^{12}}{\gamma^{11}} + i\frac{1}{\gamma^{11}\|\mathcal{F}\|} = \frac{\gamma_{12}}{\gamma_{22}} + i\frac{\|\mathcal{F}\|}{\gamma_{22}} \tag{8.62}$$

satisfies the conditions (8.60) and (8.61). Moreover,

$$|\tau| = \sqrt{\frac{\gamma_{12}^2}{\gamma_{22}^2} + \frac{\|\mathcal{F}\|^2}{\gamma_{22}^2}} = \frac{1}{\sqrt{\gamma_{22}}}\sqrt{\frac{\gamma_{21}\gamma_{12} + \|\mathcal{F}\|}{\gamma_{22}}} = \frac{1}{\sqrt{\gamma_{22}}}\sqrt{\gamma_{11}}. \tag{8.63}$$

An easy manipulation gives

$$A_n(w_1) \tag{8.64}$$
$$= -\frac{1}{4\pi n}\left(e^{2\pi i \tau n w_1} + e^{2\pi i \tau n w_1}\frac{e^{2\pi i \tau n}}{1 - e^{2\pi i \tau n}} + e^{2\pi i \bar{\tau} n w_1}\frac{e^{-2\pi i \bar{\tau} n}}{1 - e^{-2\pi i \bar{\tau} n}}\right).$$

In connection with (8.64) it follows that

$$\lim_{w_1 \to 0}\left(\sum_{n=1}^{\infty}\left(A_n(w_1) + \overline{A_n(w_1)}\right) + \frac{1}{4\pi}\sum_{n=1}^{\infty}\frac{1}{n}\left(e^{2\pi i \tau w_1 n} + e^{-2\pi i \bar{\tau} w_1 n}\right)\right)$$

$$= -\frac{2}{4\pi}\sum_{n=1}^{\infty}\left(\frac{1}{n}\frac{e^{2\pi i \tau n}}{1 - e^{2\pi i \tau n}} + \frac{1}{n}\frac{e^{-2\pi i \bar{\tau} n}}{1 - e^{-2\pi i \bar{\tau} n}}\right)$$

$$= \frac{1}{2\pi}\ln\prod_{n=1}^{\infty}(1 - e^{2\pi i \tau n}) + \frac{1}{2\pi}\ln\prod_{n=1}^{\infty}\overline{(1 - e^{2\pi i \tau n})}$$

$$= \frac{1}{2\pi}\ln\prod_{n=1}^{\infty}|1 - e^{2\pi i \tau n}|^2. \tag{8.65}$$

Observing the definition of $\tilde{G}(w_1, \cdot)$ we obtain

$$\lim_{w_1 \to 0}\left(\sum_{n=1}^{\infty}\left(A_n(w_1) + \overline{A_n(w_1)}\right) + \frac{1}{4\pi}\sum_{n=1}^{\infty}\frac{1}{n}\left(e^{2\pi i \tau w_1 n} + e^{-2\pi i \bar{\tau} w_1 n}\right)\right)$$

$$= \lim_{w_1 \to 0}\left(\tilde{G}(w_1, 0) - A_0(w_1) + \frac{1}{4\pi}\sum_{n=1}^{\infty}\frac{1}{n}\left(e^{2\pi i \tau w_1 n} + e^{-2\pi i \bar{\tau} w_1 n}\right)\right)$$

$$= \lim_{w_1 \to 0}\left(\tilde{G}(w_1, 0) - A_0(w_1) - \frac{1}{4\pi}\ln|1 - e^{2\pi i \tau w_1}|^2\right) \tag{8.66}$$

such that

$$\lim_{w_1 \to 0}\left(\tilde{G}(w_1, 0) - A_0(w_1) - \frac{1}{2\pi}\ln|1 - e^{2\pi i \tau w_1}|\right) = \frac{1}{2\pi}\ln|1 - e^{2\pi i \tau n}|^2. \tag{8.67}$$

Summarizing our results we find in connection with (8.57)

$$\lim_{w_1 \to 0} \left(\tilde{G}(w_1, 0) - \frac{1}{2\pi} \ln \left| 1 - e^{2\pi i \tau w_1} \right| \right) = -\frac{1}{12} \frac{\|\mathcal{F}\|}{\gamma_{22}} + \frac{1}{2\pi} \ln \prod_{n=1}^{\infty} \left| 1 - e^{2\pi i \tau n} \right|^2.$$

(8.68)

We know that

$$G\left(\Delta; x\right) = \frac{1}{2\pi} \ln |x| - \frac{1}{4\|\mathcal{F}\|} |x|^2 + H(x),$$

(8.69)

where H is of the form (8.45). This shows that

$$\lim_{x \to 0} \left(G\left(\Delta; x\right) - \frac{1}{2\pi} \ln |x| \right) = H(0) = Y_0(\Lambda).$$

(8.70)

A simple calculation yields

$$\left(1 - e^{2\pi i \tau w_1} \right) \left(1 - e^{-2\pi i \bar{\tau} w_1} \right) = 4\pi^2 |\tau|^2 w_1^2 + \ldots .$$

(8.71)

Thus it follows that

$$\lim_{w_1 \to 0} \left(\frac{1}{2\pi} \ln \left| 1 - e^{2\pi i \tau w_1} \right| - \frac{1}{2\pi} \ln(w_1) \right) = \frac{1}{2\pi} \left(\ln(2\pi) + \ln |\tau| \right).$$

(8.72)

Furthermore we see that for $x = w_1 g_1$

$$-\frac{1}{2\pi} \ln |x| = -\frac{1}{2\pi} \ln(w_1) - \frac{1}{2\pi} \ln |g_1|.$$

(8.73)

Combining (8.70),(8.72), and (8.73) we therefore obtain

$$
\begin{aligned}
Y_0(\Lambda) \quad &= \quad \lim_{w_1 \to 0} \left(\tilde{G}(w_1, 0) - \frac{1}{2\pi} \ln(w_1) - \frac{1}{2\pi} \ln |g_1| \right) \qquad (8.74) \\[2mm]
&= \quad \lim_{w_1 \to 0} \left(\tilde{G}(w_1, 0) - \frac{1}{2\pi} \ln(w_1) + \frac{1}{2\pi} \ln \left| 1 - e^{2\pi i \tau w_1} \right| \right. \\[1mm]
&\qquad \left. - \ln |g_1| - \frac{1}{2\pi} \ln \left| 1 - e^{2\pi i \tau w_1} \right| \right) \\[2mm]
&\overset{(8.72)}{=} \quad \lim_{w_1 \to 0} \left(\tilde{G}(w_1, 0) + \frac{1}{2\pi} \ln(2\pi) + \frac{1}{2\pi} |\tau| \right. \\[1mm]
&\qquad \left. - \ln |g_1| - \frac{1}{2\pi} \ln \left| 1 - e^{2\pi i \tau w_1} \right| \right).
\end{aligned}
$$

Thus, we get

$$
\begin{aligned}
Y_0(\Lambda) \quad = \quad &-\frac{1}{12} \frac{\|\mathcal{F}\|}{\gamma_{22}} + \frac{1}{2\pi} \ln \prod_{n=1}^{\infty} \left| 1 - e^{2\pi i \tau n} \right|^2 \qquad (8.75) \\[2mm]
&- \frac{1}{2\pi} \ln |g_1| + \frac{1}{2\pi} \ln(2\pi) + \frac{1}{2\pi} \ln |\tau|.
\end{aligned}
$$

Combining our results we therefore obtain

Lemma 8.9. *With* $\tau = \frac{\gamma_{12}}{\gamma_{22}} + i\frac{\|\mathcal{F}\|}{\gamma_{22}}$ *we have*

$$Y_0(\Lambda) = -\frac{1}{12}\frac{\|\mathcal{F}\|}{\gamma_{22}} + \frac{1}{2\pi}\ln\left|\frac{2\pi}{\sqrt{\gamma_{11}}}\tau\right| + \frac{1}{2\pi}\ln\prod_{n=1}^{\infty}|1 - e^{2\pi i \tau n}|^2. \qquad (8.76)$$

This lemma motivates the introduction of the *(Theta) function* Θ (see, e.g., W. Magnus et al. [1966]) by

$$\Theta(\tau) = \prod_{n=1}^{\infty}\left(1 - e^{2\pi i \tau n}\right), \quad \tau \in \mathbb{C}, \ \Im(\tau) > 0. \qquad (8.77)$$

In terms of the (Theta) function Θ Lemma 8.9 allows the following reformulation.

Lemma 8.10. *With* $\tau = \frac{\gamma_{12}}{\gamma_{22}} + i\frac{\|\mathcal{F}\|}{\gamma_{22}}$ *we have*

$$Y_0(\Lambda) = H(0) = \frac{1}{2\pi}\ln\left(\frac{2\pi}{\sqrt{\gamma_{22}}}\left|\Theta(\tau)e^{\frac{\pi i \tau}{12}}\right|^2\right). \qquad (8.78)$$

The same procedure leading to Lemma 8.10 can be done by taking $\tilde{G}(0, w_2)$ with $w_2 \in (0,1)$ fixed. In this case τ must be replaced by

$$\frac{1}{\tilde{\tau}} = \frac{\gamma_{12}}{\gamma_{11}} + i\frac{\|\mathcal{F}\|}{\gamma_{11}}. \qquad (8.79)$$

Furthermore, we have

$$\Theta\left(\frac{1}{\tilde{\tau}}\right) = \overline{\Theta\left(-\frac{1}{\tau}\right)}. \qquad (8.80)$$

This leads to the following statement.

Lemma 8.11.

$$Y_0(\Lambda) = \frac{1}{2\pi}\ln\left(\frac{2\pi}{\sqrt{\gamma_{11}}}\left|\Theta\left(-\frac{1}{\tau}\right)e^{-\frac{\pi i}{12\tau}}\right|^2\right). \qquad (8.81)$$

Because of $|\tau| = \sqrt{\frac{\gamma_{11}}{\gamma_{22}}}$ (cf. (8.63)) we are therefore able to recover the well known functional equation of the Theta function.

Theorem 8.3. *The (Theta) function* Θ *satisfies the functional equation*

$$|\tau\Theta^2(\tau)| = \left|\Theta\left(-\frac{1}{\tau}\right)e^{-\frac{\pi i}{12}\left(\tau + \frac{1}{\tau}\right)}\right|^2. \qquad (8.82)$$

Both Lemma 8.10 and Lemma 8.11 hold true for every lattice $\Lambda \subset \mathbb{R}^2$. Therefore they are also valid for the inverse lattice Λ^{-1}. In this case we obtain instead of the coefficient $Y_0(\Lambda)$ related to the lattice Λ, the corresponding coefficient $Y_0(\Lambda^{-1})$ related to the inverse lattice Λ^{-1}.

The reformulation of Lemma 8.10 then reads as follows.

Lemma 8.12. *With*

$$-\frac{1}{\tau} = \frac{\gamma^{12}}{\gamma^{22}} + i\frac{1}{\gamma^{22}\|\mathcal{F}\|} = -\frac{\gamma_{12}}{\gamma_{11}} + i\frac{\|\mathcal{F}\|}{\gamma_{11}} \tag{8.83}$$

we have

$$Y_0(\Lambda^{-1}) = \frac{1}{2\pi}\ln\left(\frac{2\pi}{\sqrt{\gamma^{22}}}\left|\Theta\left(-\frac{1}{\tau}\right)e^{-\frac{\pi i}{12\tau}}\right|^2\right) \tag{8.84}$$

or

$$Y_0(\Lambda^{-1}) = \frac{1}{2\pi}\ln\left(\frac{2\pi}{\sqrt{\gamma^{11}}}\|\mathcal{F}\||\tau|\left|\Theta\left(\tau\right)e^{\frac{\pi i\tau}{12}}\right|^2\right). \tag{8.85}$$

Because of $|\tau| = \sqrt{\gamma_{11}/\gamma_{22}}$ we finally obtain a relation between the coefficients $Y_0(\Lambda)$ and $Y_0(\Lambda^{-1})$.

Lemma 8.13. *If Λ is a lattice in $\mathbb{R}^2$, then*

$$Y_0(\Lambda^{-1}) = Y_0(\Lambda) + \frac{1}{2\pi}\ln\|\mathcal{F}\|. \tag{8.86}$$

In particular, for $\Lambda = \mathbb{Z}^2$, we have (see C. Müller [1954a])

$$Y_0(\Lambda^{-1}) = Y_0(\Lambda) = \frac{1}{2\pi}\ln\left(2\pi\Theta^2(i)\right) + \frac{\pi}{6}, \tag{8.87}$$

where

$$\Theta^2(i) = \frac{1}{2}e^{\frac{\pi}{12}}\pi^{-\frac{3}{4}}\Gamma\left(\frac{1}{4}\right). \tag{8.88}$$

Lemma 8.14. *For $\Lambda = \mathbb{Z}^2$*

$$Y_0(\Lambda^{-1}) = Y_0(\Lambda) = \frac{1}{2\pi}\ln\left(\frac{1}{2}\pi^{-\frac{1}{2}}\Gamma^2\left(\frac{1}{4}\right)\right). \tag{8.89}$$

It remains to express the coefficients $Y^\wedge(n,j;\Lambda)$, $n = 1,2,\ldots$, $j = 1,2$, of the Fourier expansion of H in certain terms known in the theory of complex analysis (see C. Müller [1954a]). For that purpose we use the Cartesian notation $x = x_1\epsilon^1 + x_2\epsilon^2$ in $\mathbb{R}^2$ as well as the complex notation $z = x_1 + ix_2$ of $\mathbb{C}$ in parallel. We set

$$U(x_1,x_2) = \frac{\partial^2}{\partial x_1^2}G(\Delta;x_1,x_2) \tag{8.90}$$

and

$$V(x_1,x_2) = -\frac{\partial^2}{\partial x_1\partial x_2}G(\Delta;x_1,x_2). \tag{8.91}$$

Then, for all $x \notin \Lambda$, the function $F = U + iV$ is holomorphic (note that, because of $\Delta G(\Delta;\cdot) = -\|\mathcal{F}\|^{-1}$, the Cauchy–Riemann differential equations are satisfied for all $x \notin \Lambda$). Moreover, we have in the neighborhood of the origin

$$\left(\frac{\partial^2}{\partial x_1^2} - i\frac{\partial^2}{\partial x_1 \partial x_2}\right) \ln \sqrt{x_1^2 + x_2^2} = \frac{x_2^2 - x_1^2 + 2ix_1x_2}{(x_1^2 + x_2^2)^2} \qquad (8.92)$$

and

$$-\frac{1}{z^2} = -\frac{(\bar{z})^2}{(z\bar{z})^2} = \frac{x_2^2 - x_1^2 + 2ix_1x_2}{(x_1^2 + x_2^2)^2}, \qquad (8.93)$$

for $(x_1, x_2)^T \neq (0,0)^T$. In other words, we have

Lemma 8.15.

$$\frac{\partial^2}{\partial x_1^2} G(\Delta; \cdot) - i \frac{\partial^2}{\partial x_1 \partial x_2} G(\Delta; \cdot) \qquad (8.94)$$

is a holomorphic function showing in all lattice points of Λ a singularity of the form

$$-\frac{1}{2\pi}\frac{1}{z^2}. \qquad (8.95)$$

From the classical theory of elliptic functions (see, e.g., K. Knopp [1971], H. Rademacher [1973]), we are therefore led to conclude

Lemma 8.16.

$$\frac{\partial^2}{\partial x_1^2} G(\Delta; \cdot) - i \frac{\partial^2}{\partial x_1 \partial x_2} G(\Delta; \cdot) = -\frac{1}{2\pi}\mathcal{P}, \qquad (8.96)$$

where $\mathcal{P}$ is the (Weierstraß) $\mathcal{P}$-function (see, e.g., K. Knopp [1971]).

Now it is known in complex analysis that

$$|x|^{2n} \left(Y^\wedge(2n, 1; \Lambda)\cos(2n\varphi) + Y^\wedge(2n, 2; \Lambda)\sin(2n\varphi)\right) \qquad (8.97)$$

$$= \frac{1}{2}\left(Y^\wedge(2n, 1; \Lambda) - iY^\wedge(2n, 2; \Lambda)\right)z^{2n}$$

$$+ \frac{1}{2}\left(Y^\wedge(2n, 1; \Lambda) + iY^\wedge(2n, 2; \Lambda)\right)\bar{z}^{2n}.$$

Furthermore, it can be readily deduced that

$$\left(\frac{\partial^2}{\partial x_1^2} - i\frac{\partial^2}{\partial x_1 \partial x_2}\right)\left(\frac{1}{2}(Y^\wedge(2n, 1; \Lambda) - i\, Y^\wedge(2n, 2; \Lambda))\, z^{2n}\right)$$

$$+ \left(\frac{\partial^2}{\partial x_1^2} - i\frac{\partial^2}{\partial x_1 \partial x_2}\right)\left(\frac{1}{2}(Y^\wedge(2n, 1; \Lambda)) + i\, Y^\wedge(2n, 2; \Lambda))\, \bar{z}^{2n}\right)$$

$$= \frac{2n(2n-1)}{2}\left(Y^\wedge(2n, 1; \Lambda) - i\, Y^\wedge(2n, 2; \Lambda)\right)z^{2n} \qquad (8.98)$$

such that

$$-\frac{1}{2\pi}\frac{1}{z^2} + \frac{1}{2\sqrt{\pi}\|\mathcal{F}\|}\sum_{n=1}^{\infty}\binom{2n}{2}\left(Y^\wedge(2n, 1; \Lambda) - i\, Y^\wedge(2n, 2; \Lambda)\right)z^{2n-2}$$

$$= -\frac{1}{2\pi}\mathcal{P}(z). \qquad (8.99)$$

This leads us to the conclusion that all expansion coefficients of the two-dimensional lattice function $G(\Delta; \cdot)$ are expressible by known values.

8.4 Lattice Function in Terms of Spherical Harmonics

For dimensions $q \geq 3$ we omit the presentations of one-dimensional reductions of the Λ-lattice function (as presented in the preceding subsection). Instead, we want to discuss the behavior of an iterated Λ-lattice function in the neighborhood of lattice points by use of spherical harmonics. Because of the periodicity it suffices to study the (punctured) neighborhood of the origin (cf. R. Wienkamp [1958]).

Representation in Terms of Spherical Harmonics

For $x \in \overline{\mathbb{B}^q_{\rho_0,\rho_1}}$, $0 < \rho_0 \leq \rho_1 < \inf_{x \in \mathcal{F}} |x|$, we consider the separation of $G(\Delta^k; \cdot)$ into radial and angular sum components

$$G\left(\Delta^k; r\xi\right) = \sum_{n=0}^{\infty} g_n^{q,k}(r) Y_n^{q,k}(\xi), \quad k \in \mathbb{N}, \tag{8.100}$$

with

$$Y_n^{q,k}(\xi) = \sum_{j=1}^{N(q;n)} (Y_n^{q,k})^\wedge(n,j) Y_{n,j}(q;\xi), \tag{8.101}$$

where, as usual, $x = r\xi$, $r = |x|$, $\xi \in \mathbb{S}^{q-1}$, and the family

$$\{Y_{n,j}(q;\cdot)\}_{\substack{n=0,1,\dots \\ j=1,\dots,N(q;n)}} \tag{8.102}$$

forms an $L^2(\mathbb{S}^{q-1})$-orthonormal system of spherical harmonics. Clearly, the series on the right side of (8.100) is absolutely and uniformly convergent on the specified domain $\overline{\mathbb{B}^q_{\rho_0,\rho_1}}$ (note that $\xi \mapsto G(\Delta^k; r\xi)$, $\xi \in \mathbb{S}^{q-1}$, $r \in [\rho_0, \rho_1]$, is infinitely often differentiable such that the convergence of the series expansion is guaranteed by Lemma 6.18). From the recursion relation for iterated Λ-lattice functions it follows for all $x \in \overline{\mathbb{B}^q_{\rho_0,\rho_1}}$ that

$$\Delta_x \left(\sum_{n=0}^{\infty} g_n^{q,k}(r) Y_n^{q,k}(\xi) \right) = \sum_{n=0}^{\infty} g_n^{q,k-1}(r) Y_n^{q,k-1}(\xi). \tag{8.103}$$

Therefore, the separation of the Laplace operator into radial and angular parts, the orthogonality of spherical harmonics, and the uniform convergence of the series (8.100) on $\overline{\mathbb{B}^q_{\rho_0,\rho_1}}$ imply the recursion relation

$$r^{1-q} \frac{d}{dr} r^{q-1} \frac{d}{dr} g_n^{q,k}(r) - \frac{n(n+q-2)}{r^2} g_n^{q,k}(r) = g_n^{q,k-1}(r). \tag{8.104}$$

In other words, for all integers $k \geq 1$, the coefficients constituting the solution $g_n^{q,k}$ of the differential equation (8.104) become calculable by elementary operations.

The leading coefficient functions for the recursion process can be characterized as follows: For $n > 0$ and $k \geq 1$, $g_n^{q,k}$ allows a representation of the form

$$g_n^{q,k}(r) = a_{n,k-1}^{q,k} r^{n+2(k-1)} + \ldots + a_{n,1}^{q,k} r^{n+2} + r^n. \qquad (8.105)$$

For $n = 0$ and $k \geq 1$ we distinguish two cases, namely q even and q odd:
(i) If q is odd, then

$$g_0^{q,k}(r) \;=\; a_{0,k-1}^{q,k} r^{2(k-1)} + \ldots + a_{0,1}^{q,k} r^2 + 1 \qquad (8.106)$$
$$+ \, b^{q,k} r^{2k-q} + c^{q,k} r^{2k}.$$

(ii) If q is even, then we are confronted with the representation (8.105) if $k < \frac{q}{2}$. However, if q is even with $k \geq \frac{q}{2}$, then

$$g_0^{q,k}(r) \;=\; a_{0,k-1}^{q,k} r^{2(k-1)} + \ldots + a_{0,1}^{q,k} r^2 + a_{0,0}^{q,k} \qquad (8.107)$$
$$+ \, b^{q,k} r^{2(k-\frac{q}{2})} \ln(r) + c^{q,k} r^{2k}.$$

The recursive determination of the coefficients occurring in (8.105), (8.106), and (8.107) is straightforward, but rather technical; hence, it will be omitted here. Nevertheless it should be kept in mind that all coefficients $g_{n,k}^{q,k}$ in (8.100) are explicitly determinable (cf. R. Wienkamp [1958]).

Remark 8.3. *In particular, the two-dimensional representation (8.39) and (8.45) in terms of circular harmonics*

$$G(\Delta; r\xi) = \frac{1}{2\pi} \ln(r) - \frac{1}{4\|\mathcal{F}\|} r^2 + Y_0 + \sum_{n=1}^{\infty} r^n Y_n(\xi), \quad \xi \in \mathbb{S}^1 \qquad (8.108)$$

becomes compatible with the (two-dimensional) spherical harmonic expansion (8.100). In fact, we have

$$G(\Delta; r\xi) = g_0^{2,1}(r) Y_0^{2,1} + \sum_{n=1}^{\infty} g_n^{2,1}(r) Y_n^{2,1}(\xi), \qquad (8.109)$$

with

$$Y_n^{2,1}(\xi) = \sum_{j=1}^{N(2;n)} (Y_n^{2,1})^{\wedge}(n,j) Y_{n,j}(2;\xi), \qquad (8.110)$$

where

$$g_0^{2,1}(r) = a_{0,0}^{2,1} + b^{2,1} \ln(r) + c^{2,1} r^2 \tag{8.111}$$

$$Y_0^{2,1} = Y_0, \tag{8.112}$$

$$a_{0,0}^{2,1} = 1, \tag{8.113}$$

$$b^{2,1} = \frac{Y_0^{-1}}{2\pi}, \tag{8.114}$$

$$c^{2,1} = -\frac{Y_0^{-1}}{4\|\mathcal{F}\|}, \tag{8.115}$$

$$g_n^{2,1}(r) = r^n, \tag{8.116}$$

$$Y_n^{2,1}(\xi) = Y_n(\xi), \quad \xi \in \mathbb{S}^1 \tag{8.117}$$

(note that the quantities $b^{2,1}, c^{2,1}$, and $Y_n^{2,1}$, $n \in \mathbb{N}_0$, are dependent on the two-dimensional lattice Λ under consideration).

9

Euler Summation on Regular Regions

CONTENTS

9.1 Euler Summation Formula for the Iterated Laplace Operator 270
    Euler Summation for the Laplace Operator 270
    Euler Summation for the Iterated Laplace Operator 272
    Euler Summation to Certain Boundary Conditions 274
    Euler Summation to Dirichlet Conditions 275
    Euler Summation to Homogeneous Boundary Conditions 277
9.2 Lattice Point Discrepancy Involving the Laplace Operator 278
    Homogeneous Boundary Conditions 279
    (Poly)Harmonicity Under Boundary Conditions 280
    Constant Weights 281
9.3 Zeta Function and Lattice Function 282
    Two-Dimensional Theory 282
    Multi-Dimensional Theory 288
9.4 Euler Summation Formulas for Iterated Helmholtz Operators 294
    Euler Summation for the Helmholtz Operator 294
    Euler Summation for Iterated Helmholtz Operators 296
9.5 Lattice Point Discrepancy Involving the Helmholtz Operator 299
    Application to Periodical Polynomials 300

In the first part of this chapter we generalize the Euler summation formula to the multi-dimensional case. In fact, we give its formulation for the (iterated) Laplace operator Δ^m, arbitrary lattices $\Lambda \subset \mathbb{R}^q$, and regular regions $\mathcal{G} \subset \mathbb{R}^q$. The essential tools are the Λ-lattice function for the Laplace operator and its constituting properties (as introduced in the last chapter). In the second part of the chapter we go over to the Euler summation formula with respect to an iterated Helmholtz operator. In addition, the particular structure of Euler summation formulas is used to relate Λ-lattice point discrepancies to certain expressions involving elliptic partial differential operators such as iterated Laplace and Helmholtz operators.

The composition of this chapter is as follows: based on the properties of the multi-dimensional Λ-lattice function for iterated Laplace operators, Section 9.1 develops the Euler summation formula by partial integration, i.e., by use of the second Green formula for the Laplacian. Within the polyharmonic framework, the Euler summation formula is formulated under Dirichlet/Neumann boundary conditions for regular regions, e.g., "potatoes". In Section 9.2, Euler summation formulas with respect to the Laplace operator lead to representations of the so–called lattice point discrepancy. Remainder terms in the polyharmonic context are listed for the discrepancies. By use of Euler summation formulas interrelations between the Zeta function and (the spherical harmonic coefficients of) the (iterated) Λ-lattice function are worked out in Section 9.3. Kronecker's limit relation is established. For simplicity, we first discuss the theory of the Zeta function within the two-dimensional framework (in Section 9.3), thereby taking special advantage of complex analysis. Subsequently, the general q-dimensional theory is under discussion (in Section 9.3). The chapter ends with the explanation of (lattice point-generated) Euler summation formulas for iterated Helmholtz operators and regular regions (in Section 9.4). The discrepancy expressions are briefly analyzed in the polymetaharmonic framework.

9.1 Euler Summation Formula for the Iterated Laplace Operator

We begin with the Euler summation formula with respect to the Laplace operator, arbitrary lattices, and regular regions in $\mathbb{R}^q$, $q \geq 2$.

Euler Summation for the Laplace Operator

Let Λ be an arbitrary lattice in $\mathbb{R}^q$. Suppose that $\mathcal{G} \subset \mathbb{R}^q$ is a regular region. Let F be a function of class $\mathrm{C}^{(2)}(\overline{\mathcal{G}}), \overline{\mathcal{G}} = \mathcal{G} \cup \partial\mathcal{G}$.

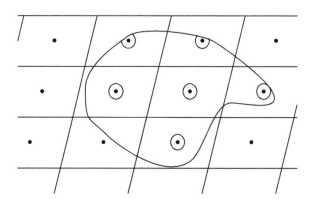

FIGURE 9.1
The geometric situation of Euler summation in Equation (9.1).

Then, for every (sufficiently small) $\varepsilon > 0$, the Second Green Theorem gives

(see Figure 9.1)

$$\int_{\substack{x \in \overline{\mathcal{G}} \\ x \notin \mathbb{B}_{\varepsilon}^{q}+\Lambda}} \{F(x)\,(\Delta G\,(\Delta; x)) - G\,(\Delta; x)\,(\Delta F(x))\}\ dV(x)$$

$$= \int_{\substack{x \in \partial \mathcal{G} \\ x \notin \mathbb{B}_{\varepsilon}^{q}+\Lambda}} \left\{ F(x) \left(\frac{\partial}{\partial \nu} G\,(\Delta; x) \right) - G\,(\Delta; x) \left(\frac{\partial F}{\partial \nu}(x) \right) \right\}\ dS(x)$$

$$+ \sum_{\substack{g \in \overline{\mathcal{G}} \\ g \in \Lambda}} \int_{\substack{|x-g|=\varepsilon \\ x \in \overline{\mathcal{G}}}} \left\{ F(x) \left(\frac{\partial}{\partial \nu} G\,(\Delta; x) \right) - G\,(\Delta; x) \left(\frac{\partial F}{\partial \nu}(x) \right) \right\}\ dS(x),$$

$$(9.1)$$

where ν is the outer (unit) normal field. Observing the differential equation (Condition *(ii)* of Definition 8.1) we get

$$\int_{\substack{x \in \overline{\mathcal{G}} \\ x \notin \mathbb{B}_{\varepsilon}^{q}+\Lambda}} F(x)\Delta G\,(\Delta; x)\ dV(x) = -\frac{1}{\|\mathcal{F}\|} \int_{\substack{x \in \overline{\mathcal{G}} \\ x \notin \mathbb{B}_{\varepsilon}^{q}+\Lambda}} F(x)\ dV(x). \qquad (9.2)$$

Hence, on passing to the limit $\varepsilon \to 0$ and observing the characteristic singularity of the Λ-lattice function (i.e., Condition *(iii)* of Definition 8.1) we obtain in connection with Lemma 6.1

Theorem 9.1. *(Lattice Point-Generated Euler Summation Formula for the Laplace Operator Δ) Let Λ be an arbitrary lattice in $\mathbb{R}^{q}$. Suppose that $\mathcal{G} \subset \mathbb{R}^{q}$ is a regular region. Let F be twice continuously differentiable on $\overline{\mathcal{G}}$, $\overline{\mathcal{G}} = \mathcal{G} \cup \partial \mathcal{G}$. Then*

$$\sideset{}{'}\sum_{\substack{g \in \overline{\mathcal{G}} \\ g \in \Lambda}} F(g) = \frac{1}{\|\mathcal{F}\|} \int_{\mathcal{G}} F(x)\,dV(x) \qquad (9.3)$$

$$+ \int_{\mathcal{G}} G\,(\Delta; x)\,\Delta F(x)\ dV(x)$$

$$+ \int_{\partial \mathcal{G}} \left\{ F(x) \left(\frac{\partial}{\partial \nu} G\,(\Delta; x) \right) - G\,(\Delta; x) \left(\frac{\partial F}{\partial \nu}(x) \right) \right\}\ dS(x),$$

where

$$\sideset{}{'}\sum_{\substack{g \in \overline{\mathcal{G}} \\ g \in \Lambda}} F(g) = \sum_{\substack{g \in \overline{\mathcal{G}} \\ g \in \Lambda}} \alpha(g)\,F(g) \qquad (9.4)$$

and $\alpha(g)$ is the solid angle subtended at $g \in \overline{\mathcal{G}}$ by the surface $\partial \mathcal{G}$.

This formula provides a comparison between the integral over a regular region $\mathcal{G}$ and the sum over all functional values of the twice continuously differentiable function F in lattice points $g \in \overline{\mathcal{G}}$ under explicit knowledge of the remainder term in integral form.

Remark 9.1. *The formula for the Laplacian Δ (Theorem 9.1) is an immediate generalization to the multi-dimensional case of the one-dimensional Euler summation formula as presented in Chapter 4, where $G(\Delta; \cdot)$ takes the role of the Bernoulli polynomial of degree 2.*

Euler Summation for the Iterated Laplace Operator

Provided that the function F is of class $C^{(2k+2)}\left(\overline{\mathcal{G}}\right)$, $k \in \{1,\ldots,m-1\}$, on $\overline{\mathcal{G}} = \mathcal{G} \cup \partial\mathcal{G}$, $\mathcal{G}$ regular region, we get from the Second Green Theorem by aid of the differential equation (8.31)

$$\int_{\substack{x \in \overline{\mathcal{G}} \\ x \notin \mathbb{B}_\varepsilon^q + \Lambda}} G\left(\Delta^{k+1}; x\right)\left(\Delta^{k+1} F(x)\right)\, dV(x) \tag{9.5}$$

$$-\int_{\substack{x \in \overline{\mathcal{G}} \\ x \notin \mathbb{B}_\varepsilon^q + \Lambda}} \left(\Delta G\left(\Delta^{k+1}; x\right)\right) \Delta^k F(x)\, dV(x)$$

$$= \int_{\substack{x \in \partial\mathcal{G} \\ x \notin \mathbb{B}_\varepsilon^q + \Lambda}} G\left(\Delta^{k+1}; x\right)\left(\frac{\partial}{\partial\nu}\Delta^k F(x)\right)\, dS(x)$$

$$-\int_{\substack{x \in \partial\mathcal{G} \\ x \notin \mathbb{B}_\varepsilon^q + \Lambda}} \left(\frac{\partial}{\partial\nu}G\left(\Delta^{k+1}; x\right)\right) \Delta^k F(x)\, dS(x)$$

$$+\sum_{\substack{g \in \overline{\mathcal{G}} \\ g \in \Lambda}} \int_{\substack{|x-g|=\varepsilon \\ x \in \overline{\mathcal{G}}}} G\left(\Delta^{k+1}; x\right)\left(\frac{\partial}{\partial\nu}\Delta^k F(x)\right)\, dS(x)$$

$$-\sum_{\substack{g \in \overline{\mathcal{G}} \\ g \in \Lambda}} \int_{\substack{|x-g|=\varepsilon \\ x \in \overline{\mathcal{G}}}} \left(\frac{\partial}{\partial\nu}G\left(\Delta^{k+1}; x\right)\right) \Delta^k F(x)\, dS(x)$$

for every (sufficiently small) $\varepsilon > 0$. From classical potential theory (see, e.g., O.D. Kellogg [1929]) we know that the integrals over all hyperspheres around the lattice points tend to 0 as $\varepsilon \to 0$. This leads to the recursion formula

$$\int_{\mathcal{G}} G\left(\Delta^{k+1}; x\right) \Delta^{k+1} F(x)\, dV(x) \tag{9.6}$$

$$= \int_{\mathcal{G}} G\left(\Delta^k; x\right) \Delta^k F(x)\, dV(x)$$

$$+ \int_{\partial\mathcal{G}} G\left(\Delta^{k+1}; x\right)\left(\frac{\partial}{\partial\nu}\Delta^k F(x)\right)\, dS(x)$$

$$- \int_{\partial\mathcal{G}} \left(\frac{\partial}{\partial\nu}G\left(\Delta^{k+1}; x\right)\right) \Delta^k F(x)\, dS(x).$$

From (9.6) we easily obtain for $F \in C^{(2m)} \left(\overline{\mathcal{G}} \right)$, $m \in \mathbb{N}$,

$$\int_{\mathcal{G}} G\left(\Delta; x\right) \Delta F(x) \, dV(x) \tag{9.7}$$

$$= \int_{\mathcal{G}} G\left(\Delta^m; x\right) \Delta^m F(x) \, dV(x)$$

$$+ \sum_{k=1}^{m-1} \int_{\partial\mathcal{G}} \left(\frac{\partial}{\partial\nu} G\left(\Delta^{k+1}; x\right) \right) \Delta^k F(x) \, dS(x)$$

$$- \sum_{k=1}^{m-1} \int_{\partial\mathcal{G}} G\left(\Delta^{k+1}; x\right) \left(\frac{\partial}{\partial\nu} \Delta^k F(x) \right) \, dS(x).$$

In connection with the Euler summation formula (Theorem 9.1) we therefore obtain the Euler summation formula with respect to the operator Δ^m.

Theorem 9.2. *(Euler Summation Formula for the Operator Δ^m, $m \in \mathbb{N}$) Let $\mathcal{G} \subset \mathbb{R}^q$ be a regular region. Suppose that F is of class $C^{(2m)} \left(\overline{\mathcal{G}} \right)$, $\overline{\mathcal{G}} = \mathcal{G} \cup \partial\mathcal{G}$. Then,*

$$\sum_{\substack{g \in \overline{\mathcal{G}} \\ g \in \Lambda}}{}' F(g) = \frac{1}{\|\mathcal{F}\|} \int_{\mathcal{G}} F(x) \, dV(x) \tag{9.8}$$

$$+ \int_{\mathcal{G}} G\left(\Delta^m; x\right) \Delta^m F(x) \, dV(x)$$

$$+ \sum_{k=0}^{m-1} \int_{\partial\mathcal{G}} \left(\frac{\partial}{\partial\nu} G\left(\Delta^{k+1}; x\right) \right) \Delta^k F(x) \, dS(x)$$

$$- \sum_{k=0}^{m-1} \int_{\partial\mathcal{G}} G\left(\Delta^{k+1}; x\right) \left(\frac{\partial}{\partial\nu} \Delta^k F(x) \right) \, dS(x).$$

Replacing the lattice Λ by a point lattice $\Lambda + \{x\}$ based on $x \in \mathbb{R}^q$ we obtain as multi-dimensional counterpart to the extended Euler summation formula (Corollary 4.8) the following identity.

Theorem 9.3. *Let Λ be an arbitrary lattice in $\mathbb{R}^q$. Let $\mathcal{G} \subset \mathbb{R}^q$ be a regular region, $\overline{\mathcal{G}} = \mathcal{G} \cup \partial\mathcal{G}$. Suppose that F is a member of class $C^{(2m)} \left(\overline{\mathcal{G}} \right)$, $m \in \mathbb{N}$.*

Then, for every $x \in \mathbb{R}^q$,

$$\sum_{\substack{g+x\in\overline{\mathcal{G}} \\ g\in\Lambda}} {}' F(g+x) = \frac{1}{\|\mathcal{F}\|} \int_{\mathcal{G}} F(y) \, dV(y) \tag{9.9}$$

$$+ \int_{\mathcal{G}} G\left(\Delta^m; x-y\right) \Delta^m F(y) \, dV(y)$$

$$+ \sum_{k=0}^{m-1} \int_{\partial\mathcal{G}} \left(\frac{\partial}{\partial\nu} G\left(\Delta^{k+1}; x-y\right)\right) \left(\Delta^k F(y)\right) \, dS(y)$$

$$- \sum_{k=0}^{m-1} \int_{\partial\mathcal{G}} G\left(\Delta^{k+1}; x-y\right) \left(\frac{\partial}{\partial\nu}\Delta^k F(y)\right) \, dS(y),$$

where

$$\sum_{\substack{g+x\in\overline{\mathcal{G}} \\ g\in\Lambda}} {}' F(g+x) = \sum_{\substack{g+x\in\overline{\mathcal{G}} \\ g\in\Lambda}} \alpha(g+x) \, F(g+x). \tag{9.10}$$

Euler Summation to Certain Boundary Conditions

Let Λ be an arbitrary lattice in $\mathbb{R}^q$. Let $\mathcal{G}$ be a regular region in $\mathbb{R}^q$. Then, for all $F \in \mathrm{C}^{(2m)}\left(\overline{\mathcal{G}}\right), \overline{\mathcal{G}} = \mathcal{G} \cup \partial\mathcal{G}, \ m \in \mathbb{N}$,

$$\sum_{\substack{g\in\overline{\mathcal{G}} \\ g\in\Lambda}} {}' F(g) = \frac{1}{\|\mathcal{F}\|} \int_{\mathcal{G}} F(x) \, dV(x) \tag{9.11}$$

$$+ \int_{\mathcal{G}} \Delta^m G\left(\Delta^{2m}; x\right) \left(\Delta^m F(x)\right) \, dV(x)$$

$$+ \int_{\partial\mathcal{G}} \mathcal{B}^{(m)}_{(q)} \left[F(x), G\left(\Delta^{2m}; x\right)\right] \, dS(x),$$

where the "boundary term" explicitly reads as follows

$$\int_{\partial\mathcal{G}} \mathcal{B}^{(m)}_{(q)} \left[F(x), G\left(\Delta^{2m}; x\right)\right] \, dS(x) \tag{9.12}$$

$$= \sum_{k=0}^{m-1} \int_{\partial\mathcal{G}} \left(\frac{\partial}{\partial\nu}\Delta^{2m-(k+1)} G\left(\Delta^{2m}; x\right)\right) \Delta^k F(x) \, dS(x)$$

$$- \sum_{k=0}^{m-1} \int_{\partial\mathcal{G}} \Delta^{2m-(k+1)} G\left(\Delta^{2m}; x\right) \left(\frac{\partial}{\partial\nu}\Delta^k F(x)\right) \, dS(x).$$

Easy to handle in practical applications are summation formulas for which the *"boundary term"* (9.12) vanishes. We list two examples:

(HBC) Suppose that $F \in C^{(2m)}(\overline{\mathcal{G}})$ satisfies the homogeneous boundary conditions *(HBC)*

$$\Delta^k F | \partial \mathcal{G} = 0, \quad \frac{\partial}{\partial \nu} \Delta^k F | \partial \mathcal{G} = 0, \quad k = 0, \ldots, m-1. \tag{9.13}$$

Then we have

$$\int_{\partial \mathcal{G}} \mathcal{B}_{(q)}^{(m)} \left[F(x), \Delta^m G\left(\Delta^{2m}; x \right) \right] \, dS(x) = 0. \tag{9.14}$$

(PBC) Suppose that F is of class $C_\Lambda^{(2m)}(\mathbb{R}^q)$. Then, because of the Λ-periodical boundary condition *(PBC)*,

$$\int_{\partial \mathcal{F}} \mathcal{B}_{(q)}^{(m)} \left[F(x), G\left(\Delta^{2m}; x \right) \right] \, dS(x) = 0. \tag{9.15}$$

Euler Summation to Dirichlet Conditions

For simplicity, let $\mathcal{G}$ be a regular region in $\mathbb{R}^q$ such that its boundary $\partial \mathcal{G}$ does not contain a lattice point of Λ (if this is not the case we can modify the Euler summation formula by (a finitely often application of) the Third Green Theorem with respect to the lattice points on the boundary so that the following approach still remains valid for the modified formula). Suppose that $B_k : \partial \mathcal{G} \to \mathbb{R}$, $k = 0, \ldots, m-1$, are given continuous functions. We are interested in a Euler summation formula of the form (9.11), (9.12) corresponding to Dirichlet's boundary conditions

$$\Delta^k F | \partial \mathcal{G} = B_k,$$

$k = 0, \ldots, m-1$. For that purpose we understand $G^{(2m)} \in C^{(4m)}(\overline{\mathcal{G}})$, $m \in \mathbb{N}$, to be the solution of the boundary-value problem

$$\Delta^{2m} G^{(2m)} | \mathcal{G} = 0 \tag{9.16}$$

such that

$$\Delta^k G^{(2m)} | \partial \mathcal{G} = G\left(\Delta^{2m-k}; \cdot \right) \Big| \partial \mathcal{G}, \tag{9.17}$$

$k = 0, \ldots, 2m - 1$. Since $G^{(2m)}$ is a member of the class $\mathrm{C}^{(4m)}\left(\overline{\mathcal{G}}\right)$, it follows from the Extended Second Green Theorem that

$$\int_{\mathcal{G}} \underbrace{\Delta^{2m} G^{(2m)}(x)}_{=0} F(x) \, dV(x) \tag{9.18}$$

$$= \int_{\mathcal{G}} \left(\Delta^m G^{(2m)}(x)\right) \Delta^m F(x) \, dV(x)$$

$$+ \sum_{k=0}^{m-1} \int_{\partial \mathcal{G}} \left(\frac{\partial}{\partial \nu} \Delta^{2m-(k+1)} G^{(2m)}(x)\right) \Delta^k F(x) \, dS(x)$$

$$- \sum_{k=0}^{m} \int_{\partial \mathcal{G}} \Delta^{2m-(k+1)} G^{(2m)}(x) \left(\frac{\partial}{\partial \nu} \Delta^k F(x)\right) \, dS(x).$$

In connection with (9.11) and (9.12) this leads us to the following modified *Euler summation formula under Dirichlet's conditions.*

Theorem 9.4. *Let* Λ *be an arbitrary lattice in* $\mathbb{R}^q$. *Let* $\mathcal{G} \subset \mathbb{R}^q$ *be a regular region such that* $\partial \mathcal{G} \cap \Lambda = \emptyset$. *Let* $F \in \mathrm{C}^{(2m)}\left(\overline{\mathcal{G}}\right)$, $\overline{\mathcal{G}} = \mathcal{G} \cup \partial \mathcal{G}$, *satisfy the boundary conditions*

$$\Delta^k F|\partial \mathcal{G} = B_k,$$

$k = 0, \ldots, m - 1$. *Furthermore, suppose that* $G^{(2m)} \in \mathrm{C}^{(4m)}\left(\overline{\mathcal{G}}\right)$ *solves the boundary-value problem*

$$\Delta^{2m} G^{(2m)}|\mathcal{G} = 0, \quad \Delta^k G^{(2m)}|\partial \mathcal{G} = G(\Delta^{2m-k}; \, \cdot \,)|\partial \mathcal{G}, \tag{9.19}$$

$k = 0, \ldots, 2m - 1$. *Then*

$$\sum_{\substack{g \in \mathcal{G} \\ g \in \Lambda}} F(g) = \frac{1}{\|\mathcal{F}\|} \int_{\mathcal{G}} F(x) \, dV(x) \tag{9.20}$$

$$+ \int_{\mathcal{G}} \left(\Delta^m D^{(2m)}(x)\right) \Delta^m F(x) \, dV(x)$$

$$- \sum_{k=0}^{m-1} \int_{\partial \mathcal{G}} B_k(x) \left(\frac{\partial}{\partial \nu} \Delta^{2m-(k+1)} D^{(2m)}(x)\right) \, dS(x),$$

where

$$D^{(2m)} = G(\Delta^{2m}; \, \cdot \,) - G^{(2m)}. \tag{9.21}$$

Of course, Theorem 9.4 is a reformulation of the Euler summation formula corresponding to given "Dirichlet boundary" conditions $\Delta^k F|\partial \mathcal{G}$, $k = 0, \ldots, m-1$, at the price of the additional introduction of the function $G^{(2m)}$.

Remark 9.2. *The functions* $\frac{\partial}{\partial \nu} \Delta^k F|\partial \mathcal{G}$, $k = 0, \ldots, m-1$, *can be used in the same manner to handle "Neumann's boundary values". For the Neumann case, however, the construction of* $G^{(2m)}$ *must be modified accordingly.*

Euler Summation to Homogeneous Boundary Conditions

Theorem 9.4 can be used to formulate homogeneous boundary conditions on $\partial \mathcal{G}$. For this purpose, for each $k = 0, \ldots, m - 1$, we define functions $W^{(k)} \in C^{(2m)}(\overline{\mathcal{G}})$, $k = 0, \ldots, m - 1$, such that,

$$
\begin{aligned}
\Delta^{k+1} W^{(k)}|\mathcal{G} &= 0, & (9.22) \\
\Delta^k W^{(k)}|\partial \mathcal{G} &= B_k, & (9.23) \\
W^{(k)}|\partial \mathcal{G} = \Delta W^{(k)}|\partial \mathcal{G} = \ldots = \Delta^{k-1} W^{(k)}|\partial \mathcal{G} &= 0 & (9.24)
\end{aligned}
$$

(note that, for $k = 0$, Condition (9.24) must be omitted). The boundary-value problem is equivalent to the totality of subproblems

$$
\begin{aligned}
\Delta V^{(1)}|\mathcal{G} &= 0, & (9.25) \\
V^{(1)}|\partial \mathcal{G} &= B_k, & (9.26) \\
\Delta V^{(2)}|\mathcal{G} &= V^{(1)}, & (9.27) \\
V^{(2)}|\partial \mathcal{G} &= 0, & (9.28)
\end{aligned}
$$

$$
\begin{aligned}
&\vdots & \vdots \quad & \vdots \\
\Delta V^{(k)}|\mathcal{G} &= V^{(k-1)}, & (9.29) \\
V^{(k)}|\partial \mathcal{G} &= 0, & (9.30) \\
\Delta W^{(k)}|\mathcal{G} &= V^{(k)}, & (9.31) \\
W^{(k)}|\partial \mathcal{G} &= 0, & (9.32)
\end{aligned}
$$

where each of the subproblems can be treated separately.

If we choose $F - \sum_{k=0}^{m-1} W^{(k)}$ instead of F in Theorem 9.4 we obtain the following (modified) *Euler summation formula corresponding to homogeneous boundary conditions.*

Corollary 9.1. *For $k = 0, \ldots, m - 1$, let $W^{(k)} \in C^{(2m)}(\overline{\mathcal{G}})$ be defined by (9.22), (9.23), and (9.24). Then under the assumptions of Theorem 9.4,*

$$
\sum_{\substack{g \in \mathcal{G} \\ g \in \Lambda}} \left(F(g) - \sum_{k=0}^{m-1} W^{(k)}(g) \right)
$$

$$
= \frac{1}{\|\mathcal{F}\|} \int_{\mathcal{G}} \left(F(x) - \sum_{k=0}^{m-1} W^{(k)}(x) \right) dV(x)
$$

$$
+ \int_{\mathcal{G}} \left(\Delta^m D^{(2m)}(x) \right) \Delta^m \left(F(x) - \sum_{k=0}^{m-1} W^{(k)}(x) \right) dV(x).
$$

9.2 Lattice Point Discrepancy Involving the Laplace Operator

Our approach to Euler summation canonically provides the concept of the lattice point discrepancy for a (weight) function F defined on $\overline{\mathcal{G}}$ with $\mathcal{G} \subset \mathbb{R}^q$ being regular.

Definition 9.1. *Let Λ be an arbitrary lattice in $\mathbb{R}^q$. Let $\mathcal{G}$ be a regular region in $\mathbb{R}^q$. Suppose that F is of class $\mathrm{C}^{(0)}(\overline{\mathcal{G}})$, $\overline{\mathcal{G}} = \mathcal{G} \cup \partial\mathcal{G}$. The difference $P(F; \overline{\mathcal{G}})$ given by*

$$P(F; \overline{\mathcal{G}}) = \underset{\substack{g \in \overline{\mathcal{G}} \\ g \in \Lambda}}{\sum}{}' F(g) - \frac{1}{\|\mathcal{F}\|} \int_{\mathcal{G}} F(x)\, dV(x) \tag{9.33}$$

is called the Λ-lattice point discrepancy of F in $\overline{\mathcal{G}}$.

The Euler summation formula (Theorem 9.2) enables us to relate the Λ-lattice point discrepancy of F in $\overline{\mathcal{G}}$ to iterated Laplace operators, however, under the additional assumption that F is of class $\mathrm{C}^{(2m)}(\overline{\mathcal{G}})$, $m \in \mathbb{N}$.

Theorem 9.5. *Let Λ be an arbitrary lattice in $\mathbb{R}^q$. Let $\mathcal{G}$ be a regular region in $\mathbb{R}^q$. Suppose that F is of class $\mathrm{C}^{(2m)}(\overline{\mathcal{G}})$, $m \in \mathbb{N}$, $\overline{\mathcal{G}} = \mathcal{G} \cup \partial\mathcal{G}$. Then*

$$
\begin{aligned}
P(F; \overline{\mathcal{G}}) &= \int_{\mathcal{G}} G(\Delta^m; x)\, \Delta^m F(x)\, dV(x) \tag{9.34}\\
&+ \sum_{k=0}^{m-1} \int_{\partial\mathcal{G}} \left(\frac{\partial}{\partial\nu} G\left(\Delta^{k+1}; x\right) \right) \left(\Delta^k F(x) \right)\, dS(x)\\
&- \sum_{k=0}^{m-1} \int_{\partial\mathcal{G}} G\left(\Delta^{k+1}; x\right) \left(\frac{\partial}{\partial\nu} \Delta^k F(x) \right)\, dS(x).
\end{aligned}
$$

In accordance with our approach we are able to distinguish different representations for the Λ-lattice point discrepancy of F in $\overline{\mathcal{G}}$:

Remark 9.3. *If F is a polyharmonic function of class $\mathrm{C}^{(2m)}(\overline{\mathcal{G}})$, $m \in \mathbb{N}$ (i.e., $\Delta^m F = 0$), then the Λ-lattice point discrepancy of F in $\overline{\mathcal{G}}$ consists exclusively of surface integrals*

$$
\begin{aligned}
P(F; \overline{\mathcal{G}}) &= \sum_{k=0}^{m-1} \int_{\partial\mathcal{G}} \left(\frac{\partial}{\partial\nu} G\left(\Delta^{k+1}; x\right) \right) \left(\Delta^k F(x) \right)\, dS(x) \tag{9.35}\\
&- \sum_{k=0}^{m-1} \int_{\partial\mathcal{G}} G\left(\Delta^{k+1}; x\right) \left(\frac{\partial}{\partial\nu} \Delta^k F(x) \right)\, dS(x).
\end{aligned}
$$

If F is a function of class $\mathrm{C}^{(2m)}(\overline{\mathcal{G}})$, $m \in \mathbb{N}$, satisfying the homogeneous boundary conditions (9.13), then the Λ-lattice point discrepancy of F in $\overline{\mathcal{G}}$ is a volume integral

$$P(F;\overline{\mathcal{G}}) = \int_{\mathcal{G}} G\left(\Delta^m;x\right) \, \Delta^m F(x) \, dV(x). \tag{9.36}$$

Theorem 9.4 enables us to reduce the number of integrals on the right side of (9.34), however, under additional assumptions on the weight function F and the geometry $\mathcal{G}$.

Theorem 9.6. *Let $\mathcal{G} \subset \mathbb{R}^q$ be a regular region such that $\partial\mathcal{G} \cap \Lambda = \emptyset$. Let $F \in \mathrm{C}^{(2m)}\left(\overline{\mathcal{G}}\right)$, $\overline{\mathcal{G}} = \mathcal{G} \cup \partial\mathcal{G}$, satisfy the boundary conditions*

$$\Delta^k F|\partial\mathcal{G} = B_k,$$

$k = 0,\ldots,m-1$. Furthermore, suppose that $G^{(2m)} \in \mathrm{C}^{(4m)}\left(\overline{\mathcal{G}}\right)$ solves the boundary-value problem

$$\Delta^{2m} G^{(2m)}|\mathcal{G} = 0, \qquad \Delta^k G^{(2m)}|\partial\mathcal{G} = G(\Delta^{2m-k}; \,\cdot\,)|\partial\mathcal{G}, \tag{9.37}$$

$k = 0,\ldots,2m-1$. Then

$$\begin{aligned}
P(F;\overline{\mathcal{G}}) \;=\;& \int_{\mathcal{G}} \left(\Delta^m D^{(2m)}(x)\right) \Delta^m F(x) \, dV(x) \\
&- \sum_{k=0}^{m-1} \int_{\partial\mathcal{G}} B_k(x) \left(\frac{\partial}{\partial\nu} \Delta^{2m-(k+1)} D^{(2m)}(x)\right) dS(x),
\end{aligned}$$

where

$$D^{(2m)} = G(\Delta^{2m}; \,\cdot\,) - G^{(2m)}. \tag{9.38}$$

Theorem 9.6 yields two important specifications caused by different features, namely homogeneous boundary conditions and polyharmonicity under certain boundary conditions.

Homogeneous Boundary Conditions

First we obtain from Theorem 9.6

Corollary 9.2. *(Homogeneous Boundary Conditions) Let $\mathcal{G} \subset \mathbb{R}^q$ be a regular region such that $\partial\mathcal{G} \cap \Lambda = \emptyset$. Let $F \in \mathrm{C}^{(2m)}\left(\overline{\mathcal{G}}\right)$, $\overline{\mathcal{G}} = \mathcal{G} \cup \partial\mathcal{G}$, satisfy the boundary conditions*

$$\Delta^k F|\partial\mathcal{G} = 0,$$

$k = 0,\ldots,m-1$. Furthermore, suppose that $G^{(2m)} \in \mathrm{C}^{(4m)}\left(\overline{\mathcal{G}}\right)$ solves the boundary-value problem

$$\Delta^{2m} G^{(2m)}|\mathcal{G} = 0, \qquad \Delta^k G^{(2m)}|\partial\mathcal{G} = G(\Delta^{2m-k}; \,\cdot\,)|\partial\mathcal{G}, \tag{9.39}$$

$k = 0,\ldots,2m-1$. Then (9.34) reduces to the volume integral

$$P(F;\overline{\mathcal{G}}) = \int_{\mathcal{G}} \left(\Delta^m D^{(2m)}(x)\right) \Delta^m F(x) \, dV(x).$$

Specializing to the case $m = 1$ we find

Corollary 9.3. *(Homogeneous Boundary Condition) Let $\mathcal{G} \subset \mathbb{R}^q$ be a regular region such that $\partial\mathcal{G} \cap \Lambda = \emptyset$. Let $F \in C^{(2)}\left(\overline{\mathcal{G}}\right), \overline{\mathcal{G}} = \mathcal{G} \cup \partial\mathcal{G}$, satisfy the boundary condition*

$$F|\partial\mathcal{G} = 0.$$

Furthermore, suppose that $G^{(2)} \in C^{(4)}\left(\overline{\mathcal{G}}\right)$ solves the boundary-value problem

$$\Delta^2 G^{(2)}|\mathcal{G} = 0, \qquad \Delta^k G^{(2)}|\partial\mathcal{G} = G(\Delta^{2-k};\,\cdot\,)|\partial\mathcal{G}, \tag{9.40}$$

$k = 0, 1$. *Then*

$$P(F;\overline{\mathcal{G}}) = \int_{\mathcal{G}} \left(\Delta D^{(2)}(x)\right) \Delta F(x)\; dV(x).$$

(Poly)Harmonicity Under Boundary Conditions

Second we obtain from Theorem 9.6

Corollary 9.4. *(Polyharmonicity) Let $\mathcal{G} \subset \mathbb{R}^q$ be a regular region such that $\partial\mathcal{G} \cap \Lambda = \emptyset$. Let $F \in C^{(2m)}\left(\overline{\mathcal{G}}\right), \overline{\mathcal{G}} = \mathcal{G} \cup \partial\mathcal{G}$, satisfy the boundary-value problem*

$$\Delta^m F|\mathcal{G} = 0, \qquad \Delta^k F|\partial\mathcal{G} = B_k, \tag{9.41}$$

$k = 0, \ldots, m - 1$. *Furthermore, suppose that $G^{(2m)} \in C^{(4m)}\left(\overline{\mathcal{G}}\right)$ solves the boundary-value problem*

$$\Delta^{2m} G^{(2m)}|\mathcal{G} = 0, \qquad \Delta^k G^{(2m)}|\partial\mathcal{G} = G(\Delta^{2m-k};\,\cdot\,)|\partial\mathcal{G}, \tag{9.42}$$

$k = 0, \ldots, 2m - 1$. *Then (9.34) reduces to the boundary integrals*

$$P(F;\overline{\mathcal{G}}) = \sum_{k=0}^{m-1} \int_{\partial\mathcal{G}} B_k(x) \left(\frac{\partial}{\partial\nu} \Delta^{2m-(k+1)} D^{(2m)}(x)\right)\; dS(x).$$

Corollary 9.5. *(Harmonicity) Let $\mathcal{G} \subset \mathbb{R}^q$ be a regular region such that $\partial\mathcal{G} \cap \Lambda = \emptyset$. Let $F \in C^{(2)}\left(\overline{\mathcal{G}}\right), \overline{\mathcal{G}} = \mathcal{G} \cup \partial\mathcal{G}$, satisfy the Dirichlet boundary-value problem*

$$\Delta F|\mathcal{G} = 0, \qquad F|\partial\mathcal{G} = B_0. \tag{9.43}$$

Furthermore, suppose that $G^{(2)} \in C^{(4)}\left(\overline{\mathcal{G}}\right)$ solves the boundary-value problem

$$\Delta^2 G^{(2)}|\mathcal{G} = 0, \qquad \Delta^k G^{(2)}|\partial\mathcal{G} = G(\Delta^{2-k};\,\cdot\,)|\partial\mathcal{G}, \tag{9.44}$$

$k = 0, 1$. *Then*

$$P(F;\overline{\mathcal{G}}) = \int_{\partial\mathcal{G}} B_0(x) \left(\frac{\partial}{\partial\nu} \Delta D^{(2)}(x)\right)\; dS(x).$$

Constant Weights

The total number of lattice points inside and on the boundary of $\overline{\mathcal{G}} \subset \mathbb{R}^q$ can be easily determined (by especially taking constant weight $F = 1$) in the Euler summation formula (cf. Theorem 9.2):

$$\sum_{\substack{g \in \overline{\mathcal{G}} \\ g \in \Lambda}}{}' 1 = \frac{1}{\|\mathcal{F}\|} \underbrace{\int_{\substack{x \in \overline{\mathcal{G}} \\ x \in \mathbb{R}^q}} dV(x)}_{=\|\mathcal{G}\|} + \int_{\substack{x \in \partial \mathcal{G} \\ x \in \mathbb{R}^q}} \frac{\partial}{\partial \nu} G\left(\Delta; x\right) \, dS(x). \tag{9.45}$$

Thus, the Λ-lattice point discrepancy $P(\overline{\mathcal{G}}) = P(1; \overline{\mathcal{G}})$ of $F = 1$ in $\overline{\mathcal{G}}$ can be represented in the form

$$P(\overline{\mathcal{G}}) = \int_{\substack{x \in \partial \mathcal{G} \\ x \in \mathbb{R}^q}} \frac{\partial}{\partial \nu} G\left(\Delta; x\right) \, dS(x). \tag{9.46}$$

Example 9.1. *For the lattice* $\Lambda = \mathbb{Z}^2$ *and the circle* $\mathbb{B}^2_{\sqrt{N}} \subset \mathbb{R}^2$ *around the origin with radius* $\sqrt{N}$ *we especially know from Lemma 5.6 that*

$$\begin{aligned} P\left(\overline{\mathbb{B}^2_{\sqrt{N}}}\right) &= \#_{\mathbb{Z}^2}\left(\overline{\mathbb{B}^2_{\sqrt{N}}}\right) - \pi N - \frac{1}{2} \sum_{\substack{|g|^2 = N \\ g \in \mathbb{Z}^2}} 1 \tag{9.47} \\ &= \#_{\mathbb{Z}^2}\left(\overline{\mathbb{B}^2_{\sqrt{N}}}\right) - \pi N + O(N^\varepsilon) \end{aligned}$$

for every $\varepsilon > 0$. Moreover, the well known formula (5.44) of C.F. Gauß [1801] enables us to conclude

$$\#_{\mathbb{Z}^2}\left(\overline{\mathbb{B}^2_{\sqrt{N}}}\right) - \pi N = O(N^{\frac{1}{2}}); \tag{9.48}$$

hence, it is clear that we have in (rough) estimation

$$P\left(\overline{\mathbb{B}^2_{\sqrt{N}}}\right) = \int_{\substack{|x| = \sqrt{N} \\ x \in \mathbb{R}^q}} \frac{\partial}{\partial \nu} G\left(\Delta; x\right) \, dS(x) = O(N^{\frac{1}{2}}). \tag{9.49}$$

Even better, Lemma 5.6 enables us to write

$$P\left(\overline{\mathbb{B}^2_{\sqrt{N}}}\right) = \int_{\substack{|x| = \sqrt{N} \\ x \in \mathbb{R}^q}} \frac{\partial}{\partial \nu} G\left(\Delta; x\right) \, dS(x) = O\left(N^{\frac{1}{4} + \varepsilon_2}\right), \tag{9.50}$$

where $\varepsilon_2 \leq \frac{1}{4}$ is a positive number for which

$$\#_{\mathbb{Z}^2}\left(\overline{\mathbb{B}^2_{\sqrt{N}}}\right) - \pi N = O\left(N^{\frac{1}{4} + \varepsilon_2}\right) \tag{9.51}$$

is valid (for more details the reader is referred to Section 5.3).

9.3   Zeta Function and Lattice Function

In Section 4.3 we discussed the one-dimensional relations between the Riemann Zeta function and the $\mathbb{Z}$-lattice function for Δ. Next we are interested in the close relationship between the (Epstein) Zeta function (cf. P.S. Epstein [1903, 1907]) and the Λ-lattice function for (iterated) Laplace operators. The essential tool is the lattice point generated Euler summation formula.

Two-Dimensional Theory

We apply the (two-dimensional) Euler summation formula to the *two-dimensional Zeta function* $\zeta(\cdot;\Lambda)$ defined by

$$\zeta(s;\Lambda) = \sum_{\substack{|g|>0 \\ g\in\Lambda}} \frac{1}{|g|^s} = \lim_{N\to\infty} \sum_{\substack{0<|g|\leq N \\ g\in\Lambda}} \frac{1}{|g|^s}, \tag{9.52}$$

where $s \in \mathbb{C}$ satisfies $\Re(s) > 2$. Following our approach of Euler summation (see also C. Müller [1954a,b]) we have to consider the auxiliary function $F : \mathbb{R}^2 \setminus \{0\} \to \mathbb{C}$ given by

$$F(x) = \frac{1}{|x|^s}, \quad \Re(s) > 2. \tag{9.53}$$

The Euler summation formula (with $\lambda = 0$, i.e., for the Laplacian Δ) then yields

$$\sideset{}{'}\sum_{\substack{\rho\leq|g|\leq N \\ g\in\Lambda}} \frac{1}{|g|^s} = \frac{1}{\|\mathcal{F}\|} \int_{\substack{\rho\leq|x|\leq N \\ x\in\mathbb{R}^2}} \frac{1}{|x|^s}\, dV(x) \tag{9.54}$$

$$+ s^2 \int_{\substack{\rho\leq|x|\leq N \\ x\in\mathbb{R}^2}} G(\Delta;x)\, \frac{1}{|x|^{s+2}}\, dV(x)$$

$$+ \int_{\substack{|x|=N \\ x\in\mathbb{R}^2}} \left\{ \frac{1}{N^s}\left(\frac{\partial}{\partial\nu} G(\Delta;x)\right) + s\frac{1}{N^{s+1}} G(\Delta;x) \right\} dS(x)$$

$$+ \int_{\substack{|x|=\rho \\ x\in\mathbb{R}^2}} \left\{ \frac{1}{\rho^s}\left(\frac{\partial}{\partial\nu} G(\Delta;x)\right) - s\frac{1}{\rho^{s+1}} G(\Delta;x) \right\} dS(x),$$

where $\rho > 0$ is chosen in such a way that $\mathbb{B}_\rho^2$ does not contain any point of the lattice Λ except $g = 0$ and ν is the outward directed (unit) normal field. Observing the representation (see (8.45))

$$G(\Delta;x) = \frac{1}{2\pi} \ln|x| - \frac{1}{4\|\mathcal{F}\|}|x|^2 + H(x) \tag{9.55}$$

we immediately obtain

$$\int_{\substack{|x|=\rho \\ x\in\mathbb{R}^2}} G\left(\Delta;x\right)\,dS(x) = \rho\ln\rho - \frac{\pi}{2\|\mathcal{F}\|}\rho^3 + 2\pi Y_0(\Lambda)\rho \qquad (9.56)$$

and

$$\int_{\substack{|x|=\rho \\ x\in\mathbb{R}^2}} \frac{\partial}{\partial\nu}G\left(\Delta;x\right)\,dS(x) = -1 + \frac{\pi}{\|\mathcal{F}\|}\rho^2, \qquad (9.57)$$

where ν is pointing into the exterior of $\mathbb{B}^2_\rho$. From Section 5.3 and (9.49) we are able to derive that

$$\int_{|x|=N} \frac{\partial}{\partial\nu}G\left(\Delta;x\right)\,dS(x) = O\left(N^{\frac{1}{2}+2\varepsilon_2}\right), \quad N\to\infty, \qquad (9.58)$$

where $\varepsilon_2 \le \frac{1}{4}$ is a positive number for which $P\left(\overline{\mathbb{B}^2_{\sqrt{N}}}\right) = O\left(N^{\frac{1}{4}+\varepsilon_2}\right)$ is valid (see (9.51)). From the identity

$$\int_{\substack{|x|=N \\ x\in\mathbb{R}^2}} \frac{\partial}{\partial\nu}G\left(\Delta;x\right)\,dS(x) = N\frac{d}{dN}\frac{1}{N}\int_{\substack{|x|=N \\ x\in\mathbb{R}^2}} G\left(\Delta;x\right)\,dS(x), \qquad (9.59)$$

that is valid for all N with $|g| \ne N$, $g \in \Lambda$, we are able to deduce the (rough) estimate

$$\int_{\substack{|x|=N \\ x\in\mathbb{R}^2}} G\left(\Delta;x\right)\,dS(x) = O(N^2), \quad N\to\infty. \qquad (9.60)$$

Letting N tend to infinity we therefore get for $s \in \mathbb{C}$ with $\Re(s) > 2$

$$\zeta(s;\Lambda) = \sum_{\substack{\rho<|g| \\ g\in\Lambda}} \frac{1}{|g|^s} = \frac{2\pi}{\|\mathcal{F}\|}\frac{1}{s-2}\rho^{-s+2} \qquad (9.61)$$

$$+ s^2\int_{\substack{\rho\le|x| \\ x\in\mathbb{R}^2}} \frac{1}{|x|^{s+2}}G\left(\Delta;x\right)\,dV(x)$$

$$-\rho^{-s}(1 + s\ln\rho - 2\pi s\, Y_0(\Lambda)) + \frac{\pi}{\|\mathcal{F}\|}\rho^{-s+2}\left(1 + \frac{s}{2}\right).$$

Hence, the series (9.61) represents a holomorphic function for all $s \in \mathbb{C}$ with $\Re(s) > 2$. For all $N > \sigma$ we obtain

$$\int_{\substack{N-\sigma\le|x|\le N+\sigma \\ x\in\mathbb{R}^2}} G\left(\Delta;x\right)\,dV(x) \qquad (9.62)$$

$$= -\frac{1}{4\pi^2\sqrt{\|\mathcal{F}\|}}\sum_{\substack{|h|>0 \\ h\in\Lambda^{-1}}} \frac{1}{|h|^2}\int_{\substack{N-\sigma\le|x|\le N+\sigma \\ x\in\mathbb{R}^2}} \overline{\Phi_h(x)}\,dV(x)$$

$$= -\frac{1}{2\pi\|\mathcal{F}\|}\sum_{\substack{|h|>0 \\ h\in\Lambda^{-1}}} \frac{1}{|h|^2}\int_{N-\sigma}^{N+\sigma} r J_0(2\pi|h|r)\,dr$$

(note that, in accordance with our notation, $J_0(r) = J_0(2; r)$). Since the two-dimensional Λ-lattice function $G(\Delta; \cdot)$ has a logarithmic singularity, we are able to see that

$$\lim_{\substack{\sigma \to 0 \\ \sigma > 0}} \frac{1}{2\sigma} \int_{\substack{N-\sigma \le |x| \le N+\sigma \\ x \in \mathbb{R}^2}} G(\Delta; x) \, dV(x) = \int_{\substack{|x|=N \\ x \in \mathbb{R}^2}} G(\Delta; x) \, dS(x). \tag{9.63}$$

From the well known estimate $|J_0(r)| \le \frac{C}{\sqrt{r}}$ of the theory of Bessel functions we get

$$\left| \frac{1}{2\sigma} \int_{N-\sigma}^{N+\sigma} r J_0(2\pi|h|r) dr \right| \le \frac{C}{2\sigma} \frac{1}{\sqrt{2\pi|h|}} \int_{N-\sigma}^{N+\sigma} \sqrt{r} \, dr. \tag{9.64}$$

Hence, with N fixed and $\sigma < \sigma_0$, we are able to find a constant $D > 0$ such that

$$\left| \frac{1}{2\sigma} \int_{N-\sigma}^{N+\sigma} r J_0(2\pi|h|r) dr \right| \le \frac{D}{\sqrt{|h|}}. \tag{9.65}$$

It follows that

$$\lim_{\substack{\sigma \to 0 \\ \sigma > 0}} \frac{1}{2\pi\|\mathcal{F}\|} \sum_{\substack{|h|>0 \\ h \in \Lambda^{-1}}} \frac{1}{|h|^2} \frac{1}{2\sigma} \int_{N-\sigma}^{N+\sigma} r J_0(2\pi|h|r) \, dr \tag{9.66}$$

$$= \frac{1}{2\pi\|\mathcal{F}\|} \sum_{\substack{|h|>0 \\ h \in \Lambda^{-1}}} \frac{N}{|h|^2} J_0(2\pi|h|N).$$

Summarizing our results we obtain the following identity.

Lemma 9.1. *For $N > 0$,*

$$\int_{\substack{|x|=N \\ x \in \mathbb{R}^2}} G(\Delta; x) \, dS(x) = \frac{2\pi N}{\|\mathcal{F}\|} \sum_{\substack{|h|>0 \\ h \in \Lambda^{-1}}} \frac{1}{-\Delta^\wedge(h)} J_0(2\pi|h|N). \tag{9.67}$$

In the same way we find the following result.

Lemma 9.2. *For $N > 0$ and $\lambda \in \mathbb{R}$,*

$$\int_{\substack{|x|=N \\ x \in \mathbb{R}^2}} G(\Delta+\lambda; x) \, dS(x) = \frac{2\pi N}{\|\mathcal{F}\|} \sum_{\substack{(\Delta+\lambda)^\wedge(h)\neq 0 \\ h \in \Lambda^{-1}}} \frac{1}{-(\Delta+\lambda)^\wedge(h)} J_0(2\pi|h|N). \tag{9.68}$$

From Lemma 9.1 we are immediately able to deduce that

$$\int_{\substack{|x|=N \\ x \in \mathbb{R}^2}} G(\Delta; x) \, dS(x) = O(\sqrt{N}), \quad N \to \infty. \tag{9.69}$$

Therefore the integral

$$\int_{\rho \le |x|} \frac{1}{|x|^{s+2}} G\left(\Delta; x\right) \, dV(x) = \int_{\rho}^{\infty} \frac{1}{r^{s+2}} \int_{\substack{|x|=r \\ x \in \mathbb{R}^2}} G\left(\Delta; x\right) \, dS(x) \, dr \quad (9.70)$$

converges for all $s \in \mathbb{C}$ with $\Re(s) \ge s_0 > -\frac{1}{2}$ uniformly with respect to s such that $\zeta(s; \Lambda)$, in connection with (9.61), can be extended to the half plane $\{s \in \mathbb{C} | \Re(s) > -\frac{1}{2}\}$. In this half plane $\zeta(\cdot, \Lambda)$ is a holomorphic function except for $s = 2$, where it has a pole of order 1. Thus, $\zeta(\cdot, \Lambda)$ is a meromorphic function in the half plane $\{s \in \mathbb{C} | \Re(s) > -\frac{1}{2}\}$ showing a pole of order 1 at $s = 2$. Furthermore, the integral

$$\int_{\rho}^{1} \frac{1}{r^{s+2}} \int_{\substack{|x|=r \\ x \in \mathbb{R}^2}} G\left(\Delta; x\right) \, dS(x) \quad (9.71)$$

converges uniformly as ρ tends to 0, provided that $\Re(s) \le s_1 < 0$. Hence, we find the following lemma.

Lemma 9.3. *For $s \in \mathbb{C}$ with $-\frac{1}{2} < \Re(s) < 0$,*

$$\zeta(s; \Lambda) = s^2 \int_{0}^{\infty} \frac{1}{r^{s+2}} \left(\int_{\substack{|x|=r \\ x \in \mathbb{R}^2}} G\left(\Delta; x\right) \, dS(x) \right) dr. \quad (9.72)$$

The theory of Bessel functions enables us to find a constant A such that $|J_0(r)| \le \frac{A}{\sqrt[3]{r}}$ for all $r > 0$. This leads to the following estimates.

Lemma 9.4. *There exists a constant $C > 0$ such that for $N \in (0,1]$ and all $R > 0$*

$$\left| \frac{2\pi N}{\|\mathcal{F}\|} \sum_{\substack{0 < |h|^2 \le R \\ h \in \Lambda^{-1}}} \frac{1}{-4\pi^2 h^2} J_0(2\pi|h|N) \right| \le C N^{\frac{3}{4}} \quad (9.73)$$

and for $N > 1$ and all $R > 0$

$$\left| \frac{2\pi N}{\|\mathcal{F}\|} \sum_{\substack{0 < |h|^2 \le R \\ h \in \Lambda^{-1}}} \frac{1}{-4\pi^2 h^2} J_0(2\pi|h|N) \right| \le C N^{\frac{1}{2}}. \quad (9.74)$$

For $N \in [C_1, C_2], C_2 \ge C_1 > 0$, we have

$$\lim_{R \to \infty} \left(\frac{2\pi N}{\|\mathcal{F}\|} \sum_{\substack{0 < |h|^2 \le R \\ h \in \Lambda^{-1}}} \frac{1}{-4\pi^2 h^2} J_0(2\pi|h|N) \right) = \int_{\substack{|x|=N \\ x \in \mathbb{R}^2}} G\left(\Delta; x\right) \, dS(x). \quad (9.75)$$

The well known Lebesgue theorem guarantees

Lemma 9.5. *For all $s \in \mathbb{C}$ with $-\frac{1}{2} < \Re(s) < -\frac{1}{4}$*

$$\zeta(s;\Lambda) = \lim_{R \to \infty} \frac{2\pi}{\|\mathcal{F}\|} \sum_{\substack{0<|h|^2 \leq R \\ h \in \Lambda^{-1}}} \frac{s^2}{-4\pi^2 h^2} \int_0^\infty \frac{r}{r^{s+2}} J_0(2\pi|h|r) \, dr$$

$$= s^2 \int_0^\infty \frac{1}{r^{s+2}} \int_{\substack{|x|=r \\ x \in \mathbb{R}^2}} G(\Delta; x) \, dS(x) \, dr. \tag{9.76}$$

By aid of the formula (see also G.N. Watson [1944])

$$\int_0^\infty \frac{1}{r^{s+1}} J_0(2\pi|h|r) \, dr = (2\pi|h|)^s \int_0^\infty \frac{J_0(r)}{r^{s+1}} \, dr \tag{9.77}$$

$$= \frac{1}{2}(\pi|h|)^s \frac{\Gamma\left(-\frac{s}{2}\right)}{\Gamma\left(\frac{s}{2}+1\right)}$$

we get for all $s \in \mathbb{C}$ with $-\frac{1}{2} < \Re(s) < -\frac{1}{4}$

$$\zeta(s;\Lambda) = -\frac{s^2}{4}\pi^{s-1}\frac{\Gamma\left(-\frac{s}{2}\right)}{\Gamma\left(\frac{s}{2}+1\right)}\frac{1}{\|\mathcal{F}\|} \sum_{\substack{|h|>0 \\ h \in \Lambda^{-1}}} \frac{1}{|h|^{2-s}} \tag{9.78}$$

$$= \pi^{s-1}\frac{\Gamma\left(1-\frac{s}{2}\right)}{\Gamma\left(\frac{s}{2}\right)}\frac{1}{\|\mathcal{F}\|} \sum_{\substack{|h|>0 \\ h \in \Lambda^{-1}}} \frac{1}{|h|^{2-s}}.$$

Summarizing our considerations we finally arrive at the following theorem, which includes the functional equation of the Zeta function.

Theorem 9.7. *(Functional Equation of the Two-Dimensional Zeta Function) The Zeta function $\zeta(\cdot;\Lambda)$, given by*

$$\zeta(s;\Lambda) = \sum_{\substack{|g|>0 \\ g \in \Lambda}} \frac{1}{|g|^s}, \quad s \in \mathbb{C}, \ \Re(s) > 2, \tag{9.79}$$

can be extended as a meromorphic function with the pole

$$\frac{2\pi}{\|\mathcal{F}\|}\frac{1}{s-2} \tag{9.80}$$

to the whole complex plane $\mathbb{C}$. Moreover, $\zeta(\cdot;\Lambda)$ satisfies the functional equation

$$\zeta(s;\Lambda) = \pi^{s-1}\frac{\Gamma\left(1-\frac{s}{2}\right)}{\Gamma\left(\frac{s}{2}\right)}\frac{1}{\|\mathcal{F}\|}\zeta(2-s;\Lambda^{-1}). \tag{9.81}$$

Proof. First $\zeta(\cdot;\Lambda)$ is meromorphic in the half plane $\{s \in \mathbb{C}| \Re(s) > -\frac{1}{2}\}$ with a pole of the form $2\pi\|\mathcal{F}\|^{-1}(s-2)^{-1}$. The functional equation holds true for all $s \in \mathbb{C}$ with $-\frac{1}{2} < \Re(s) < -\frac{1}{4}$. The function $\zeta(2-\cdot;\Lambda^{-1})$ is meromorphic in the half plane $\Re(s) < \frac{5}{2}$. Since the functional equation is valid in the aforementioned strip, it holds true everywhere. $\qquad\square$

Next we come to the two-dimensional *Kronecker limit formula*. To this end Theorem 9.7 shows that (9.61) also holds true for all $s \in \mathbb{C}$ with $\Re(s) > -\frac{1}{2}$. Therefore, we obtain

$$\zeta(0; \Lambda) = -1 \qquad (9.82)$$

and

$$\zeta'(0; \Lambda) = -2\pi \, Y_0(\Lambda). \qquad (9.83)$$

Letting $s = 2 + t$ we get from Theorem 9.7

$$\zeta(2 + t; \Lambda) = -\frac{2\pi}{\|\mathcal{F}\|} \pi^t \frac{\Gamma\left(1 - \frac{t}{2}\right)}{\Gamma\left(1 + \frac{t}{2}\right)} \zeta(-t; \Lambda^{-1}). \qquad (9.84)$$

In the neighborhood of the point $t = 0$ we have

$$\zeta(-t; \Lambda^{-1}) = -1 + 2\pi Y_0(\Lambda^{-1})t + \dots \qquad (9.85)$$

and

$$\pi^t \frac{\Gamma\left(1 - \frac{t}{2}\right)}{\Gamma\left(1 + \frac{t}{2}\right)} = 1 + (\ln \pi - \Gamma'(1))t + \dots, \qquad (9.86)$$

such that, because of $-\Gamma'(1) = C$ (C is the Euler constant), we find

$$\zeta(2 + t; \Lambda) = \frac{2\pi}{\|\mathcal{F}\|} \left(\frac{1}{t} + (2\pi Y_0(\Lambda^{-1}) + \ln(\pi) + C) + \dots \right). \qquad (9.87)$$

This gives us

Theorem 9.8. *(Kronecker's Limit Formula)*

$$\lim_{s \to 2} \left(\zeta(s; \Lambda) - \frac{2\pi}{\|\mathcal{F}\|} \frac{1}{s - 2} \right) = \frac{2\pi}{\|\mathcal{F}\|} (\ln \pi + C - 2\pi Y_0(\Lambda) - \ln \|\mathcal{F}\|). \qquad (9.88)$$

By virtue of (9.61) we obtain

$$\lim_{s \to 2} \left(\zeta(s; \Lambda) - \frac{2\pi}{\|\mathcal{F}\|} \frac{1}{s - 2} \right) \qquad (9.89)$$

$$= \lim_{s \to 2} \left(\frac{2\pi}{\|\mathcal{F}\|} \frac{\rho^{-s+2} - 1}{s - 2} + \frac{2\pi}{\|\mathcal{F}\|} - \rho^{-2}(1 + \ln \rho - 4\pi Y_0(\Lambda)) \right.$$

$$\left. + 4 \int_{\substack{\rho \le |x| \\ x \in \mathbb{R}^2}} G(\Delta; x) \frac{1}{|x|^4} \, dV(x) \right)$$

$$\qquad (9.90)$$

such that

$$\lim_{s \to 2} \left(\zeta(s; \Lambda) - \frac{2\pi}{\|\mathcal{F}\|} \frac{1}{s - 2} \right) \qquad (9.91)$$

$$= \frac{2\pi}{\|\mathcal{F}\|} (1 - \ln \rho) - \rho^{-2}(1 + 2 \ln \rho - 4\pi Y_0(\Lambda))$$

$$+ 4 \int_{\substack{\rho \le |x| \\ x \in \mathbb{R}^2}} G(\Delta; x) \frac{1}{|x|^4} \, dV(x).$$

In connection with the Euler summation formula (9.61) we find

$$
\lim_{N\to\infty} \left(\sum_{\substack{0<|g|\le N\\ g\in\Lambda}}{}' \frac{1}{|g|^2} - \frac{1}{\|\mathcal{F}\|} \int_{\substack{\rho\le|x|\le N\\ x\in\mathbb{R}^2}} \frac{1}{|x|^2}\, dV(x) \right) \tag{9.92}
$$

$$
= \frac{2\pi}{\|\mathcal{F}\|} - \rho(1+2\ln\rho - 4\pi Y_0(\Lambda)) + 4\int_{\substack{\rho\le|x|\\ x\in\mathbb{R}^2}} G\left(\Delta;x\right)\frac{1}{|x|^4}\, dV(x).
$$

Comparing (9.92) with (9.89) we finally get the following limit relation.

Theorem 9.9.

$$
\lim_{N\to\infty} \left(\sum_{\substack{0\le|g|\le N\\ g\in\Lambda}}{}' \frac{1}{|g|^2} - \frac{2\pi}{\|\mathcal{F}\|}\ln(N) \right) = \frac{2\pi}{\|\mathcal{F}\|}\left(\ln\pi + C - 2\pi Y_0(\Lambda) - \ln\|\mathcal{F}\|\right),
$$
$$\tag{9.93}$$

where C is the Euler constant (i.e., $C = -\Gamma'(1)$).

Multi-Dimensional Theory

We start with some preparatory material (see, e.g., P.S. Epstein [1903, 1907], R. Wienkamp [1958], and the literature therein) concerning the *Zeta function* $\zeta_n^q(\cdot;\eta;\Lambda)$ *of dimension $q \ge 3$ and degree n* defined by

$$
s \mapsto \zeta_n^q(s;\eta;\Lambda) = \sum_{\substack{|g|>0\\ g\in\Lambda}} \frac{1}{|g|^s}P_n\left(q;\frac{g}{|g|}\cdot\eta\right), \tag{9.94}
$$

where $s \in \mathbb{C}$ satisfies $\Re(s) > q$, η is an arbitrary, but fixed, element of the unit sphere $\mathbb{S}^{q-1}$, and $P_n(q;\cdot)$ is the Legendre polynomial of degree n and dimension q (cf. (3.86)).

For each positive value $\rho < \inf_{x\in\mathcal{F}}|x|$, sufficiently large positive N, and for all $s \in \mathbb{C}$ with $\Re(s) > q$, the Euler summation formula (Theorem 9.2) gives us in terms of the auxiliary function F defined by

$$
x \mapsto F(x) = \frac{1}{|x|^s}P_n\left(q;\frac{x}{|x|}\cdot\eta\right), \quad \eta\in\mathbb{S}^{q-1},\ n\in\mathbb{N}_0,\ x\ne 0, \tag{9.95}
$$

the identity

$$\sideset{}{'}\sum_{\substack{\rho \leq |g| \leq N \\ g \in \Lambda}} F(g) \quad = \quad \frac{1}{\|\mathcal{F}\|} \int_{\substack{\rho \leq |x| \leq N \\ x \in \mathbb{R}^q}} F(x)\, dV(x) \tag{9.96}$$

$$+ \int_{\substack{\rho \leq |x| \leq N \\ x \in \mathbb{R}^q}} G\left(\Delta^m; x\right)\, \left(\Delta^m F(x)\right)\, dV(x)$$

$$+ \sum_{k=0}^{m-1} \int_{\substack{|x|=N \\ x \in \mathbb{R}^q}} \left(\frac{\partial}{\partial \nu} G\left(\Delta^{k+1}; x\right)\right) \left(\Delta^k F(x)\right)\, dS(x)$$

$$- \sum_{k=0}^{m-1} \int_{\substack{|x|=N \\ x \in \mathbb{R}^q}} G\left(\Delta^{k+1}; x\right) \left(\frac{\partial}{\partial \nu} \Delta^k F(x)\right)\, dS(x)$$

$$+ \sum_{k=0}^{m-1} \int_{\substack{|x|=\rho \\ x \in \mathbb{R}^q}} \left(\frac{\partial}{\partial \nu} G\left(\Delta^{k+1}; x\right)\right) \left(\Delta^k F(x)\right)\, dS(x)$$

$$- \sum_{k=0}^{m-1} \int_{\substack{|x|=\rho \\ x \in \mathbb{R}^q}} G\left(\Delta^{k+1}; x\right) \left(\frac{\partial}{\partial \nu} \Delta^k F(x)\right)\, dS(x),$$

where $m \in \mathbb{N}$ is chosen such that $m > q/2$ and ν is the outward unit normal to $\mathbb{B}^q_{\rho,N}$. First we want to calculate the second integral on the right side of (9.96). It is not difficult to see that

$$\Delta_x F(x) = (s+n)(s-n-q+2)\frac{F(x)}{|x|^2},\quad x \neq 0, \tag{9.97}$$

such that

$$\Delta_x^m F(x) = A_{s,n,m}\frac{F(x)}{|x|^{2m}},\quad x \neq 0, \tag{9.98}$$

where

$$A_{s,n,m} = (s+n)(s+n+2)\cdot\ldots\cdot(s+n+2(m-1))(s-n+2-q)\cdot\ldots\cdot(s-n+2m-q). \tag{9.99}$$

For $m > q/2$ we find in connection with the absolutely and uniformly convergent bilinear expansion of the Λ-lattice function in $\mathbb{R}^q$

$$\int_{\substack{\rho \leq |x| \leq N \\ x \in \mathbb{R}^q}} G\left(\Delta^m; x\right)\, \Delta^m F(x)\, dV(x) \tag{9.100}$$

$$= \frac{A_{s,n,m}}{\|\mathcal{F}\|} \sum_{\substack{|h|>0 \\ h \in \Lambda^{-1}}} \frac{1}{(-4\pi^2 h^2)^m} \int_{\substack{\rho \leq |x| \leq N \\ x \in \mathbb{R}^q}} \frac{e^{2\pi i x \cdot h}\, P_n\left(q; \eta \cdot \frac{x}{|x|}\right)}{|x|^{s+2m}}\, dV(x).$$

From the theory of Bessel functions we obtain after some elementary calculations

$$\int_{\substack{\rho \le |x| \le N \\ x \in \mathbb{R}^q}} G\left(\Delta^m; x\right) \, \Delta^m F(x) \, dV(x) \tag{9.101}$$

$$= A_{s,n,m} \frac{i^n \|\mathbb{S}^{q-1}\| \Gamma(\frac{q}{2})}{2^{\frac{2-q}{2}} \|\mathcal{F}\|} \sum_{\substack{|h|>0 \\ h \in \Lambda^{-1}}} \frac{1}{(2\pi|h|)^{q-s}} P_n\left(q; \frac{h}{|h|} \cdot \eta\right) \int_{2\pi|h|\rho}^{2\pi|h|N} \frac{J_{n+\frac{q-2}{2}}(r)}{r^{s+2m-\frac{q}{2}}} \, dr.$$

Remembering the asymptotic behavior of the Bessel function we see that the last series converges uniformly with respect to ρ and N for all $s \in \mathbb{C}$ with $-2m + \frac{q}{2} + \frac{1}{2} < \Re(s) < \min(0; -2m + q + n)$. Consequently, with $s \in \mathbb{C}$ indicated as before, it follows that

$$\int_{\mathbb{R}^q} G\left(\Delta^m; x\right) \, \Delta^m F(x) \, dV(x) \tag{9.102}$$

$$= A_{s,n,m} \frac{i^n \|\mathbb{S}^{q-1}\| \, \Gamma(\frac{q}{2})}{2^{\frac{2-q}{2}} \|\mathcal{F}\|} \sum_{\substack{|h|>0 \\ h \in \Lambda^{-1}}} \frac{1}{(2\pi|h|)^{q-s}} P_n\left(q; \frac{h}{|h|} \cdot \eta\right) \int_0^\infty \frac{J_{n+\frac{q-2}{2}}(r)}{r^{s+2m-\frac{q}{2}}} \, dr.$$

In connection with the formula (see, e.g., W. Magnus et al. [1966])

$$\int_0^\infty \frac{J_\nu(r)}{r^{\nu+1-\mu}} \, dr = \frac{\Gamma(\frac{\mu}{2}) 2^{\mu-\nu-1}}{\Gamma(\nu - \frac{1}{2}\mu + 1)}, \quad 0 < \mu < \frac{3}{2} + \nu, \tag{9.103}$$

this allows us to formulate

Lemma 9.6. *For all $s \in \mathbb{C}$ with*

$$-2m + \frac{q+1}{2} < \Re(s) < \min(0, -2m + q + n) \tag{9.104}$$

we have

$$\int_{\mathbb{R}^q} G\left(\Delta^m; x\right) \left(\Delta_x^m \frac{1}{|x|^s} P_n\left(q; \frac{x}{|x|} \cdot \eta\right)\right) \, dV(x) \tag{9.105}$$

$$= \frac{i^n}{\|\mathcal{F}\|} \pi^{s-\frac{q}{2}} \frac{\Gamma(\frac{q+n-s}{2})}{\Gamma(\frac{n+s}{2})} \sum_{\substack{|h|>0 \\ h \in \Lambda^{-1}}} \frac{1}{|h|^{q-s}} P_n\left(q; \frac{h}{|h|} \cdot \eta\right).$$

In connection with (9.96) and Lemma 9.6 we are able to show that the sum

$$\sum_{k=0}^{m-1} A_{s,n,k} \int_{\substack{|x|=N \\ x \in \mathbb{R}^q}} \left(\frac{\partial}{\partial \nu_x} G\left(\Delta^{k+1}; x\right)\right) \frac{F(x)}{|x|^{2k}} - G\left(\Delta^{k+1}; x\right) \left(\frac{\partial}{\partial \nu_x} \frac{F(x)}{|x|^{2k}}\right) dS(x) \tag{9.106}$$

is convergent as $N \to \infty$ provided that $\Re(s) > q$. Even more, for $\Re(s) > q$, the integrals (9.106) tend to zero as $N \to \infty$. Thus it follows for $\Re(s) > q$ that

$$\lim_{N \to \infty} \sum_{\substack{\rho \leq |g| \leq N \\ g \in \Lambda}}{}' \frac{1}{|g|^s} P_n \left(q; \frac{g}{|g|} \cdot \eta \right) \tag{9.107}$$

$$= \begin{cases} \frac{\|\mathbb{S}^{q-1}\|}{\|\mathcal{F}\|} \frac{\rho^{q-s}}{s-q} & , \quad n = 0 \\ 0 & , \quad n > 0 \end{cases}$$

$$+ A_{s,n,m} \int_{\substack{\rho \leq |x| \\ x \in \mathbb{R}^q}} G\left(\Delta^m; x\right) \frac{F(x)}{|x|^{2m}} \, dV(x)$$

$$+ \sum_{k=0}^{m-1} A_{s,n,k} \int_{\substack{|x|=\rho \\ x \in \mathbb{R}^q}} \left(\frac{\partial}{\partial \nu} G\left(\Delta^{k+1}; x\right) \right) \frac{F(x)}{|x|^{2k}} \, dS(x)$$

$$- \sum_{k=0}^{m-1} A_{s,n,k} \int_{\substack{|x|=\rho \\ x \in \mathbb{R}^q}} G\left(\Delta^{k+1}; x\right) \left(\frac{\partial}{\partial \nu} \frac{F(x)}{|x|^{2k}} \right) dS(x).$$

Consequently, the right side of (9.107) shows that $\zeta_n^q(\cdot; \eta; \Lambda)$ can be continued by the left side of (9.107) to the half plane $\Re(s) > -2m + \frac{q}{2} + \frac{1}{2}$. It remains to investigate the sums

$$\sum_{k=0}^{m-1} A_{s,n,k} \int_{\substack{|x|=\rho \\ x \in \mathbb{R}^q}} \left(\frac{\partial}{\partial \nu} G(\Delta^{k+1}; x) \right) \frac{F(x)}{|x|^{2m}} \, dS(x) \tag{9.108}$$

$$- \sum_{k=0}^{m-1} A_{s,n,k} \int_{\substack{|x|=\rho \\ x \in \mathbb{R}^q}} G\left(\Delta^{k+1}; x\right) \left(\frac{\partial}{\partial \nu} \frac{F(x)}{|x|^{2m}} \right) dS(x).$$

Because of the singularity behavior of the Λ-lattice function at 0, the sums (9.108) tend to 0 as $\rho \to 0$, provided that $s \in \mathbb{C}$ with $\Re(s) < -2m + q$. Thus, for all values $s \in \mathbb{C}$ with $-2m + \frac{q}{2} + \frac{1}{2} < \Re(s) < \min(0, -2m + q + n)$, $\zeta_n^q(\cdot; \eta; \Lambda)$ admits the representation

$$\zeta_n^q(s; \eta; \Lambda) = \frac{i^n \pi^{s - \frac{q}{2}} \Gamma\left(\frac{n+q-s}{2}\right)}{\|\mathcal{F}\| \Gamma\left(\frac{s+n}{2}\right)} \sum_{|h|>0} \frac{1}{|h|^{q-s}} P_n \left(q; \frac{h}{|h|} \cdot \eta \right). \tag{9.109}$$

Summarizing our results we therefore obtain

Theorem 9.10. *(Functional Equation of the q-Dimensional Zeta Function)*
For $n > 0$, the Zeta function $\zeta_n^q(\cdot; \eta; \Lambda)$ of dimension q and degree n defined by

$$s \mapsto \zeta_n^q(s; \eta; \Lambda) = \sum_{\substack{|g|>0 \\ g \in \Lambda}} \frac{1}{|g|^s} P_n \left(q; \frac{g}{|g|} \cdot \eta \right), \quad \Re(s) > q, \tag{9.110}$$

admits a holomorphic continuation that represents an entire function in $\mathbb{C}$.

For $n = 0$, the continuation of $\zeta_n^q(\cdot; \eta; \Lambda)$, i.e., $\zeta_0^q(\cdot; \eta; \Lambda)$, is a meromorphic *function showing the single pole*

$$\frac{\|\mathbb{S}^{q-1}\|}{\|\mathcal{F}\|}\frac{1}{s-q}. \tag{9.111}$$

For $n \in \mathbb{N}_0$, $\zeta_n^q(\cdot; \eta; \Lambda)$ satisfies the functional equation

$$\zeta_n^q(s; \eta; \Lambda) = \frac{i^n}{\|\mathcal{F}\|}\, \pi^{s-\frac{q}{2}}\, \frac{\Gamma(\frac{n+q-s}{2})}{\Gamma(\frac{s+n}{2})}\, \zeta_n^q(q - s; \eta; \Lambda^{-1}). \tag{9.112}$$

Proof. $\zeta_n^q(\cdot; \eta; \Lambda)$ is for $\Re(s) > -2m + \frac{q}{2} + \frac{1}{2}$ (except for $s = q$ in case of $n = 0$) holomorphic. According to (9.109) the functional equation holds true for all $s \in \mathbb{C}$ with $-2m + \frac{q}{2} + \frac{1}{2} < \Re(s) < \min(0, -2m + q + n)$. $\zeta_n^q(q - \cdot; \eta; \Lambda^{-1})$ is holomorphic (except for $s = q$ in the case of $n = 0$) for all $s \in \mathbb{C}$ with $\Re(s) < 2m + \frac{q}{2} - \frac{1}{2}$. As the functional equation is valid in the aforementioned strip, it is valid everywhere. $\qquad\square$

Remark 9.4. $\zeta_0^q(\cdot; \eta; \Lambda)$ *is independent of $\eta \in \mathbb{S}^{q-1}$; hence, we simply write* $\zeta_0^q(\cdot; \Lambda)$ *instead of* $\zeta_0^q(\cdot; \eta; \Lambda)$. *Furthermore,* $\zeta_0^2(\cdot; \eta; \Lambda)$ *coincides with* $\zeta(\cdot; \Lambda)$ *as discussed in Section 9.3.*

Next our aim is to formulate an analogue of the q-dimensional *Kronecker limit formula*. To this end, we turn over to $n = 0$ and even dimensions, i.e., $q = 2l, l > 1$, and observe that, for $l < m$ and $\rho > 0$,

$$
\begin{aligned}
\zeta_0^{2l}(s; \Lambda) &= \frac{\|\mathbb{S}^{2l-1}\|\rho^{2l-1}}{\|\mathcal{F}\|(s - 2l)} + A_{s,0,m} \int_{\substack{\rho < |x| \\ x \in \mathbb{R}^{2l}}} G\left(\Delta^m; x\right)\frac{1}{|x|^{s+2m}}\, dV(x) \\
&\quad + \sum_{k=0}^{m-1} A_{s,0,k}\frac{1}{\rho^{s+2m}} \int_{\substack{|x|=\rho \\ x \in \mathbb{R}^{2l}}} \frac{\partial}{\partial\nu} G\left(\Delta^{k+1}; x\right)\, dS(x) \\
&\quad - \sum_{k=0}^{m-1} A_{s,0,k}\frac{s + 2k}{\rho^{s+2k+1}} \int_{\substack{|x|=\rho \\ x \in \mathbb{R}^{2l}}} G\left(\Delta^{k+1}; x\right)\, dS(x). \tag{9.113}
\end{aligned}
$$

By an easy calculation we find

$$(-1)^{k+1} \int_{\substack{|x|=\rho \\ x \in \mathbb{R}^{2l}}} G\left(\Delta^{k+1}; x\right)\, dS(x) \tag{9.114}$$

$$= \rho^{2l-1}\|\mathbb{S}^{2l-1}\|\left\{\frac{b^{2l,k+1}}{\rho^{2l-2(k+1)}} + Y_0^{2l,k+1} h^{2l,k+1}(\rho)\right\}$$

and

$$(-1)^{k+1} \int_{\substack{|x|=\rho \\ x \in \mathbb{R}^{2l}}} \frac{\partial}{\partial\nu} G\left(\Delta^{k+1}; x\right)\, dS(x) \tag{9.115}$$

$$= \rho^{2l-1}\|\mathbb{S}^{2l-1}\|\left\{\frac{b^{2l,k+1}(2l - 2k - 2)}{\rho^{2l-2(k+1)}} + Y_0^{2l,k+1}\frac{d}{d\rho} h^{2l,k+1}(\rho)\right\},$$

where $h^{2l,k+q}(\rho)$ can be expressed in terms of the coefficients (8.105), (8.106), and (8.107)

$$h^{2l,k+1}(\rho) = g_0^{2l,k+1}(\rho) - \frac{b^{2l,k+1}}{\rho^{2l-2(k+1)}}. \tag{9.116}$$

This allows us to write

$$\zeta_0^{2l}(0,\Lambda) = \lim_{\rho \to 0} \int_{\substack{|x|=\rho \\ x \in \mathbb{R}^{2l}}} \frac{\partial}{\partial \nu} G(\Delta; x) \, dS(x) = -1. \tag{9.117}$$

Furthermore,

$$
\begin{aligned}
\frac{d}{ds} \zeta_0^{2l}(s;\Lambda) \;=\; & -\frac{\|\mathbb{S}^{2l-1}\|}{\|\mathcal{F}\|} \left(\frac{1}{(s-2l)^2} + \frac{\ln(\rho)}{s-2l} \right) \rho^{2l-s} \\
& + \frac{d}{ds} \left(A_{s,0,m} \int_{\substack{\rho \le |x| \\ x \in \mathbb{R}^{2l}}} G(\Delta^m; x) \frac{1}{|x|^{s+2m}} \, dV(x) \right) \\
& + \frac{d}{ds} \left(\sum_{\nu=l}^{m-1} A_{s,0,\nu} \frac{1}{\rho^{s+2\nu}} \int_{\substack{|x|=\rho \\ x \in \mathbb{R}^{2l}}} \frac{\partial}{\partial \nu} G(\Delta^{k+1}; x) \, dS(x) \right) \\
& - \frac{d}{ds} \left(\sum_{\nu=l}^{m-1} A_{s,0,\nu} \frac{s+2\nu}{\rho^{2+2\nu+1}} \int_{\substack{|x|=\rho \\ x \in \mathbb{R}^{2l}}} G(\Delta^{k+1}; x) \, dS(x) \right) \\
& + \frac{d}{ds} \left(\sum_{\nu=0}^{l-1} A_{s,0,\nu} \frac{1}{\rho^{s+2\nu}} \int_{\substack{|x|=\rho \\ x \in \mathbb{R}^{2l}}} \frac{\partial}{\partial \nu} G(\Delta^{k+1}; x) \, dS(x) \right) \\
& - \frac{d}{ds} \left(\sum_{\nu=0}^{l=1} A_{s,0,\nu} \frac{s+2\nu}{\rho^{s+2\nu+1}} \int_{\substack{|x|=\rho \\ x \in \mathbb{R}^{2l}}} G(\Delta^{k+1}; x) \, dS(x) \right).
\end{aligned}
\tag{9.118}
$$

The second, third, and fourth term of (9.118) is zero at $s = 0$. It therefore remains to deal with the last two terms of (9.118). In fact, an easy calculation shows that for $\rho \to 0$

$$\frac{d}{ds} \zeta_0^{2l}(0;\Lambda) = \|\mathbb{S}^{2l-1}\|(2l-2)2^{2(l-1)}\left((l-1)!\right)^2 Y_0^{2l,l}, \quad l > 1 \tag{9.119}$$

(for the case $l = 1$ see (9.83)). Moreover, in analogy to (9.84), it is not hard to deduce from the functional equation of the Zeta function that

$$
\begin{aligned}
\zeta_0^{2l}(2l+t;\Lambda) \;=\; & \frac{1}{\|\mathcal{F}\|} \pi^{t+l} \frac{\Gamma(-\frac{t}{2})}{\Gamma(\frac{2l+t}{2})} \zeta_0^{2l}\left(t; \Lambda^{-1}\right) \tag{9.120} \\
\;=\; & -\frac{1}{\|\mathcal{F}\|} \pi^{t+l} \frac{2}{t} \frac{\Gamma(1-\frac{t}{2})}{\Gamma(l+\frac{t}{2})} \zeta_0^{2l}\left(-t; \Lambda^{-1}\right).
\end{aligned}
$$

In connection with

$$\zeta_0^{2l}\left(-t; \Lambda^{-1}\right) = -1 - \|\mathbb{S}^{2l-1}\| (2l-2)\left((l-1)!\right)^2 Y_0^{2l,l} t + \cdots \tag{9.121}$$

and

$$\pi^t \frac{\Gamma(1 - \frac{t}{2})}{\Gamma(l + \frac{t}{2})} = \frac{1}{\Gamma(l)} + \frac{t}{2\Gamma(l)} \left(2\ln(\pi) - \Gamma'(1) - \frac{\Gamma'(l)}{\Gamma(l)} \right) + \dots \qquad (9.122)$$

we therefore obtain

$$\zeta_0^{2l}(2l + t; \Lambda) = \frac{2\pi^l}{\|\mathcal{F}\|} \left(\frac{1}{t\Gamma(l)} + \frac{1}{2\Gamma(l)} \left(-\Gamma'(1) - \frac{\Gamma'(l)}{\Gamma(l)} \right) + 2\ln(\pi) \right.$$
$$\left. + \frac{\|\mathbb{S}^{2l-1}\|(2l - 2)}{\Gamma(l)\, 2^{2-2l}} \left((l - 1)! \right)^2 Y_0^{2l,l} + \dots \right). \qquad (9.123)$$

Thus we finally arrive at the following result (note that the case $l = 1$ is already known from Theorem 9.8).

Theorem 9.11. *(Kronecker's Limit Formula). For $l > 1$ we have*

$$\lim_{s \to 2l} \left(\zeta_0^{2l}(s; \Lambda) - \frac{\|\mathbb{S}^{2l-1}\|}{\|\mathcal{F}\|(s - 2l)} \right) \qquad (9.124)$$
$$= \frac{\|\mathbb{S}^{2l-1}\|}{\|\mathcal{F}\|} \left(\ln(\pi) - \frac{\Gamma'(1)}{2} - \frac{\Gamma'(l)}{2\Gamma(l)} + \frac{\|\mathbb{S}^{2l-1}\|}{2^{2-2l}} (2l - 2) \left((l - 1)! \right)^2 Y_0^{2l,l} \right).$$

9.4 Euler Summation Formulas for Iterated Helmholtz Operators

In analogy to the procedure for the Laplace operator we prove the multi-dimensional Euler summation formula for Helmholtz operators.

Euler Summation for the Helmholtz Operator

Let Λ be an arbitrary lattice in $\mathbb{R}^q$. Suppose that $\mathcal{G} \subset \mathbb{R}^q$ is a regular region. Let F be a function of class $C^{(2)}(\overline{\mathcal{G}})$, $\overline{\mathcal{G}} = \mathcal{G} \cup \partial\mathcal{G}$. Then, the Second Green Theorem gives for every (sufficiently small) $\varepsilon > 0$ (see Figure 9.1),

$$\int_{\substack{x \in \overline{\mathcal{G}} \\ x \notin \mathbb{B}_\varepsilon^q + \Lambda}} \left\{ F(x) \left((\Delta + \lambda) G(\Delta + \lambda; x) \right) - G(\Delta + \lambda; x) \left((\Delta + \lambda) F(x) \right) \right\} \, dV(x)$$

$$= \int_{\substack{x \in \partial\mathcal{G} \\ x \notin \mathbb{B}_\varepsilon^q + \Lambda}} \left\{ F(x) \left(\frac{\partial}{\partial \nu} G(\Delta + \lambda; x) \right) - G(\Delta + \lambda; x) \left(\frac{\partial F}{\partial \nu}(x) \right) \right\} \, dS(x)$$

$$+ \sum_{\substack{g \in \overline{\mathcal{G}} \\ g \in \Lambda}} \int_{\substack{|x - g| = \varepsilon \\ x \in \overline{\mathcal{G}}}} \left\{ F(x) \left(\frac{\partial}{\partial \nu} G(\Delta + \lambda; x) \right) - G(\Delta + \lambda; x) \left(\frac{\partial F}{\partial \nu}(x) \right) \right\} \, dS(x)$$

$$\qquad (9.125)$$

with an arbitrary but fixed parameter $\lambda \in \mathbb{R}$, where ν is the outer (unit) normal field. Observing the differential equation (i.e., Condition *(ii)* of Definition 8.1) we get

$$\int_{\substack{x \in \overline{\mathcal{G}} \\ x \notin \mathbb{B}_{\varepsilon}^{q} + \Lambda}} F(x)\, (\Delta + \lambda)G\,(\Delta + \lambda; x)\ dV(x) \tag{9.126}$$

$$= -\frac{1}{\sqrt{\|\mathcal{F}\|}} \int_{\substack{x \in \overline{\mathcal{G}} \\ x \notin \mathbb{B}_{\varepsilon}^{q} + \Lambda}} \sum_{\substack{(\Delta + \lambda)^{\wedge}(h) = 0 \\ h \in \Lambda^{-1}}} F(x)\overline{\Phi_h(x)}\ dV(x),$$

where the sum on the right side is to be extended over all $h \in \Lambda^{-1}$ with $(\Delta + \lambda)^{\wedge}(h) = 0$, i.e., $4\pi^2 h^2 = \lambda$ (note that, in the case of $4\pi^2 h^2 \neq \lambda$ for all $h \in \Lambda^{-1}$, the right side is assumed to be zero). Letting $\varepsilon \to 0$ and observing the characteristic singularity of the Λ-lattice function (Condition *(iii)* of Definition 8.1) we obtain in connection with the potential theoretical results of Lemma 6.1 the following summation formula.

Theorem 9.12. *(Euler Summation Formula for the Operator $\Delta + \lambda$, $\lambda \in \mathbb{R}$) Let Λ be an arbitrary lattice in $\mathbb{R}^q$. Let $\mathcal{G} \subset \mathbb{R}^q$ be a regular region. Let F be twice continuously differentiable on $\overline{\mathcal{G}}$, $\overline{\mathcal{G}} = \mathcal{G} \cup \partial\mathcal{G}$.*

Then, for $\lambda \notin \mathrm{Spect}_{\Delta}(\Lambda)$, we have

$$\sideset{}{'}\sum_{\substack{g \in \overline{\mathcal{G}} \\ g \in \Lambda}} F(g) \tag{9.127}$$

$$= \int_{\mathcal{G}} G\,(\Delta + \lambda; x)\ (\Delta + \lambda)F(x)\ dV(x)$$

$$+ \int_{\partial\mathcal{G}} \left\{ F(x) \left(\frac{\partial}{\partial \nu} G\,(\Delta + \lambda; x) \right) - G\,(\Delta + \lambda; x) \left(\frac{\partial F}{\partial \nu}(x) \right) \right\}\ dS(x),$$

while, for $\lambda \in \mathrm{Spect}_{\Delta}(\Lambda)$, we have

$$\sideset{}{'}\sum_{\substack{g \in \overline{\mathcal{G}} \\ g \in \Lambda}} F(g) = \frac{1}{\sqrt{\|\mathcal{F}\|}} \sum_{\substack{(\Delta + \lambda)^{\wedge}(h) = 0 \\ h \in \Lambda^{-1}}} \int_{\mathcal{G}} F(x)\overline{\Phi_h(x)}\ dV(x) \tag{9.128}$$

$$+ \int_{\mathcal{G}} G\,(\Delta + \lambda; x)\ (\Delta + \lambda)F(x)\ dV(x)$$

$$+ \int_{\partial\mathcal{G}} \left\{ F(x) \left(\frac{\partial}{\partial \nu} G\,(\Delta + \lambda; x) \right) - G\,(\Delta + \lambda; x) \left(\frac{\partial F}{\partial \nu}(x) \right) \right\}\ dS(x),$$

where $\frac{\partial}{\partial \nu}$ denotes the derivative in the direction of the outer normal ν on $\partial\mathcal{G}$ and

$$(\Delta + \lambda)^{\wedge}(h) = 4\pi^2 h^2 - \lambda, \quad h \in \Lambda^{-1}.$$

The difference between the two cases (9.127) and (9.128) shows that we have to expect some kind of resonance phenomena in the summation procedures in case of eigenvalues. Later on, several examples will bring out more clearly this structure of the summation formula.

From an algorithmic point of view it should be noted that the formulas express a sum in terms of integrals over $\mathcal{G}$ and its boundary $\partial\mathcal{G}$ involving the derivatives of the function F up to the second order.

Remark 9.5. *The particular case $\lambda = 0$, i.e., the Euler summation formula corresponding to the Laplacian, turns out to be a "resonance case". Indeed, the formula for the Laplacian Δ is an immediate multi-dimensional generalization of the one-dimensional Euler summation formula as presented in Chapter 4, where $G(\Delta;\cdot)$ takes the multi-dimensional role of the (negative) Bernoulli polynomial of degree 2.*

Euler Summation for Iterated Helmholtz Operators

Let us assume that the function F is of class $C^{(2k+2)}\left(\overline{\mathcal{G}}\right)$, $k \in \{1,\ldots,m-1\}$. Let $\mathcal{G} \subset \mathbb{R}^q$ be a regular region, $\overline{\mathcal{G}} = \mathcal{G} \cup \partial\mathcal{G}$. Let Λ be an arbitrary lattice in $\mathbb{R}^q$. Then we get the following identity from the Second Green Theorem by aid of the differential equation (8.31)

$$
\int_{\substack{x\in\overline{\mathcal{G}} \\ x\notin B_\varepsilon^q + \Lambda}} G\left((\Delta+\lambda)^{k+1};x\right)\left((\Delta+\lambda)^{k+1}F(x)\right)\,dV(x) \tag{9.129}
$$

$$
- \int_{\substack{x\in\overline{\mathcal{G}} \\ x\notin B_\varepsilon^q + \Lambda}} \left((\Delta+\lambda)\,G\left((\Delta+\lambda)^{k+1};x\right)\right)(\Delta+\lambda)^k F(x)\,dV(x)
$$

$$
= \int_{\substack{x\in\partial\mathcal{G} \\ x\notin B_\varepsilon^q + \Lambda}} G\left((\Delta+\lambda)^{k+1};x\right)\left(\frac{\partial}{\partial\nu}(\Delta+\lambda)^k F(x)\right)\,dS(x)
$$

$$
- \int_{\substack{x\in\partial\mathcal{G} \\ x\notin B_\varepsilon^q + \Lambda}} \left(\frac{\partial}{\partial\nu}G\left((\Delta+\lambda)^{k+1};x\right)\right)(\Delta+\lambda)^k F(x)\,dS(x)
$$

$$
+ \sum_{\substack{g\in\overline{\mathcal{G}} \\ g\in\Lambda}} \int_{\substack{|x-g|=\varepsilon \\ x\in\overline{\mathcal{G}}}} G\left((\Delta+\lambda)^{k+1};x\right)\left(\frac{\partial}{\partial\nu}(\Delta+\lambda)^k F(x)\right)\,dS(x)
$$

$$
- \sum_{\substack{g\in\overline{\mathcal{G}} \\ g\in\Lambda}} \int_{\substack{|x-g|=\varepsilon \\ x\in\overline{\mathcal{G}}}} \left(\frac{\partial}{\partial\nu}G\left((\Delta+\lambda)^{k+1};x\right)\right)(\Delta+\lambda)^k F(x)\,dS(x)
$$

for every (sufficiently small) $\varepsilon > 0$. The integrals over all hyperspheres around the lattice points tend to 0 as $\varepsilon \to 0$. This leads to the recursion formula

$$\int_{\mathcal{G}} G\left((\Delta + \lambda)^{k+1}; x\right) (\Delta + \lambda)^{k+1} F(x) \, dV(x) \tag{9.130}$$

$$= \int_{\mathcal{G}} G\left((\Delta + \lambda)^{k}; x\right) (\Delta + \lambda)^{k} F(x) \, dV(x)$$

$$+ \int_{\partial \mathcal{G}} G\left((\Delta + \lambda)^{k+1}; x\right) \left(\frac{\partial}{\partial \nu} (\Delta + \lambda)^{k} F(x)\right) \, dS(x)$$

$$- \int_{\partial \mathcal{G}} \left(\frac{\partial}{\partial \nu} G\left((\Delta + \lambda)^{k+1}; x\right)\right) (\Delta + \lambda)^{k} F(x) \, dS(x).$$

Consequently, for $F \in \mathrm{C}^{(2m)}\left(\overline{\mathcal{G}}\right)$, we find the identity

$$\int_{\mathcal{G}} G(\Delta + \lambda; x) (\Delta + \lambda) \, F(x) \, dV(x) \tag{9.131}$$

$$= \int_{\mathcal{G}} G\left((\Delta + \lambda)^{m}; x\right) (\Delta + \lambda)^{m} F(x) \, dV(x)$$

$$+ \sum_{k=1}^{m-1} \int_{\partial \mathcal{G}} \left(\frac{\partial}{\partial \nu} G\left((\Delta + \lambda)^{k+1}; x\right)\right) (\Delta + \lambda)^{k} F(x) \, dS(x)$$

$$- \sum_{k=1}^{m-1} \int_{\partial \mathcal{G}} G\left((\Delta + \lambda)^{k+1}; x\right) \left(\frac{\partial}{\partial \nu} (\Delta + \lambda)^{k} F(x)\right) \, dS(x).$$

In connection with the Euler summation formula (i.e., Theorem 9.12) we therefore obtain the Euler summation formula with respect to the operator $(\Delta + \lambda)^{m}$.

Theorem 9.13. *(Euler Summation Formula for the Iterated Helmholtz Operator $(\Delta + \lambda)^{m}$, $\lambda \in \mathbb{R}$, $m \in \mathbb{N}$) Let Λ be an arbitrary lattice in $\mathbb{R}^{q}$. Let $\mathcal{G} \subset \mathbb{R}^{q}$ be a regular region. Suppose that F is of class $\mathrm{C}^{(2m)}\left(\overline{\mathcal{G}}\right), \overline{\mathcal{G}} = \mathcal{G} \cup \partial \mathcal{G}$.*

Then, for all $\lambda \in \mathbb{R}$,

$$\sideset{}{'}\sum_{\substack{g \in \overline{\mathcal{G}} \\ g \in \Lambda}} F(g) = \frac{1}{\sqrt{\|\mathcal{F}\|}} \sum_{\substack{(\Delta+\lambda)^{\wedge}(h)=0 \\ h \in \Lambda^{-1}}} \int_{\mathcal{G}} F(x) \overline{\Phi_{h}(x)} \, dV(x) \tag{9.132}$$

$$+ \int_{\mathcal{G}} G((\Delta + \lambda)^{m}; x) \, (\Delta + \lambda)^{m} F(x) \, dV(x)$$

$$+ \sum_{k=0}^{m-1} \int_{\partial \mathcal{G}} \left(\frac{\partial}{\partial \nu} G\left((\Delta + \lambda)^{k+1}; x\right)\right) (\Delta + \lambda)^{k} F(x) \, dS(x)$$

$$- \sum_{k=0}^{m-1} \int_{\partial \mathcal{G}} G\left((\Delta + \lambda)^{k+1}; x\right) \left(\frac{\partial}{\partial \nu} (\Delta + \lambda)^{k} F(x)\right) \, dS(x),$$

where $\sum_{(\Delta+\lambda)^{\wedge}(h)=0} \cdots$ only occurs if $\lambda \in \mathrm{Spect}_{\Delta}(\Lambda)$.

The close relation of these identities to the one-dimensional Euler summation formulas is best seen by specializing the "wave number" $\lambda \in \mathbb{R}$. In the case $\lambda = 0$, the multi-dimensional formula (9.132) takes the following already known form.

Corollary 9.6. *(Euler Summation Formula for the Operator Δ^m, $m \in \mathbb{N}$) Let Λ be an arbitrary lattice in $\mathbb{R}^q$. Let $\mathcal{G} \subset \mathbb{R}^q$ be a regular region. Suppose that F is of class $\mathrm{C}^{(2m)}\left(\overline{\mathcal{G}}\right), \overline{\mathcal{G}} = \mathcal{G} \cup \partial\mathcal{G}$. Then*

$$\sideset{}{'}\sum_{\substack{g \in \overline{\mathcal{G}} \\ g \in \Lambda}} F(g) \quad = \quad \frac{1}{\|\mathcal{F}\|} \int_{\mathcal{G}} F(x)\, dV(x) \tag{9.133}$$

$$+ \int_{\mathcal{G}} G\left(\Delta^m; x\right)\, \Delta^m F(x)\, dV(x)$$

$$+ \sum_{k=0}^{m-1} \int_{\partial\mathcal{G}} \left(\frac{\partial}{\partial\nu} G\left(\Delta^{k+1}; x\right)\right) \Delta^k F(x)\, dS(x)$$

$$- \sum_{k=0}^{m-1} \int_{\partial\mathcal{G}} G\left(\Delta^{k+1}; x\right) \left(\frac{\partial}{\partial\nu}\Delta^k F(x)\right)\, dS(x).$$

The generalization of the Euler summation formula to Helmholtz operators $(\Delta + \lambda)^m$, $\lambda \in \mathbb{R}$, opens new perspectives to subtle questions of the convergence of multi-dimensional series and the integration of multi-variate functions because the Helmholtz operator $(\Delta + \lambda)^m$, $\lambda \in \mathbb{R}$, can be closely related to alternating or oscillating properties of the summands or integrands, respectively.

For the series convergence in one dimension direct techniques are well known, which are particularly suited for verifying the convergence of an alternating series. In multi-dimensional summation problems the situation is different because the one-dimensional concept of the alternating or oscillating series is not directly applicable in most cases. According to our approach, the choice of the "wave number" $\lambda \in \mathbb{R}$ within the operator $(\Delta+\lambda)^m$ may be used to adapt the summation formula to oscillating properties of the summands constituting the series. These aspects will be of great advantage in deriving special identities in analytic theory of numbers. The "Hardy–Landau" identities of Section 10.4 and its multi-dimensional extensions of Section 10.5 are examples of this procedure.

Remark 9.6. *Of course, the Euler summation formulas can be formulated to more general differential operators (see V.K. Ivanow [1963], W. Freeden [1982]), W. Freeden, J. Fleck [1987]). For example, a summation formula can be formulated for hyperbolic differential operators. Our work, however, is restricted to (iterated) Helmholtz operators and to arbitrary lattices. Hence, all summation formulas are included which are based on (iterations of) arbitrary* *second order elliptic differential operators with constant coefficients. In consequence, spherical summation and/or integration procedures are particularly reflected by our approach.*

9.5 Lattice Point Discrepancy Involving the Helmholtz Operator

We start with a canonical extension of the Λ-lattice point discrepancy.

Definition 9.2. *Let Λ be a lattice in $\mathbb{R}^q$. Let $\mathcal{G}$ be a regular region in $\mathbb{R}^q$. Suppose that F is of class $\mathrm{C}^{(0)}(\overline{\mathcal{G}})$, $\overline{\mathcal{G}} = \mathcal{G} \cup \partial \mathcal{G}$. The difference $P^\lambda (F; \overline{\mathcal{G}})$, $\lambda \in \mathbb{R}$, given by*

$$P^\lambda \left(F; \overline{\mathcal{G}}\right) = \sideset{}{'}\sum_{\substack{g \in \overline{\mathcal{G}} \\ g \in \Lambda}} F(g) - \frac{1}{\sqrt{\|\mathcal{F}\|}} \sum_{\substack{(\Delta+\lambda)^\wedge(h)=0 \\ h \in \Lambda^{-1}}} \int_{\mathcal{G}} F(y) \overline{\Phi_h(y)} \, dV(y) \quad (9.134)$$

is called the Λ-lattice point discrepancy of F in $\overline{\mathcal{G}}$ with respect to $\Delta + \lambda$.

Remark 9.7. *By convention, the Λ-lattice point discrepancy of F in $\overline{\mathcal{G}}$ with respect to Δ, i.e. $P^0 \left(F; \overline{\mathcal{G}}\right)$, coincides with $P\left(F; \overline{\mathcal{G}}\right)$.*

From the Euler summation formula (cf. Theorem 9.3) involving iterated Helmholtz operators we get for the Λ-lattice point discrepancy for a (weight) function $F \in \mathrm{C}^{(2m)}\left(\overline{\mathcal{G}}\right)$, $m \in \mathbb{N}$, $\overline{\mathcal{G}} = \mathcal{G} \cup \partial \mathcal{G}$, $\mathcal{G}$ regular region in $\mathbb{R}^q$,

Theorem 9.14. *Let Λ be a lattice in $\mathbb{R}^q$. Let $\mathcal{G}$ be a regular region in $\mathbb{R}^q$. Suppose that F is of class $\mathrm{C}^{(2m)}(\overline{\mathcal{G}})$, $m \in \mathbb{N}$, $\overline{\mathcal{G}} = \mathcal{G} \cup \partial \mathcal{G}$. Then*

$$P^\lambda \left(F; \overline{\mathcal{G}}\right) = \int_{\mathcal{G}} G\left((\Delta+\lambda)^m; x\right) \, (\Delta+\lambda)^m F(x) \, dV(x) \quad (9.135)$$

$$+ \sum_{k=0}^{m-1} \int_{\partial \mathcal{G}} \left(\frac{\partial}{\partial \nu} G\left((\Delta+\lambda)^{k+1}; x\right)\right) (\Delta+\lambda)^k \, F(x) \, dS(x)$$

$$- \sum_{k=0}^{m-1} \int_{\partial \mathcal{G}} G\left((\Delta+\lambda)^{k+1}; x\right) \left(\frac{\partial}{\partial \nu} (\Delta+\lambda)^k \, F(x)\right) \, dS(x).$$

Homogeneous boundary conditions and polymetaharmonicity can be treated analogously to the case of iterated Laplace operators (see Section 9.2). We omit these considerations. We mention only some typical examples.

Application to Periodical Polynomials

For $\lambda = 4\pi^2 b^2$, $b \in \mathbb{R}^q$, the Λ-lattice point discrepancy of $e^{2\pi i b \cdot}$ in $\overline{\mathcal{G}}$, $\mathcal{G}$ regular region in $\mathbb{R}^q$,

$$P^{4\pi^2 b^2}\left(e^{2\pi i b \cdot}; \overline{\mathcal{G}}\right) = \sideset{}{'}\sum_{\substack{g \in \overline{\mathcal{G}} \\ g \in \Lambda}} e^{2\pi i b \cdot g} - \frac{1}{\|\mathcal{F}\|} \sum_{\substack{|h| = |b| \\ h \in \Lambda^{-1}}} \int_{\mathcal{G}} e^{2\pi i (b-h) \cdot y} \, dV(y) \quad (9.136)$$

reads

$$P^{4\pi^2 b^2}\left(e^{2\pi i b \cdot}; \overline{\mathcal{G}}\right) = \int_{\partial\mathcal{G}} \left(\frac{\partial}{\partial\nu} G\left(\Delta + 4\pi^2 b^2; x\right)\right) e^{2\pi i b \cdot x} \, dS(x) \quad (9.137)$$
$$- \int_{\partial\mathcal{G}} G\left(\Delta + 4\pi^2 b^2; x\right) \left(\frac{\partial}{\partial\nu} e^{2\pi i (b \cdot x)}\right) \, dS(x).$$

Note that the second sum on the right side of (9.136) occurs only if $4\pi b^2 \in \mathrm{Spect}_\Delta(\Lambda)$. In particular, for the ball $\overline{\mathbb{B}}_R^q$, we find with the aid of (6.454)

$$P^{4\pi^2 b^2}\left(e^{2\pi i b \cdot}; \overline{\mathbb{B}}_R^q\right) \quad (9.138)$$
$$= \sideset{}{'}\sum_{\substack{|g| \leq R \\ g \in \Lambda}} e^{2\pi i b \cdot g} - \frac{\|\mathbb{S}^{q-1}\| \, R^q}{\|\mathcal{F}\|} \sum_{\substack{|h| = |b| \\ h \in \Lambda^{-1}}} \frac{J_1(q; 2\pi |b - h| R)}{2\pi |b - h| R}$$

with

$$P^{4\pi^2 b^2}\left(e^{2\pi i b \cdot}; \overline{\mathbb{B}}_R^q\right) = \int_{\substack{|x| = R \\ x \in \mathbb{R}^q}} \left(\frac{\partial}{\partial\nu} G\left(\Delta + 4\pi^2 b^2; x\right)\right) e^{2\pi i (b \cdot x)} \, dS(x)$$
$$- \int_{\substack{|x| = R \\ x \in \mathbb{R}^q}} G\left(\Delta + 4\pi^2 b^2; x\right) \left(\frac{\partial}{\partial\nu} e^{2\pi i b \cdot x}\right) \, dS(x).$$
$$(9.139)$$

Example 9.2. *For* $q = 2, R > 0, b \in \mathbb{R}^2$, *we are left with the following situation:*

If $4\pi^2 b^2 \notin \mathrm{Spect}_\Delta(\Lambda)$, *then*

$$P^{4\pi^2 b^2}\left(e^{2\pi i b \cdot}; \overline{\mathbb{B}}_R^2\right) = \sideset{}{'}\sum_{\substack{|g| \leq R \\ g \in \Lambda}} e^{2\pi i b \cdot g} \quad (9.140)$$
$$= \int_{\substack{|x| = R \\ x \in \mathbb{R}^q}} \left(\frac{\partial}{\partial\nu} G\left(\Delta + 4\pi^2 b^2; x\right)\right) e^{2\pi i (b \cdot x)} \, dS(x)$$
$$- \int_{\substack{|x| = R \\ x \in \mathbb{R}^q}} G\left(\Delta + 4\pi^2 b^2; x\right) \left(\frac{\partial}{\partial\nu} e^{2\pi i b \cdot x}\right) \, dS(x).$$

If $4\pi^2 b^2 \in \mathrm{Spect}_\Delta(\Lambda)$, *then*

$$P^{4\pi^2 b^2}\left(e^{2\pi i b \cdot}; \overline{\mathbb{B}_R^2}\right) \tag{9.141}$$

$$= \sum_{\substack{|g| \le R \\ g \in \Lambda}}{}' e^{2\pi i b \cdot g} - \frac{R}{\|\mathcal{F}\|} \sum_{\substack{|h| = |b| \\ h \in \Lambda^{-1}}} \frac{J_1(2; 2\pi |b - h| R)}{|b - h|}$$

$$= \int_{\substack{|x| = R \\ x \in R^q}} \left(\frac{\partial}{\partial \nu} G\left(\Delta + 4\pi^2 b^2; x\right)\right) e^{2\pi i (b \cdot x)} \, dS(x)$$

$$- \int_{\substack{|x| = R \\ x \in R^q}} G\left(\Delta + 4\pi^2 b^2; x\right) \left(\frac{\partial}{\partial \nu} e^{2\pi i b \cdot x}\right) \, dS(x).$$

Later on, the integrals on the right side of (9.140) and (9.141), respectively, are shown to be expressible as series expansions in terms of Bessel functions.

10

Lattice Point Summation

CONTENTS

10.1 Integral Asymptotics for (Iterated) Lattice Functions 304
10.2 Convergence Criteria and Theorems 308
10.3 Lattice Point-Generated Poisson Summation Formula 312
10.4 Classical Two-Dimensional Hardy–Landau Identity 314
10.5 Multi-Dimensional Hardy–Landau Identities 317

In this chapter we are concerned with lattice point summation. Multi-dimensional Euler summation formulas with respect to (iterated) Helmholtz operators are applied to lattice point sums inside spheres $\mathbb{S}_N^{q-1}$, $q \geq 2$, which provide interesting results particularly when N tends toward infinity. Convergence theorems are formulated for multi-dimensional lattice point series, thereby adapting the iterated Helmholtz operator $(\Delta + \lambda)^m$, $\lambda \in \mathbb{R}$, to the specific (oscillating) properties of the weight function under consideration. Essential ingredients in this context are integral estimates with respect to (iterated) Λ-lattice functions for Helmholtz operators (as proposed by W. Freeden [1975, 1978a] for the two-dimensional case, C. Müller, W. Freeden [1980] for the higher-dimensional case). Limits of lattice point sums are expressed in terms of integrals involving the Λ-lattice function for the chosen Helmholtz operator. Multi-dimensional analogues to the one-dimensional Poisson formula (see Section 4.4 as well as Section 4.5) are formulated in the lattice point context of Euclidean spaces $\mathbb{R}^q$, $q \geq 2$. Their special application to alternating series of the Hardy–Landau type are studied in more detail.

The layout of this chapter can be described briefly as follows: Section 10.1 is concerned with the proof of asymptotic integral estimates of (iterated) Λ-lattice functions for Helmholtz operators. The resulting asymptotic integral relations are used in Section 10.2 to derive sufficient criteria for the (spherically understood) convergence of lattice point sums. Multi-dimensional convergence theorems for weighted lattice point sums are developed from (asymptotic) Euler summation formulas. The occurring (infinite) series are expressed in terms of integrals involving the Λ-lattice function for $(\Delta + \lambda)^m$, $\lambda \in \mathbb{R}$, $m \in \mathbb{N}$. Next, the parameter $\lambda \in \mathbb{R}$ is adaptively utilized to enforce spherical convergence of the Poisson summation formula in $\mathbb{R}^q$ (cf. Section 10.3) by adaptation of the operator $(\Delta + \lambda)^m$, $\lambda \in \mathbb{R}$, $m \in \mathbb{N}$, to the specific properties of the (oscillating) summand (i.e., weight function) under consideration. As a first application of our convergence criteria, the classical two-dimensional Hardy–Landau identity is provided in Section 10.4. Finally, Section 10.5 presents theoretical extensions of the Hardy–Landau identity to higher dimensions based on our metaharmonically oriented framework.

10.1 Integral Asymptotics for (Iterated) Lattice Functions

We apply the multi-dimensional Euler summation formula (see Theorem 9.3) especially to sums of the spherical type

$$\sum_{g \in \Lambda} \ldots = \lim_{N \to \infty} \sum_{\substack{|g| \leq N \\ g \in \Lambda}} \ldots . \tag{10.1}$$

Let F be a $2m$-times $(m \in \mathbb{N})$ continuously differentiable function in $\overline{\mathbb{B}_N^q}$. Then, for each value $\lambda \in \mathbb{R}$, we have

$$\sum_{\substack{|g| \leq N \\ g \in \Lambda}} F(g) \;=\; \frac{1}{\sqrt{\|\mathcal{F}\|}} \sum_{\substack{(\Delta+\lambda)^\wedge (h)=0 \\ h \in \Lambda^{-1}}} \int_{\substack{|x| \leq N \\ x \in \mathbb{R}^q}} F(x)\overline{\Phi_h(x)}\, dV(x) \tag{10.2}$$

$$+ \int_{\substack{|x| \leq N \\ x \in \mathbb{R}^q}} G\left((\Delta+\lambda)^m ; x\right)\,(\Delta+\lambda)^m F(x)\, dV(x)$$

$$+ R_{(q)}^{(m)}(N),$$

where the sum $\Sigma_{(\Delta+\lambda)^\wedge (h)=0}$ on the right side of (10.2) occurs only if $\lambda \in \mathrm{Spect}_\Delta(\Lambda)$; i.e., λ is an eigenvalue of Λ. The *"remainder term"* is given by

$$R_{(q)}^{(m)}(N) \;=\; \frac{1}{2} \sum_{\substack{|g|=N \\ g \in \Lambda}} F(g) \tag{10.3}$$

$$+ \sum_{k=0}^{m-1} \int_{\substack{|x|=N \\ x \in \mathbb{R}^q}} \left(\frac{\partial}{\partial \nu} G\left((\Delta+\lambda)^{k+1}; x\right)\right) (\Delta+\lambda)^k F(x)\, dS(x)$$

$$- \sum_{k=0}^{m-1} \int_{\substack{|x|=N \\ x \in \mathbb{R}^q}} G\left((\Delta+\lambda)^{k+1}; x\right) \left(\frac{\partial}{\partial \nu} (\Delta+\lambda)^k F(x)\right) dS(x).$$

Questions of the convergence as $N \to \infty$ require estimates of the remainder term (10.3). The following results play an important part in this respect.

Theorem 10.1. *For all lattices $\Lambda \subset \mathbb{R}^q$ and all real numbers λ the estimates*

$$\int_{\substack{|x|=N \\ x \in \mathbb{R}^q}} |G\left(\Delta+\lambda; x\right)|\, dS_{(q-1)}(x) = O(N^{q-1}), \; N \to \infty, \tag{10.4}$$

and

$$\int_{\substack{|x|=N \\ x \in \mathbb{R}^q}} \left| \frac{\partial}{\partial \nu} G\left(\Delta + \lambda; x\right) \right| \, dS_{(q-1)}(x) = O(N^{q-1}), \ N \to \infty, \tag{10.5}$$

hold true.

Proof. The proof is essentially based on the material provided by Section 6.3. We start by remembering that there is a positive constant E dependent on the lattice Λ, such that $|g - g'| \geq E$ holds for all points $g, g' \in \Lambda$ with $g \neq g'$. We set

$$\delta \leq \frac{1}{2} \, \min(1, E). \tag{10.6}$$

Then we observe that, for each $\lambda \in \mathbb{R}$, there exists a constant C such that the estimates

$$\begin{array}{rcll}
|G\left(\Delta + \lambda; x\right)| & \leq & C \left| \ln|x - g| \right| & , \quad q = 2, \\
|G\left(\Delta + \lambda; x\right)| & \leq & C \, |x - g|^{2-q} & , \quad q \geq 3
\end{array} \tag{10.7}$$

and

$$\begin{array}{rcll}
\left| \nabla_x G\left(\Delta + \lambda; x\right) - \frac{1}{2\pi} \frac{x-g}{|x-g|^2} \right| & \leq & C \frac{1}{|x-g|} & , \quad q = 2, \\
\left| \nabla_x G\left(\Delta + \lambda; x\right) - \frac{1}{(q-2)\|\mathbb{S}^{q-1}\|} \frac{x-g}{|x-g|^q} \right| & \leq & C \frac{1}{|x-g|^{q-2}} & , \quad q \geq 3
\end{array} \tag{10.8}$$

hold uniformly in $\overline{\mathbb{B}_\delta^q(g)}$. We denote the distance of $x \in \mathbb{R}^q$ to the lattice Λ by

$$D(x; \Lambda) = \text{dist}(x; \Lambda) = \min_{g \in \Lambda} |x - g|. \tag{10.9}$$

It is clear that there is a constant B (depending on λ and δ) such that the estimates

$$\begin{array}{rcl}
|G\left(\Delta + \lambda; x\right)| & \leq & B, \\
|\nabla_x G\left(\Delta + \lambda; x\right)| & \leq & B
\end{array} \tag{10.10} \tag{10.11}$$

are valid for all $x \in \mathbb{R}^q$ with $D(x; \Lambda) = \text{dist}(x; \Lambda) \geq \delta$. Moreover, by Theorem 5.5, we are led to deduce that

$$\#_\Lambda \left(\overline{\mathbb{B}_{N+\delta}^q} \setminus \mathbb{B}_{N-\delta}^q \right) = \sum_{\substack{N-\delta \leq |g| \leq N+\delta \\ g \in \Lambda}} 1 = O\left(N^{q-1}\right) \tag{10.12}$$

for δ (fixed) and $N \to \infty$. Thus it follows that

$$\sum_{\substack{|g|=N \\ g \in \Lambda}} 1 = O\left(N^{q-1}\right), \quad N \to \infty. \tag{10.13}$$

We use the different results for the cases $D(x;\Lambda) > \delta$ and $D(x;\Lambda) \leq \delta$ to get an estimate for the integrals (10.4) and (10.5)

$$\int_{\substack{|x|=N \\ x \in \mathbb{R}^q}} |G(\Delta + \lambda; x)| \, dS_{(q-1)}(x) \tag{10.14}$$

$$= \int_{\substack{|x|=N \\ D(x;\Lambda)>\delta \\ x \in \mathbb{R}^q}} |G(\Delta + \lambda; x)| \, dS_{(q-1)}(x)$$

$$+ \int_{\substack{|x|=N \\ D(x;\Lambda)\leq\delta \\ x \in \mathbb{R}^q}} |G(\Delta + \lambda; x)| \, dS_{(q-1)}(x).$$

From (10.10) it follows that

$$\int_{\substack{|x|=N \\ D(x;\lambda)>\delta \\ x \in \mathbb{R}^q}} |G(\Delta + \lambda; x)| \, dS_{(q-1)}(x) \leq B \, \|\mathbb{S}^{q-1}\| \, N^{q-1}. \tag{10.15}$$

Because of the characteristic singularity of $G(\Delta + \lambda; \cdot)$, for dimensions $q \geq 3$, the estimate

$$\int_{\substack{|x|=N \\ D(x;\lambda)\leq\delta \\ x \in \mathbb{R}^q}} |G(\Delta + \lambda; x)| \, dS_{(q-1)}(x) \tag{10.16}$$

$$= O\left(\left(\#_\Lambda \left(\overline{\mathbb{B}^q_{N+\delta}} \setminus \mathbb{B}^q_{N-\delta}\right)\right) \int_{\substack{|x|=N \\ |x-g|\leq\delta \\ x \in \mathbb{R}^q}} \frac{1}{|x - g|^{q-2}} \, dS_{(q-1)}(x) \right)$$

is valid for $N \to \infty$. In connection with (10.12) and Lemma 6.6 we therefore find

$$\int_{\substack{|x|=N \\ D(x;\lambda)\leq\delta \\ x \in \mathbb{R}^q}} |G(\Delta + \lambda; x)| \, dS_{(q-1)}(x) = O\left(N^{q-1}\right) \tag{10.17}$$

for $q \geq 3$ and $N \to \infty$ (note that the case $q = 2$ can be verified by the same arguments observing the logarithmic singularity (cf. Lemma 6.7)). This establishes the proof of the first part of Theorem 10.1 for $q \geq 2$.

Concerning the second part of Theorem 10.1 we again split the integral such that

$$\int_{\substack{|x|=N \\ x \in \mathbb{R}^q}} \left| \frac{\partial}{\partial\nu} G(\Delta + \lambda; x) \right| \, dS_{(q-1)}(x) \tag{10.18}$$

$$= \int_{\substack{|x|=N \\ D(x;\Lambda)>\delta \\ x \in \mathbb{R}^q}} \left| \frac{\partial}{\partial\nu} G(\Delta + \lambda; x) \right| \, dS_{(q-1)}(x)$$

$$+ \int_{\substack{|x|=N \\ D(x;\Lambda)\leq\delta \\ x \in \mathbb{R}^q}} \left| \frac{\partial}{\partial\nu} G(\Delta + \lambda; x) \right| \, dS_{(q-1)}(x).$$

Then, by virtue of (10.11) and Lemma 6.8, we find

$$\int_{\substack{|x|=N \\ x \in \mathbb{R}^q}} \left| \frac{\partial}{\partial \nu} G\left(\Delta + \lambda; x\right) \right| \, dS_{(q-1)}(x) \tag{10.19}$$

$$= O\left(N^{q-1}\right) + O\left(\#_\Lambda \left(\overline{\mathbb{B}^q_{N+\delta}} \setminus \mathbb{B}^q_{N-\delta}\right)\right)$$

$$= O\left(N^{q-1}\right)$$

for $N \to \infty$. Altogether, this is the desired result. $\qquad \square$

Similar results are obtainable for the Λ-lattice functions $G\left((\Delta + \lambda)^k; \cdot\right)$, $k \in \mathbb{N}$. Since each iteration reduces the order of the singularity by two, $G\left((\Delta + \lambda)^k; \cdot\right)$ is continuous for $k > \frac{q}{2}$ and continuously differentiable for $k > \frac{q}{2} + 1$. The estimates

$$\int_{\substack{|x|=N \\ x \in \mathbb{R}^q}} \left| G\left((\Delta + \lambda)^k; x\right) \right| \, dS_{(q-1)}(x) = O\left(N^{q-1}\right) \tag{10.20}$$

and

$$\int_{\substack{|x|=N \\ x \in \mathbb{R}^q}} \left| \frac{\partial}{\partial \nu} G\left((\Delta + \lambda)^k; x\right) \right| \, dS_{(q-1)}(x) = O\left(N^{q-1}\right) \tag{10.21}$$

for $N \to \infty$, therefore, are obvious for all $k > \frac{q}{2} + 1$. For the intermediate cases $k \in (1, \frac{q}{2} + 1]$ we use Lemma 6.6 and estimate the integrals

$$\int_{\substack{|x|=N \\ x \in \mathbb{R}^q}} \ldots = \int_{\substack{|x|=N \\ D(x;\Lambda) > \delta \\ x \in \mathbb{R}^q}} \ldots + \int_{\substack{|x|=N \\ D(x;\Lambda) \le \delta \\ x \in \mathbb{R}^q}} \ldots \tag{10.22}$$

in the same way as described above by aid of Lemma 6.8.

This finally justifies the results of

Theorem 10.2. *For all lattices* $\Lambda \subset \mathbb{R}^q$, *all numbers* $\lambda \in \mathbb{R}$, *and all positive integers* k *the* Λ-*lattice functions for iterated Helmholtz operators satisfy the asymptotic estimates*

$$\int_{\substack{|x|=N \\ x \in \mathbb{R}^q}} \left| G\left((\Delta + \lambda)^k; x\right) \right| \, dS_{(q-1)}(x) = O\left(N^{q-1}\right) \tag{10.23}$$

and

$$\int_{\substack{|x|=N \\ x \in \mathbb{R}^q}} \left| \frac{\partial}{\partial \nu} G\left((\Delta + \lambda)^k; x\right) \right| \, dS_{(q-1)}(x) = O\left(N^{q-1}\right), \tag{10.24}$$

$N \to \infty$.

10.2 Convergence Criteria and Theorems

In the sequel, the properties developed for Λ-lattice functions for iterated Helmholtz operators are used to formulate convergence theorems for multi-dimensional (alternating) series (see Theorem 4.10 for a one-dimensional analogue), where the Euler summation formula is the key structure in our context. To this end we first introduce two subspaces of $\mathrm{C}^{(2m)}(\mathbb{R}^q)$, which indicate the sufficient criteria for the convergence derivable from our multi-dimensional approach:

Definition 10.1. *Let $m \in \mathbb{N}$, $\varepsilon > 0$, and $\lambda \in \mathbb{R}$, be given values. Then the spaces $\mathrm{CP}_1^{(2m)}(\lambda; \mathbb{R}^q)$ and $\mathrm{CP}_2^{(2m)}(\varepsilon, \lambda; \mathbb{R}^q)$, respectively, are defined as follows:*

(i) $\mathrm{CP}_1^{(2m)}(\lambda; \mathbb{R}^q)$ is the space of all functions $H \in \mathrm{C}^{(2m)}(\mathbb{R}^q)$ such that the asymptotic relations

$$(\Delta_x + \lambda)^k H(x) = o\left(|x|^{1-q}\right), \quad |x| \to \infty, \tag{10.25}$$

$$\left|\nabla_x (\Delta_x + \lambda)^k H(x)\right| = o\left(|x|^{1-q}\right), \quad |x| \to \infty \tag{10.26}$$

are valid for $k = 0, \dots, m-1$ (note that the case $m = 1$ is independent of $\lambda \in \mathbb{R}$).

(ii) $\mathrm{CP}_2^{(2m)}(\varepsilon, \lambda; \mathbb{R}^q)$ is the space of all functions $H \in \mathrm{C}^{(2m)}(\mathbb{R}^q)$ such that

$$(\Delta_x + \lambda)^m H(x) = O\left(|x|^{-(q+\varepsilon)}\right), \quad |x| \to \infty. \tag{10.27}$$

We begin our discussion of the remainder term (10.3) with the boundary terms.

Lemma 10.1. *For given $m \in \mathbb{N}$ and $\lambda \in \mathbb{R}$, assume that the function F is of class $\mathrm{CP}_1^{(2m)}(\lambda; \mathbb{R}^q)$. Then, for $N \to \infty$, we have*

$$R_{(q)}^{(m)}(N) = \frac{1}{2} \sum_{\substack{|g|=N \\ g \in \Lambda}} F(g) \tag{10.28}$$

$$+ \sum_{k=0}^{m-1} \int_{\substack{|x|=N \\ x \in \mathbb{R}^q}} \left(\frac{\partial}{\partial \nu} G\left((\Delta + \lambda)^{k+1}; x\right)\right) (\Delta + \lambda)^k F(x)\, dS(x)$$

$$- \sum_{k=0}^{m-1} \int_{\substack{|x|=N \\ x \in \mathbb{R}^q}} G\left((\Delta + \lambda)^{k+1}; x\right) \left(\frac{\partial}{\partial \nu} (\Delta + \lambda)^k F(x)\right) dS(x)$$

$$= o(1).$$

Proof. If F is assumed to be a member of class $\mathrm{CP}_1^{(2m)}(\lambda;\mathbb{R}^q)$, then

$$\frac{1}{2} \sum_{\substack{|g|=N \\ g\in\Lambda}} F(g) = o\left(N^{1-q} \sum_{\substack{|g|=N \\ g\in\Lambda}} 1\right) = o(1), \quad N \to \infty. \tag{10.29}$$

Furthermore, we are allowed to conclude

$$\sum_{k=0}^{m-1} \int_{\substack{|x|=N \\ x\in\mathbb{R}^q}} \left(\frac{\partial}{\partial\nu} G\left((\Delta+\lambda)^{k+1};x\right)\right) \left((\Delta+\lambda)^k F(x)\right) \, dS(x)$$

$$- \sum_{k=0}^{m-1} \int_{\substack{|x|=N \\ x\in\mathbb{R}^q}} \left(G\left((\Delta+\lambda)^{k+1};x\right)\right) \left(\frac{\partial}{\partial\nu} (\Delta+\lambda)^k F(x)\right) \, dS(x)$$

$$= o\left(N^{1-q} \left(\sum_{k=0}^{m-1} \int_{\substack{|x|=N \\ x\in\mathbb{R}^q}} \left|G\left((\Delta+\lambda)^{k+1};x\right)\right| \right.\right. \tag{10.30}$$

$$\left.\left. + \left|\frac{\partial}{\partial\nu} G\left((\Delta+\lambda)^{k+1};x\right)\right| \, dS(x)\right)\right)$$

for $N \to \infty$. In connection with Theorem 10.2 we therefore find

$$\sum_{k=0}^{m-1} \int_{\substack{|x|=N \\ x\in\mathbb{R}^q}} \left(\frac{\partial}{\partial\nu} G\left((\Delta+\lambda)^{k+1};x\right)\right) ((\Delta+\lambda)^k F(x)) \, dS(x)$$

$$- \sum_{k=0}^{m-1} \int_{\substack{|x|=N \\ x\in\mathbb{R}^q}} \left(G\left((\Delta+\lambda)^{k+1};x\right)\right) \left(\frac{\partial}{\partial\nu} (\Delta+\lambda)^k F(x)\right) \, dS(x)$$

$$= o(1) \tag{10.31}$$

for $N \to \infty$. Collecting all details we obtain the promised result of Lemma 10.1. $\qquad\square$

Next we come to the discussion of the volume integral in (10.2) involving the derivative $(\Delta+\lambda)^m F$.

Lemma 10.2. *For given $m \in \mathbb{N}$, $\varepsilon > 0$, and $\lambda \in \mathbb{R}$, assume that the function $F \in \mathrm{C}^{(2m)}(\mathbb{R}^q)$ is of class $\mathrm{CP}_2^{(2m)}(\varepsilon,\lambda;\mathbb{R}^q)$. Then the integral*

$$\int_{\mathbb{R}^q} G\left((\Delta+\lambda)^m;x\right) (\Delta+\lambda)^m F(x) \, dV(x) \tag{10.32}$$

is absolutely convergent.

Proof. From Theorem 10.2 we are immediately able to guarantee with suitable positive constants M, N

$$\left| \int_{\substack{M \leq |x| \leq N \\ x \in \mathbb{R}^q}} G\left((\Delta + \lambda)^m ; x\right) (\Delta + \lambda)^m F(x) \, dV(x) \right| \tag{10.33}$$

$$= O\left(\int_M^N \frac{1}{(1+r)^{q+\varepsilon}} \left(\int_{\substack{|x|=r \\ x \in \mathbb{R}^q}} \left| G\left((\Delta + \lambda)^m ; x\right) \right| \, dS(x) \right) dr \right)$$

$$= O\left(\int_M^N \frac{r^{q-1}}{(1+r)^{q+\varepsilon}} \, dr \right).$$

Consequently if F is of class $\mathrm{CP}_2^{(2m)}(\varepsilon, \lambda; \mathbb{R}^q)$, the absolute convergence of the integral (10.32) is guaranteed. $\qquad\square$

Combining Lemma 10.1 and Lemma 10.2 we obtain as a first consequence from (10.2) and (10.3)

Theorem 10.3. *Let Λ be an arbitrary lattice in $\mathbb{R}^q$. For given $m \in \mathbb{N}$, $\varepsilon > 0$, and $\lambda \in \mathbb{R}$, assume that F is a member of $\mathrm{CP}_1^{(2m)}(\lambda; \mathbb{R}^q) \cap \mathrm{CP}_2^{(2m)}(\varepsilon, \lambda; \mathbb{R}^q)$. Then, the limit*

$$\lim_{N \to \infty} \left(\sum_{\substack{|g| \leq N \\ g \in \Lambda}} F(g) - \frac{1}{\sqrt{\|\mathcal{F}\|}} \sum_{\substack{(\Delta+\lambda)^\wedge(h)=0 \\ h \in \Lambda^{-1}}} \int_{\substack{|x| \leq N \\ x \in \mathbb{R}^q}} F(x) \overline{\Phi_h(x)} \, dV(x) \right) \tag{10.34}$$

exists, and we have the limit relation

$$\lim_{N \to \infty} \left(\sum_{\substack{|g| \leq N \\ g \in \Lambda}} F(g) - \frac{1}{\sqrt{\|\mathcal{F}\|}} \sum_{\substack{(\Delta+\lambda)^\wedge(h)=0 \\ h \in \Lambda^{-1}}} \int_{\substack{|x| \leq N \\ x \in \mathbb{R}^q}} F(x) \overline{\Phi_h(x)} \, dV(x) \right)$$

$$= \int_{\mathbb{R}^q} G\left((\Delta + \lambda)^m ; x \right) (\Delta + \lambda)^m F(x) \, dV(x), \tag{10.35}$$

where the sum $\sum_{(\Delta+\lambda)^\wedge(h)=0} \cdots$ has to be extended over all $h \in \Lambda^{-1}$ satisfying $(\Delta+\lambda)^\wedge(h) = 0$.

In the case that $\lambda \notin \mathrm{Spect}_\Delta(\Lambda)$, the sum $\sum_{(\Delta+\lambda)^\wedge(h)=0} \cdots$ is always understood to be zero, and we simply have

$$\lim_{N \to \infty} \left(\sum_{\substack{|g| \leq N \\ g \in \Lambda}} F(g) \right) = \int_{\mathbb{R}^q} G\left((\Delta + \lambda)^m ; x\right) (\Delta + \lambda)^m F(x) \, dV(x). \tag{10.36}$$

Theorem 10.3 demonstrates that the convergence of multi-dimensional sums (understood in spherical summation)

$$\sum_{g \in \Lambda} F(g) = \lim_{N \to \infty} \sum_{|g| \leq N} F(g) \tag{10.37}$$

is closely related to the spectrum $\mathrm{Spect}_\Delta(\Lambda)$. As a matter of fact, for the class of non-eigenvalues, i.e., $\lambda \notin \mathrm{Spect}_\Delta(\Lambda)$, Theorem 10.3 immediately guarantees the convergence of the infinite series on the left side of (10.36).

In order to ensure the convergence of the series (10.37) for an eigenvalue $\lambda \in \mathrm{Spect}_\Delta(\Lambda)$, however, another subspace $\mathrm{CP}_3^{(2m)}(\lambda; \Lambda; \mathbb{R}^q)$ of $\mathrm{C}^{(2m)}(\mathbb{R}^q)$ has to come into play.

Definition 10.2. *Let $m \in \mathbb{N}$ and $\lambda \in \mathbb{R}$ be given values such that $\lambda \in \mathrm{Spect}_\Delta(\Lambda)$. Then the subspace $\mathrm{CB}_3^{(2m)}(\lambda; \Lambda; \mathbb{R}^q)$ of $\mathrm{C}^{(2m)}(\mathbb{R}^q)$ is defined as follows:*

(iii) $\mathrm{CB}_3^{(2m)}(\lambda; \mathbb{R}^q)$ is the space of all functions $H \in \mathrm{C}^{(2m)}(\mathbb{R}^q)$ such that the integrals

$$\int_{\mathbb{R}^q} H(x)\overline{\Phi_h(x)} \, dV(x) \tag{10.38}$$

exist for all $h \in \Lambda^{-1}$ with $(\Delta)^\wedge(h) = 4\pi h^2 = \lambda$, i.e., $(\Delta + \lambda)^\wedge(h) = 0$, in the (spherical) sense

$$\int_{\mathbb{R}^q} H(x)\overline{\Phi_h(x)} \, dV(x) = \lim_{N \to \infty} \int_{\substack{|x| \leq N \\ x \in \mathbb{R}^q}} H(x)\overline{\Phi_h(x)} \, dV(x). \tag{10.39}$$

In fact, for a value $\lambda \in \mathrm{Spect}_\Delta(\Lambda)$ in connection with the structure of $\mathrm{CP}_3^{(2m)}(\lambda; \Lambda; \mathbb{R}^q)$, we are able to formulate the following convergence theorem.

Theorem 10.4. *Let Λ be an arbitrary lattice in $\mathbb{R}^q$. For given $m \in \mathbb{N}$, $\varepsilon > 0$, and $\lambda \in \mathbb{R}$, suppose that $F \in \mathrm{C}^{(2m)}(\mathbb{R}^q)$, $m \in \mathbb{N}$, is a member of the class $\mathrm{CP}_1^{(2m)}(\lambda, \mathbb{R}^q) \cap \mathrm{CP}_2^{(2m)}(\varepsilon, \lambda, \mathbb{R}^q) \cap \mathrm{CP}_3^{(2m)}(\lambda; \Lambda; \mathbb{R}^q)$. Then the series*

$$\sum_{g \in \Lambda} F(g) = \lim_{N \to \infty} \sum_{\substack{|g| \leq N \\ g \in \Lambda}} F(g) \tag{10.40}$$

is convergent.

More explicitly, for a value $\lambda \notin \mathrm{Spect}_\Delta(\Lambda)$ and a function $F \in \mathrm{C}^{(2m)}(\mathbb{R}^q)$ belonging to $\mathrm{CP}_1^{(2m)}(\lambda, \mathbb{R}^q) \cap \mathrm{CP}_2^{(2m)}(\varepsilon, \lambda, \mathbb{R}^q)$, we have

$$\sum_{g \in \Lambda} F(g) = \int_{\mathbb{R}^q} G\left((\Delta + \lambda)^m ; x\right) (\Delta + \lambda)^m F(x) \, dV(x), \tag{10.41}$$

whereas, for a value $\lambda \in \text{Spect}_\Delta(\Lambda)$ and a function $F \in C^{(2m)}(\mathbb{R}^q)$ belonging to $\text{CP}_1^{(2m)}(\lambda, \mathbb{R}^q) \cap \text{CP}_2^{(2m)}(\varepsilon, \lambda, \mathbb{R}^q) \cap \text{CP}_3^{(2m)}(\lambda; \Lambda; \mathbb{R}^q)$, we have

$$\sum_{g \in \Lambda} F(g) = \frac{1}{\sqrt{\|\mathcal{F}\|}} \sum_{\substack{(\Delta+\lambda)^\wedge(h)=0 \\ h \in \Lambda^{-1}}} \int_{\mathbb{R}^q} F(x) \overline{\Phi_h(x)} \, dV(x) \tag{10.42}$$

$$+ \int_{\mathbb{R}^q} G\left((\Delta+\lambda)^m ; x\right) (\Delta+\lambda)^m F(x) \, dV(x).$$

10.3   Lattice Point-Generated Poisson Summation Formula

The convergence conditions (in Theorem 10.4) enable us to derive multi-dimensional analogues of the Poisson summation formula. As, for $m \in \mathbb{N}$ with $m > \frac{q}{2}$, the Λ-lattice function $G\left((\Delta+\lambda)^m ; \cdot\right)$ permits an absolutely and uniformly convergent Fourier series in $\mathbb{R}^q$, Lebesgue's theorem allows us to interchange summation and integration such that

$$\int_{\mathbb{R}^q} G\left((\Delta+\lambda)^m ; x\right) (\Delta+\lambda)^m F(x) \, dV(x) \tag{10.43}$$

$$= \frac{1}{\sqrt{\|\mathcal{F}\|}} \sum_{\substack{(\Delta+\lambda)^\wedge(h)\neq 0 \\ h \in \Lambda^{-1}}} \frac{1}{-((\Delta+\lambda)^m)^\wedge(h)} \int_{\mathbb{R}^q} (\Delta+\lambda)^m F(x) \overline{\Phi_h(x)} \, dV(x)$$

holds for a function $F \in \text{CP}_1^{(2m)}(\lambda; \mathbb{R}^q) \cap \text{CP}_2^{(2m)}(\varepsilon, \lambda; \mathbb{R}^q)$. Moreover, by observation of $F \in C_1^{(2m)}(\lambda; \mathbb{R}^q)$, repeated application of the Second Green Theorem yields

$$\int_{\mathbb{R}^q} \overline{\Phi_h(x)} (\Delta+\lambda)^m F(x) \, dV(x) \tag{10.44}$$

$$= \int_{\mathbb{R}^q} \left((\Delta+\lambda)^m \overline{\Phi_h(x)}\right) F(x) \, dV(x)$$

$$= -\left((\Delta+\lambda)^m\right)^\wedge(h) \int_{\mathbb{R}^q} \overline{\Phi_h(x)} F(x) \, dV(x).$$

Inserting (10.44) into (10.43) we therefore find

$$\int_{\mathbb{R}^q} G\left((\Delta+\lambda)^m ; x\right) (\Delta+\lambda)^m F(x) \, dV(x) \tag{10.45}$$

$$= \frac{1}{\sqrt{\|\mathcal{F}\|}} \sum_{\substack{(\Delta+\lambda)^\wedge(h)\neq 0 \\ h \in \Lambda^{-1}}} \int_{\mathbb{R}^q} F(x) \overline{\Phi_h(x)} \, dV(x).$$

This finally leads to the *multi-dimensional Poisson summation formula.*

Theorem 10.5. *Let Λ be an arbitrary lattice in the Euclidean space $\mathbb{R}^q$. If, for $\varepsilon > 0$ and $\lambda \in \mathbb{R}$, the function $F \in C^{(2m)}(\mathbb{R}^q)$, $m > \frac{q}{2}$, is a member of class $\mathrm{CP}_1^{(2m)}(\lambda; \mathbb{R}^q) \cap \mathrm{CP}_2^{(2m)}(\varepsilon, \lambda; \mathbb{R}^q)$, then*

$$\lim_{N \to \infty} \left(\sum_{\substack{|g| \leq N \\ g \in \Lambda}} F(g) - \frac{1}{\sqrt{\|\mathcal{F}\|}} \sum_{\substack{(\Delta+\lambda)^\wedge(h)=0 \\ h \in \Lambda^{-1}}} \int_{\substack{|x| \leq N \\ x \in \mathbb{R}^q}} F(x) \overline{\Phi_h(x)} \, dV(x) \right)$$
$$= \frac{1}{\sqrt{\|\mathcal{F}\|}} \sum_{\substack{(\Delta+\lambda)^\wedge(h) \neq 0 \\ h \in \Lambda^{-1}}} \int_{\mathbb{R}^q} F(x) \overline{\Phi_h(x)} \, dV(x). \tag{10.46}$$

More explicitly, if $\lambda \notin \mathrm{Spect}_\Delta(\Lambda)$ and $F \in \mathrm{CP}_1^{(2m)}(\lambda; \mathbb{R}^q) \cap \mathrm{CP}_2^{(2m)}(\varepsilon, \lambda; \mathbb{R}^q)$, then

$$\sum_{g \in \Lambda} F(g) = \frac{1}{\sqrt{\|\mathcal{F}\|}} \sum_{h \in \Lambda^{-1}} \int_{\mathbb{R}^q} F(x) \overline{\Phi_h(x)} \, dV(x). \tag{10.47}$$

If $\lambda \in \mathrm{Spect}_\Delta(\Lambda)$ and F is of class $\mathrm{CP}_1^{(2m)}(\lambda; \mathbb{R}^q) \cap \mathrm{CP}_2^{(2m)}(\varepsilon, \lambda; \mathbb{R}^q) \cap \mathrm{CP}_3^{(2m)}(\lambda; \Lambda; \mathbb{R}^q)$, then

$$\sum_{g \in \Lambda} F(g) = \frac{1}{\sqrt{\|\mathcal{F}\|}} \sum_{h \in \Lambda^{-1}} \int_{\mathbb{R}^q} F(x) \overline{\Phi_h(x)} \, dV(x). \tag{10.48}$$

The sum on the left side of the identities (10.47) and (10.48) is not necessarily absolutely convergent in $\mathbb{R}^q$, so the process of summation must be specified. Following our approach the convergence of the series on the left side in Theorem 10.5 is understood in the spherical sense (10.1).

Remark 10.1. *The "wave number" $\lambda \in \mathbb{R}$ reflects the specific character of the subspaces $\mathrm{CP}_1^{(2m)}(\lambda; \mathbb{R}^q)$, $\mathrm{CP}_2^{(2m)}(\varepsilon, \lambda; \mathbb{R}^q)$, and $\mathrm{CP}_3^{(2m)}(\lambda; \Lambda; \mathbb{R}^q)$ of $C^{(2m)}(\mathbb{R}^q)$. In fact, $\lambda \in \mathbb{R}$ can be adapted specifically to the oscillating properties of a function under consideration. In consequence, our sufficient conditions to establish the validity of the Poisson summation formula are particularly suited for the discussion of multi-dimensional alternating series.*

Case-by-case studies within the framework of Hardy–Landau (alternating) identities are given in the next sections.

10.4   Classical Two-Dimensional Hardy–Landau Identity

Observing the sinc-representation (6.477) in terms of Bessel functions

$$
\mathrm{sinc}(2\pi h R) = J_0(3; 2\pi h R) == \sqrt{\frac{\pi}{2}}\,\frac{J_{\frac{1}{2}}(2; 2\pi h R)}{\sqrt{2\pi h R}} = \frac{J_1(1; 2\pi h R)}{2\pi h R}
$$

$$(10.49)$$

the *one-dimensional Hardy–Landau identity* (see (4.242)) can be rewritten in the form

$$
\sideset{}{'}\sum_{\substack{|g|\leq R \\ g\in\mathbb{Z}}} 1 = 2R\sum_{h\in\mathbb{Z}} \frac{J_1(1; 2\pi h R)}{2\pi h R}. \tag{10.50}
$$

In what follows we are interested in the classical *two-dimensional Hardy–Landau identity*. Our work starts with the proof of the two-dimensional identity for arbitrary lattices Λ, however, in strict verification of the sufficient convergence criteria as proposed in Section 10.2. In Section 10.5, we turn over to multi-dimensional extensions of Hardy–Landau identities on arbitrary lattices $\Lambda \subset \mathbb{R}^q$.

The point of departure is the infinitely often differentiable function F_R (see (6.474)) given by

$$
F_R(x) = \frac{J_1(2; 2\pi|x|R)}{|x|} = \frac{J_1(2\pi|x|R)}{|x|}, \quad x \in \mathbb{R}^2, \tag{10.51}
$$

where R is a positive number and, in accordance with our theory of Bessel functions.

Our approach is essentially based on the differential equation (10.52) relating the Bessel function J_1 of order 1 to the Bessel function J_2 of order 2 (of course, both of dimension $q = 2$)

$$
\Delta_x \frac{J_1(2\pi|x|R)}{|x|} + 4\pi^2 R^2 \frac{J_1(2\pi|x|R)}{|x|} = 4\pi R \frac{J_2(2\pi|x|R)}{|x|^2}. \tag{10.52}
$$

This equation indicates to use the "wave number" $\lambda = 4\pi^2 R^2$ as an adaptive parameter to the oscillating properties of the function F_R as defined by (10.51). In fact, the well known asymptotic estimates known from the theory of Bessel functions (Lemma 6.37) tell us that

$$
\frac{J_1(2\pi|x|R)}{|x|} = O\left(|x|^{-\frac{3}{2}}\right) \tag{10.53}
$$

and

$$
\left|\nabla_x \frac{J_1(2\pi|x|R)}{|x|}\right| = O\left(|x|^{-\frac{3}{2}}\right) \tag{10.54}
$$

are valid for $|x| \to \infty$. Furthermore, we see from (10.52) that

$$(\Delta_x + 4\pi^2 R^2)\frac{J_1(2\pi|x|R)}{|x|} = O\left(|x|^{-\frac{5}{2}}\right) \tag{10.55}$$

for $|x| \to \infty$. Hence, for the particular choice

$$\varepsilon = \frac{1}{2}, \quad \lambda = 4\pi^2 R^2, \tag{10.56}$$

the function F_R belongs to $\mathrm{CP}_1^{(2m)}(\lambda; \mathbb{R}^2) \cap \mathrm{CP}_1^{(2m)}(\varepsilon, \lambda; \mathbb{R}^2)$, where $m \in \mathbb{N}$ can be taken arbitrarily. Theorem 10.3, therefore, enables us to conclude that the limit

$$\lim_{N\to\infty}\left(\sum_{\substack{|g|\le N \\ g\in\Lambda}}\frac{J_1(2\pi|g|R)}{|g|} - \frac{2\pi}{\|\mathcal{F}\|}\sum_{\substack{|h|=R \\ h\in\Lambda^{-1}}}\int_0^N J_1(2\pi r R)J_0(2\pi|h|r)\ dr\right) \tag{10.57}$$

exists in the indicated spherical sense. Moreover, in connection with (10.52) we have

$$\lim_{N\to\infty}\left(\sum_{\substack{|g|\le N \\ g\in\Lambda}}\frac{J_1(2\pi|g|R)}{|g|} - \frac{2\pi}{\|\mathcal{F}\|}\sum_{\substack{|h|=R \\ h\in\Lambda^{-1}}}\int_0^N J_1(2\pi r R)J_0(2\pi|h|r)\ dr\right) \tag{10.58}$$

$$= 4\pi R \int_0^\infty \frac{J_2(2\pi r R)}{r^2}\left(\int_{\substack{|x|=r \\ x\in\mathbb{R}^2}} G\left(\Delta + 4\pi^2 R^2; x\right)\ dS(x)\right)\ dr.$$

The last integral can be rewritten in the form

$$4\pi R \int_0^\infty \frac{J_2(2\pi r R)}{r^2}\int_{\substack{|x|=r \\ x\in\mathbb{R}^2}} G\left(\Delta + 4\pi^2 R^2; x\right)\ dS(x)\ dr \tag{10.59}$$

$$= -\frac{2\pi}{\|\mathcal{F}\|}\sum_{\substack{|h|\ne R \\ h\in\Lambda^{-1}}}\frac{4\pi R}{-4\pi^2 h^2 + 4\pi^2 R^2}\int_0^\infty \frac{J_2(2\pi r R)}{r}J_0(2\pi|h|R)\ dr.$$

Observing the identity

$$\frac{4\pi R}{4\pi^2 h^2 - 4\pi^2 R^2}\int_0^\infty J_2(2\pi r R)J_0(2\pi|h|r)\frac{dr}{r} = \int_0^\infty J_1(2\pi r R)J_0(2\pi|h|r)\ dr \tag{10.60}$$

we get (cf. Section 7.4)

$$
\lim_{N \to \infty} \left(\sum_{\substack{|g| \le N \\ g \in \Lambda}} \frac{J_1(2\pi|g|R)}{|g|} - \frac{2\pi}{\|\mathcal{F}\|} \sum_{\substack{|h|=R \\ h \in \Lambda^{-1}}} \int_0^N J_1(2\pi r R) J_0(2\pi|h|r)\, dr \right)
$$
$$
= \frac{2\pi}{\|\mathcal{F}\|} \sum_{\substack{|h| \ne R \\ h \in \Lambda^{-1}}} \int_0^\infty J_1(2\pi r R) J_0(2\pi|h|r)\, dr.
$$

(10.61)

Now, for arbitrary $R > 0$, the Hankel transform of discontinuous integrals (see (7.139)) guarantees the existence of all occuring integrals such that

$$
\lim_{N \to \infty} \sum_{\substack{|g| \le N \\ g \in \Lambda}} \frac{J_1(2\pi|g|R)}{|g|} = \frac{2\pi}{\|\mathcal{F}\|} \sum_{\substack{|h|=R \\ h \in \Lambda^{-1}}} \int_0^\infty J_1(2\pi r R) J_0(2\pi|h|r)\, dr
$$
$$
+ \frac{2\pi}{\|\mathcal{F}\|} \sum_{\substack{|h| \ne R \\ h \in \Lambda^{-1}}} \int_0^\infty J_1(2\pi r R) J_0(2\pi|h|r)\, dr.
$$

(10.62)

Even more, the theory of discontinuous integrals (see Lemma 7.11) yields

$$
2\pi R \int_0^\infty J_1(2\pi r R) J_0(2\pi|h|r)\, dr = \begin{cases} 1 & , \quad |h| < R \\ \frac{1}{2} & , \quad |h| = R \\ 0 & , \quad |h| > R. \end{cases}
$$

(10.63)

Hence, we find

$$
\lim_{N \to \infty} \sum_{\substack{|g| \le N \\ g \in \Lambda}} \frac{J_1(2\pi|g|R)}{|g|} = \frac{2\pi}{\|\mathcal{F}\|} \sum_{h \in \Lambda^{-1}} \int_0^\infty J_1(2\pi r R) J_0(2\pi|h|r)\, dr
$$
$$
= \frac{1}{\|\mathcal{F}\| R} \sum_{\substack{|h| \le R \\ h \in \Lambda^{-1}}} {}' 1.
$$

(10.64)

Summarizing our results we therefore obtain

Theorem 10.6. *For arbitrary, but fixed radius $R > 0$, and arbitrary lattices $\Lambda \subset \mathbb{R}^2$*

$$
\lim_{N \to \infty} \sum_{\substack{|g| \le N \\ g \in \Lambda}} \frac{J_1(2\pi|g|R)}{|g|} = \frac{1}{\|\mathcal{F}\| R} \sum_{\substack{|h| \le R \\ h \in \Lambda^{-1}}} {}' 1.
$$

(10.65)

Replacing $\Lambda \subset \mathbb{R}^2$ by its inverse lattice $\Lambda^{-1} \subset \mathbb{R}^2$ we finally get the *Hardy–Landau identity* in its canonical form

$$
\sum_{\substack{|g| \le R \\ g \in \Lambda}} {}' 1 = \lim_{N \to \infty} \frac{R}{\|\mathcal{F}\|} \sum_{\substack{|h| \le N \\ h \in \Lambda^{-1}}} \frac{J_1(2\pi|h|R)}{|h|}.
$$

(10.66)

Observing $J_1(r) = \frac{r}{2} + \ldots$ we are able to rewrite the identity (10.66) in its standard form comparing the number of lattice points inside a circle around the origin of radius R with the area of the circle under explicit specification of the remaining Hardy–Landau series.

Corollary 10.1. *(Classical Two-Dimensional Hardy–Landau Identity) For all positive numbers R and for each lattice $\Lambda \subset \mathbb{R}^2$*

$$\sum_{\substack{|g| \leq R \\ g \in \Lambda}}{}' 1 = \frac{\pi}{\|\mathcal{F}\|} R^2 + \lim_{N \to \infty} \frac{R}{\|\mathcal{F}\|} \sum_{\substack{0 < |h| \leq N \\ h \in \Lambda^{-1}}} \frac{J_1(2\pi|h|R)}{|h|} . \tag{10.67}$$

As already mentioned, for $\Lambda = \mathbb{Z}^2$, the identity (10.67) is found in G.H. Hardy [1915], E. Landau [1915], G.H. Hardy, E. Landau [1924].

Remark 10.2. *Using the well known asymptotic relation from the theory of Bessel functions (see Lemma 6.38)*

$$J_1(2\pi|h|R) = \frac{(|h|R)^{-\frac{1}{2}}}{\pi} \cos\left(2\pi|h|R - \frac{3\pi}{4}\right) + O\left((|h|R)^{-\frac{3}{2}}\right) \tag{10.68}$$

we are able to state that, apart from an additive remainder term, the Hardy–Landau series

$$\frac{R}{\|\mathcal{F}\|} \sum_{\substack{0 < |h| \\ h \in \Lambda^{-1}}} \frac{J_1(2\pi|h|R)}{|h|} \tag{10.69}$$

is equal to

$$\frac{R^{\frac{1}{2}}}{\pi\|\mathcal{F}\|} \sum_{\substack{0 < |h| \\ h \in \Lambda^{-1}}} \frac{\cos\left(2\pi|h|R - \frac{3\pi}{4}\right)}{|h|^{\frac{3}{2}}} . \tag{10.70}$$

In a certain sense, all investigations involving the classical two-dimensional Hardy–Landau identity, which have been done during the past century, are based on (partial sums of) the series (10.70).

10.5 Multi-Dimensional Hardy–Landau Identities

Next we apply the sufficient convergence criteria (i.e., Theorem 10.5) to higher-dimensional extensions of the Hardy–Landau identity (Corollary 10.1). As the point of departure we choose a product function consisting of a radial and an angular weight function, i.e., the infinitely often differentiable function $F_R : \mathbb{R}^q \to \mathbb{R}$, $R > 0$ fixed, given by

$$F_R(x) = \frac{J_\nu(q; 2\pi|x|R)}{|x|^\nu} H_n(q; x), \tag{10.71}$$

where $H_n(q; \cdot)$ is a homogeneous, harmonic polynomial of degree $n \in \mathbb{N}_0$ in q dimensions (as introduced in Section 6.3) and $J_\nu(q; \cdot)$ is the Bessel function of order $\nu > 0$ and dimension q (as introduced in Section 6.7).

Expanding $J_\nu(q; 2\pi r R)$ in powers of its argument we have

$$\frac{J_\nu(q; 2\pi r R)}{r^\nu} = \frac{(\pi R)^\nu \Gamma\left(\frac{q}{2}\right)}{\Gamma\left(\nu + \frac{q}{2}\right)} \left(1 - \frac{(\pi r R)^2}{1!(\nu + \frac{q}{2})} + - \cdots \right). \tag{10.72}$$

For $n \in \mathbb{N}$ and $r \to \infty$ the standard estimates of the theory of Bessel functions (i.e., Lemma 6.37) yield the relation

$$
\begin{aligned}
J_\nu(q; 2\pi R r) &= J_n(q; 2\pi R r) \, \cos\left((\nu - n)\frac{\pi}{2}\right) \tag{10.73} \\
&\quad + J_{n+1}(q; 2\pi R r) \, \sin\left((\nu - n)\frac{\pi}{2}\right) \\
&\quad + O\left(r^{-\frac{q+1}{2}}\right).
\end{aligned}
$$

Under the assumption

$$q \geq 2, \ \nu - n > \frac{q-1}{2}, \ \nu > 0, \ n \in \mathbb{N}_0, \tag{10.74}$$

the theory of Bessel functions implies the asymptotic relations

$$
\begin{aligned}
(\Delta + \lambda)^k F_R(x) &= o(|x|^{1-q}), \ |x| \to \infty, \tag{10.75} \\
|\nabla(\Delta + \lambda)^k F_R(x)| &= o(|x|^{1-q}), \ |x| \to \infty
\end{aligned}
$$

for each non-negative integer k and all numbers $\lambda \in \mathbb{R}$. Moreover, from the differential equation of the Bessel function we are able to deduce that

$$\left| (\Delta + 4\pi^2 R^2) \frac{J_\nu(q; 2\pi |x| R)}{|x|^\nu} \right| = O\left(\frac{J_{\nu+1}(q; 2\pi |x| R)}{|x|^{\nu+1}} \right) \tag{10.76}$$

for $|x| \to \infty$. Thus, in view of the assumption (10.74), we obtain the estimate

$$(\Delta + \lambda) F_R(x) = O\left(|x|^{-(q+\varepsilon)} \right), \ |x| \to \infty, \tag{10.77}$$

under the special choice

$$\lambda = 4\pi^2 R^2, \quad \varepsilon = \nu - n - \frac{q-1}{2} > 0. \tag{10.78}$$

Altogether, Theorem 10.5 leads to

Lemma 10.3. *Under the assumption (10.74), $F_R \in C^{(\infty)}(\mathbb{R}^q)$ given by (10.71) belongs to $\mathrm{CP}_1^{(2m)}(\lambda; \mathbb{R}^q) \cap \mathrm{CP}_2^{(2m)}(\varepsilon, \lambda; \mathbb{R}^q)$, where $m \in \mathbb{N}$ can be* *chosen arbitrarily, such that*

$$\lim_{N\to\infty}\left(\sum_{\substack{|g|\leq N \\ g\in\Lambda}} F_R(g) \quad - \quad \frac{1}{\sqrt{\|\mathcal{F}\|}} \sum_{\substack{|h|=R \\ h\in\Lambda^{-1}}} \int_{\substack{|x|\leq N \\ x\in\mathbb{R}^q}} F_R(x)\overline{\Phi_h(x)}\,dV(x)\right)$$

$$= \frac{1}{\sqrt{\|\mathcal{F}\|}} \sum_{\substack{|h|\neq R \\ h\in\Lambda^{-1}}} \int_{\mathbb{R}^q} F_R(x)\overline{\Phi_h(x)}\,dV(x). \qquad (10.79)$$

For values R with $4\pi^2 R^2 \notin \mathrm{Spect}_\Delta(\Lambda)$, F_R already allows the identity

Lemma 10.4. *For all R with $4\pi^2 R^2 \notin \mathrm{Spect}_\Delta(\Lambda)$*

$$\lim_{N\to\infty} \sum_{\substack{|g|\leq N \\ g\in\Lambda}} F_R(g) = \frac{1}{\sqrt{\|\mathcal{F}\|}} \sum_{h\in\Lambda^{-1}} \int_{\mathbb{R}^q} F_R(x)\overline{\Phi_h(x)}\,dV(x). \qquad (10.80)$$

It remains to investigate the integrals on the right side of (10.80). Introducing standard polar coordinates we find

$$\frac{1}{\sqrt{\|\mathcal{F}\|}} \int_{\mathbb{R}^q} F_R(x)\overline{\Phi_h(x)}\,dV(x) \qquad (10.81)$$

$$= \frac{i^n}{\|\mathcal{F}\|} \|\mathbb{S}^{q-1}\|\, Y_n\left(q; \frac{h}{|h|}\right) \int_0^\infty J_\nu(q; 2\pi rR) J_n(q; 2\pi|h|r)\, r^{n-\nu+q-1}\,dr.$$

The theory of discontinuous integrals gives detailed information on the convergence of all terms

$$\int_0^\infty J_\nu(q; 2\pi rR) J_n(q; 2\pi|h|r) r^{n-\nu+q-1}\,dr. \qquad (10.82)$$

From (10.73) we deduce that the integrand is asymptotically equal to

$$r^{n-\nu+q-1}\, J_n(q; 2\pi|h|r) J_n(q; 2\pi Rr) \cos\left((\nu-n)\frac{\pi}{2}\right) \qquad (10.83)$$

$$= r^{n-\nu+q-1}\, J_n(q; 2\pi|h|r) J_{n+1}(q; 2\pi Rr) \sin\left((\nu-n)\frac{\pi}{2}\right)$$

$$+ O\left(r^{n-\nu-1}\right).$$

Hence, under the following conditions imposed on the values ν, n

(i) $4\pi^2 R^2 \notin \mathrm{Spect}_\Delta(\Lambda)$, *i.e.*, $R \neq |h|$ *for all* $h \in \Lambda^{-1}$:

$$\nu - n > 0, \quad \nu > 0. \qquad (10.84)$$

(ii) $4\pi^2 R^2 \in \mathrm{Spect}_\Delta(\Lambda)$, *i.e.*, $R = |h|$ *for some* $h \in \Lambda^{-1}$:

$$\nu - n > 1, \quad \nu > 0, \qquad (10.85)$$

or

$$\nu - n > 0, \quad \nu - n \ \text{odd integer,} \tag{10.86}$$

the integrals (10.82) are convergent. Moreover, simultaneously to the case $4\pi^2 R^2 \notin \text{Spect}_\Delta(\Lambda)$, the condition of the space $\text{CP}_1^{(2m)}(4\pi^2 R^2; \mathbb{R}^q)$ becomes transparent by (10.85) and (10.86) (more details on discontinuous integral theory can be found in the monograph by G.N. Watson [1944]).

In accordance with our approach we therefore have to distinguish two cases, namely $4\pi^2 R^2 \notin \text{Spect}_\Delta(\Lambda)$ and $4\pi^2 R^2 \in \text{Spect}_\Delta(\Lambda)$:

First we deal with the case $4\pi^2 R^2 \notin \text{Spect}_\Delta(\Lambda)$: under the assumptions (10.74) and (10.84), i.e., $\nu - n > \frac{q-1}{2}$, $\nu > 0, n \in \mathbb{N}_0$, the integrals (10.82) exist. Even more, they are known from the Hankel transform of discontinuous integrals (Section 7.4)

$$\frac{1}{\sqrt{\|\mathcal{F}\|}} \int_{\mathbb{R}^q} F_R(x) \overline{\Phi_h(x)} \, dV(x) \tag{10.87}$$

$$= \begin{cases} \dfrac{2i^n}{\|\mathbb{S}^{q-1}\|\|\mathcal{F}\|} \dfrac{\pi^{\nu-n}}{R^{2n-\nu+q}\Gamma(\nu-n)} \left(1 - \left(\dfrac{|h|}{R}\right)^2\right)^{\nu-n-1} H_n(q;h) & , \quad |h| < R, \\[3mm] 0 & , \quad |h| > R. \end{cases}$$

Therefore, as a first remarkable result, we obtain from (10.79)

Theorem 10.7. *Let Λ be an arbitrary lattice in $\mathbb{R}^q$. Then, for all numbers R with $4\pi^2 R^2 \notin \text{Spect}_\Delta(\Lambda)$ and for all values ν, n with $\nu - n > \frac{q-1}{2}$ and $\nu > 0$, $n \in \mathbb{N}_0$ (see (10.78)), we have*

$$\frac{2i^n \pi^{\nu-n}}{\|\mathbb{S}^{q-1}\|\|\mathcal{F}\|R^{2n-\nu+q}\Gamma(\nu-n)} \sum_{\substack{|h|<R \\ h\in\Lambda^{-1}}} \left(1 - \left(\frac{|h|}{R}\right)^2\right)^{\nu-n-1} H_n(q;h)$$

$$= \sum_{g\in\Lambda} \frac{J_\nu(q; 2\pi|g|R)}{|g|^\nu} H_n(q;g). \tag{10.88}$$

If n is an odd integer, both sides of the last identity are zero for reasons of symmetry. Therefore, the identity (10.88) is interesting only for even integers.

Next we turn over to the case $4\pi^2 R^2 \in \text{Spect}_\Delta(\Lambda)$: for all values ν, n with

$\nu - n > \frac{q-1}{2}$ and $\nu > 0$, $n \in \mathbb{N}_0$ the identity (10.79) leads to the limit relation

$$
\lim_{N \to \infty} \left(\sum_{\substack{|g| \leq N \\ g \in \Lambda}} \frac{J_\nu(q; 2\pi|g|R)}{|g|^\nu} H_n(q; g) \right.
$$

(10.89)

$$
\left. - \frac{1}{\|\mathcal{F}\|} \sum_{\substack{|h| = R \\ h \in \Lambda^{-1}}} \int_{\substack{|x| \leq N \\ x \in \mathbb{R}^q}} \frac{J_\nu(q; 2\pi|x|R)}{|x|^\nu} H_n(q; x) e^{-2\pi i h \cdot x} \, dV(x) \right)
$$

$$
= \frac{2i^n \pi^{\nu-n}}{\|\mathbb{S}^{q-1}\| \|\mathcal{F}\| R^{2n-\nu+q} \Gamma(\nu - n)} \sum_{\substack{|h| < R \\ h \in \Lambda^{-1}}} \left(1 - \left(\frac{|h|}{R} \right)^2 \right)^{\nu-n-1} H_n(q; h).
$$

Therefore, for values R with $4\pi^2 R^2 \in \mathrm{Spect}_\Delta(\Lambda)$, we obtain by combination of (10.84), (10.85), and (10.86)

Theorem 10.8. *Let Λ be an arbitrary lattice in $\mathbb{R}^q$. Let ν, n satisfy one of the following assumptions:*

$$
\nu - n > \max\left(1, \frac{q-1}{2} \right), \quad \nu > 0, \; n \in \mathbb{N}_0,
$$

(10.90)

or

$$
\nu - n > \frac{q-1}{2}, \quad \nu > 0, \; n \in \mathbb{N}_0, \; \nu - n \text{ is an odd integer.}
$$

(10.91)

Then the following q-dimensional analogue to the Hardy–Landau identity holds true:

$$
\frac{2i^n \pi^{\nu-n}}{\|\mathbb{S}^{q-1}\| \|\mathcal{F}\| R^{2n-\nu+q} \Gamma(\nu - n)} \sum_{\substack{|h| \leq R \\ h \in \Lambda^{-1}}}{}' \left(1 - \left(\frac{|h|}{R} \right)^2 \right)^{\nu-n-1} H_n(q; h)
$$

$$
= \sum_{g \in \Lambda} \frac{J_\nu(q; 2\pi|g|R)}{|g|^\nu} H_n(q; g).
$$

(10.92)

Replacing the inverse lattice $\Lambda^{-1} \subset \mathbb{R}^q$ by the lattice $\Lambda \subset \mathbb{R}^q$ we find

Corollary 10.2. *Under the assumptions of Theorem 10.8*

$$
\frac{2i^n \pi^{\nu-n}}{\|\mathbb{S}^{q-1}\| R^{2n-\nu+q} \Gamma(\nu - n)} \sum_{\substack{|g| \leq R \\ g \in \Lambda}}{}' \left(1 - \left(\frac{|g|}{R} \right)^2 \right)^{\nu-n-1} H_n(q; g)
$$

$$
= \frac{1}{\|\mathcal{F}\|} \sum_{h \in \Lambda^{-1}} \frac{J_\nu(q; 2\pi|h|R)}{|h|^\nu} H_n(q; h).
$$

(10.93)

Remark 10.3. *The remaining cases $q = 2$, $\frac{1}{2} < \nu - n < 1$, in dependence on $4\pi^2 R^2 \notin \operatorname{Spect}_\triangle(\Lambda)$ and $4\pi^2 R^2 \in \operatorname{Spect}_\triangle(\Lambda)$, respectively, are not discussed at this place. They are of a particular significance in the motivation of planar non-uniform lattice point distribution (see Section 14.1).*

We conclude this section with some important two-dimensional examples.

Example 10.1. *For $q = 2$ the homogeneous harmonic polynomials of degree $n \in \mathbb{N}_0$ are linear combinations of the functions*

$$(x_1, x_2)^T \mapsto (x_1 + ix_2)^n + (x_1 - ix_2)^n, \tag{10.94}$$

and

$$(x_1, x_2)^T \mapsto i^{-1} \left((x_1 + ix_2)^n - (x_1 - ix_2)^n \right). \tag{10.95}$$

Taking $\Lambda \subset \mathbb{R}^2$ simply as the unit lattice $\mathbb{Z}^2$ and setting as homogeneous harmonic polynomial

$$(x_1, x_2)^T \mapsto H_n(2; x) = H_n(2; x_1, x_2) = (x_1 + ix_2)^n, \quad n \in \mathbb{N}_0, \tag{10.96}$$

we are able to deduce from Corollary 10.2 with $\nu = 4k + 1$, $n = 4k, k \in \mathbb{N}_0$, the following identity (see also W. Freeden [1975, 1978a]):

$$\sum_{(n_1, n_2)^T \in \mathbb{Z}^2} \frac{J_{4k+1}(2\pi \sqrt{n_1^2 + n_2^2}\, R)}{\sqrt{n_1^2 + n_2^2}^{\,4k+1}} (n_1 + in_2)^{4k} \tag{10.97}$$

$$= \frac{1}{R^{\,4k+1}} \sum_{\substack{n_1^2 + n_2^2 \leq R^2 \\ (n_1, n_2)^T \in \mathbb{Z}^2}}{}' (n_1 + in_2)^{4k}.$$

Example 10.2. *For $k = 0$ we recognize the classical two-dimensional Hardy–Landau identity in its original form*

$$\sum_{(n_1, n_2)^T \in \mathbb{Z}^2} \frac{J_1(2\pi \sqrt{n_1^2 + n_2^2}R)}{\sqrt{n_1^2 + n_2^2}} = \frac{1}{R} \sum_{\substack{n_1^2 + n_2^2 \leq R^2 \\ (n_1, n_2)^T \in \mathbb{Z}^2}}{}' 1. \tag{10.98}$$

11

Lattice Ball Summation

CONTENTS

11.1 Lattice Ball-Generated Euler Summation Formulas 324
11.2 Lattice Ball Discrepancy Involving the Laplacian 328
11.3 Convergence Criteria and Theorems 331
11.4 Lattice Ball-Generated Poisson Summation Formula 337
11.5 Multi-Dimensional Hardy–Landau Identities 338

In this book we distinguish two kinds of summation over lattices (illustrated by Figure 11.1), namely lattice point-generated Euler summation formulas (Section 9.1) and lattice ball-generated Euler summation formulas (Section 11.1).

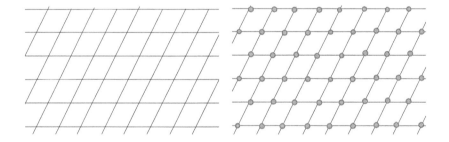

FIGURE 11.1
Lattice points (left), lattice balls (right).

The first kind of formulas are based on the constituting properties of the Λ-lattice functions for (iterated) Helmholtz operators, while the second kind uses ball-averaged integral variants of the Λ-lattice functions. It is not surprising that an integral average implies a smoothing effect; hence, the lattice ball summation offers a much better convergence behavior for the Fourier series of the ball-averaged integral variants of the Λ-lattice functions over infinite lattices than the lattice point summation.

This chapter is organized as follows: at the beginning (in Section 11.1) we develop ball-averaged variants of the Euler summation formula. In fact, we formulate the averages of the Λ-lattice function for (iterated) Laplace operators Δ^m, arbitrary lattices $\Lambda \subset \mathbb{R}^q$, and regular regions $\mathcal{G} \subset \mathbb{R}^q$. Based on the properties of the multi-dimensional ball-averaged Λ-lattice functions, modifications of the Euler summation formulas are developed just by use of the Second Green Theorem. Section 11.2 explains why the lattice ball-generated Euler summation formulas canonically produce representations of lattice ball discrepancies, which can be related to the lattice point-generated discrepancies (as discussed in Section 9.2). Multi-dimensional convergence theorems for weighted lattice ball sums are developed from (asymptotic) ball-averaged Euler summation formulas in Section 11.3. The occurring series are expressed in terms of integrals involving the ball-averaged Λ-lattice function for iterated Helmholtz operators $(\Delta + \lambda)^m$, $\lambda \in \mathbb{R}$, $m \in \mathbb{N}$, and arbitrary lattices $\Lambda \in \mathbb{R}^q$. Once more, a remarkable feature is that the "wave number" $\lambda \in \mathbb{R}$ can be chosen in close adaptation of the iterated Helmholtz operator $(\Delta + \lambda)^m$, $m \in \mathbb{N}$, to the specific properties of the (oscillating) weight function under consideration. The problem of modifying the already specified criteria known from the lattice point-generated Poisson formula to the lattice ball-generated case is undertaken within spherical summation. Finally, Section 11.5 gives a number of straightforward extensions of the Hardy–Landau identities within the lattice ball-generated framework.

11.1   Lattice Ball-Generated Euler Summation Formulas

In the previous chapter we used the (pointwise) Euler summation formula corresponding to the Laplace operator to sum up values of a given function on a set of lattice points inside and on the boundary surface of a regular region $\mathcal{G} \subset \mathbb{R}^q$. Now, the Euler summation formula corresponding to the Laplace operator is formulated for a sum of "mean values over (small) balls" with centers located at the lattice points of Λ. Consequently, our considerations are based on the so–called Λ-lattice τ-ball function, which is nothing else than an average of the Λ-lattice function over a (small) ball around lattice points.

Definition 11.1. *Let Λ be an arbitrary lattice in $\mathbb{R}^q$. For sufficiently small $\tau > 0$ (i.e., $\tau \in \mathbb{R}$ with $0 < \tau < \inf_{x \in \partial \mathcal{F}} |x|$), let $G_\tau(\Delta; \cdot)$ be the function defined by*

$$G_\tau(\Delta; x) = \frac{1}{\|\mathbb{B}_\tau^q\|} \int_{\substack{|y| \leq \tau \\ y \in \mathbb{R}^q}} G(\Delta; x - y) \, dV(y), \quad x \in \mathbb{R}^q \backslash \Lambda, \qquad (11.1)$$

where

$$\|\mathbb{B}_\tau^q\| = \int_{\substack{|y| \leq \tau \\ y \in \mathbb{R}^q}} dV(y) = \frac{\pi^{\frac{q}{2}}}{\Gamma\left(\frac{q}{2} + 1\right)} \tau^q. \qquad (11.2)$$

Then, $G_\tau(\Delta; \cdot)$ is called the Λ-lattice τ-ball function in $\mathbb{R}^q$ for Δ (sometimes briefly called lattice ball function in $\mathbb{R}^q$ for Δ).

From the definition of the Λ-lattice function (Definition 8.1) it is clear that the Λ-lattice τ-ball function in $\mathbb{R}^q$ for Δ is Λ-periodical, i.e.,

$$G_\tau(\Delta; x) = G_\tau(\Delta; x + g), \qquad x \in \mathbb{R}^q \backslash \Lambda, \; g \in \Lambda. \tag{11.3}$$

Lemma 8.4 provides an interesting result for the Laplace derivative of the τ-mean (introduced in Definition 11.1).

Theorem 11.1. *Let* $G_\tau(\Delta; \cdot)$ *be the Λ-lattice τ-ball function in $\mathbb{R}^q$ for Δ. Then, for all $x \in \mathcal{F}$, we have*

$$\Delta_x G_\tau(\Delta; x) = \begin{cases} \frac{1}{\|\mathbb{B}^q_\tau\|} - \frac{1}{\|\mathcal{F}\|}, & |x| \leq \tau, \\ -\frac{1}{\|\mathcal{F}\|}, & |x| > \tau. \end{cases} \tag{11.4}$$

Theorem 11.1 can be rewritten by use of the *characteristic lattice ball function* $\mathcal{X}_{\overline{\mathbb{B}^q_\tau}+\Lambda} : \mathbb{R}^q \to \mathbb{R}$ given by

$$\mathcal{X}_{\overline{\mathbb{B}^q_\tau}+\Lambda}(x) = \begin{cases} 1, & x \in \bigcup_{g \in \Lambda} \overline{\mathbb{B}^q_\tau(g)} \\ 0, & \text{otherwise.} \end{cases} \tag{11.5}$$

This yields the following representation of $\Delta G_\tau(\Delta; \cdot)$:

$$\Delta_x G_\tau(\Delta; x) = \frac{1}{\|\mathbb{B}^q_\tau\|} \left(\mathcal{X}_{\overline{\mathbb{B}^q_\tau}+\Lambda}(x) - \frac{\|\mathbb{B}^q_\tau\|}{\|\mathcal{F}\|} \right), \quad x \in \mathbb{R}^q. \tag{11.6}$$

The representation (11.6) is used to construct a variant of the Euler summation formula based on the Λ-lattice τ-ball function in $\mathbb{R}^q$ for Δ. Let $\mathcal{G}$ be a regular region in $\mathbb{R}^q$. Furthermore, let F be a function of class $C^{(2)}(\overline{\mathcal{G}})$, $\overline{\mathcal{G}} = \mathcal{G} \cup \partial\mathcal{G}$. Then partial integration (i.e., the Second Green Theorem) yields for every (sufficiently small) $\varepsilon > 0$,

$$\int_{\substack{x \in \overline{\mathcal{G}} \\ x \notin \mathbb{B}^q_\varepsilon+\Lambda}} \{F(x)(\Delta G_\tau(\Delta; x)) - G_\tau(\Delta; x)(\Delta F(x))\} \; dV(x)$$

$$= \int_{\substack{x \in \partial\mathcal{G} \\ x \notin \mathbb{B}^q_\varepsilon+\Lambda}} \left\{ F(x) \left(\frac{\partial}{\partial\nu} G_\tau(\Delta; x) \right) - G_\tau(\Delta; x) \left(\frac{\partial}{\partial\nu} F(x) \right) \right\} \; dS(x)$$

$$+ \sum_{\substack{g \in \overline{\mathcal{G}} \\ g \in \Lambda}} \int_{\substack{|x-g|=\varepsilon \\ x \in \overline{\mathcal{G}}}} \left\{ F(x) \left(\frac{\partial}{\partial\nu} G_\tau(\Delta; x) \right) - G_\tau(\Delta; x) \left(\frac{\partial}{\partial\nu} F(x) \right) \right\} \; dS(x),$$

$$\tag{11.7}$$

where - as always - ν is the outer (unit) normal field. By observation of the differential equation (11.6) we get for $\varepsilon \to 0$

$$\int_\mathcal{G} \left\{ F(x) \frac{1}{\|\mathbb{B}^q_\tau\|} \left(\mathcal{X}_{\overline{\mathbb{B}^q_\tau}+\Lambda}(x) - \frac{\|\mathbb{B}^q_\tau\|}{\|\mathcal{F}\|} \right) - G_\tau(\Delta; x) \; \Delta F(x) \right\} \; dV(x)$$

$$= \int_{\partial\mathcal{G}} \left\{ F(x) \left(\frac{\partial}{\partial\nu} G_\tau(\Delta; x) \right) - G_\tau(\Delta; x) \left(\frac{\partial}{\partial\nu} F(x) \right) \right\} \; dS(x) \,.$$

$$\tag{11.8}$$

In terms of the characteristic lattice ball function (see (11.5)) this leads to

Theorem 11.2. *(Lattice Ball-Generated Euler Summation Formula for the Laplace Operator Δ) Let Λ be an arbitrary lattice in $\mathbb{R}^q$. Let F be of class $C^{(2)}(\overline{\mathcal{G}})$, $\overline{\mathcal{G}} = \mathcal{G} \cup \partial\mathcal{G}$. Then, for sufficiently small $\tau > 0$, we have*

$$\frac{1}{\|\mathbb{B}_\tau^q\|} \int_\mathcal{G} \mathcal{X}_{\overline{\mathbb{B}_\tau^q} + \Lambda}(x) F(x) \, dV(x) = \frac{1}{\|\mathcal{F}\|} \int_\mathcal{G} F(x) \, dV(x) \qquad (11.9)$$

$$+ \int_\mathcal{G} G_\tau(\Delta; x) \, \Delta F(x) \, dV(x)$$

$$+ \int_{\partial\mathcal{G}} \left\{ F(x) \left(\frac{\partial}{\partial\nu} G_\tau(\Delta; x) \right) - G_\tau(\Delta; x) \left(\frac{\partial}{\partial\nu} F(x) \right) \right\} \, dS(x).$$

In fact, Theorem 11.2 is the lattice ball counterpart of the Euler summation formula related to the Λ-lattice ball function in $\mathbb{R}^q$ and the Laplace operator Δ.

Next we go over to *iterated lattice ball functions in $\mathbb{R}^q$*. Their definition is straightforward.

Definition 11.2. *For sufficiently small $\tau > 0$, let the function $G_\tau(\Delta^k; \cdot) : \mathbb{R}^q \to \mathbb{R}$, $k \in \mathbb{N}$, be defined by*

$$G_\tau(\Delta^k; x) = \frac{1}{\|\mathbb{B}_\tau^q\|} \int_{\substack{|y| \le \tau \\ y \in \mathbb{R}^q}} G(\Delta^k; x - y) \, dV_{(q)}(y), \quad x \in \mathbb{R}^q \backslash \Lambda, \qquad (11.10)$$

Then, $G_\tau(\Delta^k; \cdot)$ is called the Λ-lattice τ-ball function in $\mathbb{R}^q$ for Δ^k.

Again, it can be readily seen that the Λ-lattice τ-ball function in $\mathbb{R}^q$ for Δ^k is Λ-periodical, i.e.,

$$G_\tau(\Delta^k; x) = G_\tau(\Delta^k; x + g), \quad x \in \mathbb{R}^q \backslash \Lambda, \ g \in \Lambda. \qquad (11.11)$$

Observing the identity (see Lemma 6.30)

$$\int_{\substack{|x| \le \tau \\ x \in \mathbb{R}^q}} e^{-2\pi i h \cdot x} \, dV(x) = \|\mathbb{S}^{q-1}\| \, \tau^q \, \frac{J_1(q; 2\pi|h|\tau)}{2\pi|h|\tau} \qquad (11.12)$$

and using the bilinear expansion of $G(\Delta^k; \cdot)$ we are able to write down the (formal) Fourier series representation of $G_\tau(\Delta^k; \cdot)$, namely

$$\frac{\|\mathbb{S}^{q-1}\|}{\|\mathbb{B}_\tau^q\|} \frac{\tau^q}{\|\mathcal{F}\|} \sum_{\substack{((\Delta)^k)^\wedge(h) \ne 0 \\ h \in \Lambda^{-1}}} \frac{J_1(q; 2\pi\tau|h|)}{2\pi\tau|h|} \frac{e^{2\pi i h \cdot x}}{-((\Delta)^k)^\wedge(h)}. \qquad (11.13)$$

For $k > \frac{q-1}{4}$, the series (11.13) converges absolutely and uniformly in $\mathbb{R}^q$, so that $G_\tau(\Delta^k; \cdot)$ is continuous in $\mathbb{R}^q$, and we have

$$G_\tau(\Delta^k; x) = \frac{\|\mathbb{S}^{q-1}\|}{\|\mathbb{B}_\tau^q\|} \frac{\tau^q}{\|\mathcal{F}\|} \sum_{\substack{|h| > 0 \\ h \in \Lambda^{-1}}} \frac{J_1(q; 2\pi\tau|h|)}{2\pi\tau|h|} \frac{e^{2\pi i h \cdot x}}{(-4\pi^2 h^2)^k}. \qquad (11.14)$$

Next we want to extend Theorem 11.2 to iterated operators Δ^m, $m \in \mathbb{N}$, in canonical way. Suppose that F is $(2k+2)$-times, $k \in \{1, \ldots, m-1\}$, continuously differentiable in $\overline{\mathcal{G}}$. Then the Second Green Theorem yields

$$\int_{\mathcal{G}} G_\tau \left(\Delta^{k+1}; x \right) \Delta^{k+1} F(x) \, dV_{(q)}(x) \tag{11.15}$$

$$= \int_{\mathcal{G}} G_\tau \left(\Delta^k; x \right) \Delta^k F(x) \, dV_{(q)}(x)$$

$$+ \int_{\partial\mathcal{G}} G_\tau \left(\Delta^{k+1}; x \right) \left(\frac{\partial}{\partial\nu} \Delta^k F(x) \right) dS_{(q-1)}(x)$$

$$- \int_{\partial\mathcal{G}} \left(\frac{\partial}{\partial\nu} G_\tau \left(\Delta^{k+1}; x \right) \right) \Delta^k F(x) \, dS_{(q-1)}(x).$$

Summing up over all k from 1 to $m-1$ we find in the well known way known from the lattice point theory

$$\int_{\mathcal{G}} G_\tau \left(\Delta^m; x \right) \Delta^m F(x) \, dV_{(q)}(x) \tag{11.16}$$

$$= \int_{\mathcal{G}} G_\tau \left(\Delta \; ; x \right) \Delta F(x) \, dV_{(q)}(x)$$

$$+ \sum_{k=1}^{m-1} \int_{\partial\mathcal{G}} G_\tau \left(\Delta^{k+1}; x \right) \left(\frac{\partial}{\partial\nu} \Delta^k F(x) \right) dS_{(q-1)}(x)$$

$$- \sum_{k=1}^{m-1} \int_{\partial\mathcal{G}} \left(\frac{\partial}{\partial\nu} G_\tau \left(\Delta^{k+1}; x \right) \right) \Delta^k F(x) \, dS_{(q-1)}(x).$$

Combining this result with Theorem 11.2 we obtain

Theorem 11.3. *(Lattice Ball-Generated Euler Summation Formula for Iterated Laplace Operators Δ^m, $m \in \mathbb{N}$) Let Λ be an arbitrary lattice in $\mathbb{R}^q$. Let F be of class $C^{(2m)} \left(\overline{\mathcal{G}} \right)$, $m \in \mathbb{N}$, with $\mathcal{G}$ being a regular region. Then, for sufficiently small $\tau > 0$, we have*

$$\frac{1}{\|\mathbb{B}^q_\tau\|} \int_{\mathcal{G}} \mathcal{X}_{\overline{\mathbb{B}^q_\tau}+\Lambda}(x) F(x) \, dV(x) = \frac{1}{\|\mathcal{F}\|} \int_{\mathcal{G}} F(x) \, dV(x) \tag{11.17}$$

$$+ \int_{\mathcal{G}} G_\tau \left(\Delta^m; x \right) \Delta^m F(x) \, dV(x)$$

$$+ \sum_{k=0}^{m-1} \int_{\partial\mathcal{G}} \left(\frac{\partial}{\partial\nu} G_\tau \left(\Delta^{k+1}; x \right) \right) \Delta^k F(x) \, dS(x)$$

$$- \sum_{k=0}^{m-1} \int_{\partial\mathcal{G}} G_\tau \left(\Delta^{k+1}; x \right) \left(\frac{\partial}{\partial\nu} \Delta^k F(x) \right) dS(x).$$

The last identity is the Euler summation formula related to the Λ-lattice ball function in $\mathbb{R}^q$ for the iterated Laplacian Δ^m, $m \in \mathbb{N}$.

Remark 11.1. *The lattice ball-generated Euler summation formula for the iterated Laplace operators, i.e., Theorem 11.3, can be obtained by application of the τ-mean (introduced in Definition 11.1) to the lattice point-generated Euler summation formula (Theorem 9.3), too.*

11.2   Lattice Ball Discrepancy Involving the Laplacian

Lattice ball discrepancies can be introduced in a way analogous to lattice point discrepancies (see Section 9.5).

Definition 11.3. *Let Λ be a lattice in $\mathbb{R}^q$. Let $\mathcal{G}$ be a regular region in $\mathbb{R}^q$. Suppose that F is of class $\mathrm{C}^{(0)}(\overline{\mathcal{G}})$, $\overline{\mathcal{G}} = \mathcal{G} \cup \partial \mathcal{G}$. The difference $P_\tau(F; \overline{\mathcal{G}})$ given by*

$$P_\tau(F; \overline{\mathcal{G}}) = \frac{1}{\|\mathbb{B}_\tau^q\|} \int_{\mathcal{G}} \mathcal{X}_{\overline{\mathbb{B}_\tau^q} + \Lambda}(x) F(x)\, dV(x) - \frac{1}{\|\mathcal{F}\|} \int_{\mathcal{G}} F(x)\, dV(x) \quad (11.18)$$

is called the Λ-lattice τ-ball discrepancy of F in $\overline{\mathcal{G}}$.

From Theorem 11.3 we immediately obtain for the Λ-lattice τ-ball discrepancy of F in $\overline{\mathcal{G}}$

$$P_\tau(F; \overline{\mathcal{G}}) = \int_{\mathcal{G}} G_\tau\left(\Delta^m; x\right) \Delta^m F(x)\, dV(x) \quad (11.19)$$

$$+ \sum_{k=0}^{m-1} \int_{\partial \mathcal{G}} \left(\frac{\partial}{\partial \nu} G_\tau\left(\Delta^{k+1}; x\right)\right) \Delta^k F(x)\, dS(x)$$

$$- \sum_{k=0}^{m-1} \int_{\partial \mathcal{G}} G_\tau\left(\Delta^{k+1}; x\right) \left(\frac{\partial}{\partial \nu} \Delta^k F(x)\right) dS(x),$$

provided that F is of class $\mathrm{C}^{(2m)}(\overline{\mathcal{G}})$, $m \in \mathbb{N}$.

For a constant weight function, i.e., $F = 1$, the corresponding Λ-lattice τ-ball discrepancy of F in $\overline{\mathcal{G}}$ reads with $P_\tau(\overline{\mathcal{G}}) = P_\tau(1; \overline{\mathcal{G}})$

$$P_\tau(\overline{\mathcal{G}}) = \frac{1}{\|\mathbb{B}_\tau^q\|} \int_{\mathcal{G}} \mathcal{X}_{\overline{\mathbb{B}_\tau^q} + \Lambda}(x)\, dV(x) - \underbrace{\frac{1}{\|\mathcal{F}\|} \int_{\mathcal{G}} dV(x)}_{= \frac{\|\mathcal{G}\|}{\|\mathcal{F}\|}}. \quad (11.20)$$

From (11.19) it follows that

$$P_\tau(\overline{\mathcal{G}}) = \int_{\partial \mathcal{G}} \frac{\partial}{\partial \nu} G_\tau(\Delta; x)\, dS(x). \quad (11.21)$$

In other words, the Λ-lattice τ-ball discrepancy can be described by the boundary integral over the normal derivative of the Λ-lattice τ-ball function.

The formula (with $\mathcal{G} = \mathbb{B}^q_{\sqrt{N}}$, $q \geq 2$)

$$\frac{1}{\|\mathbb{B}^q_\tau\|} \int_{\substack{|x| \leq \sqrt{N} \\ x \in \mathbb{R}^q}} \mathcal{X}_{\overline{\mathbb{B}^q_\tau} + \Lambda}(x) \, dV(x) = \frac{1}{\|\mathcal{F}\|} \frac{\pi^{\frac{q}{2}}}{\Gamma(\frac{q}{2}+1)} N^{\frac{q}{2}} \tag{11.22}$$

$$+ \int_{\substack{|x| = \sqrt{N} \\ x \in \mathbb{R}^q}} \frac{\partial}{\partial \nu} G_\tau(\Delta; x) \, dS(x)$$

is a τ-weighted counterpart to (9.45). The second term on the right side of (11.22) can be expressed in terms of Bessel functions. Indeed, from (6.182), (6.427), and (6.429) we readily get

$$\int_{\substack{|x| = \sqrt{N} \\ x \in \mathbb{R}^q}} \frac{\partial}{\partial \nu} e^{2\pi i h \cdot x} \, dS_{(q-1)}(x) = -2\pi |h| \ N^{\frac{q-1}{2}} \ \|\mathbb{S}^{q-1}\| \, J_1(q; 2\pi |h| \sqrt{N}).$$

$$\tag{11.23}$$

Therefore, by use of (11.12) and (11.23), we are able to conclude that for sufficiently small $\tau > 0$

$$\int_{\substack{|x| = \sqrt{N} \\ x \in \mathbb{R}^q}} \frac{\partial}{\partial \nu_x} G_\tau(\Delta; x) \, dS(x) \tag{11.24}$$

$$= \frac{\|\mathbb{S}^{q-1}\|^2}{\|\mathbb{B}^q_\tau\|} \frac{N^{\frac{q-1}{2}} \tau^{q+1}}{\|\mathcal{F}\|} \sum_{\substack{|h| > 0 \\ h \in \Lambda^{-1}}} \frac{J_1\left(q; 2\pi\sqrt{N}|h|\right) J_1\left(q; 2\pi\tau|h|\right)}{4\pi^2 |h|^2},$$

such that

$$\frac{1}{\|\mathbb{B}^q_\tau\|} \int_{\substack{|x| \leq \sqrt{N} \\ x \in \mathbb{R}^q}} \mathcal{X}_{\overline{\mathbb{B}^q_\tau} + \Lambda}(x) \, dV(x) = \frac{1}{\|\mathcal{F}\|} \frac{\pi^{\frac{q}{2}}}{\Gamma(\frac{q}{2}+1)} N^{\frac{q}{2}} \tag{11.25}$$

$$+ \frac{\|\mathbb{S}^{q-1}\|^2}{\|\mathbb{B}^q_\tau\|} \frac{N^{\frac{q-1}{2}} \tau^{q+1}}{\|\mathcal{F}\|} \sum_{\substack{|h| > 0 \\ h \in \Lambda^{-1}}} \frac{J_1\left(q; 2\pi\sqrt{N}|h|\right) J_1\left(q; 2\pi\tau|h|\right)}{4\pi^2 |h|^2}.$$

Observing the asymptotic relation for the Bessel function $J_1(q; \cdot)$ we are able to see that the series on the right side of (11.24) is absolutely convergent in $\mathbb{R}^q$. Thus, for sufficiently small $\tau > 0$, we finally find in the lattice ball nomenclature

$$P_\tau\left(\overline{\mathbb{B}^q_{\sqrt{N}}}\right) = \int_{\substack{|x| = \sqrt{N} \\ x \in \mathbb{R}^2}} \frac{\partial}{\partial \nu_x} G_\tau(\Delta; x) \, dS(x) = O\left(N^{\frac{q-1}{4}}\right). \tag{11.26}$$

Summarizing our results we therefore obtain

Lemma 11.1. *The Λ-lattice τ-ball discrepancy in $\mathbb{B}^q_{\sqrt{N}}$ satisfies the asymptotic relation*

$$\frac{1}{\|\mathbb{B}^q_\tau\|} \int_{\substack{|x| \leq \sqrt{N} \\ x \in \mathbb{R}^q}} \mathcal{X}_{\overline{\mathbb{B}^q_\tau + \Lambda}}(x) \, dV(x) = \frac{1}{\|\mathcal{F}\|} \frac{\pi^{\frac{q}{2}}}{\Gamma(\frac{q}{2}+1)} N^{\frac{q}{2}} + O\left(N^{\frac{q-1}{4}}\right), \quad (11.27)$$

$N \to \infty$.

Even more, going over to the limit $\tau \to 0$ we are led to

$$P\left(\overline{\mathbb{B}^q_{\sqrt{N}}}\right) = \lim_{\substack{\tau \to 0 \\ \tau > 0}} \frac{\|S^{q-1}\|^2}{\|\mathbb{B}^q_\tau\|} \frac{N^{\frac{q-1}{2}} \tau^{q+1}}{\|\mathcal{F}\|} \sum_{\substack{|h| > 0 \\ h \in \Lambda^{-1}}} \frac{J_1\left(q; 2\pi\sqrt{N}|h|\right) J_1\left(q; 2\pi\tau|h|\right)}{4\pi^2 |h|^2},$$

$$(11.28)$$

such that one could argue that the lattice point discrepancy $P\left(\overline{\mathbb{B}^q_{\sqrt{N}}}\right)$ would behave like the lattice ball discrepancy $P_\tau\left(\overline{\mathbb{B}^q_{\sqrt{N}}}\right)$ for $N \to \infty$.

Table 11.1, however, gives the already known answer. It shows the comparison of the $\mathbb{Z}^q$-lattice ball discrepancy and the $\mathbb{Z}^q$-lattice point discrepancy for the dimensions $q \geq 2$. The values ε_q, $\varepsilon_q \leq \frac{q-1}{4}$, $q = 2, 3, 4$, respectively, represent positive values known from the asymptotic relations of Section 5.5.

$q = 2$	$P_\tau\left(\overline{\mathbb{B}^2_{\sqrt{N}}}\right) = O\left(N^{\frac{1}{4}}\right)$	$P\left(\overline{\mathbb{B}^2_{\sqrt{N}}}\right) = O\left(N^{\frac{1}{4}+\varepsilon_2}\right)$
$q = 3$	$P_\tau\left(\overline{\mathbb{B}^3_{\sqrt{N}}}\right) = O\left(N^{\frac{1}{2}}\right)$	$P\left(\overline{\mathbb{B}^3_{\sqrt{N}}}\right) = O\left(N^{\frac{1}{2}+\varepsilon_3}\right)$
$q = 4$	$P_\tau\left(\overline{\mathbb{B}^4_{\sqrt{N}}}\right) = O\left(N^{\frac{3}{4}}\right)$	$P\left(\overline{\mathbb{B}^4_{\sqrt{N}}}\right) = O\left(N^{1+\varepsilon_4}\right)$
$q \geq 5$	$P_\tau\left(\overline{\mathbb{B}^q_{\sqrt{N}}}\right) = O\left(N^{\frac{q-1}{4}}\right)$	$P\left(\overline{\mathbb{B}^q_{\sqrt{N}}}\right) = O\left(N^{\frac{q-2}{2}}\right)$

TABLE 11.1
The $\mathbb{Z}^q$-lattice ball discrepancy and the $\mathbb{Z}^q$-lattice point discrepancy.

In more detail, for all dimensions $q \geq 5$, we are able to formulate the following result for the $\mathbb{Z}^q$-lattice ball discrepancy and the $\mathbb{Z}^q$-lattice point discrepancy, respectively.

Theorem 11.4. *For $q \geq 5$ and sufficiently small (fixed) $\tau > 0$, the $\mathbb{Z}^q$-lattice ball discrepancy*

$$P_\tau\left(\overline{\mathbb{B}^q_{\sqrt{N}}}\right) = \frac{1}{\|\mathbb{B}^q_\tau\|} \int_{\substack{|x| \leq \sqrt{N} \\ x \in \mathbb{R}^q}} \mathcal{X}_{\overline{\mathbb{B}^q_\tau} + \mathbb{Z}^q}(x) \, dV(x) - \frac{\pi^{\frac{q}{2}}}{\Gamma(\frac{q}{2} + 1)} N^{\frac{q}{2}} \qquad (11.29)$$

can be expressed in the form

$$P_\tau\left(\overline{\mathbb{B}^q_{\sqrt{N}}}\right) = \int_{\substack{|x| = \sqrt{N} \\ x \in \mathbb{R}^q}} \frac{\partial}{\partial\nu} G_\tau\left(\Delta; x\right) \, dS(x), \qquad (11.30)$$

where $P_\tau\left(\overline{\mathbb{B}^q_{\sqrt{N}}}\right)$ satisfies the estimate

$$P_\tau\left(\overline{\mathbb{B}^q_{\sqrt{N}}}\right) = O\left(N^{\frac{q-1}{4}}\right), \quad N \to \infty. \qquad (11.31)$$

For comparison, the $\mathbb{Z}^q$-lattice point discrepancy

$$P\left(\overline{\mathbb{B}^q_{\sqrt{N}}}\right) = \sum_{\substack{|g| \leq \sqrt{N} \\ g \in \mathbb{Z}^q}}{}' 1 \quad - \quad \frac{\pi^{\frac{q}{2}}}{\Gamma(\frac{q}{2} + 1)} N^{\frac{q}{2}} \qquad (11.32)$$

allows the representation

$$P\left(\overline{\mathbb{B}^q_{\sqrt{N}}}\right) = \int_{\substack{|x| = \sqrt{N} \\ x \in \mathbb{R}^q}} \frac{\partial}{\partial\nu} G\left(\Delta; x\right) \, dS(x), \qquad (11.33)$$

where $P\left(\overline{\mathbb{B}^q_{\sqrt{N}}}\right)$ satisfies the estimate

$$P\left(\overline{\mathbb{B}^q_{\sqrt{N}}}\right) = O\left(N^{\frac{q-2}{2}}\right), \quad N \to \infty. \qquad (11.34)$$

11.3 Convergence Criteria and Theorems

The Λ-*lattice τ-ball function* for the operator $(\Delta + \lambda)^m$, $\lambda \in \mathbb{R}$, $m \in \mathbb{N}$, can be defined in a straightforward way as the ball-averaged counterpart of the corresponding Λ-lattice function. The corresponding Euler summation formula is straightforward.

Theorem 11.5. *Let Λ be an arbitrary lattice in $\mathbb{R}^q$. Let F be of class $\mathrm{C}^{(2m)}\left(\overline{\mathcal{G}}\right)$, $m \in \mathbb{N}$, with $\mathcal{G} \subset \mathbb{R}^q$ being a regular region. Then, for each sufficiently small $\tau > 0$ and each $\lambda \in \mathbb{R}$, we have*

$$\frac{1}{\|\mathbb{B}_\tau^q\|} \int_{\mathcal{G}} \mathcal{X}_{\overline{\mathbb{B}_\tau^q}+\Lambda}(x)\, F(x)\, dV(x) \tag{11.35}$$

$$= \frac{\|\mathbb{S}^{q-1}\|}{\|\mathbb{B}_\tau^q\|} \frac{\tau^q}{\sqrt{\|\mathcal{F}\|}} \sum_{\substack{(\Delta+\lambda)^\wedge(h)=0 \\ h \in \Lambda^{-1}}} \frac{J_1(q; 2\pi\tau|h|)}{2\pi\tau|h|} \int_{\mathcal{G}} F(y)\, \overline{\Phi_h(y)}\, dV(y)$$

$$+ \int_{\mathcal{G}} G_\tau\left((\Delta+\lambda)^m; x\right) (\Delta+\lambda)^m F(x)\, dV(x)$$

$$+ \sum_{k=0}^{m-1} \int_{\partial\mathcal{G}} \left(\frac{\partial}{\partial\nu} G_\tau\left((\Delta+\lambda)^{k+1}; x\right)\right) (\Delta+\lambda)^k F(x)\, dS(x)$$

$$- \sum_{k=0}^{m-1} \int_{\partial\mathcal{G}} G_\tau\left((\Delta+\lambda)^{k+1}; x\right) \left(\frac{\partial}{\partial\nu}(\Delta+\lambda)^k F(x)\right) dS(x),$$

where the finite sum on the right side of (11.35) occurs only if $\lambda \in \mathrm{Spect}_\Delta(\Lambda)$.

As in the lattice point theory (cf. Section 10.2) we start from a specialization of the Euler summation formula (i.e., Theorem 11.5) to balls $\mathbb{B}_N^q$ in order to derive convergence criteria as N tends toward infinity.

Corollary 11.1. *Let Λ be an arbitrary lattice in $\mathbb{R}^q$. Let F be of class $\mathrm{C}^{(2m)}\left(\overline{\mathbb{B}_N^q}\right)$, $m \in \mathbb{N}$, $N > 0$. Then, for all sufficiently small $\tau > 0$ and all $\lambda \in \mathbb{R}$, we have*

$$\frac{1}{\|\mathbb{B}_\tau^q\|} \int_{\substack{|x| \le N \\ x \in \mathbb{R}^q}} \mathcal{X}_{\overline{\mathbb{B}_\tau^q}+\Lambda}(x)\, F(x)\, dV(x) \tag{11.36}$$

$$= \frac{\|\mathbb{S}^{q-1}\|}{\|\mathbb{B}_\tau^q\|} \frac{\tau^q}{\sqrt{\|\mathcal{F}\|}} \sum_{\substack{(\Delta+\lambda)^\wedge(h)=0 \\ h \in \Lambda^{-1}}} \frac{J_1(q; 2\pi\tau|h|)}{2\pi\tau|h|} \int_{\substack{|x| \le N \\ x \in \mathbb{R}^q}} F(y)\, \overline{\Phi_h(y)}\, dV(y)$$

$$+ \int_{\substack{|x| \le N \\ x \in \mathbb{R}^q}} G_\tau\left((\Delta+\lambda)^m; x\right) (\Delta+\lambda)^m F(x)\, dV(x)$$

$$+ \sum_{k=0}^{m-1} \int_{\substack{|x|=N \\ x \in \mathbb{R}^q}} \left(\frac{\partial}{\partial\nu} G_\tau\left((\Delta+\lambda)^{k+1}; x\right)\right) (\Delta+\lambda)^k F(x)\, dS(x)$$

$$- \sum_{k=0}^{m-1} \int_{\substack{|x|=N \\ x \in \mathbb{R}^q}} G_\tau\left((\Delta+\lambda)^{k+1}; x\right) \left(\frac{\partial}{\partial\nu}(\Delta+\lambda)^k F(x)\right) dS(x).$$

Remark 11.2. *For $\lambda = 0$, Theorem 11.5 coincides with Theorem 11.3.*

In combination with our asymptotic relations known from the metaharmonic theory Corollary 11.1 enables us to derive convergence theorems for *lattice ball series* with radii $\tau > 0$ arbitrary, but fixed. The procedure is similar to the lattice point situation.

Definition 11.4. *Let $m \in \mathbb{N}$, $\varepsilon > 0$, and $\lambda \in \mathbb{R}$, be given values. Then the spaces $\mathrm{CB}_1^{(2m)}(\lambda; \mathbb{R}^q)$ and $\mathrm{CB}_2^{(2m)}(\varepsilon, \lambda; \mathbb{R}^q)$, respectively, are defined as follows:*

(i) $\mathrm{CB}_1^{(2m)}(\lambda; \mathbb{R}^q)$ is the space of all functions $H \in \mathrm{C}^{(2m)}(\mathbb{R}^q)$ such that the asymptotic relations

$$(\Delta + \lambda)^k H(x) = o\left(|x|^{\frac{1-q}{2}}\right), \quad |x| \to \infty, \tag{11.37}$$

$$|\nabla(\Delta + \lambda)^k H(x)| = o\left(|x|^{\frac{1-q}{2}}\right), \quad |x| \to \infty \tag{11.38}$$

are valid for $k = 0, \ldots, m-1$ (note that, for $m = 1$, the dependence on $\lambda \in \mathbb{R}$ is not reflected by (11.37), (11.38)).

(ii) $\mathrm{CB}_2^{(2m)}(\varepsilon, \lambda; \mathbb{R}^q)$ is the space of all functions $H \in \mathrm{C}^{(2m)}(\mathbb{R}^q)$ such that the asymptotic relation

$$(\Delta + \lambda)^m H(x) = O\left(|x|^{-\frac{q+1}{2} - \varepsilon}\right), \quad |x| \to \infty, \tag{11.39}$$

holds true.

The application of (11.37), (11.38) of Definition 11.4 to the boundary terms in Corollary 11.1 leads to the following asymptotic statement.

Lemma 11.2. *For given $m \in \mathbb{N}$ and $\lambda \in \mathbb{R}$, suppose that $F \in \mathrm{C}^{(k)}(\mathbb{R}^q)$ with $k \geq 2m + \frac{q}{2}$ belongs to $\mathrm{CB}_1^{(2m)}(\lambda; \mathbb{R}^q)$. Then*

$$\sum_{k=0}^{m-1} \int_{\substack{|x|=N \\ x \in \mathbb{R}^q}} \left(\frac{\partial}{\partial\nu} G_\tau\left((\Delta+\lambda)^{k+1};x\right)\right) (\Delta+\lambda)^k F(x)\, dS(x) \tag{11.40}$$

$$- \sum_{k=0}^{m-1} \int_{\substack{|x|=N \\ x \in \mathbb{R}^q}} \left(G_\tau\left((\Delta+\lambda)^{k+1};x\right)\right) \left(\frac{\partial}{\partial\nu}(\Delta+\lambda)^k F(x)\right) dS(x)$$

$$= o(1) \quad N \to \infty.$$

Proof. Observing the Fourier expansion for $G_\tau\left((\Delta+\lambda)^k; \cdot\right)$, $\tau > 0$ sufficiently small, we are left with absolutely and uniformly convergent series expansions involving integrals of the form

$$\int_{\substack{|x|=N \\ x \in \mathbb{R}^q}} \left(\frac{\partial}{\partial\nu} e^{2\pi i h \cdot x}\right) (\Delta+\lambda)^k F(x)\, dS(x) \tag{11.41}$$

and

$$\int_{\substack{|x|=N \\ x \in \mathbb{R}^q}} e^{2\pi i h \cdot x} \left(\frac{\partial}{\partial\nu}(\Delta+\lambda)^k F(x)\right) dS(x) \tag{11.42}$$

for $k = 0, \ldots, m-1$ and $N \to \infty$, respectively. From the metaharmonic theory (i.e., Corollary 6.7) we are able to deduce that the integrals of the type (11.41) are of the order

$$2\pi i |h| N^{q-1} \left(\frac{1}{|h|N} \right)^{\frac{q-1}{2}} \tag{11.43}$$

$$\times \left(i^{\frac{1-q}{2}} e^{2\pi i |h| N} (\Delta + \lambda)^k F \left(N \frac{h}{|h|} \right) \right.$$

$$\left. + i^{\frac{q-1}{2}} e^{-2\pi i |h| N} (\Delta + \lambda)^k F \left(-N \frac{h}{|h|} \right) \right)$$

$$+ o \left(\left(\frac{1}{|h|N} \right)^{\frac{q-1}{2}} \right),$$

while the integrals of the type (11.42) show an asymptotic behavior of the order

$$N^{q-1} \left(\frac{1}{|h|N} \right)^{\frac{q-1}{2}} \tag{11.44}$$

$$\times \left(i^{\frac{1-q}{2}} e^{2\pi i |h| N} \frac{\partial}{\partial \nu} (\Delta + \lambda)^k F \left(N \frac{h}{|h|} \right) \right.$$

$$\left. - i^{\frac{q-1}{2}} e^{-2\pi i |h| N} \frac{\partial}{\partial \nu} (\Delta + \lambda)^k F \left(-N \frac{h}{|h|} \right) \right)$$

$$+ o \left(\left(\frac{1}{|h|N} \right)^{\frac{q-1}{2}} \right).$$

Under the assumption of F we therefore see that the left side of (11.40) is of the order

$$N^{\frac{q-1}{2}} \sum_{k=0}^{m-1} \sum_{\substack{4\pi^2 h^2 \neq \lambda \\ h \in \Lambda^{-1}}} \frac{J_1(q; 2\pi |h| \tau)}{|h|^{\frac{q-1}{2}}} \frac{1}{|4\pi^2 h^2 - \lambda|^{k+1}} o \left(N^{\frac{1-q}{2}} \right), \quad N \to \infty. \tag{11.45}$$

This yields the desired asymptotic order $o(1)$.   □

The asymptotic estimate (11.39) in Definition 11.4 applied to the volume integral in Corollary 11.1 leads to the following statement.

Lemma 11.3. *For given* $m \in \mathbb{N}$, $\varepsilon > 0$, $\lambda \in \mathbb{R}$, *assume that* $F \in C^{(k)}(\mathbb{R}^q)$, $k \geq 2m + \frac{q}{2} + 1$, *belongs to* $CB_2^{(2m)}(\varepsilon, \lambda; \mathbb{R}^q)$. *Then*

$$\lim_{N \to \infty} \int_{\substack{|x| \leq N \\ x \in \mathbb{R}^q}} G_\tau \left((\Delta + \lambda)^m ; x \right) (\Delta + \lambda)^m F(x) \, dV(x) \tag{11.46}$$

is absolutely convergent.

Proof. For suitable positive constants M, N, apart from a (multiplicative) constant, the volume integral

$$\int_{\substack{M \leq |x| \leq N \\ x \in \mathbb{R}^q}} G_\tau \left((\Delta + \lambda)^m ; x \right) (\Delta + \lambda)^m F(x) \, dV(x) \tag{11.47}$$

is equal to

$$\sum_{\substack{4\pi^2 h^2 \neq \lambda \\ h \in \Lambda^{-1}}} \frac{1}{(\lambda - 4\pi^2 h^2)^m} \frac{J_1(q; 2\pi|h|\tau)}{2\pi|h|\tau} \int_{\substack{M \leq |x| \leq N \\ x \in \mathbb{R}^q}} e^{2\pi i h \cdot x} (\Delta + \lambda)^m F(x) \, dV(x).$$

$$\tag{11.48}$$

Introducing polar coordinates $x = r\xi$, $r = |x|$, $\xi \in \mathbb{S}^{q-1}$, we get

$$\int_{\substack{M \leq |x| \leq N \\ x \in \mathbb{R}^q}} e^{2\pi i (h \cdot x)} (\Delta + \lambda)^m F(x) \, dV(x) \tag{11.49}$$

$$= \int_M^N r^{q-1} \int_{\mathbb{S}^{q-1}} e^{2\pi i |h| r \left(\xi \cdot \frac{h}{|h|} \right)} (\Delta + \lambda)^m F(r\xi) \, dS(\xi) \, dr.$$

The occurring integral on the right side of (11.49) can be handled in the context of the metaharmonic theory by virtue of Corollary 6.7, provided that $F \in C^{(k)}(\mathbb{R}^q)$, $k \geq 2m + \frac{q}{2} + 1$, $m \in \mathbb{N}$, is of class $\text{CB}_2^{(2m)}(\varepsilon, \lambda; \mathbb{R}^q)$:

$$\int_{\substack{M \leq |x| \leq N \\ x \in \mathbb{R}^q}} G_\tau \left((\Delta + \lambda)^m ; x \right) (\Delta + \lambda)^m F(x) \, dV(x) \tag{11.50}$$

$$= O \left(\int_M^N r^{q-1} \frac{1}{r^{\frac{q-1}{2}}} \frac{1}{r^{\frac{q+1}{2} + \varepsilon}} \, dr \sum_{h \in \Lambda^{-1}} \frac{1}{|h|^{2m+q}} \right).$$

Consequently, the convergence of (11.48) and, hence, the convergence of (11.46) become obvious for $N \to \infty$. $\square$

Combining Lemma 11.2 and Lemma 11.3 we therefore obtain the following limit relation.

Theorem 11.6. *Let Λ be an arbitrary lattice in $\mathbb{R}^q$. For given $m \in \mathbb{N}$, $\varepsilon > 0$, and $\lambda \in \mathbb{R}$, assume that $F \in C^{(k)}(\mathbb{R}^q)$, $k \geq 2m + \frac{q}{2} + 1$, belongs to the class $\text{CB}_1^{(2m)}(\varepsilon, \lambda; \mathbb{R}^q) \cap \text{CB}_2^{(2m)}(\varepsilon, \lambda; \mathbb{R}^q)$. Then the limit*

$$\lim_{N \to \infty} \left(\frac{1}{\|\mathbb{B}_\tau^q\|} \int_{\substack{|x| \leq N \\ x \in \mathbb{R}^q}} \mathcal{X}_{\overline{\mathbb{B}_\tau^q} + \Lambda}(x) \, F(x) \, dV(x) \right.$$

$$\left. - \frac{\|\mathbb{S}^{q-1}\|}{\|\mathbb{B}_\tau^q\|} \frac{\tau^q}{\sqrt{\|\mathcal{F}\|}} \sum_{\substack{(\Delta + \lambda)^\wedge(h) = 0 \\ h \in \Lambda^{-1}}} \frac{J_1(q; 2\pi|h|\tau)}{2\pi\tau|h|} \int_{\substack{|x| \leq N \\ x \in \mathbb{R}^q}} F(x) \overline{\Phi_h(x)} \, dV(x) \right)$$

exists, and we have

$$
\lim_{N \to \infty} \left(\frac{1}{\|\mathbb{B}_\tau^q\|} \int_{\substack{|x| \leq N \\ x \in \mathbb{R}^q}} \mathcal{X}_{\overline{\mathbb{B}_\tau^q} + \Lambda}(x) \, F(x) \, dV(x) \right.
$$

$$
\left. - \frac{\|\mathbb{S}^{q-1}\|}{\|\mathbb{B}_\tau^q\|} \frac{\tau^q}{\sqrt{\|\mathcal{F}\|}} \sum_{\substack{(\Delta + \lambda)^\wedge (h) = 0 \\ h \in \Lambda^{-1}}} \frac{J_1(q; 2\pi |h| \tau)}{2\pi \tau |h|} \int_{\substack{|x| \leq N \\ x \in \mathbb{R}^q}} F(x) \overline{\Phi_h(x)} \, dV(x) \right)
$$

$$
= \int_{\mathbb{R}^q} G_\tau \left((\Delta + \lambda)^m ; x \right) \, (\Delta + \lambda)^m \, F(x) \, dV(x).
$$

Theorem 11.6 demonstrates that the convergence of multi-dimensional lattice ball sums (understood in spherical convergence)

$$
\frac{1}{\|\mathbb{B}_\tau^q\|} \int_{\mathbb{R}^q} \mathcal{X}_{\overline{\mathbb{B}_\tau^q} + \Lambda}(x) F(x) \, dV(x) = \lim_{N \to \infty} \frac{1}{\|\mathbb{B}_\tau^q\|} \int_{\substack{|x| \leq N \\ x \in \mathbb{R}^q}} \mathcal{X}_{\overline{\mathbb{B}_\tau^q} + \Lambda}(x) F(x) \, dV(x)
$$

$$
\tag{11.51}
$$

is closely related to the spectrum $\mathrm{Spect}_\Delta(\Lambda)$. As in the case of lattice point sums, by looking at the class of non-eigenvalues, i.e., $\lambda \notin \mathrm{Spect}_\Delta(\Lambda)$, we see that Theorem 11.6 immediately assures the convergence of the infinite "lattice ball series".

Corollary 11.2. *If $\varepsilon > 0$, $\lambda \notin \mathrm{Spect}_\Delta(\Lambda)$ and $F \in C^{(k)}(\mathbb{R}^q)$, $k \geq 2m + \frac{q}{2} + 1$, is a member of class $\mathrm{CB}_1^{(2m)}(\varepsilon, \lambda; \mathbb{R}^q) \cap \mathrm{CB}_2^{(2m)}(\varepsilon, \lambda; \mathbb{R}^q)$, then*

$$
\lim_{N \to \infty} \frac{1}{\|\mathbb{B}_\tau^q\|} \int_{\substack{|x| \leq N \\ x \in \mathbb{R}^q}} \mathcal{X}_{\overline{\mathbb{B}_\tau^q} + \Lambda}(x) \, F(x) \, dV(x)
$$

$$
\tag{11.52}
$$

$$
= \int_{\mathbb{R}^q} G_\tau \left((\Delta + \lambda)^m ; x \right) \, (\Delta + \lambda)^m \, F(x) \, dV(x).
$$

However, in order to discuss a value $\lambda \in \mathrm{Spect}_\Delta(\Lambda)$, we remember Definition 10.2.

Theorem 11.7. *Let Λ be an arbitrary lattice in $\mathbb{R}^q$. For given $m \in \mathbb{N}$, $\varepsilon > 0$, and $\lambda \in \mathbb{R}$ such that $\lambda \in \mathrm{Spect}_\Delta(\Lambda)$, assume that the function $F \in C^{(k)}(\mathbb{R}^q)$, $k \geq 2m + \frac{q}{2} + 1$, belongs to the class $\mathrm{CB}_1^{(2m)}(\lambda; \mathbb{R}^q) \cap \mathrm{CB}_2^{(2m)}(\varepsilon, \lambda; \mathbb{R}^q) \cap \mathrm{CP}_3^{(2m)}(\lambda; \Lambda; \mathbb{R}^q)$. Then*

$$
\frac{1}{\|\mathbb{B}_\tau^q\|} \int_{\mathbb{R}^q} \mathcal{X}_{\overline{\mathbb{B}_\tau^q} + \Lambda}(x) \, F(x) \, dV(x) = \lim_{N \to \infty} \frac{1}{\|\mathbb{B}_\tau^q\|} \int_{\substack{|x| \leq N \\ x \in \mathbb{R}^q}} \mathcal{X}_{\mathbb{B}_\tau^q + \Lambda}(x) \, F(x) \, dV(x)
$$

$$
\tag{11.53}
$$

is convergent.

Explaining our results in more detail we get the following statements:

for $\lambda \notin \mathrm{Spect}_\Delta(\Lambda)$ and $F \in \mathrm{C}^{(k)}(\mathbb{R}^q)$, $k \geq 2m + \frac{q}{2} + 1$, belonging to the class $\mathrm{CB}_1^{(2m)}(\lambda; \mathbb{R}^q) \cap \mathrm{CB}_2^{(2m)}(\varepsilon, \lambda; \mathbb{R}^q)$, we have

$$\frac{1}{\|\mathbb{B}_\tau^q\|} \int_{\mathbb{R}^q} \mathcal{X}_{\overline{\mathbb{B}_\tau^q} + \Lambda}(x)\, F(x)\, dV(x) = \int_{\mathbb{R}^q} G_\tau\left((\Delta + \lambda)^m; x\right)\, (\Delta + \lambda)^m\, F(x)\, dV(x),$$

(11.54)

while, for $\lambda \in \mathrm{Spect}_\Delta(\Lambda)$ and $F \in \mathrm{C}^{(k)}(\mathbb{R}^q)$, $k \geq 2m + \frac{q}{2} + 1$, belonging to the class $\mathrm{CB}_1^{(2m)}(\lambda; \mathbb{R}^q) \cap \mathrm{CB}_2^{(2m)}(\varepsilon, \lambda; \mathbb{R}^q) \cap \mathrm{CP}_3^{(2m)}(\lambda; \Lambda; \mathbb{R}^q)$, we have

$$\frac{1}{\|\mathbb{B}_\tau^q\|} \int_{\mathbb{R}^q} \mathcal{X}_{\overline{\mathbb{B}_\tau^q} + \Lambda}(x)\, F(x)\, dV(x) \qquad (11.55)$$

$$= \frac{\|\mathbb{S}^{q-1}\|}{\|\mathbb{B}_\tau^q\|} \frac{\tau^q}{\sqrt{\|\mathcal{F}\|}} \sum_{\substack{(\Delta + \lambda)^\wedge(h) = 0 \\ h \in \Lambda^{-1}}} \frac{J_1(q; 2\pi|h|\tau)}{2\pi\tau|h|} \int_{\mathbb{R}^q} F(x)\overline{\Phi_h(x)}\, dV(x)$$

$$+ \int_{\mathbb{R}^q} G_\tau\left((\Delta + \lambda)^m; x\right)\, (\Delta + \lambda)^m\, F(x)\, dV(x).$$

11.4 Lattice Ball-Generated Poisson Summation Formula

For all integers $m > \frac{q-1}{4}$, the convergence conditions (of Theorem 11.7) enable us to derive multi-dimensional analogues of the Poisson summation formula. As the function $G_\tau((\Delta + \lambda)^m; \cdot)$, $\tau > 0$ fixed and sufficiently small, permits an absolutely and uniformly convergent Fourier expansion for $m > \frac{q-1}{4}$, we are allowed to interchange summation and integration such that

$$\int_{\mathbb{R}^q} G_\tau\left((\Delta + \lambda)^m; x\right)\, (\Delta + \lambda)^m\, F(x)\, dV(x) \qquad (11.56)$$

$$= \frac{\|\mathbb{S}^{q-1}\|}{\|\mathbb{B}_\tau^q\|} \frac{\tau^q}{\sqrt{\|\mathcal{F}\|}} \sum_{\substack{(\Delta + \lambda)^\wedge(h) \neq 0 \\ h \in \Lambda^{-1}}} \frac{1}{-((\Delta + \lambda)^m)^\wedge(h)}$$

$$\times \frac{J_1(q; 2\pi|h|\tau)}{2\pi\tau|h|} \int_{\mathbb{R}^q} (\Delta + \lambda)^m\, F(x)\, \overline{\Phi_h(x)}\, dV(x)$$

provided that the function $F \in \mathrm{C}^{(k)}(\mathbb{R}^q)$, $k \geq 2m + \frac{q}{2} + 1$, is of the class $\mathrm{CB}_1^{(2m)}(\lambda; \mathbb{R}^q) \cap \mathrm{CB}_2^{(2m)}(\varepsilon, \lambda; \mathbb{R}^q) \cap \mathrm{CP}_3^{(2m)}(\lambda; \Lambda; \mathbb{R}^q)$. In addition, the Second Green Theorem yields

$$\int_{\mathbb{R}^q} \overline{\Phi_h(x)}\, (\Delta + \lambda)^m\, F(x)\, dV(x) \qquad (11.57)$$

$$= -((\Delta + \lambda)^m)^\wedge(h) \int_{\mathbb{R}^q} \overline{\Phi_h(x)}\, F(x)\, dV(x).$$

Inserting (11.57) into (11.56) we obtain the *lattice ball generated multi-dimensional Poisson summation formula* in the Euclidean space $\mathbb{R}^q$.

Theorem 11.8. *Let Λ be an arbitrary lattice in $\mathbb{R}^q$. For given $\varepsilon > 0$, $\lambda \in \mathbb{R}$, suppose that $F \in C^{(k)}(\mathbb{R}^q)$, $k \geq 2m + \frac{q}{2} + 1 > q + \frac{1}{2}$, is an element of the class* $\mathrm{CB}_1^{(2m)}(\lambda; \mathbb{R}^q) \cap \mathrm{CB}_2^{(2m)}(\varepsilon, \lambda; \mathbb{R}^q) \cap \mathrm{CP}_3^{(2m)}(\lambda; \Lambda; \mathbb{R}^q)$, *then*

$$\frac{1}{\|\mathbb{B}_\tau^q\|} \int_{\mathbb{R}^q} \mathcal{X}_{\overline{\mathbb{B}_\tau^q} + \Lambda}(x)\, F(x)\, dV(x) \tag{11.58}$$

$$= \frac{\|\mathbb{S}^{q-1}\|}{\|\mathbb{B}_\tau^q\|} \frac{\tau^q}{\sqrt{\|\mathcal{F}\|}} \sum_{h \in \Lambda^{-1}} \frac{J_1(q; 2\pi|h|\tau)}{2\pi|h|\tau} \int_{\mathbb{R}^q} F(x)\, \overline{\Phi_h(x)}\, dV(x).$$

11.5 Multi-Dimensional Hardy–Landau Identities

It is not surprising that the sufficient convergence criteria developed for the lattice ball-generated Poisson summation formula (i.e., Theorem 11.8) offer new variants of multi-dimensional Hardy–Landau identities. It turns out that Hardy–Landau lattice ball sums can be formulated under weaker asymptotic assumptions imposed on the occurring parameters of the weight functions than their corresponding lattice point counterparts, but at the cost of additional Bessel terms forcing the convergence of the Hardy–Landau series.

In the sequel, the multi-dimensional convergence criteria leading to Theorem 11.8 are applied to the infinitely often differentiable function $F_R : \mathbb{R}^q \to \mathbb{R}$, $R > 0$ fixed, given by (cf. (10.71))

$$F_R(x) = \frac{J_\nu(q; 2\pi|x|R)}{|x|^\nu}\, H_n(q; x), \tag{11.59}$$

where $H_n(q; \cdot)$ is a homogeneous, harmonic polynomial of degree n in q dimensions, i.e., $H_n(q; x) = |x|^n Y_n(q; \xi)$, $r = |x|$, $\xi \in \mathbb{S}^{q-1}$, and $J_\nu(q; \cdot)$ is the Bessel function of order $\nu > 0$ and dimension q.

We restrict ourselves to the relations (for comparison with the lattice point relations see (10.74))

$$q \geq 2, \quad \nu - n > 0, \quad \nu > 0, \quad n \in \mathbb{N}_0. \tag{11.60}$$

Then, under the conditions (11.60), it is not difficult to see from the well known estimates of Bessel functions that

$$(\Delta + \lambda)^k F_R(x) \;=\; o\left(|x|^{\frac{1-q}{2}}\right), \quad |x| \to \infty, \tag{11.61}$$

$$\left|\nabla(\Delta + \lambda)^k F_R(x)\right| \;=\; o\left(|x|^{\frac{1-q}{2}}\right), \quad |x| \to \infty$$

hold for all integers k and all values $\lambda \in \mathbb{R}$. From the differential equation of the Bessel function we already know (see (10.76)) that

$$(\Delta + 4\pi^2 R^2) \frac{J_\nu(q; 2\pi|x|R)}{|x|^\nu} = O\left(\frac{J_{\nu+1}(q; 2\pi|x|R)}{|x|^{\nu+1}}\right) \tag{11.62}$$

for $|x| \to \infty$; hence, in connection with (11.60), we obtain with $\varepsilon = \nu - n > 0$ and $\lambda = 4\pi^2 R^2$ the estimate

$$(\Delta + \lambda) F_R(x) = O\left(|x|^{-\frac{q+1}{2}-\varepsilon}\right), \quad |x| \to \infty. \tag{11.63}$$

Lemma 11.4. *Under the particular choice*

$$\lambda = 4\pi^2 R^2, \quad \varepsilon = \nu - n > 0, \tag{11.64}$$

$F_R \in C^{(\infty)}(\mathbb{R}^q)$ *belongs to* $CB_1^{(2m)}(\lambda; \mathbb{R}^q) \cap CB_2^{(2m)}(\varepsilon, \lambda; \mathbb{R}^q)$ *(with $m \in \mathbb{N}$ being arbitrary), and we have*

$$\lim_{N \to \infty} \left(\frac{1}{\|\mathbb{B}_\tau^q\|} \int_{\substack{|x| \le N \\ x \in \mathbb{R}^q}} \mathcal{X}_{\mathbb{B}_\tau^q + \Lambda}(x)\, F_R(x)\, dV(x) \right. \tag{11.65}$$

$$\left. - \frac{\|\mathbb{S}^{q-1}\|\, \tau^q}{\|\mathbb{B}_\tau^q\|\sqrt{\|\mathcal{F}\|}} \sum_{\substack{(\Delta+\lambda)^\wedge(h)=0 \\ h \in \Lambda^{-1}}} \frac{J_1(q; 2\pi|h|\tau)}{2\pi\tau|h|} \int_{\substack{|x| \le N \\ x \in \mathbb{R}^q}} F_R(x)\, \overline{\Phi_h(x)}\, dV(x) \right)$$

$$= \int_{\mathbb{R}^q} G_\tau\left((\Delta + \lambda)^m; x\right) (\Delta + \lambda)^m F_R(x)\, dV(x).$$

In accordance with our approach we have to distinguish two cases, namely $4\pi^2 R^2 \notin \mathrm{Spect}_\Delta(\Lambda)$ and $4\pi^2 R^2 \in \mathrm{Spect}_\Delta(\Lambda)$:

First we are concerned with the case $4\pi^2 R^2 \notin \mathrm{Spect}_\Delta(\Lambda)$: the identity (11.65) of Lemma 11.4 serves as the point of departure to formulate

Lemma 11.5. *For all R with $4\pi^2 R^2 \notin \mathrm{Spect}_\Delta(\Lambda)$ and under the conditions (11.64)*

$$\frac{1}{\|\mathbb{B}_\tau^q\|} \int_{x \in \mathbb{R}^q} \mathcal{X}_{\overline{\mathbb{B}_\tau^q} + \Lambda}(x)\, F_R(x)\, dV(x) \tag{11.66}$$

$$= \frac{\|\mathbb{S}^{q-1}\|}{\|\mathbb{B}_\tau^q\|} \frac{\tau^q}{\sqrt{\|\mathcal{F}\|}} \sum_{h \in \Lambda^{-1}} \frac{J_1(q; 2\pi|h|\tau)}{2\pi\tau|h|} \int_{x \in \mathbb{R}^q} F_R(x) \overline{\Phi_h(x)}\, dV(x).$$

As is well known, under the assumption (10.84), the discontinuous integrals (10.82) exist and their values are known from the Hankel transform of discontinuous integrals (see (10.87)). Thus we are able to formulate

Theorem 11.9. *Let Λ be an arbitrary lattice in $\mathbb{R}^q$. Then, for all real numbers R with $4\pi^2 R^2 \notin \mathrm{Spect}_\Delta(\Lambda)$, we have*

$$\frac{2 i^n \pi^{\nu-n}}{R^{2n-\nu+q} \Gamma(\nu-n)} \frac{1}{\|\mathbb{B}_\tau^q\|} \frac{\tau^q}{\|\mathcal{F}\|} \tag{11.67}$$

$$\times \sum_{\substack{|h|<R \\ h \in \Lambda^{-1}}} \frac{J_1(q; 2\pi|h|\tau)}{2\pi\tau|h|} \left(1 - \left(\frac{|h|}{R}\right)^2\right)^{\nu-n-1} H_n(q; h)$$

$$= \frac{1}{\|\mathbb{B}_\tau^q\|} \int_{x \in \mathbb{R}^q} X_{\overline{\mathbb{B}_\tau^q}+\Lambda}(x) \frac{J_\nu(q; 2\pi|x|R)}{|x|^\nu} H_n(q; x) \, dV(x).$$

Replacing the lattice $\Lambda \subset \mathbb{R}^q$ by its inverse lattice $\Lambda^{-1} \subset \mathbb{R}^q$ we find

$$\frac{2 i^n \pi^{\nu-n}}{R^{2n-\nu+q} \Gamma(\nu-n)} \frac{1}{\|\mathbb{B}_\tau^q\|} \tau^q \tag{11.68}$$

$$\times \sum_{\substack{|g|<R \\ h \in \Lambda^{-1}}} \frac{J_1(q; 2\pi|g|\tau)}{2\pi\tau|g|} \left(1 - \left(\frac{|g|}{R}\right)^2\right)^{\nu-n-1} H_n(q; g)$$

$$= \frac{1}{\|\mathcal{F}\|} \frac{1}{\|\mathbb{B}_\tau^q\|} \int_{x \in \mathbb{R}^q} X_{\overline{\mathbb{B}_\tau^q}+\Lambda^{-1}}(x) \frac{J_\nu(q; 2\pi|x|R)}{|x|^\nu} H_n(q; x) \, dV(x).$$

Next we discuss the case $4\pi^2 R^2 \in \mathrm{Spect}_\Delta(\Lambda)$: again, the identity (11.65) of Lemma 11.4 serves as the point of departure.

Theorem 11.10. *Let Λ be an arbitrary lattice in $\mathbb{R}^q$. Then, for values R with $4\pi^2 R^2 \in \mathrm{Spect}_\Delta(\Lambda)$, we have*

$$\lim_{N \to \infty} \left(\frac{1}{\|\mathbb{B}_\tau^q\|} \int_{\substack{|x|\leq N \\ x \in \mathbb{R}^q}} X_{\overline{\mathbb{B}_\tau^q}+\Lambda}(x) \, F_R(x) \, dV(x) \right. \tag{11.69}$$

$$\left. - \frac{\|\mathbb{S}^{q-1}\|}{\|\mathbb{B}_\tau^q\|} \frac{\tau^q}{\|\mathcal{F}\|} \sum_{\substack{|h|=R \\ h \in \Lambda^{-1}}} \frac{J_1(q; 2\pi|h|\tau)}{2\pi|h|\tau} \int_{\substack{|x|\leq N \\ x \in \mathbb{R}^q}} F_R(x) e^{-2\pi i h \cdot x} \, dV(x) \right)$$

$$= \frac{2 i^n \pi^{\nu-n} \tau^q}{\|\mathbb{B}_\tau^q\| \|\mathcal{F}\| R^{2n-\nu+q} \Gamma(\nu-n)}$$

$$\times \sum_{\substack{|h|<R \\ h \in \Lambda^{-1}}} \frac{J_1(q; 2\pi|h|\tau)}{2\pi|h|\tau} \left(1 - \left(\frac{|h|}{R}\right)^2\right)^{\nu-n-1} H_n(q; h).$$

Combining the already known results (10.84), (10.85), and (10.86) about the existence of the discontinuous integrals with the assumptions (11.64) of Theorem 11.10 we are able to come to the following conclusion.

Theorem 11.11. *Under the validity of one of the assumptions imposed on the values $\nu > 0$, $n \in \mathbb{N}_0$,*

$$\nu - n > 1 \tag{11.70}$$

or

$$\nu - n > 0, \quad \nu - n \ \text{odd integer} \tag{11.71}$$

the following q-dimensional analogue to the Hardy–Landau identity holds true

$$\frac{2i^n \pi^{\nu-n} \tau^q}{\|\mathbb{B}^q_\tau\| \ \|\mathcal{F}\| R^{2n-\nu+q} \Gamma (\nu - n)} \tag{11.72}$$

$$\times \sum_{\substack{|h|\leq R \\ h\in\Lambda^{-1}}}{}' \frac{J_1(q; 2\pi|h|\tau)}{2\pi|h|\tau} \left(1 - \left(\frac{|h|}{R}\right)^2\right)^{\nu-n-1} H_n(q; h)$$

$$= \lim_{N\to\infty} \frac{1}{\|\mathbb{B}^q_\tau\|} \int_{\substack{|x|\leq N \\ x\in\mathbb{R}^q}} \mathcal{X}_{\overline{\mathbb{B}^q_\tau}+\Lambda}(x) \frac{J_\nu(q; 2\pi|x|R)}{|x|^\nu} H_n(q; x) \, dV(x).$$

Replacing the lattice $\Lambda \subset \mathbb{R}^q$ by its inverse lattice $\Lambda^{-1} \subset \mathbb{R}^q$ we get

$$\frac{2i^n \pi^{\nu-n} \tau^q}{\|\mathbb{B}^q_\tau\| R^{2n-\nu+q} \Gamma (\nu - n)} \tag{11.73}$$

$$\times \sum_{\substack{|g|\leq R \\ g\in\Lambda^{-1}}}{}' \frac{J_1(q; 2\pi|g|\tau)}{2\pi|g|\tau} \left(1 - \left(\frac{|g|}{R}\right)^2\right)^{\nu-n-1} H_n(q; g)$$

$$= \frac{1}{\|\mathcal{F}\|} \lim_{N\to\infty} \frac{1}{\|\mathbb{B}^q_\tau\|} \int_{\substack{|x|\leq N \\ x\in\mathbb{R}^q}} \mathcal{X}_{\overline{\mathbb{B}^q_\tau}+\Lambda^{-1}}(x) \frac{J_\nu(q; 2\pi|x|R)}{|x|^\nu} H_n(q; x) \, dV(x).$$

Remark 11.3. *The remaining cases $q \geq 2$, $0 < \nu - n < 1$, in dependence on $4\pi^2 R^2 \notin \mathrm{Spect}_\Delta(\Lambda)$ and $4\pi^2 R^2 \in \mathrm{Spect}_\Delta(\Lambda)$, respectively, are not discussed in more detail here. They play a particular part in the motivation of non-uniform lattice ball distributions (see Section 14.1).*

Example 11.1. *For $q = 2$ and $\nu = 1, n = 0$ we find*

$$\frac{1}{\pi\tau} \sum_{\substack{|g|\leq R \\ g\in\Lambda}}{}' J_1(2\pi|g|\tau)|g| \tag{11.74}$$

$$= \frac{R}{\|\mathcal{F}\|} \lim_{N\to\infty} \frac{1}{\pi\tau^2} \int_{\substack{|x|\leq N \\ x\in\mathbb{R}^q}} \mathcal{X}_{\overline{\mathbb{B}^2_\tau}+\Lambda^{-1}}(x) \frac{J_1(2\pi|x|R)}{|x|} \, dV(x) \, ,$$

which may be regarded as the direct lattice ball counterpart of the classical Hardy–Landau identity (see Corollary 10.1).

Remark 11.4. *From (11.74) in connection with the estimate $J_1(r) = \frac{r}{2} + \ldots$ we obtain*

$$\sum_{\substack{|g| \leq R \\ g \in \Lambda}}{}' = \lim_{\substack{\tau \to 0 \\ \tau > 0}} \lim_{N \to \infty} \frac{R}{\|\mathcal{F}\|} \frac{1}{\pi \tau^2} \int_{\substack{|x| \leq N \\ x \in \mathbb{R}^q}} \mathcal{X}_{\mathbb{B}_\tau^2 + \Lambda - 1}(x) \, \frac{J_1(2\pi |x| R)}{|x|} \, dV(x) \, , \quad (11.75)$$

where we know from Section 10.4 that the limits on the right side of (11.75) can be interchanged, hence, leading to the classical Hardy–Landau identity (10.67).

12

Poisson Summation on Regular Regions

CONTENTS

12.1 Theta Function and Gauß–Weierstraß Summability 344
      Theta Function and Its Functional Equation 345
      Gauß–Weierstraß Summability on Regular Regions 346
12.2 Convergence Criteria for the Poisson Series 350
      Non-Negative Expansion Coefficients 352
      Homogeneous Boundary Weights 352
      Weighted Landau Formulas 354
12.3 Generalized Parseval Identity 355
12.4 Minkowski's Lattice Point Theorem 359

In this chapter we are concerned with the concept of *periodization under non-specified geometry*, that is to say, the formulation of the Poisson summation formula for regular regions $\mathcal{G} \subset \mathbb{R}^q$ such as "potatoes" (see Figure 12.1).

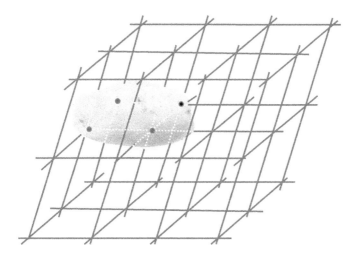

FIGURE 12.1
Lattice points of a 3D-lattice inside a regular region (such as a "potato").

The obvious problem is that the consistency of periodization realized in the Poisson summation formula cannot be detected in the same way for, e.g., potato-like regions as for the canonical "geometries" like the fundamental cell of a lattice or the whole Euclidean space $\mathbb{R}^q$. This is the reason why we are interested in establishing the Poisson summation formula on regular regions within the framework of simply structured summability, e.g., by Gauß–Weierstraß means. The point of departure is the q-dimensional functional equation of the Theta function. The essential tool for our approach is the Fourier inversion formula in the Gauß–Weierstraß context (cf. Section 7.4). Particular emphasis is laid on suitable concepts to overcome the summability imposed on the Poisson summation formula, for instance by requiring non-negative expansion coefficients or certain homogeneous boundary conditions.

The layout of the chapter can be described as follows: Section 12.1 investigates the lattice point Poisson summation formula for continuous functions on regular regions in $\mathbb{R}^q$, $q \geq 3$, however, in the Gauß–Weierstraß framework. Different strategies are studied to avoid the Gauß–Weierstraß summability (cf. Section 12.2). Vanishing weights on spheres are considered in more detail. The metaharmonic apparatus of Section 6.7 serves as the essential tool. Section 12.3 gives a formulation of the Parseval identity for regular regions and its specialization to lattice cells as well as to balls. As a particular application we present in Section 12.4 (an alternative proof of) the well known Minkowski Theorem of the geometry of numbers.

12.1   Theta Function and Gauß–Weierstraß Summability

The one-dimensional variant of the Theta function (see Definition 4.3) is reflected in an adequate way by the following multi-dimensional counterpart.

Definition 12.1. *For arbitrary points $x, y \in \mathbb{R}^q$, and arbitrary lattices $\Lambda \subset \mathbb{R}^q$ we call $\vartheta_n^{(q)}(\,\cdot\,;x,y;\Lambda)$ given by*

$$\vartheta_n^{(q)}(\sigma;x,y;\Lambda) = \sum_{g\in\Lambda} e^{-\pi\sigma|g-x|^2} H_n(q;g-x)\, e^{2\pi i g\cdot y}, \ \sigma \in \mathbb{C}, \ \Re(\sigma) > 0, \ (12.1)$$

the Theta function of degree n and dimension q.

Remark 12.1. *Clearly, $\vartheta_n^{(q)}(\,\cdot\,;x,y;\Lambda)$ is dependent on the homogeneous harmonic polynomial $H_n : x \mapsto H_n(q;x) = |x|^n Y_n(q;\xi)$, $x \in \mathbb{R}^q$, $x = |x|\xi$, $\xi \in \mathbb{S}^{q-1}$, of degree n and dimension q, where $Y_n(q;\cdot)$ is a member of class Harm_n.*

Theta Function and Its Functional Equation

For the function $F : \mathbb{R}^q \to \mathbb{C}$ given by

$$F(z) = e^{-\pi\sigma|z-x|^2} H_n(q; z-x) \, e^{2\pi i z \cdot y}, \quad z \in \mathbb{R}^q, \tag{12.2}$$

we obtain from the Poisson summation formula (e.g., Theorem 10.5)

$$\sum_{g \in \Lambda} F(g) = \frac{1}{\|\mathcal{F}\|} \sum_{h \in \Lambda^{-1}} \int_{\mathbb{R}^q} F(z) \, e^{2\pi i h \cdot z} \, dV(z). \tag{12.3}$$

When (12.3) is explicitly written out we are confronted with the following identity:

$$\vartheta_n^{(q)}(\sigma; x, y; \Lambda) \tag{12.4}$$

$$= \sum_{g \in \Lambda} e^{-\pi\sigma|g-x|^2} H_n(q; g-x) \, e^{2\pi i g \cdot y}$$

$$= \frac{1}{\|\mathcal{F}\|} \sum_{h \in \Lambda^{-1}} e^{2\pi i x \cdot (y+h)} \int_{\mathbb{R}^q} e^{-\pi\sigma|z-x|^2} H_n(q; z-x) e^{2\pi i(z-x)\cdot(y+h)} \, dV(z).$$

Introducing the polar coordinates $z - x = r\eta$ and $y + h = \rho\xi$ with $\eta, \xi \in \mathbb{S}^{q-1}$, we get

$$\vartheta_n^{(q)}(\sigma; x, y; \Lambda) \tag{12.5}$$

$$= \frac{1}{\|\mathcal{F}\|} \sum_{h \in \Lambda^{-1}} e^{2\pi i x \cdot (\rho\xi)}$$

$$\times \int_0^\infty \left(\int_{\mathbb{S}^{q-1}} Y_n(q; \eta) e^{2\pi i r \rho \xi \cdot \eta} \, dS(\eta) \right) e^{-\pi\sigma r^2} r^{n+q-1} \, dr.$$

The formula of Funk and Hecke (i.e., Theorem 6.15) in connection with the integral representation (6.429) of the Bessel function gives

$$\int_{\mathbb{S}^{q-1}} Y_n(q; \eta) e^{2\pi i r \rho \xi \cdot \eta} \, dS(\eta) = i^n \, \|\mathbb{S}^{q-1}\| \, J_n(q; 2\pi r\rho) \, Y_n(q; \xi). \tag{12.6}$$

Thus we get

$$\vartheta_n^{(q)}(\sigma; x, y; \Lambda) = \frac{i^n \|\mathbb{S}^{q-1}\|}{\|\mathcal{F}\|} \sum_{h \in \Lambda^{-1}} e^{2\pi i x \cdot (y+h)} \, Y_n\left(q; \frac{y+h}{|y+h|} \right) \tag{12.7}$$

$$\times \int_0^\infty e^{-\pi\sigma r^2} J_n(q; 2\pi r\rho) r^{n+q-1} \, dr.$$

The integral on the right side of (12.7) leads back to technicalities known from the theory of Bessel functions. In fact, Lemma 7.2 yields

$$\vartheta_n^{(q)}(\sigma; x, y; \Lambda) \tag{12.8}$$

$$= \sum_{g \in \Lambda} e^{-\pi\sigma|g-x|^2} H_n(q; g-x) \, e^{2\pi i g \cdot y}$$

$$= \frac{i^n \|\mathbb{S}^{q-1}\|}{2\|\mathcal{F}\|} \frac{\Gamma\left(\frac{q}{2}\right)}{\pi^{\frac{q}{2}} \sigma^{n+\frac{q}{2}}} e^{2\pi i x \cdot y} \sum_{h \in \Lambda^{-1}} e^{-\frac{\pi}{\sigma}|h+y|^2} H_n(q; h+y) \, e^{2\pi i h \cdot x}.$$

With $\|\mathbb{S}^{q-1}\| = 2\pi^{\frac{q}{2}}\left(\Gamma\left(\frac{q}{2}\right)\right)^{-1}$ this yields the *functional equation of the Theta function of degree n and dimension q.*

Theorem 12.1. *For all values $\sigma \in \mathbb{C}$ with $\Re(\sigma) > 0$ the Theta function $\vartheta_n^{(q)}(\cdot;x,y;\Lambda)$ is holomorphic, and we have*

$$\vartheta_n^{(q)}(\sigma;x,y;\Lambda) = \frac{i^n}{\|\mathcal{F}\|}e^{2\pi i x\cdot y}\sigma^{-n-\frac{q}{2}}\,\vartheta_n^{(q)}\left(\frac{1}{\sigma};-y,x;\Lambda^{-1}\right). \qquad (12.9)$$

Gauß–Weierstraß Summability on Regular Regions

Let Λ be an arbitrary lattice in $\mathbb{R}^q$. Let $\mathcal{G}$ be a regular region in $\mathbb{R}^q$. Suppose that x is an arbitrary but fixed point of the Euclidean space $\mathbb{R}^q$. *An important question is in what respect the infinite series*

$$\sum_{h\in\Lambda^{-1}}\int_{\mathcal{G}}F(y)\overline{\Phi_h(y)}\,dV(y)\,\Phi_h(x) \qquad (12.10)$$

can be "summed" to the finite sum

$$\sum_{\substack{g+x\in\overline{\mathcal{G}}\\g\in\Lambda}}{}^{'}F(g+x) = \sum_{\substack{g+x\in\overline{\mathcal{G}}\\g\in\Lambda}}\alpha(g+x)\,F(g+x)\;? \qquad (12.11)$$

The basic idea is to identify (12.10) and (12.11) in the sense of the Gauß–Weierstraß summability. In consequence, our main result is based on Lemma 7.4. It shows that the Gauß kernel is the Fourier transform of the Weierstraß kernel, and vice versa.

To be more concrete, for all $\tau \in \mathbb{R}$ with $\tau > 0$ and $x \in \mathbb{R}^q$ we easily obtain from the functional equation of the Theta function in $\mathbb{R}^q$ (cf. Theorem 12.1)

$$\frac{1}{\|\mathcal{F}\|}\vartheta_0^{(q)}\left(\pi\tau;0,y-x;\Lambda^{-1}\right) = (\pi\tau)^{\frac{q}{2}}\,\vartheta_0^{(q)}\left(\frac{1}{\pi\tau};x-y;0;\Lambda\right) \qquad (12.12)$$

the identity

$$\frac{1}{\|\mathcal{F}\|}\int_{\mathcal{G}}\vartheta_0^{(q)}\left(\pi\tau;0,y-x;\Lambda^{-1}\right)\,F(y)\,dV(y) \qquad (12.13)$$

$$= (\pi\tau)^{\frac{q}{2}}\int_{\mathcal{G}}\vartheta_0^{(q)}\left(\frac{1}{\pi\tau};x-y;0;\Lambda\right)\,F(y)\,dV(y).$$

Explicitly written out in series notation we have

$$\int_{\mathcal{G}}\sum_{h\in\Lambda^{-1}}e^{-\tau\pi^2 h^2}F(y)\overline{\Phi_h(y)}\,dV(y)\,\Phi_h(x) \qquad (12.14)$$

$$= (\tau\pi)^{-\frac{q}{2}}\int_{\mathcal{G}}\sum_{g\in\Lambda}e^{-\frac{|y-(g+x)|^2}{\tau}}F(y)\,dV(y).$$

for all $\tau > 0$ and all $x \in \mathbb{R}^q$, for all regular regions $\mathcal{G}$ as well as for all (weight) functions F of class $\mathrm{C}^{(0)}(\overline{\mathcal{G}})$. Interchanging sums and integrals we get

$$(\tau\pi)^{-\frac{q}{2}} \sum_{g\in\Lambda} \int_{\mathcal{G}} e^{-\frac{|y-(g+x)|^2}{\tau}} F(y) \, dV(y) \tag{12.15}$$

$$= \sum_{h\in\Lambda^{-1}} e^{-\tau\pi^2 h^2} \int_{\mathcal{G}} F(y)\overline{\Phi_h(y)} \, dV(y) \, \Phi_h(x).$$

After these preparations we are able to prove (cf. W. Freeden, P. Hermann [1985])

Theorem 12.2. *(Poisson Summation Formula in Gauß–Weierstraß Summability) Let $\mathcal{G} \subset \mathbb{R}^q$ be a regular region. Suppose that F is a member of class $\mathrm{C}^{(0)}(\overline{\mathcal{G}})$, $\overline{\mathcal{G}} = \mathcal{G}\cup\partial\mathcal{G}$. Then, for all $x \in \mathbb{R}^q$ and all $\tau \in \mathbb{R}$ with $\tau > 0$, the series*

$$\sum_{h\in\Lambda^{-1}} e^{-\tau\pi^2 h^2} \int_{\mathcal{G}} F(y)\overline{\Phi_h(y)} \, dV(y) \, \Phi_h(x) \tag{12.16}$$

is convergent.

Moreover, for all $x \in \mathbb{R}^q$, we have the summation formula

$$\sideset{}{'}\sum_{\substack{g+x\in\overline{\mathcal{G}} \\ g\in\Lambda}} F(g+x) = \lim_{\substack{\tau\to 0 \\ \tau>0}} \sum_{h\in\Lambda^{-1}} e^{-\tau\pi^2 h^2} \int_{\mathcal{G}} F(y)\overline{\Phi_h(y)} \, dV(y) \, \Phi_h(x). \tag{12.17}$$

Proof. For given $x \in \mathbb{R}^q$, take a sufficiently large positive constant R such that $|x+y| \leq R$ for all $y \in \overline{\mathcal{G}}$ and $|y-(g+x)|^2 \geq \frac{1}{2}|g+x|^2$ for all $g \in \Lambda$ with $|g+x| > R$. Then, because of the continuity of F, we are able to see that

$$(\tau\pi)^{-\frac{q}{2}} \sum_{\substack{|g+x|>R \\ g\in\Lambda}} \int_{\mathcal{G}} e^{-\frac{|y-(g+x)|^2}{\tau}} F(y) \, dV(y) \tag{12.18}$$

$$= O\left((\tau\pi)^{-\frac{q}{2}} \sum_{\substack{|g+x|\geq R \\ g\in\Lambda}} e^{\frac{-|g+x|^2}{2\tau}}\right)$$

$$= O\left((\tau\pi)^{-\frac{q}{2}} \int_{\substack{|x+y|\geq R \\ y\in\mathbb{R}^q}} e^{\frac{-|x+y|^2}{2\tau}} \, dV(y)\right)$$

for $\tau \to 0$. Moreover, there exists a positive constant $T > 0$ such that

$\mathrm{dist}(g, \overline{\mathcal{G}}) \geq T$ holds for all $g \notin \overline{\mathcal{G}}$. Thus,

$$\sum_{\substack{g+x \notin \overline{\mathcal{G}} \\ |g+x| \leq R \\ g \in \Lambda}} (\tau\pi)^{-\frac{q}{2}} \int_{\mathcal{G}} e^{-\frac{|y-(g+x)|^2}{\tau}} F(y) \, dV(y) \tag{12.19}$$

$$= O\left((\tau\pi)^{-\frac{q}{2}} \sum_{\substack{g+x \notin \overline{\mathcal{G}} \\ |g+x| \leq R \\ g \in \Lambda}} \int_{\mathcal{G}} e^{-\frac{|y-(g+x)|^2}{\tau}} \, dV(y) \right)$$

$$= O\left((\tau\pi)^{-\frac{q}{2}} \left(\int_{\substack{|y| \geq T \\ y \in \mathbb{R}^q}} e^{-\frac{|y|^2}{\tau}} \, dV(y) \right) \sum_{\substack{g+x \notin \overline{\mathcal{G}} \\ |g+x| \leq R \\ g \in \Lambda}} 1 \right)$$

$$= O\left((\tau\pi)^{-\frac{q}{2}} \int_{\substack{|y| \geq T \\ y \in \mathbb{R}^q}} e^{-\frac{|y|^2}{\tau}} \, dV(y) \right)$$

for $\tau \to 0$. Summarizing our results we therefore obtain

$$\sum_{\substack{g+x \notin \overline{\mathcal{G}} \\ g \in \Lambda}} (\tau\pi)^{-\frac{q}{2}} \int_{\mathcal{G}} e^{-\frac{|y-(g+x)|^2}{\tau}} F(y) \, dV(y) \tag{12.20}$$

$$= O\left(\int_{\frac{R^2}{2\tau}}^{\infty} e^{-r} r^{\frac{q}{2}-1} dr + \int_{\frac{T^2}{\tau}}^{\infty} e^{-r} r^{\frac{q}{2}-1} dr \right)$$

$$= o(1).$$

As is well known, for every $\rho > 0$,

$$(\tau\pi)^{-\frac{q}{2}} \int_{\mathbb{B}_\rho^q} e^{-\frac{|y|^2}{\tau}} \, dV(y) = \|\mathbb{S}^{q-1}\| \, \pi^{-\frac{q}{2}} \int_0^\rho e^{-\frac{r^2}{\tau}} r^{q-1} \, dr \tag{12.21}$$

$$= \frac{1}{2} \|\mathbb{S}^{q-1}\| \, \pi^{-\frac{q}{2}} \int_0^{\frac{\rho^2}{\tau}} e^{-s} s^{\frac{q}{2}-1} \, ds.$$

We have $\frac{1}{2} \|\mathbb{S}^{q-1}\| \, \pi^{-\frac{q}{2}} = \left(\Gamma(\frac{q}{2}) \right)^{-1}$. Thus it can be deduced from (12.21) that

$$\lim_{\substack{\tau \to 0 \\ \tau > 0}} (\tau\pi)^{-\frac{q}{2}} \int_{\mathbb{B}_\rho^q} e^{-\frac{|y|^2}{\tau}} \, dV(y) = 1. \tag{12.22}$$

In the same way it is not hard to verify that

$$\lim_{\substack{\tau \to 0 \\ \tau > 0}} (\tau \pi)^{-\frac{q}{2}} \int_{\substack{y \in \overline{\mathbb{B}_\rho^q(x)} \\ y \in \overline{\mathcal{G}}}} e^{-\frac{|y-x|^2}{\tau}} \, dV(y) = \alpha(x), \quad x \in \mathbb{R}^q, \tag{12.23}$$

where $\alpha(x)$ is the solid angle at $x \in \mathbb{R}^q$ subtended by the surface $\partial \mathcal{G}$. Even more general, for $\tau \to 0$, the *Fourier inversion formula in the Gauß–Weierstraß framework* (as known from Lemma 7.7) shows that

$$\lim_{\tau \to 0} (\tau \pi)^{-\frac{q}{2}} \int_{\mathcal{G}} e^{-\frac{|y-(g+x)|^2}{\tau}} F(y) \, dV(y) = \alpha(g+x) F(g+x), \tag{12.24}$$

where $\alpha(g + x)$ is the solid angle subtended at $g + x \in \overline{\mathcal{G}}$ by the surface $\partial \mathcal{G}$. It follows that

$$\lim_{\tau \to 0} (\tau \pi)^{-\frac{q}{2}} \sum_{\substack{g+x \in \overline{\mathcal{G}} \\ g \in \Lambda}} \int_{\mathcal{G}} e^{-\frac{|y-(g+x)|^2}{\tau}} F(y) \, dV(y) = \sum_{\substack{g+x \in \overline{\mathcal{G}} \\ g \in \Lambda}} \alpha(g+x) F(g+x).$$

$$\tag{12.25}$$

Combining the relations (12.15), (12.24), and (12.25) we get the assertion of Theorem 12.2. □

In fact, Theorem 12.2 is the identification of (12.10) and (12.11) in the sense of the Gauß–Weierstraß summability.

Remark 12.2. *An analogous result follows in the sense of the Abel–Poisson summability (note that the Abel kernel is the Fourier transform of the Poisson kernel, and vice versa).*

From Theorem 12.2 we immediately obtain by setting $x = 0$

Corollary 12.1. *Let Λ be an arbitrary lattice in $\mathbb{R}^q$. Let $\mathcal{G}$ be a regular region. Suppose that F is a member of class $C^{(0)}(\overline{\mathcal{G}})$, $\overline{\mathcal{G}} = \mathcal{G} \cup \partial \mathcal{G}$. Then*

$$\sideset{}{'}\sum_{\substack{g \in \overline{\mathcal{G}} \\ g \in \Lambda}} F(g) = \frac{1}{\|\mathcal{F}\|} \lim_{\substack{\tau \to 0 \\ \tau > 0}} \sum_{h \in \Lambda^{-1}} e^{-\tau \pi^2 h^2} \int_{\mathcal{G}} F(y) e^{-2\pi i h \cdot y} \, dV(y) \,. \tag{12.26}$$

Recognizing the nomenclature of lattice point discrepancies

$$P(F; \overline{\mathcal{G}}) = \sideset{}{'}\sum_{\substack{g \in \overline{\mathcal{G}} \\ g \in \Lambda}} F(g) - \frac{1}{\|\mathcal{F}\|} \int_{\mathcal{G}} F(x) \, dV(x) \tag{12.27}$$

in the framework of Gauß–Weierstraß summability we get

$$P(F; \overline{\mathcal{G}}) = \lim_{\substack{\tau \to 0 \\ \tau > 0}} \frac{1}{\sqrt{\|\mathcal{F}\|}} \sum_{\substack{h \neq 0 \\ h \in \Lambda}} e^{-\tau \pi^2 h^2} \int_{\mathcal{G}} F(y) \overline{\Phi_h(y)} \, dV(y). \tag{12.28}$$

Remark 12.3. *For balls* $\overline{\mathbb{B}^q_{\sqrt{N}}}$, $q \geq 2$, $N > 0$, *the lattice point discrepancy*

$$P\left(\overline{\mathbb{B}^q_{\sqrt{N}}}\right) = \sum_{\substack{|g| \leq \sqrt{N} \\ g \in \Lambda}}' 1 - \frac{1}{\|\mathcal{F}\|} \frac{\pi^{\frac{q}{2}}}{\Gamma(\frac{q}{2}+1)} N^{\frac{q}{2}} \tag{12.29}$$

$$= \int_{\substack{|x|=\sqrt{N} \\ x \in \mathbb{R}^q}} \frac{\partial}{\partial \nu} G(\Delta; x) \, dS(x)$$

satisfies the limit relations

$$P\left(\overline{\mathbb{B}^q_{\sqrt{N}}}\right) = \lim_{\substack{\tau \to 0 \\ \tau \to 0}} \frac{1}{\|\mathcal{F}\|} \sum_{\substack{h \neq 0 \\ h \in \Lambda}} e^{-\tau \pi^2 h^2} \int_{\substack{|x|=\sqrt{N} \\ x \in \mathbb{R}^q}} e^{-2\pi i h \cdot y} \, dV(y) \tag{12.30}$$

$$= \frac{\|\mathbb{S}^{q-1}\| \, N^{\frac{q}{2}}}{\|\mathcal{F}\|} \lim_{\substack{\tau \to 0 \\ \tau \to 0}} \sum_{\substack{h \neq 0 \\ h \in \Lambda}} e^{-\tau \pi^2 h^2} \frac{J_1(q; 2\pi |g| \sqrt{N})}{2\pi |g| \sqrt{N}},$$

where the theory of Bessel functions informs us that

$$N^{\frac{q}{2}} \frac{J_1(q; 2\pi |g| \sqrt{N})}{2\pi |g| \sqrt{N}} = O\left(N^{\frac{q-1}{4}}\right), \quad N \to \infty. \tag{12.31}$$

Needless to say, the estimate (12.31) does not mean that the asymptotic relation

$$P\left(\overline{\mathbb{B}^q_{\sqrt{N}}}\right) = O\left(N^{\frac{q-1}{4}}\right), \quad N \to \infty, \tag{12.32}$$

is valid (see Theorem 11.4 for a clarification involving lattice ball means).

12.2 Convergence Criteria for the Poisson Series

In the literature we know some attempts to overcome the Gauß–Weierstraß summability. Next we follow these techniques in order to establish the process of periodization for regular regions via Theorem 12.2.

Theorem 12.3. *Let Λ be an arbitrary lattice in $\mathbb{R}^q$. Let $\mathcal{G} \subset \mathbb{R}^q$ be a regular region. Suppose that F is of class $C^{(0)}(\overline{\mathcal{G}})$. Furthermore, assume that*

$$\lim_{N \to \infty} \sum_{\substack{|h| \leq N \\ h \in \Lambda^{-1}}} F_{\mathcal{G}}^{\wedge}(h) \tag{12.33}$$

is convergent, where the expansion coefficients $F_{\mathcal{G}}^{\wedge}(h)$ are given by

$$F_{\mathcal{G}}^{\wedge}(h) = \int_{\mathcal{G}} F(y) e^{-2\pi i h \cdot y} \, dV(y). \tag{12.34}$$

Then

$$\sideset{}{'}\sum_{\substack{g\in\bar{\mathcal{G}} \\ g\in\Lambda}} F(g) = \frac{1}{\|\mathcal{F}\|} \lim_{N\to\infty} \sum_{\substack{|h|\leq N \\ h\in\Lambda^{-1}}} F_{\mathcal{G}}^{\wedge}(h). \tag{12.35}$$

Proof. For given $\varepsilon > 0$, the convergence of (12.33) enables us to specify a constant $M(= M(\varepsilon))$ such that

$$\left| \sum_{\substack{|h|\leq N'' \\ h\in\Lambda^{-1}}} F_{\mathcal{G}}^{\wedge}(h) - \sum_{\substack{|h|\leq N' \\ h\in\Lambda^{-1}}} F_{\mathcal{G}}^{\wedge}(h) \right| \leq \varepsilon \tag{12.36}$$

for all N', N'' with $N'' \geq N' \geq M$. Now, define $S(N)$, $N \geq N'$, by

$$S(N) = \sum_{\substack{|h|\leq N \\ h\in\Lambda^{-1}}} F_{\mathcal{G}}^{\wedge}(h) - \sum_{\substack{|h|\leq N' \\ h\in\Lambda^{-1}}} F_{\mathcal{G}}^{\wedge}(h). \tag{12.37}$$

Then it follows with $\hat{\tau} = \tau\pi^2$ that

$$\sum_{\substack{N'<|h|\leq N \\ h\in\Lambda^{-1}}} e^{-\hat{\tau}h^2} F_{\mathcal{G}}^{\wedge}(h) = \int_{N'}^{N} e^{-\hat{\tau}r^2} \, dS(r) \tag{12.38}$$

$$= e^{-\hat{\tau}N^2} S(N) - \int_{N'}^{N} S(r) \, d\left(e^{-\hat{\tau}r^2}\right),$$

such that

$$\left| \sum_{\substack{N'<|h|\leq N \\ h\in\Lambda^{-1}}} e^{-\hat{\tau}h^2} F_{\mathcal{G}}^{\wedge}(h) \right| = O\left(e^{-\hat{\tau}N^2}\varepsilon - \varepsilon\int_{N'}^{N} d\left(e^{-\hat{\tau}r^2}\right)\right) = O(\varepsilon) \tag{12.39}$$

for all sufficiently small $\hat{\tau} > 0$. Thus, under the assumption of Theorem 12.3, the series

$$\lim_{N\to\infty} \frac{1}{\|\mathcal{F}\|} \sum_{\substack{|h|\leq N \\ h\in\Lambda^{-1}}} e^{-\tau\pi h^2} F_{\mathcal{G}}^{\wedge}(h) \tag{12.40}$$

is uniformly convergent with respect to τ. Hence, we are allowed to interchange the limits $\tau \to 0$ and $N \to \infty$ such that

$$\lim_{\tau\to 0} \lim_{N\to\infty} \frac{1}{\|\mathcal{F}\|} \sum_{\substack{|h|\leq N \\ h\in\Lambda^{-1}}} e^{-\tau\pi h^2} F_{\mathcal{G}}^{\wedge}(h) = \lim_{N\to\infty} \frac{1}{\|\mathcal{F}\|} \sum_{\substack{|h|\leq N \\ h\in\Lambda^{-1}}} F_{\mathcal{G}}^{\wedge}(h). \tag{12.41}$$

This proves the assertion of Theorem 12.3. $\qquad\square$

Theorem 12.3 allows us to formulate the following remarkable consequences.

Non-Negative Expansion Coefficients

Corollary 12.2. *Assume that there exists an integer M such that for all $h \in \Lambda^{-1}$ with $|h| > M$ the integrals $F_{\mathcal{G}}^{\wedge}(h)$ are non-negative. Then $\sum_{h \in \Lambda^{-1}} F_{\mathcal{G}}^{\wedge}(h)$ is convergent, and we have*

$$\sideset{}{'}\sum_{\substack{g \in \overline{\mathcal{G}} \\ g \in \Lambda}} F(g) = \frac{1}{\|\mathcal{F}\|} \sum_{h \in \Lambda^{-1}} \int_{\mathcal{G}} F(y) e^{-2\pi i h \cdot y} \, dV(y). \tag{12.42}$$

Proof. From the first part of Theorem 12.2 it is known that the series

$$\sum_{h \in \Lambda^{-1}} e^{-\tau \pi^2 h^2} \int_{\mathcal{G}} F(y) \overline{\Phi_h(y)} \, dV(y) \; \Phi_h(x) \tag{12.43}$$

exists for all $x \in \mathbb{R}^q$ and all $\tau \in \mathbb{R}$ with $\tau > 0$. Therefore, for all sufficiently large N', $N \geq M$, and all $\tau > 0$ the relation

$$\sum_{\substack{N' \leq |h| \leq N \\ h \in \Lambda^{-1}}} e^{-\tau \pi^2 h^2} \underbrace{\int_{\mathcal{G}} F(y) e^{-2\pi i h \cdot y} \, dV(y)}_{= F_{\mathcal{G}}^{\wedge}(h)} \leq C \tag{12.44}$$

holds true; hence,

$$\sum_{\substack{N' \leq |h| \leq N \\ h \in \Lambda^{-1}}} \underbrace{\int_{\mathcal{G}} F(y) e^{-2\pi i h \cdot y} \, dV(y)}_{= F_{\mathcal{G}}^{\wedge}(h)} \leq C \tag{12.45}$$

is valid for all $N', N > M$. Thus, the series

$$\sum_{h \in \Lambda^{-1}} \int_{\mathcal{G}} F(y) e^{-2\pi i h \cdot y} \, dV(y) \tag{12.46}$$

is absolutely convergent. Hence, the assertion of Corollary 12.2 follows from Theorem 12.3. $\qquad\square$

Homogeneous Boundary Weights

A rough manifestation of sufficient criteria for the validity of the Poisson summation formula on regular regions $\mathcal{G} \subset \mathbb{R}^q$ is

Lemma 12.1. *Let Λ be an arbitrary lattice in $\mathbb{R}^q$. Let $\mathcal{G}$ be a regular region in $\mathbb{R}^q$. Suppose that $F \in C^{(2m)}(\overline{\mathcal{G}})$, $m > \frac{q}{2}$, satisfies the homogeneous boundary conditions*

$$\Delta^k F \mid \partial \mathcal{G} = 0, \quad 0 \leq k \leq m - 1, \tag{12.47}$$

and

$$\nabla \Delta^k F \mid \partial \mathcal{G} = 0, \quad 0 \le k \le m - 2 \tag{12.48}$$

(note that (12.48) should be omitted if $m = 1$). Then the series

$$\sum_{h \in \Lambda^{-1}} \int_{\mathcal{G}} F(y) e^{-2\pi i h \cdot y} \, dV(y) \tag{12.49}$$

is convergent, and we have

$$\sum_{\substack{g \in \overline{\mathcal{G}} \\ g \in \Lambda}} F(g) = \frac{1}{\|\mathcal{F}\|} \sum_{h \in \Lambda^{-1}} \int_{\mathcal{G}} F(y) e^{-2\pi i h \cdot y} \, dV(y). \tag{12.50}$$

Proof. From the Extended Second Green Theorem (cf. Theorem 6.3) we obtain

$$\int_{\mathcal{G}} F(y) e^{-2\pi i x \cdot y} \, dV(y) \tag{12.51}$$

$$= \frac{1}{(-4\pi^2 x^2)^m} \int_{\mathcal{G}} (\Delta^m F(y)) \, e^{-2\pi i x \cdot y} \, dV(y)$$

$$- \sum_{k=0}^{m-1} \frac{1}{(-4\pi^2 x^2)^{k+1}} \int_{\partial \mathcal{G}} \left(\frac{\partial}{\partial \nu} \Delta^k F(y) \right) e^{-2\pi i x \cdot y} \, dS(y)$$

$$+ \sum_{k=0}^{m-1} \frac{1}{(-4\pi^2 x^2)^{k+1}} \int_{\partial \mathcal{G}} (\Delta^k F(y)) \frac{\partial}{\partial \nu} e^{-2\pi i x \cdot y} \, dS(y)$$

for all $x \in \mathbb{R}^q$, $x \ne 0$, such that under the homogeneous boundary assumptions imposed on F we get

$$\int_{\mathcal{G}} F(y) \, e^{-2\pi i x \cdot y} \, dV(y) \tag{12.52}$$

$$= \frac{1}{(-4\pi^2 x^2)^m} \int_{\mathcal{G}} (\Delta^m F(y)) \, e^{-2\pi i x \cdot y} \, dV(y)$$

$$- \frac{1}{(-4\pi^2 x^2)^m} \int_{\partial \mathcal{G}} \left(\frac{\partial}{\partial \nu} \Delta^{m-1} F(y) \right) \, e^{-2\pi i x \cdot y} \, dS(y).$$

Consequently, because of the continuity properties of the weight function F and its derivatives $\Delta^m F$, $\nabla \Delta^{m-1} F$ on $\overline{\mathcal{G}}$, it follows that

$$\int_{\mathcal{G}} F(y) \, e^{-2\pi i x \cdot y} \, dV(y) = O \left(\frac{1}{|x|^{2m}} \right), \quad |x| \to \infty. \tag{12.53}$$

In consequence, under the assumption $2m > q$, we are able to guarantee the absolute convergence of the series (12.49); hence from Theorem 12.3, we get the required result. □

Lemma 12.1 opens a way out of (Gaussian) summability in higher dimensions for lattice point sums under the assumption of certain vanishing weights on the boundary. Even better, in the spherical case, the polymetaharmonic theory (in particular, Theorem 6.30) helps us to reduce the number of homogeneous conditions in Lemma 12.1.

Theorem 12.4. *Suppose that F is of class $C^{(k)}\left(\overline{\mathbb{B}_R^q}\right)$, $R > 0$, such that $k > 2m + \frac{q}{2} > \frac{3q}{2}$. Moreover, assume that F satisfies the homogeneous conditions*

$$\nabla \Delta^l F \mid \partial \mathbb{B}_R^q = 0, \quad 0 \le l \le \frac{q-3}{4}, \tag{12.54}$$

and

$$\Delta^l F \mid \partial \mathbb{B}_R^q = 0, \quad 0 \le l \le \frac{q-1}{4}. \tag{12.55}$$

Then the series (12.49)

$$\sum_{h \in \Lambda^{-1}} \int_{\mathbb{B}_R^q} F(y) \, e^{-2\pi i h \cdot y} \, dV(y) \tag{12.56}$$

is convergent, and the identity (12.50)

$$\sum_{\substack{g \in \overline{\mathbb{B}_R^q} \\ g \in \Lambda}} F(g) = \frac{1}{\|\mathcal{F}\|} \sum_{h \in \Lambda^{-1}} \int_{\mathbb{B}_R^q} F(y) \, e^{-2\pi i h \cdot y} \, dV(y) \tag{12.57}$$

holds true.

Proof. Theorem 12.4 follows immediately from (12.51) by consequent application of the asymptotic relations known from the polymetaharmonic theory, i.e., Theorem 6.30. □

Weighted Landau Formulas

Next we list some examples which are direct consequences of Theorem 12.4.

Example 12.1. *For $R > 0$, $p \in \mathbb{N}$ with $p > \frac{q-1}{2}$, and $F \in C^{(k)}\left(\overline{\mathbb{B}_R^q}\right)$ with $k > 2m + \frac{q}{2} > \frac{3q}{2}$ we have*

$$\sum_{\substack{|g| \le R \\ g \in \Lambda}} (R^2 - g^2)^p \, F(g) = \frac{1}{\|\mathcal{F}\|} \sum_{h \in \Lambda^{-1}} \int_{\substack{|x| \le R \\ x \in \mathbb{R}^q}} (R^2 - x^2)^p \, F(x) \, e^{-2\pi i h \cdot x} \, dV(x).$$

$$\tag{12.58}$$

Note that the factor $(R^2 - x^2)^p$ guarantees the validity of the conditions (12.54) and (12.55). In particular, taking F as a constant function, we obtain the so–called *Landau formulas* (for more details concerning constant weights see A. Walfisz [1927]).

Example 12.2. *(Landau Formulas) For $R > 0$ and $p \in \mathbb{N}$ with $p > \frac{q-1}{2}$ we have*

$$\frac{1}{p!} \sum_{\substack{|g| \le R \\ g \in \Lambda}} (R^2 - g^2)^p = \frac{\pi^{\frac{q}{2}}}{\Gamma(p + \frac{q}{2} + 1)} R^{2p+q} \tag{12.59}$$

$$+ \frac{2^p \, \|\mathbb{S}^{q-1}\|}{\|\mathcal{F}\|} R^{q+2p} \sum_{\substack{|h| > 0 \\ h \in \Lambda^{-1}}} \frac{J_{p+1}(q; 2\pi|h|R)}{(2\pi|h|R)^{p+1}}.$$

12.3 Generalized Parseval Identity

As is well known, the Parseval identity with respect to the fundamental cell of a lattice $\Lambda \subset \mathbb{R}^q$

$$\int_{\mathcal{F}} |F(y)|^2 \, dV(y) = \sum_{h \in \Lambda^{-1}} \left| \int_{\mathcal{F}} F(y) \overline{\Phi_h(y)} \, dV(y) \right|^2 \tag{12.60}$$

holds true for all $F \in \mathrm{L}^2_\Lambda(\mathbb{R}^q)$ (particularly, for all $F \in \mathrm{C}^{(0)}_\Lambda(\mathbb{R}^q)$). Of course, the Parseval identity is closely interrelated to the orthonormality of the system $\{\Phi_h\}_{h \in \Lambda^{-1}}$ with respect to the fundamental cell $\mathcal{F}$ of the lattice Λ under consideration. In other words, the Parseval identity on regular regions $\mathcal{G}$ and the Λ-orthonormality of the system $\{\Phi_h\}_{h \in \Lambda^{-1}}$ seem to be irreconcilable.

Nevertheless, in the framework of the two-dimensional Euclidean space $\mathbb{R}^2$, C. Müller [1956] noticed that

$$\int_{\mathcal{G}} \sum_{\substack{g+x \in \overline{\mathcal{G}} \\ g \in \mathbb{Z}^2}} {}' 1 \, dV(x) = \sum_{h \in \mathbb{Z}^2} \left| \int_{\mathcal{G}} e^{-2\pi i h \cdot x} \, dV(x) \right|^2 \tag{12.61}$$

holds true for all symmetrical (with respect to the origin) and convex regions $\mathcal{G} \subset \mathbb{R}^2$. An easy consequence of (12.61) is the inequality

$$\int_{\mathcal{G}} \sum_{\substack{g+x \in \overline{\mathcal{G}} \\ g \in \mathbb{Z}^2}} {}' 1 \, dV(x) \ge \|\mathcal{G}\|^2. \tag{12.62}$$

Now, under the assumption $\|\mathcal{G}\| > 4$, an essential step towards Minkowski's Theorem (cf. Theorem 5.2) can be made by considering the lattice $2\mathbb{Z}^2$. Indeed, as we will see later on, the inequality

$$\int_{\mathcal{G}} \sum_{\substack{g+x \in \overline{\mathcal{G}} \\ g \in 2\mathbb{Z}^2}} {}' 1 \, dV(x) \ge \frac{\|\mathcal{G}\|}{4} \|\mathcal{G}\| > \|\mathcal{G}\|. \tag{12.63}$$

is the key to guarantee that a symmetrical (with respect to the origin) and convex region $\mathcal{G} \subset \mathbb{R}^2$ with $\|\mathcal{G}\| > 4$ contains lattice points of $\mathbb{Z}^2$ different from the origin.

In what follows, our aim is to show that Müller's two-dimensional approach (see C. Müller [1956]) can be generalized in various ways: the Parseval identity of type (12.61) can be formulated for arbitrary lattices Λ. Variable continuous weight functions can be included instead of constant weights. Moreover, the concept of Gauß–Weierstraß summability allows its formulation in the framework of the Euclidean space $\mathbb{R}^q$. Finally, the Parseval identity is valid not only for regions "suitable" in the geometry of numbers , but also for all regular regions $\mathcal{G} \subset \mathbb{R}^q$. Needless to say, the essential tool for our consideration is Theorem 12.2.

We start our considerations with the formulation of the Parseval Identity in Gauß–Weierstraß Summability.

Theorem 12.5. *(Extended Parseval Identity in Gauß–Weierstraß Summability) Let Λ be a lattice in $\mathbb{R}^q$. Let $\mathcal{G}, \mathcal{H} \subset \mathbb{R}^q$ be regular regions. Suppose that F is of class $\mathrm{C}^{(0)}(\overline{\mathcal{G}})$ and G is of class $\mathrm{C}^{(0)}(\overline{\mathcal{H}})$, respectively. Then,*

$$\lim_{\substack{\tau \to 0 \\ \tau > 0}} \sum_{h \in \Lambda^{-1}} e^{-\tau \pi^2 h^2} \int_{\mathcal{G}} F(y) \overline{\Phi_h(y)} \, dV(y) \overline{\int_{\mathcal{H}} G(y) \overline{\Phi_h(y)} \, dV(y)}$$

$$= \int_{\mathcal{H}} \sum_{\substack{g + x \in \overline{\mathcal{G}} \\ g \in \Lambda}} {}' F(g + x) \, \overline{G(x)} \, dV(x).$$

Proof. As usual, $\overline{\mathcal{G}} - \{x\} = \{y - x \in \mathbb{R}^q \mid y \in \overline{\mathcal{G}}\}$ is the translate of $\overline{\mathcal{G}}$ by $-x$. We introduce the auxiliary function $H : \overline{\mathcal{G}} - \{x\} \to \mathbb{R}$ by

$$H(y) = F(x + y) , \qquad y \in \overline{\mathcal{G}} - \{x\}. \tag{12.64}$$

Then it is clear that H is of class $\mathrm{C}^{(0)}\left(\overline{\mathcal{G}} - \{x\}\right)$. From (12.15) we are able to deduce that

$$(\pi \tau)^{-\frac{q}{2}} \sum_{g \in \Lambda} \int_{\mathcal{G} - \{x\}} e^{-\frac{|y - g|^2}{\tau}} H(y) \, dV(y) \tag{12.65}$$

$$= \frac{1}{\|\mathcal{F}\|} \sum_{h \in \Lambda^{-1}} e^{-\tau \pi^2 h^2} \int_{\mathcal{G} - \{x\}} H(y) \, e^{-2\pi i h \cdot y} \, dV(y)$$

$$= \sum_{h \in \Lambda^{-1}} e^{-\tau \pi^2 h^2} \int_{\mathcal{G}} F(y) \overline{\Phi_h(y)} \, dV(y) \, \Phi_h(x).$$

Multiplying with $\overline{G(x)}, x \in \mathcal{H}$, and subsequent integration over $\mathcal{H}$ gives

$$(\pi\tau)^{-\frac{q}{2}} \int_{\mathcal{H}} \overline{G(x)} \left(\sum_{g \in \Lambda} \int_{\mathcal{G}-\{x\}} e^{-\frac{|y-g|^2}{\tau}} H(y) \, dV(y) \right) dV(x) \qquad (12.66)$$

$$= \sum_{h \in \Lambda^{-1}} e^{-\tau\pi^2 h^2} \int_{\mathcal{G}} F(y) \overline{\Phi_h(y)} \, dV(y) \int_{\mathcal{H}} \overline{G(x)} \Phi_h(x) \, dV(x).$$

Now, from the considerations of Section 12.1, it follows that

$$\lim_{\substack{\tau \to 0 \\ \tau > 0}} (\pi\tau)^{-\frac{q}{2}} \sum_{g \in \Lambda} \int_{\mathcal{G}-\{x\}} e^{-\frac{|y-g|^2}{\tau}} H(y) \, dV(y) \;=\; \sum_{\substack{g \in \overline{\mathcal{G}}-\{x\} \\ g \in \Lambda}} {}' H(g) \quad (12.67)$$

$$= \sum_{\substack{g+x \in \overline{\mathcal{G}} \\ g \in \Lambda}} {}' F(g+x).$$

For all $x \in \overline{\mathcal{G}}$ we find with $\gamma = \sup_{x \in \overline{\mathcal{G}}} |x|$

$$\left| (\pi\tau)^{-\frac{q}{2}} \sum_{g \in \Lambda} \int_{\mathcal{G}-\{x\}} e^{-\frac{|y-g|^2}{\tau}} F(x+y) \, dV(y) \right| \qquad (12.68)$$

$$\leq \; (\pi\tau)^{-\frac{q}{2}} \|F\|_{\mathrm{C}^{(0)}(\overline{\mathcal{G}})} \sum_{g \in \Lambda} \int_{|y| \leq 2\gamma} e^{-\frac{|y-g|^2}{\tau}} \, dV(y).$$

The series on the right side of (12.68) is convergent; hence, the expression on the left side of (12.68) is bounded. Thus there exists a constant C such that

$$\left| (\pi\tau)^{-\frac{q}{2}} \sum_{g \in \Lambda} \int_{\mathcal{G}-\{x\}} e^{-\frac{|y-g|^2}{\tau}} H(y) \, dV(y) \right| \leq C. \qquad (12.69)$$

In accordance with Lebesgue's convergence theorem we are therefore allowed to conclude from (12.66) that

$$\lim_{\substack{\tau \to 0 \\ \tau > 0}} \sum_{h \in \Lambda^{-1}} e^{-\tau\pi^2 h^2} \int_{\mathcal{G}} F(y) \overline{\Phi_h(y)} \, dV(y) \int_{\mathcal{H}} \overline{G(y)} \Phi_h(y) \, dV(y)$$

$$= \int_{\mathcal{H}} \sum_{\substack{g+x \in \overline{\mathcal{G}} \\ g \in \Lambda}} {}' F(g+x) \, \overline{G(x)} \, dV(x). \qquad (12.70)$$

Note that $x \mapsto \sum_{g+x \in \overline{\mathcal{G}}} {}' F(g+x) \, \overline{G(x)}$ constitutes (apart from the finite number of surfaces $\overline{\mathcal{H}} \cap (\partial\mathcal{G} - \{g\})$) a continuous function in $\overline{\mathcal{H}}$, where the discontinuities are finite jumps. $\qquad \square$

Example 12.3. *We choose* $G = \mathbb{B}_\rho^q$, $\mathcal{H} = \mathbb{B}_{\sqrt{N}}^q$, $\rho, N > 0$, *and* $F = 1$ *on* $\overline{\mathcal{G}}$ *as well as* $G = 1$ *on* $\overline{\mathcal{H}}$. *Then we obtain from Theorem 12.5*

$$\left\|\mathbb{S}^{q-1}\right\|^2 \rho^q N^{\frac{q}{2}} \lim_{\substack{\tau \to 0 \\ \tau > 0}} \sum_{h \in \Lambda^{-1}} e^{-\tau \pi h^2} \frac{J_1(q; 2\pi|h|\rho)}{2\pi|h|\rho} \frac{J_1(q; 2\pi|h|\sqrt{N})}{2\pi|h|\sqrt{N}}$$

$$= \int_{\mathbb{B}_{\sqrt{N}}^q} \sum_{\substack{g+x \in \mathbb{B}_\rho^q \\ g \in \Lambda}}{}' 1 \; dV(x), \tag{12.71}$$

such that

$$\left\|\mathbb{S}^{q-1}\right\|^2 \rho^q N^{\frac{q}{2}} \sum_{h \in \Lambda^{-1}} \frac{J_1(q; 2\pi|h|\rho)}{2\pi|h|\rho} \frac{J_1(q; 2\pi|h|\sqrt{N})}{2\pi|h|\sqrt{N}}$$

$$= \int_{\mathbb{B}_{\sqrt{N}}^q} \sum_{\substack{g+x \in \mathbb{B}_\rho^q \\ g \in \Lambda}}{}' 1 \; dV(x) \tag{12.72}$$

$$= \int_{\mathbb{B}_{\sqrt{N}}^q} \chi_{\overline{\mathbb{B}_\rho^q}+\Lambda}(x) \; dV(x)$$

$$= \int_{\mathbb{B}_{\sqrt{N}}^q} \chi_{\overline{\mathbb{B}_\rho^q}+\Lambda}(x) \; \chi_{\overline{\mathbb{B}_{\sqrt{N}}^q}} \; dV(x).$$

It should be noted that the identity (12.72) leads back to the concept of the lattice ball summation of Chapter 11.

By virtue of Theorem 12.5 we are led to formulate the following corollary in Gauß–Weierstraß summability:

Corollary 12.3. *(Parseval Identity in Gauß–Weierstraß Summability) Let* Λ *be a lattice in* $\mathbb{R}^q$. *Let* $\mathcal{G} \subset \mathbb{R}^q$ *be a regular region. Suppose that* F *is of class* $C^{(0)}(\overline{\mathcal{G}})$. *Then the following variant of the Parseval identity in Gauß–Weierstraß summability holds true:*

$$\lim_{\substack{\tau \to 0 \\ \tau > 0}} \sum_{h \in \Lambda^{-1}} e^{-\tau \pi^2 h^2} \left| \int_{\mathcal{G}} F(y)\overline{\Phi_h(y)} \; dV(y) \right|^2 \tag{12.73}$$

$$= \int_{\mathcal{G}} \sum_{\substack{g+x \in \overline{\mathcal{G}} \\ g \in \Lambda}}{}' F(g+x)\overline{F(x)} \; dV(x) \tag{12.74}$$

The same arguments leading to Corollary 12.2 allow us to conclude that the limit and the sum on the right of (12.73) may be interchanged. Thus we finally arrive at the following result.

Theorem 12.6. *(Parseval Identity for Regular Regions) Let Λ be a lattice in $\mathbb{R}^q$. Let $\mathcal{G} \subset \mathbb{R}^q$ be a regular region. Suppose that F is of class $\mathrm{C}^{(0)}(\overline{\mathcal{G}})$. Then we have*

$$\sum_{h \in \Lambda^{-1}} \left| \int_{\mathcal{G}} F(y) \overline{\Phi_h(y)} \, dV(y) \right|^2 \tag{12.75}$$

$$= \int_{\mathcal{G}} \sideset{}{'}\sum_{\substack{g+x \in \overline{\mathcal{G}} \\ g \in \Lambda}} F(g+x) \overline{F(x)} \, dV(x)$$

12.4 Minkowski's Lattice Point Theorem

As a preparation of Minkowski's Theorem (cf. H. Minkowski [1896] for the original reference) we remember the *equivalence of the following statements*:

(i) The region $\mathcal{G} \subset \mathbb{R}^q$ is convex (i.e., $x_1, x_2 \in \mathcal{G}$ implies that $\lambda x_1 + (1-\lambda) x_2 \in \mathcal{G}$ for all $\lambda \in [0,1]$) and symmetrical with respect to the origin (i.e., $x_1 \in \mathcal{G}$ implies $-x_1 \in \mathcal{G}$).

(ii) The region $\mathcal{G} \subset \mathbb{R}^q$ shows the property: $x_1, x_2 \in \mathcal{G}$ implies $\frac{1}{2}(x_1 - x_2) \in \mathcal{G}$.

Combining the Parseval identity (Theorem 12.3) and the equivalent statements *(i)* and *(ii)* (see also Theorem 5.2) we obtain (cf. Figure 5.8)

Theorem 12.7. *(Minkowski's Theorem) Assume that the regular region $\mathcal{G} \subset \mathbb{R}^q$ is convex and symmetrical (with respect to the origin). Moreover, suppose that*

$$\|\mathcal{G}\| > 2^q \, \|\mathcal{F}\| \, . \tag{12.76}$$

Then, $\mathcal{G}$ contains lattice points of Λ, which are different from the origin.

Proof. It is clear that

$$\sideset{}{'}\sum_{\substack{x+g \in \overline{\mathcal{G}} \\ g \in \Lambda}} 1 = \sideset{}{'}\sum_{\substack{x+g \in \overline{\mathcal{G}} \\ g \in \Lambda}} \mathcal{X}_{\overline{\mathcal{G}}}(x+g), \tag{12.77}$$

where, as always, the characteristic function $\mathcal{X}_{\overline{\mathcal{G}}}$ is given by

$$\mathcal{X}_{\overline{\mathcal{G}}}(x) = \begin{cases} 1 & , \quad x \in \overline{\mathcal{G}} \\ 0 & , \quad x \in \mathbb{R}^q \setminus \overline{\mathcal{G}}. \end{cases} \tag{12.78}$$

The application of the Parseval identity (Theorem 12.6) to (12.84) yields

$$
\int_{\mathcal{G}} \sum_{\substack{x+g\in\overline{\mathcal{G}} \\ g\in\Lambda}}{}' \mathcal{X}_{\overline{g}}(x+g)\, dV(x) \;=\; \int_{\mathcal{G}} \sum_{\substack{x+g\in\overline{\mathcal{G}} \\ g\in\Lambda}}{}' \mathcal{X}_{\overline{g}}(x+g)\, \mathcal{X}_{\overline{g}}(x)\, dV(x)
$$

$$
= \sum_{h\in\Lambda^{-1}} \left| \int_{\mathcal{G}} \mathcal{X}_{\overline{g}}(x)\overline{\Phi_h(x)}\, dV(x) \right|^2 . \tag{12.79}
$$

This demonstrates that

$$
\int_{\mathcal{G}} \sum_{\substack{x+g\in\overline{\mathcal{G}} \\ g\in\Lambda}}{}' \mathcal{X}_{\overline{g}}(x+g)\, dV(x) \geq \frac{1}{\|\mathcal{F}\|}\|\mathcal{G}\|^2. \tag{12.80}
$$

It is known that the area of the fundamental cell of the dilated lattice 2Λ is equal to $2^q\|\mathcal{F}\|$. Hence, applying the estimate (12.80) to the lattice 2Λ we get

$$
\int_{\mathcal{G}} \sum_{\substack{x+g\in\overline{\mathcal{G}} \\ g\in 2\Lambda}}{}' \mathcal{X}_{\overline{g}}(x+g)\, dV(x) \geq \frac{1}{2^q\|\mathcal{F}\|}\|\mathcal{G}\|^2. \tag{12.81}
$$

By virtue of the assumption (12.76) of Minkowski's Theorem we get

$$
\int_{\mathcal{G}} \sum_{\substack{x+g\in\overline{\mathcal{G}} \\ g\in 2\Lambda}}{}' \mathcal{X}_{\overline{g}}(x+g)\, dV(x) \geq \frac{\|\mathcal{G}\|}{2^q\|\mathcal{F}\|}\|\mathcal{G}\| > \|\mathcal{G}\|. \tag{12.82}
$$

Observing the fact that the function

$$
x \mapsto \sum_{\substack{x+g\in\overline{\mathcal{G}} \\ g\in 2\Lambda}}{}' \mathcal{X}_{\overline{g}}(x+g), \tag{12.83}
$$

apart from the finite number of surfaces $\overline{\mathcal{G}} \cap (\partial\mathcal{G} - \{2g\})$, takes only integer values, we are allowed to deduce that there exists at least one point $x \in \mathcal{G}$ such that

$$
\sum_{\substack{x+g\in\overline{\mathcal{G}} \\ g\in 2\Lambda}}{}' 1 = \sum_{\substack{x+g\in\overline{\mathcal{G}} \\ g\in 2\Lambda}}{}' \mathcal{X}_{\overline{g}}(x+g) \;\geq 2. \tag{12.84}
$$

Thus, we are led to the conclusion:

$$
x + 2g \in \mathcal{G}, \quad x + 2g' \in \mathcal{G} \tag{12.85}
$$

with $g \neq g'$ implies

$$
\frac{1}{2}((x + 2g) - (x + 2g')) = g - g' \in \mathcal{G} \tag{12.86}
$$

with $g \neq g'$. This is the desired result. $\qquad\square$

13

Poisson Summation on Planar Regular Regions

CONTENTS

13.1 Fourier Inversion Formula 362
      Formulation for Regular Regions 362
      Formulation for Circles 364
13.2 Weighted Two-Dimensional Lattice Point Identities 365
      General Geometry and Homogeneous Boundary Weight 366
      Circles and General Weights 371
      Convex "Smooth" Regions and General Weights 377
13.3 Weighted Two-Dimensional Lattice Ball Identities 379
      General Geometry and General Weights (Variant 1) 379
      General Geometry and General Weights (Variant 2) 381

Our next goal is to discuss the concept of periodization, i.e., the Poisson summation formula for two-dimensional regular regions, i.e., planar "potato slices" (cf. Figure 13.1). An essential tool is the two-dimensional Fourier inversion formula (as proposed by W. Freeden [1978a]). More concretely, Section 13.1 provides a particular treatment of the pointwise Fourier inversion formula within a (meta)harmonically oriented framework involving Bessel functions.

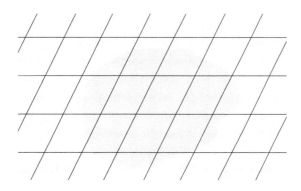

FIGURE 13.1
Lattice points of a 2D-lattice inside a regular region (such as a planar "potato slice").

It turns out that two-dimensional Poisson summation canonically leads to restrictive assumptions (see Section 13.2) such as homogeneous boundary weights or non-vanishing curvature of boundary curves. More concretely, Section 13.2 demonstrates that the two-dimensional lattice point Poisson summation formula can be verified for twice continuously differentiable functions on regular regions satisfying a homogeneous boundary condition (of Dirichlet's type), whereas its two-dimensional lattice ball (i.e., lattice circle) counterparts (see Section 13.3) are valid for all regular regions without any additional boundary condition and for all twice continuously differentiable (weight) functions. A particular case is played by the Poisson summation formula for the circle (see Subsection 13.2). This formula can be established generally for all twice continuously differentiable weight functions (again without any specification of a boundary condition).

13.1 Fourier Inversion Formula

We start this chapter with the proof of the pointwise *Fourier inversion formula for regular regions* $\mathcal{G} \subset \mathbb{R}^2$. Its proof cannot be undertaken by the Hankel transform (cf. Section 7.5) since the general geometry of a regular region does not allow a canonical decomposition into radial and angular parts. Instead, the proof is essentially based on tools of potential theory involving properties of Bessel functions.

Formulation for Regular Regions

Theorem 13.1. *Let $\mathcal{G}$ be a regular region in $\mathbb{R}^2$. Assume that F is a continuously differentiable function in $\overline{\mathcal{G}} = \mathcal{G} \cup \partial\mathcal{G}$. Then, for every $w \in \mathbb{R}^2$, the integral*

$$\int_{\mathbb{R}^2} e^{2\pi i w \cdot x} \int_{\substack{y \in \overline{\mathcal{G}} \\ y \in \mathbb{R}^2}} F(y) \, e^{-2\pi i x \cdot y} \, dV(y) \, dV(x) \tag{13.1}$$

exists in the sense

$$\int_{\mathbb{R}^2} \cdots = \lim_{N \to \infty} \int_{\substack{|x| \le N \\ x \in \mathbb{R}^2}} \cdots \,, \tag{13.2}$$

and we have

$$\int_{\mathbb{R}^2} e^{2\pi i w \cdot x} \int_{\substack{y \in \overline{\mathcal{G}} \\ y \in \mathbb{R}^2}} F(y) \, e^{-2\pi i x \cdot y} \, dV(y) \, dV(x) = \alpha(w) F(w), \tag{13.3}$$

where $\alpha(w)$, $w \in \mathbb{R}^2$, is the solid angle at w subtended by $\partial\mathcal{G}$.

Proof. The two-dimensional theory of Bessel functions tells us (for more details see Lemma 6.30) that

$$\int_{\substack{|x|\leq N \\ x\in\mathbb{R}^2}} e^{2\pi i\,((w-y)\cdot z)}\,dV(x) = 2\pi\int_0^N J_0(2\pi|y-w|r)\,dr \qquad (13.4)$$

$$= N\frac{J_1(2\pi|y-w|N)}{|y-w|}.$$

Using this result we get

$$\int_{\substack{|x|\leq N \\ x\in\mathbb{R}^2}} e^{2\pi iw\cdot x}\int_{\substack{y\in\overline{\mathcal{G}} \\ y\in\mathbb{R}^2}} F(y)\,e^{-2\pi ix\cdot y}\,dV(y)\,dV(x) \qquad (13.5)$$

$$= N\int_{\substack{y\in\overline{\mathcal{G}} \\ y\in\mathbb{R}^2}} F(y)\frac{J_1(2\pi|y-w|N)}{|y-w|}\,dV(y).$$

Observing the identities

$$\nabla_y \ln|w-y| = -\frac{w-y}{|w-y|^2}, \quad y\neq w, \qquad (13.6)$$

$$\nabla_y J_0(2\pi|w-y|N) = 2\pi N J_1(2\pi|w-y|N)\frac{w-y}{|w-y|}, \quad y\neq w, \quad (13.7)$$

we obtain by simple manipulations for $y\neq w$

$$NF(y)\frac{J_1(2\pi|w-y|N)}{|w-y|} \qquad (13.8)$$

$$= -\frac{1}{2\pi}\left(\nabla_y \ln|w-y|\right)\cdot\nabla_y\left(F(y)J_0(2\pi|w-y|N)\right)$$

$$-\frac{1}{2\pi}J_0(2\pi|w-y|N)\left(\nabla_y F(y)\cdot\frac{w-y}{|w-y|^2}\right).$$

This yields

$$N\int_{\substack{y\in\overline{\mathcal{G}} \\ y\in\mathbb{R}^2}} F(y)\frac{J_1(2\pi|w-y|N)}{|w-y|}\,dV(y) \qquad (13.9)$$

$$= -\frac{1}{2\pi}\int_{\substack{y\in\overline{\mathcal{G}} \\ y\in\mathbb{R}^2}} \left(\nabla_y \ln|w-y|\right)\cdot\nabla_y\left(F(y)J_0(2\pi|w-y|N)\right)\,dV(y)$$

$$-\frac{1}{2\pi}\int_{\substack{y\in\overline{\mathcal{G}} \\ y\in\mathbb{R}^2}} J_0(2\pi|w-y|N)\left(\nabla_y F(y)\cdot\frac{w-y}{|w-y|^2}\right)\,dV(y).$$

With the help of the well known estimate $|J_0(r)|\leq Cr^{-\frac{1}{2}}$, the second integral on the right side of (13.9) can be estimated in the form

$$-\frac{1}{2\pi}\int_{\substack{y\in\overline{\mathcal{G}} \\ y\in\mathbb{R}^2}} J_0(2\pi|w-y|N)\left(\nabla_y F(y)\right)\cdot\frac{w-y}{|w-y|^2}\,dV(y) = O\left(N^{-\frac{1}{2}}\right) \quad (13.10)$$

for $N \to \infty$. The first integral on the right side of (13.9) can be handled by potential theoretical methods. From the first Green theorem we obtain for $w \notin \overline{\mathcal{G}}$

$$-\frac{1}{2\pi} \int_{y \in \mathcal{G}} (\nabla_y \ln |w - y|) \cdot \nabla_y \left(F(y) J_0(2\pi |w - y|N) \right) \, dV(y)$$

$$= -\frac{1}{2\pi} \int_{y \in \partial \mathcal{G}} \left(\frac{\partial}{\partial \nu_y} \ln |w - y| \right) F(y) J_0(2\pi |w - y|N) \, dS(y),$$

$$(13.11)$$

whereas for $w \in \overline{\mathcal{G}}$ and sufficiently small $\rho > 0$ we are able to formulate

$$-\frac{1}{2\pi} \int_{\substack{y \in \overline{\mathcal{G}} \\ |w-y| \geq \rho}} (\nabla_y \ln |w - y|) \cdot \nabla_y \left(F(y) J_0(2\pi |w - y|N) \right) \, dV(y)$$

$$= -\frac{1}{2\pi} \int_{\substack{y \in \partial \mathcal{G} \\ |w-y| \geq \rho}} \left(\frac{\partial}{\partial \nu_y} \ln |w - y| \right) F(y) J_0(2\pi |w - y|N) \, dS(y)$$

$$-\frac{1}{2\pi} \int_{\substack{|w-y|=\rho \\ y \in \overline{\mathcal{G}}}} \left(\frac{\partial}{\partial \nu_y} \ln |w - y| \right) F(y) J_0(2\pi |w - y|N) \, dS(y).$$

$$(13.12)$$

Letting $\rho \to 0$ we therefore find by use of Lemma 6.2 and subsequent application of the standard estimate for Bessel functions

$$-\frac{1}{2\pi} \int_{\substack{y \in \overline{\mathcal{G}} \\ y \in \mathbb{R}^2}} (\nabla_y \ln |w - y|) \cdot \nabla_y \left(F(y) J_0(2\pi |w - y|N) \right) \, dV(y)$$

$$= \alpha(w) F(w) + O\left(N^{-\frac{1}{2}} \right) \qquad (13.13)$$

for $N \to \infty$. Combining (13.10) and (13.13) we get from (13.9)

$$N \int_{\substack{y \in \overline{\mathcal{G}} \\ y \in \mathbb{R}^2}} F(y) \frac{J_1(2\pi |w - y|N)}{|w - y|} \, dV(y) = \alpha(w) F(w) + O\left(N^{-\frac{1}{2}} \right). \quad (13.14)$$

This yields the desired result.   □

Formulation for Circles

Of particular interest in our lattice point theory is the Fourier inversion formula for the circle $\mathbb{B}_R^2$ around the origin with radius R (see W. Freeden [1975, 1978a])

$$\int_{\mathbb{R}^2} e^{2\pi i w \cdot x} \int_{\substack{|y| \leq R \\ y \in \mathbb{R}^2}} F(y) \, e^{-2\pi i x \cdot y} \, dV(y) \, dV(x) = \begin{cases} F(w), & |w| < R \\ \frac{1}{2} F(w), & |w| = R \\ 0, & |w| > R, \end{cases}$$

$$(13.15)$$

provided that F is a continuously differentiable function in $\overline{\mathbb{B}_R^2}$.

13.2 Weighted Two-Dimensional Lattice Point Identities

As we saw in Section 12.2, the calamity in the discussion of weighted two-dimensional lattice point sums over regular regions $\mathcal{G}$ is the proof of the convergence of the series

$$\sum_{\substack{h \in \Lambda^{-1} \\ h \in \mathbb{R}^2}} F_{\mathcal{G}}^{\wedge}(h) = \sum_{\substack{h \in \Lambda^{-1} \\ h \in \mathbb{R}^2}} \int_{\mathcal{G}} F(y) \, e^{-2\pi i h \cdot y} \, dV(y). \qquad (13.16)$$

For constant weight functions (i.e., $F = 1$) on circles $\mathcal{G}$, the convergence can be realized by estimates of the theory of Bessel functions (cf. Subsection 10.4). It seems that the convergence of the series on the right side of (13.16) for regular regions, i.e., planar "potato slices" $\mathcal{G}$, and twice continuously differentiable functions F, i.e., "general weights" on $\overline{\mathcal{G}}$, can be guaranteed only under the assumption that asymptotic procedures of the stationary phase (cf. Section 3.2) and/or asymptotic relations of the metaharmonic theory (cf. Subsection 6.7) for $|x| \to \infty$ are applicable to the curve integral

$$\int_{\partial \mathcal{G}} F(y) \frac{\partial}{\partial \nu_y} e^{-2\pi i x \cdot y} \, dS(y). \qquad (13.17)$$

In consequence, to the knowledge of the author, only extensions of the Hardy–Landau identity to regular (convex) regions with smooth boundary curves $\partial \mathcal{G}$ possessing a non-vanishing curvature are known yet (see, e.g., V.K. Ivanow [1962]).

Our interest in this chapter is a twofold generalization of the two-dimensional Hardy–Landau identity, i.e., the lattice point-generated Poisson summation formula:

- the generalization from a constant to a general, i.e., twice continuously differentiable weight function F,

- the generalization from a "circle geometry" to a "general", i.e., regular geometry $\mathcal{G}$.

Unfortunately, we do not succeed in both generalizations, and it may be questioned if it is possible at all to do so. A compromise in lattice point summation was formulated in W. Freeden [1978a]. This note shows general weights for circular geometry under spherical summation of the series (13.16).

Nevertheless, we are able to make the following progress in this work:

- the extension to "general", i.e., regular geometries, but under homogeneous "boundary weights",

- the extension to "general weights", however, restricted to convex regions allowing methods of the stationary phase for the boundary curve.

General Geometry and Homogeneous Boundary Weight

Let $\mathcal{G} \subset \mathbb{R}^2$ be a regular region. We consider the function $F_{\mathcal{G}}^{\wedge}$ defined by

$$F_{\mathcal{G}}^{\wedge}(x) = \int_{\substack{y \in \mathcal{G} \\ y \in \mathbb{R}^2}} F(y)\, e^{-2\pi i x \cdot y}\, dV(y), \quad x \in \mathbb{R}^2, \tag{13.18}$$

where F is of class $C^{(2)}(\overline{\mathcal{G}})$, $\overline{\mathcal{G}} = \mathcal{G} \cup \partial \mathcal{G}$. Obviously, $F_{\mathcal{G}}^{\wedge}$ as introduced by (13.18) is infinitely often differentiable in $\mathbb{R}^2$. Therefore, the two-dimensional Euler summation formula with respect to the iterated Laplacian Δ^m is applicable to arbitrary positive integers m.

For all positive values N and arbitrary lattices $\Lambda \subset \mathbb{R}^2$ the Euler summation formula (cf. Theorem 9.2) yields

$$\sideset{}{'}\sum_{\substack{|g| \le N \\ g \in \Lambda}} F_{\mathcal{G}}^{\wedge}(g) = \frac{1}{\|\mathcal{F}\|} \int_{\substack{|x| \le N \\ x \in \mathbb{R}^2}} F_{\mathcal{G}}^{\wedge}(x)\, dV(x) \tag{13.19}$$

$$+ \int_{\substack{|x| \le N \\ x \in \mathbb{R}^2}} G(\Delta^m; x)\, \Delta^m F_{\mathcal{G}}^{\wedge}(x)\, dV(x)$$

$$+ \sum_{k=0}^{m-1} \int_{\substack{|x| = N \\ x \in \mathbb{R}^2}} \left(\frac{\partial}{\partial \nu} G\left(\Delta^{k+1}; x\right) \right) \Delta^k F_{\mathcal{G}}^{\wedge}(x)\, dS(x)$$

$$- \sum_{k=0}^{m-1} \int_{\substack{|x| = N \\ x \in \mathbb{R}^2}} G\left(\Delta^{k+1}; x\right) \left(\frac{\partial}{\partial \nu} \Delta^k F_{\mathcal{G}}^{\wedge}(x) \right) dS(x).$$

In the sequel we use the identity (13.19) for $m \ge 2$ to establish a first lattice point manifestation of the Poisson summation formula, i.e., the "weighted Hardy–Landau identity" on a regular region $\mathcal{G} \in \mathbb{R}^2$, however, under the restrictive assumption of a homogeneous boundary weight, i.e., $F|\partial \mathcal{G} = 0$.

For all $x \in \mathbb{R}^2, x \ne 0$, and all $F \in C^{(2)}(\overline{\mathcal{G}})$, the Second Green Theorem enables us to deduce that

$$F_{\mathcal{G}}^{\wedge}(x) = -\frac{1}{4\pi^2 x^2} \int_{\substack{y \in \mathcal{G} \\ y \in \mathbb{R}^2}} e^{-2\pi i x \cdot y} \left(\Delta_y F(y) \right) dV(y) \tag{13.20}$$

$$- \frac{1}{4\pi^2 x^2} \int_{\substack{y \in \partial \mathcal{G} \\ y \in \mathbb{R}^2}} F(y) \left(\frac{\partial}{\partial \nu_y} e^{-2\pi i x \cdot y} \right) dS(y)$$

$$+ \frac{1}{4\pi^2 x^2} \int_{\substack{y \in \partial \mathcal{G} \\ y \in \mathbb{R}^2}} \left(\frac{\partial}{\partial \nu_y} F(y) \right) e^{-2\pi i x \cdot y}\, dS(y).$$

Hence, under the restriction that $F \in C^{(2)}(\overline{\mathcal{G}})$ with $F|\partial \mathcal{G} = 0$, the "critical curve integral" (13.17) vanishes in (13.20) such that the continuity of F and its derivatives $\nabla F, \Delta F$ on $\overline{\mathcal{G}}$ implies

$$F_{\mathcal{G}}^{\wedge}(x) = O\left(\frac{1}{|x|^2} \right), \quad |x| \to \infty. \tag{13.21}$$

More generally, we find the following asymptotic relations.

Lemma 13.1. *Suppose that $F \in C^{(2)}\left(\overline{\mathcal{G}}\right)$ with $F|\partial\mathcal{G} = 0$. Then $F_{\mathcal{G}}^{\wedge}$ is infinitely often differentiable in $\mathbb{R}^2$ and, for all multi-indices $\alpha = (\alpha_1, \alpha_2)^T$ with $\alpha_i \in \mathbb{N}_0$, $i = 1, 2$, we have*

$$(\nabla_x)^{\alpha} F_{\mathcal{G}}^{\wedge}(x) = O\left(\frac{1}{|x|^2}\right), \quad |x| \to \infty. \tag{13.22}$$

Furthermore, following the well known argument of C.F. Gauß [1801] (see (5.49)) we find

$$\frac{1}{2} \sum_{\substack{|g|=N \\ g\in\Lambda}} F_{\mathcal{G}}^{\wedge}(x) = O\left(N^{-2} \sum_{\substack{|g|=N \\ g\in\Lambda}} 1\right) = O\left(N^{-1}\right), \quad N \to \infty. \tag{13.23}$$

Our asymptotic integral relations known from Theorem 10.2

$$\int_{\substack{|x|=N \\ x\in\mathbb{R}^2}} \left|G(\Delta^k; x)\right| \, dS(x) = O(N) \tag{13.24}$$

and

$$\int_{\substack{|x|=N \\ x\in\mathbb{R}^2}} \left|\frac{\partial}{\partial\nu} G(\Delta^k; x)\right| \, dS(x) = O(N), \tag{13.25}$$

which hold for $N \to \infty$ and all $k \in \mathbb{N}$, play a particular role (for the proof the reader is referred to Section 10.1). In fact, the relations (13.24) and (13.25), in connection with Lemma 13.1 and (13.23), show that

$$\sum_{k=0}^{m-1} \int_{\substack{|x|=N \\ x\in\mathbb{R}^2}} \Delta^k F_{\mathcal{G}}^{\wedge}(x) \left(\frac{\partial}{\partial\nu} G\left(\Delta^{k+1}; x\right)\right) \, dS(x) \tag{13.26}$$

$$- \sum_{k=0}^{m-1} \int_{\substack{|x|=N \\ x\in\mathbb{R}^2}} G\left(\Delta^{k+1}; x\right) \left(\frac{\partial}{\partial\nu} \Delta^k F_{\mathcal{G}}^{\wedge}(x)\right) \, dS(x)$$

$$= O\left(N^{-1}\right), \quad N \to \infty.$$

From (13.19), in connection with (13.23) and (13.26), we therefore obtain

$$\sum_{\substack{|g|\leq N \\ g\in\Lambda}} F_{\mathcal{G}}^{\wedge}(x) = \frac{1}{\|\mathcal{F}\|} \int_{\substack{|x|\leq N \\ x\in\mathbb{R}^2}} F_{\mathcal{G}}^{\wedge}(x) \, dV(x) \tag{13.27}$$

$$+ \int_{\substack{|x|\leq N \\ x\in\mathbb{R}^2}} G\left(\Delta^m; x\right) \left(\Delta^m F_{\mathcal{G}}^{\wedge}(x)\right) \, dV(x)$$

$$+ O\left(N^{-1}\right), \quad N \to \infty.$$

Now, for $m \geq 2$, $G(\Delta^m; \cdot)$ is continuous in $\mathbb{R}^2$. The absolutely and uniformly convergent Fourier series of $G(\Delta^m; \cdot)$, $m \geq 2$, reads as follows

$$G\left(\Delta^m; x\right) = \frac{1}{\|\mathcal{F}\|} \sum_{\substack{|h|>0 \\ h\in\Lambda^{-1}}} \frac{e^{2\pi i h \cdot x}}{(-4\pi^2 h^2)^m}, \quad x \in \mathbb{R}^2. \tag{13.28}$$

Consequently, for $m \geq 2$, we are able to interchange integration and summation

$$\int_{\substack{|x| \leq N \\ x \in \mathbb{R}^2}} G\left(\Delta^m; x\right) \, \Delta^m F_{\mathcal{G}}^{\wedge}(x) \, dV(x) \tag{13.29}$$

$$= \frac{1}{\|\mathcal{F}\|} \sum_{\substack{|h| \neq 0 \\ h \in \Lambda^{-1}}} \frac{1}{(-4\pi^2 h^2)^m} \int_{\substack{|x| \leq N \\ x \in \mathbb{R}^2}} e^{2\pi i h \cdot x} \left(\Delta^m F_{\mathcal{G}}^{\wedge}(x)\right) \, dV(x).$$

Now, the time is ripe for the Fourier inversion formula (see Theorem 13.1). More concretely, we have for all $h \in \Lambda^{-1}$

$$\int_{\substack{|x| \leq N \\ x \in \mathbb{R}^2}} e^{2\pi i h \cdot x} \Delta^m F_{\mathcal{G}}^{\wedge}(x) \, dV(x) \tag{13.30}$$

$$= \int_{\substack{|x| \leq N \\ x \in \mathbb{R}^2}} e^{2\pi i h \cdot x} \int_{\substack{y \in \mathcal{G} \\ y \in \mathbb{R}^2}} F(y)(-4\pi^2 y^2)^m e^{-2\pi i x \cdot y} \, dV(y) \, dV(x).$$

For brevity, we introduce the auxiliary function $\tilde{F} : \overline{\mathcal{G}} \to \mathbb{R}$ by

$$\tilde{F}(y) = (-4\pi^2 y^2)^m F(y), \; y \in \overline{\mathcal{G}}. \tag{13.31}$$

Thus we are able to deduce from the inversion formula (Theorem 13.1) that, for every $\varepsilon > 0$, there exists an integer $M(= M(\varepsilon))$ such that, for all $N \geq M$,

$$\left| \int_{\substack{|x| \leq N \\ x \in \mathbb{R}^2}} e^{2\pi i \, h \cdot x} \left(\int_{\substack{y \in \mathcal{G} \\ y \in \mathbb{R}^2}} \tilde{F}(y) \, e^{-2\pi i x \cdot y} \, dV(y) \right) \, dV(x) \right| \leq \varepsilon \tag{13.32}$$

holds uniformly for all $h \in \Lambda^{-1}$ with $|h| \geq R_0$, where R_0 is chosen in such a way that $R_0 > \sup_{x \in \overline{\mathcal{G}}} |x|$. Therefore, for all $N \geq M$, we have

$$\left| \sum_{\substack{|h| \geq R_0 \\ h \in \Lambda^{-1}}} \frac{1}{(-4\pi^2 h^2)^m} \int_{\substack{|x| \leq N \\ x \in \mathbb{R}^2}} e^{2\pi i h \cdot x} \left(\Delta_x^m F_{\mathcal{G}}^{\wedge}(x)\right) \, dV(x) \right| \tag{13.33}$$

$$\leq \varepsilon C \sum_{\substack{|h| \geq R_0 \\ h \in \Lambda^{-1}}} \frac{1}{(4\pi^2 h^2)^m}$$

$$\leq \varepsilon C'.$$

In other words, for $N \to \infty$, we are therefore able to realize

$$\sum_{\substack{|g| \leq N \\ g \in \Lambda}} \int_{\substack{y \in \mathcal{G} \\ y \in \mathbb{R}^2}} F(y)\, e^{-2\pi i g \cdot y}\, dV(y) \tag{13.34}$$

$$= \frac{1}{\|\mathcal{F}\|} \int_{\substack{|x| \leq N \\ x \in \mathbb{R}^2}} \int_{\substack{y \in \mathcal{G} \\ y \in \mathbb{R}^2}} F(y)\, e^{-2\pi i x \cdot y}\, dV(y)\, dV(x)$$

$$+ \frac{1}{\|\mathcal{F}\|} \sum_{\substack{0 < |h| < R_0 \\ h \in \Lambda^{-1}}} \frac{1}{(-4\pi^2 h^2)^m}$$

$$\times \int_{\substack{|x| \leq N \\ x \in \mathbb{R}^2}} e^{2\pi i h \cdot x} \int_{\substack{y \in \mathcal{G} \\ y \in \mathbb{R}^2}} \tilde{F}(y)\, e^{-2\pi i x \cdot y}\, dV(y)\, dV(x)$$

$$+ o(1).$$

The application of the inversion formula (i.e., Theorem 13.1) to the remaining finite set of integrals, i.e., the execution of the limit $N \to \infty$, guarantees the following lattice point identity.

Theorem 13.2. *Let Λ be an arbitrary lattice in $\mathbb{R}^2$. Let $\mathcal{G} \subset \mathbb{R}^2$ be a regular region. Suppose that $F \in \mathrm{C}^{(2)}\left(\overline{\mathcal{G}}\right)$ satisfies the homogeneous boundary condition $F|\partial\mathcal{G} = 0$. Then the series*

$$\sum_{g \in \Lambda} \int_{\substack{y \in \mathcal{G} \\ y \in \mathbb{R}^2}} F(y)\, e^{-2\pi i g \cdot y}\, dV(y) \tag{13.35}$$

exists in the sense

$$\sum_{g \in \Lambda} \cdots = \lim_{N \to \infty} \sum_{\substack{|g| \leq N \\ g \in \Lambda}} \cdots , \tag{13.36}$$

and we have

$$\frac{1}{\|\mathcal{F}\|} \sum_{\substack{h \in \overline{\mathcal{G}} \\ h \in \Lambda^{-1}}} F(h) = \sum_{g \in \Lambda} \int_{\substack{y \in \mathcal{G} \\ y \in \mathbb{R}^2}} F(y)\, e^{-2\pi i g \cdot y}\, dV(y). \tag{13.37}$$

Replacing $\Lambda \subset \mathbb{R}^2$ by $\Lambda^{-1} \subset \mathbb{R}^2$ we finally obtain the following extension of the two-dimensional Poisson summation formula (i.e., the *Hardy–Landau identity on regular regions with vanishing weight on the boundary $\partial\mathcal{G}$*).

Corollary 13.1. *Let Λ be an arbitrary lattice in $\mathbb{R}^2$. Let $\mathcal{G} \subset \mathbb{R}^2$ be a regular region. Suppose that F is twice continuously differentiable on $\overline{\mathcal{G}} = \mathcal{G} \cup \partial\mathcal{G}$ with $F|\partial\mathcal{G} = 0$. Then*

$$\sum_{\substack{g \in \mathcal{G} \\ g \in \Lambda}} F(g) = \frac{1}{\sqrt{\|\mathcal{F}\|}} \sum_{h \in \Lambda^{-1}} \int_{\substack{y \in \mathcal{G} \\ y \in \mathbb{R}^2}} F(y)\, \overline{\Phi_h(y)}\, dV(y). \tag{13.38}$$

Corollary 13.1 represents a twofold advancement of the two-dimensional formulation of Theorem 12.4, viz. it weakens the order of differentiability of the weight function and it extends the geometry from the circle to general regular regions.

As an example we discuss an identity which also follows immediately by integration from the classical Hardy–Landau identity (see Corollary 10.1). Later on, this example will play a significant role in establishing the almost periodicity of lattice point discrepancies (for more details see Section 14.5).

Example 13.1. *We especially choose* $\Lambda = \mathbb{Z}^2$ *and* $F(x) = R^2 - x^2$, $x \in \overline{\mathbb{B}_R^2}$. *Clearly,* F *is an infinitely often continuously differentiable function on* $\overline{\mathbb{B}_R^2}$, *and* $F|\partial \mathbb{B}_R^2 = 0$. *Therefore we get from Corollary 13.1*

$$\sum_{\substack{|g| \leq R \\ g \in \mathbb{Z}^2}} \left(R^2 - g^2 \right) = \sum_{h \in \mathbb{Z}^2} \int_{\substack{|x| \leq R \\ x \in \mathbb{R}^2}} \left(R^2 - x^2 \right) e^{-2\pi i h \cdot x} \, dV(x). \tag{13.39}$$

The theory of Bessel functions gives

$$\sum_{\substack{|g| \leq R \\ g \in \mathbb{Z}^2}} \left(R^2 - g^2 \right) = \frac{\pi R^4}{2} + \frac{R^2}{\pi} \sum_{\substack{|h| > 0 \\ h \in \mathbb{Z}^2}} \frac{J_2(2\pi |h| R)}{|h|^2}. \tag{13.40}$$

Note that

$$\lim_{\substack{r \to 0 \\ r > 0}} \frac{R^2}{\pi} \frac{J_2(2\pi r R)}{r^2} = \frac{\pi}{2} R^4. \tag{13.41}$$

Thus we finally obtain the identity

$$\sum_{\substack{|g| \leq R \\ g \in \mathbb{Z}^2}} \left(R^2 - g^2 \right) = \frac{R^2}{\pi} \sum_{h \in \mathbb{Z}^2} \frac{J_2(2\pi |h| R)}{|h|^2}. \tag{13.42}$$

Furthermore, we are able to verify (see E. Landau [1927])

$$\int_0^{R^2} \left(\sum_{\substack{|g|^2 \leq u \\ g \in \mathbb{Z}^2}} 1 \right) du \tag{13.43}$$

$$= 1 + \sum_{n \leq R^2 - 1} \int_n^{n+1} \left(\sum_{\substack{|g|^2 \leq u \\ g \in \mathbb{Z}^2}} 1 \right) du + \int_{\lfloor R^2 \rfloor}^{R^2} \left(\sum_{\substack{|g|^2 \leq u \\ g \in \mathbb{Z}^2}} 1 \right) du$$

$$= 1 + \sum_{n \leq R^2 - 1} \left(\sum_{\substack{|g|^2 \leq n \\ g \in \mathbb{Z}^2}} 1 \right) + \left(R^2 - \lfloor R^2 \rfloor \right) \left(\sum_{\substack{|g|^2 \leq \lfloor R^2 \rfloor \\ g \in \mathbb{Z}^2}} 1 \right).$$

Partial summation leads to the identity

$$\int_0^{R^2} \left(\sum_{\substack{|g|^2 \le u \\ g \in \mathbb{Z}^2}} \right) du = \sum_{n \le R^2} \left(\sum_{\substack{|g|^2 = n \\ g \in \mathbb{Z}^2}} (R^2 - n) \right) = \sum_{\substack{|g| \le R \\ g \in \mathbb{Z}^2}} (R^2 - g^2). \quad (13.44)$$

In consequence, we find

$$\sum_{\substack{|g| \le R \\ g \in \mathbb{Z}^2}} (R^2 - g^2) = \int_0^{R^2} \left(\sum_{\substack{|g|^2 \le t \\ g \in \mathbb{Z}^2}} 1 \right) dt. \quad (13.45)$$

In connection with (13.42), the right side of (13.45) is expressible in the following way

$$\int_0^{R^2} \left(\sum_{\substack{|g|^2 \le t \\ g \in \mathbb{Z}^2}} 1 \right) dt = \frac{\pi R^4}{2} + \frac{R^2}{\pi} \sum_{\substack{|h| > 0 \\ h \in \mathbb{Z}^2}} \frac{J_2(2\pi |h| R)}{|h|^2}. \quad (13.46)$$

Combining our results we finally arrive at the identity

$$\int_0^{R^2} \left(\sum_{\substack{|g|^2 \le t \\ g \in \mathbb{Z}^2}} 1 - \pi t \right) dt = \frac{R^2}{\pi} \sum_{\substack{|h| > 0 \\ h \in \mathbb{Z}^2}} \frac{J_2(2\pi |h| R)}{|h|^2}. \quad (13.47)$$

E. Krätzel [2000] pointed out , that the identity (13.47) can be differentiated term by term leading to the Hardy–Landau identity (for R not an integer).

Circles and General Weights

Next we are going to discuss weighted lattice point sums in circles, where the weights are generated by an arbitrary twice continuously differentiable function $F : \mathbb{B}_R^2 \to \mathbb{R}$, $R > 0$ (see W. Freeden [1978a]). The essential tool is the metaharmonic theory.

In order to verify a "weighted version of the Hardy–Landau identity on circles" we again base our activities on the two-dimensional Euler summation formula for general lattices $\Lambda \subset \mathbb{R}^2$, but now on the function $F_{\mathbb{B}_R^2}^\wedge$ (cf. (13.18)) given by

$$F_{\mathbb{B}_R^2}^\wedge(x) = \int_{\substack{|y| \le R \\ y \in \mathbb{R}^2}} F(y)\, e^{-2\pi i x \cdot y}\, dV(y), \quad (13.48)$$

where F is assumed to be of class $C^{(2)} \left(\overline{\mathbb{B}_R^2} \right)$.

Remark 13.1. *A constant weight function, for example* $F = 1$, *on the circular disk* $\overline{\mathbb{B}_R^2}$ *can be handled within the theory of Bessel functions (see (11.12)) so that*

$$\int_{\substack{|y| \leq R \\ y \in \mathbb{R}^2}} e^{-2\pi i x \cdot y} \, dV(y) = R \frac{J_1(2\pi |x| R)}{|x|} = O\left(|x|^{-\frac{3}{2}}\right), \quad |x| \to \infty. \quad (13.49)$$

The two-dimensional Euler summation formula with respect to the (iterated) Laplacian Δ^m, $m \geq 2$, yields for every $N > 0$

$$\sideset{}{'}\sum_{\substack{|g| \leq N \\ g \in \Lambda}} F_{\mathbb{B}_R^2}^{\wedge}(g) = \frac{1}{\|\mathcal{F}\|} \int_{\substack{|x| \leq N \\ x \in \mathbb{R}^2}} F_{\mathbb{B}_R^2}^{\wedge}(x) \, dV(x) \quad (13.50)$$

$$+ \int_{\substack{|x| \leq N \\ x \in \mathbb{R}^2}} G\left(\Delta^m; x\right) \left(\Delta^m F_{\mathbb{B}_R^2}^{\wedge}(x)\right) \, dV(x)$$

$$+ \sum_{k=0}^{m-1} \int_{\substack{|x| = N \\ x \in \mathbb{R}^2}} \left(\frac{\partial}{\partial \nu} G\left(\Delta^{k+1}; x\right)\right) \Delta^k F_{\mathbb{B}_R^2}^{\wedge}(x) \, dS(x)$$

$$- \sum_{k=0}^{m-1} \int_{\substack{|x| = N \\ x \in \mathbb{R}^2}} G\left(\Delta^{k+1}; x\right) \left(\frac{\partial}{\partial \nu} \Delta^k F_{\mathbb{B}_R^2}^{\wedge}(x)\right) \, dS(x).$$

From the Second Green Theorem we get for all $x \in \mathbb{R}^2$ with $x \neq 0$

$$F_{\mathbb{B}_R^2}^{\wedge}(x) = -\frac{1}{4\pi^2 x^2} \int_{\substack{|y| \leq R \\ y \in \mathbb{R}^2}} e^{-2\pi i x \cdot y} \left(\Delta_y F(y)\right) \, dV(y) \quad (13.51)$$

$$- \frac{1}{4\pi^2 x^2} \int_{\substack{|y| = R \\ y \in \mathbb{R}^2}} F(y) \left(\frac{\partial}{\partial \nu_y} e^{-2\pi i x \cdot y}\right) \, dS(y)$$

$$+ \frac{1}{4\pi^2 x^2} \int_{\substack{|y| = R \\ y \in \mathbb{R}^2}} \left(\frac{\partial}{\partial \nu_y} F(y)\right) e^{-2\pi i x \cdot y} \, dS(y).$$

For $|x| \to \infty$ we therefore obtain by observation of the continuity of the first and second order derivative of F

$$F_{\mathbb{B}_R^2}^{\wedge}(x) = \frac{i}{2\pi x^2} \int_{\substack{|y| = R \\ y \in \mathbb{R}^2}} F(y) \left(\nu(y) \cdot x\right) e^{-2\pi i x \cdot y} \, dS(y) \quad (13.52)$$

$$+ O\left(\frac{1}{|x|^2}\right).$$

The integral on the right side of (13.52) can be formulated as an integral over the unit circle $\mathbb{S}^1$

$$\int_{\substack{|y| = R \\ y \in \mathbb{R}^2}} F(y) \left(\nu(y) \cdot x\right) e^{-2\pi i x \cdot y} \, dS(y) \quad (13.53)$$

$$= R|x| \int_{\eta \in \mathbb{S}^1} F(R\eta) \left(\eta \cdot \xi\right) e^{-2\pi i x \cdot R\eta} \, dS(\eta).$$

The last integral is an entire solution of the Helmholtz equation (in the variable $x \in \mathbb{R}^2$); hence, the metaharmonic theory (i.e., Theorem 6.30) provides the asymptotic estimate

$$\int_{\eta \in \mathbb{S}^1} F(R\eta) \, (\eta \cdot \xi) \, e^{-2\pi i x \cdot R\eta} \, dS(\eta) = O\left(|x|^{-\frac{1}{2}} \right) \tag{13.54}$$

for $|x| \to \infty$. In connection with (13.52) and (13.53) we therefore obtain from (13.54)

$$F_{\mathbb{B}_R^2}^{\wedge}(x) = O\left(|x|^{-\frac{3}{2}} \right), \quad |x| \to \infty. \tag{13.55}$$

More generally we find the following asymptotic relations.

Lemma 13.2. *For all multi-indices* $\alpha = (\alpha_1, \alpha_2)^T$ *with* $\alpha_i \in \mathbb{N}_0$, $i = 1, 2$, *we have*

$$(\nabla_x)^{\alpha} \, F_{\mathbb{B}_R^2}^{\wedge}(x) = O\left(|x|^{-\frac{3}{2}} \right), \quad |x| \to \infty. \tag{13.56}$$

In connection with (5.47) it follows that

$$\frac{1}{2} \sum_{\substack{|g|=N \\ g \in \Lambda}} F_{\mathbb{B}_R^2}^{\wedge}(g) = O\left(N^{-\frac{3}{2}} \sum_{\substack{|g|=N \\ g \in \Lambda}} 1 \right) = O\left(N^{-\frac{1}{2}} \right), \quad N \to \infty. \tag{13.57}$$

The relations (13.24) and (13.25) in connection with Lemma 13.2 show that

$$\sum_{k=0}^{m-1} \int_{\substack{|x|=N \\ x \in \mathbb{R}^2}} \Delta^k F_{\mathbb{B}_R^2}^{\wedge}(x) \left(\frac{\partial}{\partial \nu} G\left(\Delta^{k+1}; x \right) \right) dS(x) \tag{13.58}$$

$$- \sum_{k=0}^{m-1} \int_{\substack{|x|=N \\ x \in \mathbb{R}^2}} G\left(\Delta^{k+1}; x \right) \left(\frac{\partial}{\partial \nu} \Delta^k F_{\mathbb{B}_R^2}^{\wedge}(x) \right) dS(x)$$

$$= O\left(N^{-\frac{1}{2}} \right), \quad N \to \infty.$$

Altogether, from (13.50) in connection with (13.57) and (13.58), we are led to

$$\sum_{\substack{|g| \leq N \\ g \in \Lambda}} F_{\mathbb{B}_R^2}^{\wedge}(g) = \frac{1}{\|\mathcal{F}\|} \int_{\substack{|x| \leq N \\ x \in \mathbb{R}^2}} F_{\mathbb{B}_R^2}^{\wedge}(x) \, dV(x) \tag{13.59}$$

$$+ \int_{\substack{|x| \leq N \\ x \in \mathbb{R}^2}} G\left(\Delta^m; x \right) \left(\Delta^m F_{\mathbb{B}_R^2}^{\wedge}(x) \right) dV(x)$$

$$+ O\left(N^{-\frac{1}{2}} \right), \quad N \to \infty.$$

We use again the Fourier inversion formula (i.e., Theorem 13.1). For $h \in \Lambda^{-1}$, we have

$$\int_{\substack{|x|\leq N \\ x\in\mathbb{R}^2}} e^{2\pi ih\cdot x} \Delta_x^m F_{\mathbb{B}_R^2}^\wedge(x)\, dV(x) \tag{13.60}$$

$$= \int_{\substack{|x|\leq N \\ x\in\mathbb{R}^2}} e^{2\pi ih\cdot x} \int_{\substack{|y|\leq R \\ y\in\mathbb{R}^2}} F(y)\,(-4\pi^2 y^2)^m\, e^{-2\pi ix\cdot y}\, dV(y)\, dV(x).$$

We are able to state that for every $\varepsilon > 0$, there exists an integer $M(= M(\varepsilon))$ such that for all $N \geq M$,

$$\left| \int_{\substack{|x|\leq N \\ x\in\mathbb{R}^2}} e^{2\pi ih\cdot x} \left(\int_{\substack{|y|\leq R \\ y\in\mathbb{R}^2}} (-4\pi^2 y^2)^m F(y)\, e^{-2\pi ix\cdot y}\, dV(x) \right) dV(y) \right| \leq \varepsilon \tag{13.61}$$

holds uniformly for all $h \in \Lambda^{-1}$ with $|h| \geq R_0 > R$. Thus, for all $N \geq M$, we therefore have

$$\left| \sum_{\substack{|h|\geq R_0 \\ h\in\Lambda^{-1}}} \frac{1}{(-4\pi^2 h^2)^m} \int_{\substack{|x|\leq N \\ x\in\mathbb{R}^2}} e^{2\pi ih\cdot x} \left(\Delta_x^m F_{\mathbb{B}_R^2}^\wedge(x) \right) dV(x) \right|$$

$$\leq \varepsilon\, C \sum_{\substack{|h|\geq R_0 \\ h\in\Lambda^{-1}}} \frac{1}{(4\pi^2 h^2)^m} \tag{13.62}$$

$$\leq \varepsilon\, C'.$$

Thus, for $N \to \infty$, we get

$$\sum_{\substack{|g|\leq N \\ g\in\Lambda}} \int_{\substack{|y|\leq R \\ y\in\mathbb{R}^2}} F(y)\, e^{-2\pi ig\cdot y}\, dV(y) \tag{13.63}$$

$$= \frac{1}{\|\mathcal{F}\|} \int_{\substack{|x|\leq N \\ x\in\mathbb{R}^2}} \int_{\substack{|y|\leq R \\ y\in\mathbb{R}^2}} F(y)\, e^{-2\pi i\, x\cdot y}\, dV(y)\, dV(x)$$

$$+ \frac{1}{\|\mathcal{F}\|} \sum_{\substack{0<|h|<R_0 \\ h\in\Lambda^{-1}}} \frac{1}{(-4\pi^2 h^2)^m}$$

$$\times \int_{\substack{|x|\leq N \\ x\in\mathbb{R}^2}} e^{2\pi ih\cdot x} \int_{\substack{|y|\leq R \\ y\in\mathbb{R}^2}} (-4\pi^2 y^2)^m F(y)\, e^{-2\pi ix\cdot y}\, dV(y)\, dV(x)$$

$$+ o(1).$$

The application of the Fourier inversion formula (Theorem 13.1) to the remaining finite sum of integrals yields another interesting identity of the two-dimensional lattice point theory, namely the *Hardy–Landau identity for arbitrary weight functions on circular disks* (see W. Freeden [1975, 1978a]).

Theorem 13.3. *Let Λ be an arbitrary lattice in $\mathbb{R}^2$. Then, for every function $F \in C^{(2)}\left(\overline{\mathbb{B}_R^2}\right)$, $R > 0$, the series*

$$\sum_{g \in \Lambda} \int_{\substack{|y| \leq R \\ y \in \mathbb{R}^2}} F(y)\, e^{-2\pi i g \cdot y}\, dV(y) \tag{13.64}$$

exists in the sense

$$\sum_{g \in \Lambda} \cdots = \lim_{N \to \infty} \sum_{\substack{|g| \leq N \\ g \in \Lambda}} \cdots, \tag{13.65}$$

and we have

$$\frac{1}{\|\mathcal{F}\|} \sum_{\substack{|h| \leq R \\ h \in \Lambda^{-1}}} {}' F(h) = \sum_{g \in \Lambda} \int_{\substack{|y| \leq R \\ y \in \mathbb{R}^2}} F(y)\, e^{-2\pi i g \cdot y}\, dV(y). \tag{13.66}$$

Replacing $\Lambda \subset \mathbb{R}^2$ by $\Lambda^{-1} \subset \mathbb{R}^2$ we finally obtain the following extension of the Hardy–Landau identity.

Corollary 13.2. *Let Λ be an arbitrary lattice in $\mathbb{R}^2$. Suppose that F is twice continuously differentiable in $\overline{\mathbb{B}_R^2}$, $R > 0$. Then*

$$\sum_{\substack{|g| \leq R \\ g \in \Lambda}} {}' F(g) = \frac{1}{\sqrt{\|\mathcal{F}\|}} \sum_{h \in \Lambda^{-1}} \int_{\substack{|y| \leq R \\ y \in \mathbb{R}^2}} F(y)\, \overline{\Phi_h(y)}\, dV(y). \tag{13.67}$$

More generally we are able to verify (cf. W. Freeden [1975])

Corollary 13.3. *Let Λ be an arbitrary lattice in $\mathbb{R}^2$. Suppose that F is twice continuously differentiable in $\overline{\mathbb{B}_R^2}$, $R > 0$. Then, for $b \in \mathbb{R}^2$,*

$$\sum_{\substack{|g| \leq R \\ g \in \Lambda}} {}' e^{2\pi i b \cdot g}\, F(g) = \frac{1}{\|\mathcal{F}\|} \sum_{h \in \Lambda^{-1}} \int_{\substack{|y| \leq R \\ x \in \mathbb{R}^2}} F(x)\, e^{-2\pi i (h-b) \cdot x}\, dV(x). \tag{13.68}$$

Corollary 13.3, in connection with Lemma 6.30, yields as special case

Corollary 13.4. *For $b \in \mathbb{R}^2$*

$$\sum_{\substack{|g| \leq R \\ g \in \Lambda}} {}' e^{2\pi i b \cdot g} = \frac{R}{\|\mathcal{F}\|} \lim_{N \to \infty} \sum_{\substack{|b-h| \leq N \\ h \in \Lambda^{-1}}} \frac{J_1(2\pi |b-h| R)}{|b-h|}. \tag{13.69}$$

From Corollary 13.2 we readily obtain the following identity for weighted lattice point sums in circular rings.

Corollary 13.5. *Let Λ be an arbitrary lattice in $\mathbb{R}^2$. Suppose that F is twice continuously differentiable in $\overline{\mathbb{B}^2_{\rho,R}} = \overline{\mathbb{B}^2_R} \backslash \mathbb{B}^2_\rho$, $0 \le \rho < R$. Then*

$$\sideset{}{'}\sum_{\substack{\rho \le |g| \le R \\ g \in \Lambda}} F(g) = \frac{1}{\sqrt{\|\mathcal{F}\|}} \sum_{h \in \Lambda^{-1}} \int_{\substack{\rho \le |y| \le R \\ y \in \mathbb{R}^2}} F(y) \, \overline{\Phi_h(y)} \, dV(y), \qquad (13.70)$$

where

$$\sideset{}{'}\sum_{\substack{\rho \le |g| \le R \\ g \in \Lambda}} F(g) = \sideset{}{'}\sum_{\substack{|g| \le R \\ g \in \Lambda}} F(g) - \sideset{}{'}\sum_{\substack{|g| \le \rho \\ g \in \Lambda}} F(g). \qquad (13.71)$$

Corollary 13.2 enables us to formulate a large variety of weighted Hardy–Landau type identities. We mention the following examples for *radial or angular weight functions*.

Example 13.2. *Let Λ be a lattice in $\mathbb{R}^2$. Let $G : \overline{\mathbb{B}^2_{\rho,N}} \to \mathbb{R}$, $0 \le \rho \le N$, be a radial twice continuously differentiable function. Then*

$$\sideset{}{'}\sum_{\substack{\rho \le |g| \le N \\ g \in \Lambda}} G(|g|) = \frac{1}{\|\mathcal{F}\|} \sum_{h \in \Lambda^{-1}} \int_{\rho \le |x| \le N} G(|x|) \, e^{-2\pi i h \cdot x} \, dV(x)$$

$$= 2\pi \frac{1}{\|\mathcal{F}\|} \int_\rho^N r G(r) \, dr \qquad (13.72)$$

$$+ 2\pi \frac{1}{\|\mathcal{F}\|} \sum_{\substack{|h| \ne 0 \\ h \in \Lambda^{-1}}} \int_\rho^N r G(r) J_0(2\pi |h| r) \, dr.$$

In particular,

$$\sideset{}{'}\sum_{\substack{\rho \le |g| \le N \\ g \in \Lambda}} 1 = \frac{1}{\|\mathcal{F}\|} \sum_{h \in \Lambda^{-1}} \left(r \frac{J_1(2\pi |h| r)}{|h|} \bigg|_{r=\rho}^{r=N} \right). \qquad (13.73)$$

Example 13.3. *Let Λ be a lattice in $\mathbb{R}^2$. Let $H : \overline{\mathbb{B}^2_{\rho,N}} \to \mathbb{R}$, $0 < \rho \le N$, be an angular twice continuously differentiable function. Then*

$$\sideset{}{'}\sum_{\substack{\rho \le |g| \le N \\ g \in \Lambda}} H\left(\frac{g}{|g|}\right) = \frac{1}{\|\mathcal{F}\|} \sum_{h \in \Lambda^{-1}} \int_{\rho \le |x| \le N} H\left(\frac{x}{|x|}\right) e^{-2\pi i h \cdot x} \, dV(x)$$

$$= \frac{1}{\|\mathcal{F}\|} \left(\frac{r^2}{2} \bigg|_{r=\rho}^{r=N} \right) \int_{\mathbb{S}^1} H(\xi) \, dS(\xi) \qquad (13.74)$$

$$+ \frac{1}{\|\mathcal{F}\|} \sum_{\substack{|h| \ne 0 \\ h \in \Lambda^{-1}}} \int_\rho^N r \int_{\mathbb{S}^1} H(\xi) \, e^{-2\pi i |h| r \left(\frac{h}{|h|} \cdot \xi\right)} \, dS(\xi) \, dr.$$

Convex "Smooth" Regions and General Weights

Finally we deal with weighted lattice point sums in convex regions $\mathcal{G}$ including the origin and possessing a smooth curve $\partial\mathcal{G}$ with non-vanishing curvature (cf. V.K. Ivanow [1962]).

As we saw, the validity of a weighted Hardy-Landau type identity essentially depends on the asymptotic expansion

$$F_{\mathcal{G}}^{\wedge}(x) = \frac{i}{2\pi x^2} \int_{\partial\mathcal{G}} F(y)(\nu(y) \cdot x) \, e^{-2\pi i x \cdot y} \, dS(y) + O\left(\frac{1}{|x|^2}\right), \qquad (13.75)$$

$|x| \to \infty$. In terms of polar coordinates $x = r\xi$, $r = |x|$, $\xi \in \mathbb{S}^1$, the curve integral on the right of (13.75) can be rewritten in the form

$$\int_{\partial\mathcal{G}} F(y)(\nu(y) \cdot x) \, e^{-2\pi i x \cdot y} \, dS(y) \qquad (13.76)$$

$$= |x| \int_{\partial\mathcal{G}} F(y) \cos(\sphericalangle(\nu(y), \xi)) \, e^{-2\pi i |x| \, |y| \cos(\sphericalangle \, (\xi, \frac{y}{|y|}))} \, dS(y),$$

where $\vartheta = \sphericalangle(\nu(y), \xi), y = (y_1, y_2)^T$, is the angle between the vectors $\nu(y)$ and ξ. We fix the unit vector $\xi = (\cos\varphi, \sin\varphi)^T$, $0 \leq \varphi < 2\pi$, in the (y_1, y_2)-plane. Then we introduce a new (w_1, w_2)-coordinate system by rotating the (y_1, y_2)-plane in the positive sense with the angle φ. In doing so we are able to describe the curve $\partial\mathcal{G}$ in a parametric way by choosing as parameter the angle $\vartheta = \sphericalangle(\nu(y), \xi)$, i.e., the angle between the normal ν and the w_1-axis at the point y under consideration. Under these circumstances we have $dS = \cos(\vartheta)C(\vartheta) \, d\vartheta$, where $C(\vartheta)$ is the curvature radius of $\partial\mathcal{G}$ and $x \cdot y = |x| w_1(\vartheta)$. Consequently, the integral (13.76) can be rewritten in the form

$$\int_0^{2\pi} \tilde{F}(\vartheta) \, e^{-2\pi i |x| w_1(\vartheta)} \cos^2(\vartheta) \, C(\vartheta) \, d\vartheta \qquad (13.77)$$

where

$$\tilde{F}(\vartheta) = F(w_1(\vartheta), w_2(\vartheta)), \quad \vartheta \in [0, 2\pi). \qquad (13.78)$$

Next we introduce auxiliary functions $\check{F}, \tilde{G}$ by

$$\check{F}(\vartheta) = \tilde{F}(\vartheta) \cos^2(\vartheta) C(\vartheta), \quad \tilde{G}(\vartheta) = w_1(\vartheta). \qquad (13.79)$$

The derivatives of $\tilde{G}$ are of importance for the application of the theory of the stationary phase. We readily find for the derivative $\tilde{G}'$

$$\tilde{G}'(\vartheta) = \frac{dw_1}{dS} \frac{dS}{d\vartheta} = -C(\vartheta) \sin(\vartheta), \qquad (13.80)$$

hence, $\tilde{G}'(\vartheta) = 0$ for $\vartheta = 0, \pi, 2\pi$. The second derivative reads

$$\tilde{G}''(\vartheta) = -C'(\vartheta) \sin\vartheta - C(\vartheta) \cos(\vartheta), \qquad (13.81)$$

so that

$$\tilde{G}''(0) = \tilde{G}''(2\pi) = -C(0) < 0, \quad \tilde{G}''(\pi) = C(\pi) > 0. \tag{13.82}$$

Consequently, the method of the stationary phase (see Corollary 3.2) is applicable. It shows that

$$\int_0^{2\pi} \tilde{F}(\vartheta) \, e^{-2\pi i |x| \tilde{G}(\vartheta)} \, d\vartheta = O\left(|x|^{-\frac{1}{2}}\right), \quad |x| \to \infty. \tag{13.83}$$

Following almost literally the arguments as in Section 13.2 we therefore obtain

Theorem 13.4. *Let Λ be an arbitrary lattice in $\mathbb{R}^2$. Let $\mathcal{G} \subset \mathbb{R}^2$ be a convex region containing the origin with boundary curve ∂G such that its normal field ν is continuously differentiable and its curvature is non-vanishing. Suppose that F is of class $C^{(2)}\left(\overline{\mathcal{G}}\right)$. Then*

$$\sum_{g \in \Lambda} \int_{\mathcal{G}} F(y) \, e^{-2\pi i g \cdot y} \, dV(y) \tag{13.84}$$

exists in spherical sense, and we have

$$\sum_{g \in \Lambda} \int_{\mathcal{G}} F(y) \, e^{-2\pi i g \cdot y} \, dV(y) = \frac{1}{\|\mathcal{F}\|} \sum_{\substack{h \in \overline{\mathcal{G}} \\ h \in \Lambda^{-1}}} {}' F(h). \tag{13.85}$$

Replacing Λ by Λ^{-1} we finally obtain the following weighted version of the Hardy–Landau identity, which at the same time is a two-dimensional realization of the Poisson summation formula

Corollary 13.6. *Under the assumptions of Theorem 13.4 we have*

$$\sum_{\substack{g \in \overline{\mathcal{G}} \\ g \in \Lambda}} {}' F(g) = \frac{1}{\sqrt{\|\mathcal{F}\|}} \sum_{h \in \Lambda^{-1}} \int_{\mathcal{G}} F(y) \, \overline{\Phi_h(y)} \, dV(y). \tag{13.86}$$

Estimates for the Λ-lattice point discrepancy for constant F on $\mathcal{G}$ can be found in A. Ivic et al. [2004] and the references therein.

Corollary 13.7. *Let Λ be an arbitrary lattice in $\mathbb{R}^2$. Let $\mathcal{G}_1, \mathcal{G}_2 \subset \mathbb{R}^2$ be regions as described in Theorem 13.4 satisfying $\overline{\mathcal{G}_1} \subset \mathcal{G}_2$. Suppose that F is of class $C^{(2)}\left(\overline{\mathcal{G}_2 \backslash \mathcal{G}_1}\right)$. Then*

$$\sum_{\substack{g \in \overline{\mathcal{G}_2} \backslash \mathcal{G}_1 \\ g \in \Lambda}} {}' F(g) = \frac{1}{\sqrt{\|\mathcal{F}\|}} \sum_{h \in \Lambda^{-1}} \int_{\overline{\mathcal{G}_2} \backslash \mathcal{G}_1} F(y) \, \overline{\Phi_h(y)} \, dV(y), \tag{13.87}$$

where

$$\sum_{\substack{g \in \overline{\mathcal{G}_2} \backslash \mathcal{G}_1 \\ g \in \Lambda}} {}' F(g) = \sum_{\substack{g \in \overline{\mathcal{G}_2} \\ g \in \Lambda}} {}' F(g) - \sum_{\substack{g \in \overline{\mathcal{G}_1} \\ g \in \Lambda}} {}' F(g). \tag{13.88}$$

13.3 Weighted Two-Dimensional Lattice Ball Identities

Contrary to the weighted lattice point sums (see Section 13.2), the weighted lattice ball variants admit a twofold generalization of the two-dimensional lattice ball-generated Poisson summation formula:

- the generalization from a constant to a "general", i.e., twice continuously differentiable weight function F,

- the generalization from a "circle geometry" to a "general", i.e., regular, geometry $\mathcal{G}$.

As a matter of fact, the lattice ball concept together with our metaharmonic tools of Section 6.7 enables us to develop the Poisson summation formula for regular regions, without any restriction with respect to the boundary curve and the weight function on the boundary. Neither the convexity nor the non-zero curvature have to be prescribed. The price to be paid is an additional Bessel term to enforce the convergence of the weighted Hardy–Landau lattice ball series.

General Geometry and General Weights (Variant 1)

According to Theorem 11.2 we have for $F \in C^{(2)}\left(\overline{\mathcal{G}}\right)$ with $\mathcal{G}$ being a regular region in $\mathbb{R}^2$

$$\frac{1}{\pi\tau^2} \int_{\mathcal{G}} \mathcal{X}_{\overline{\mathbb{B}_\tau^q} + \Lambda}(x) F(x) \, dV(x) = \frac{1}{\|\mathcal{F}\|} \int_{\mathcal{G}} F(x) \, dV(x) \tag{13.89}$$

$$+ \int_{\mathcal{G}} G_\tau(\Delta; x) \, \Delta F(x) \, dV(x)$$

$$+ \int_{\partial\mathcal{G}} \left\{ F(x) \left(\frac{\partial}{\partial\nu} G_\tau(\Delta; x) \right) - G_\tau(\Delta; x) \left(\frac{\partial}{\partial\nu} F(x) \right) \right\} dS(x).$$

Inserting the absolutely and uniformly convergent Fourier series in $\mathbb{R}^2$

$$G_\tau(\Delta; x) = \frac{1}{\|\mathcal{F}\|} \sum_{\substack{|h|>0 \\ h\in\Lambda^{-1}}} \frac{J_1(2\pi\tau|h|)}{\pi\tau|h|} \frac{e^{-2\pi i h \cdot x}}{-4\pi^2 h^2}, \quad x \in \mathbb{R}^2, \tag{13.90}$$

we obtain

$$\frac{1}{\pi\tau^2}\int_{\mathcal{G}}\mathcal{X}_{\overline{\mathbb{B}_\tau^q}+\Lambda}(x)F(x)\,dV(x)-\frac{1}{\|\mathcal{F}\|}\int_{\mathcal{G}}F(x)\,dV(x) \tag{13.91}$$

$$=\frac{1}{\|\mathcal{F}\|}\sum_{\substack{|h|>0\\h\in\Lambda^{-1}}}\frac{J_1(2\pi\tau|h|)}{\pi\tau|h|}\frac{1}{-4\pi^2h^2}\int_{\mathcal{G}}e^{-2\pi ih\cdot x}\Delta F(x)\,dV(x)$$

$$+\frac{1}{\|\mathcal{F}\|}\sum_{\substack{|h|>0\\h\in\Lambda^{-1}}}\frac{J_1(2\pi\tau|h|)}{\pi\tau|h|}\frac{1}{-4\pi^2h^2}\int_{\partial\mathcal{G}}\left(\frac{\partial}{\partial\nu}e^{-2\pi ih\cdot x}\right)F(x)\,dS(x)$$

$$-\frac{1}{\|\mathcal{F}\|}\sum_{\substack{|h|>0\\h\in\Lambda^{-1}}}\frac{J_1(2\pi\tau|h|)}{\pi\tau|h|}\frac{1}{-4\pi^2h^2}\int_{\partial\mathcal{G}}e^{-2\pi ih\cdot x}\left(\frac{\partial}{\partial\nu}F(x)\right)dS(x).$$

Partial integration, i.e., the Second Green Theorem, shows that for all $h\neq0$

$$\int_{\mathcal{G}}F(x)\,e^{-2\pi ih\cdot x}\,dV(x)=\frac{1}{-4\pi^2h^2}\int_{\mathcal{G}}e^{-2\pi ih\cdot x}\Delta F(x)\,dV(x) \tag{13.92}$$

$$+\frac{1}{-4\pi^2h^2}\int_{\partial\mathcal{G}}\left(\frac{\partial}{\partial\nu}e^{-2\pi ih\cdot x}\right)F(x)\,dS(x)$$

$$-\frac{1}{-4\pi^2h^2}\int_{\partial\mathcal{G}}e^{-2\pi ih\cdot x}\left(\frac{\partial}{\partial\nu}F(x)\right)dS(x).$$

Inserting (13.92) into (13.91) we obtain the identity

$$\frac{1}{\pi\tau^2}\int_{\mathcal{G}}\mathcal{X}_{\overline{\mathbb{B}_\tau^q}+\Lambda}(x)F(x)\,dV(x) \tag{13.93}$$

$$=\frac{1}{\|\mathcal{F}\|}\sum_{h\in\Lambda^{-1}}\frac{J_1(2\pi\tau|h|)}{\pi\tau|h|}\int_{\mathcal{G}}F(x)\,e^{-2\pi ih\cdot x}\,dV(x).$$

Note that, for $F\in C^{(2)}\left(\overline{\mathcal{G}}\right)$, the Second Green Theorem yields

$$F_{\mathcal{G}}^{\wedge}(y)=\int_{\mathcal{G}}F(x)e^{-2\pi iy\cdot x}\,dV(x)=O\left(\frac{1}{|y|}\right),\quad|y|\to\infty, \tag{13.94}$$

so that the series on the right side of (13.93) is absolutely convergent.

The identity (13.93) can be interpreted as a lattice ball variant of the Poisson summation formula for an arbitrary regular region $\mathcal{G}\subset\mathbb{R}^2$ and an arbitrary weight function F of class $C^{(2)}\left(\overline{\mathcal{G}}\right)$. Even better, going over to the limit $\tau\to0$ we are able to formulate the following identity

$$\sum_{\substack{g\in\overline{\mathcal{G}}\\g\in\Lambda}}{}'F(g)=\lim_{\substack{\tau\to0\\\tau>0}}\frac{1}{\|\mathcal{F}\|}\sum_{h\in\Lambda^{-1}}\frac{J_1(2\pi\tau|h|)}{\pi\tau|h|}\int_{\mathcal{G}}F(x)\,e^{-2\pi ih\cdot x}\,dV(x). \tag{13.95}$$

We already know that, for a convex region $\mathcal{G}$ containing the origin with the boundary curve ∂G such that its normal field ν is continuously differentiable and its curvature is non-vanishing, the series and the limit in (13.95) can be interchanged; hence, Theorem 13.4 follows as the immediate result.

General Geometry and General Weights (Variant 2)

For a function $F \in C^{(2)}\left(\overline{\mathcal{G}}\right)$ with $\mathcal{G}$ being a regular region in $\mathbb{R}^2$ we continue our considerations with a second variant of the Λ-lattice ball generated Euler summation formula (cf. Theorem 11.2) applied to $F_{\mathcal{G}}^{\wedge}$ as defined by (13.18):

$$\frac{1}{\pi\tau^2} \int_{\substack{|x| \leq N \\ x \in \mathbb{R}^2}} \chi_{\overline{\mathbb{B}_\tau^2}+\Lambda}(x) F_{\mathcal{G}}^{\wedge}(x) \, dV(x) \tag{13.96}$$

$$= \frac{1}{\|F\|} \int_{\substack{|x| \leq N \\ x \in \mathbb{R}^2}} F_{\mathcal{G}}^{\wedge}(x) \, dV(x)$$

$$+ \int_{\substack{|x| \leq N \\ x \in \mathbb{R}^2}} G_\tau(\Delta; x) \Delta F_{\mathcal{G}}^{\wedge}(x) \, dV(x)$$

$$+ \int_{\substack{|x| = N \\ x \in \mathbb{R}^2}} \left(\frac{\partial}{\partial \nu} G_\tau(\Delta; x) \right) F_{\mathcal{G}}^{\wedge}(x) \, dS(x)$$

$$- \int_{\substack{|x| = N \\ x \in \mathbb{R}^2}} G_\tau(\Delta; x) \left(\frac{\partial}{\partial \nu} F_{\mathcal{G}}^{\wedge}(x) \right) \, dS(x).$$

From Lemma 13.1 we borrow the asymptotic estimate

$$(\nabla_x)^\alpha F_{\mathcal{G}}^{\wedge}(x) = O\left(\frac{1}{|x|} \right), \quad |x| \to \infty, \tag{13.97}$$

for all multi-indices $\alpha = (\alpha_1, \alpha_2)^T$ with $\alpha_i \in \mathbb{N}_0$, $i \in \{1,2\}$, provided that F is assumed to be of class $C^{(2)}\left(\overline{\mathcal{G}}\right)$ with $\mathcal{G}$ being a regular region in $\mathbb{R}^2$. For $m \in \mathbb{N}$, $G_\tau(\Delta^m; \cdot)$, $0 < \tau < \inf_{x \in \partial \mathcal{F}} |x|$, is continuous in $\mathbb{R}^2$. The absolutely and uniformly convergent series expansion of $G_\tau(\Delta; \cdot)$

$$G_\tau(\Delta; x) = \frac{1}{\|\mathcal{F}\|} \sum_{\substack{|h| > 0 \\ h \in \Lambda^{-1}}} \frac{J_1(2\pi|h|\tau)}{\pi\tau|h|} \frac{e^{-2\pi i h \cdot x}}{-4\pi^2 h^2}, \quad x \in \mathbb{R}^2, \tag{13.98}$$

enables us to write

$$\int_{\substack{|x| = N \\ x \in \mathbb{R}^2}} G_\tau(\Delta; x) \frac{\partial}{\partial \nu} F_{\mathcal{G}}^{\wedge}(x) \, dS(x) \tag{13.99}$$

$$= \frac{1}{\|\mathcal{F}\|} \frac{1}{\pi\tau} \sum_{\substack{|h| > 0 \\ h \in \Lambda^{-1}}} \frac{J_1(2\pi|h|\tau)}{-4\pi^2|h|^3} \int_{\substack{|x| = N \\ x \in \mathbb{R}^2}} e^{-2\pi i h \cdot x} \left(\frac{\partial}{\partial \nu} F_{\mathcal{G}}^{\wedge}(x) \right) dS(x)$$

and

$$\int_{\substack{|x| = N \\ x \in \mathbb{R}^2}} \left(\frac{\partial}{\partial \nu} G_\tau(\Delta; x) \right) F_{\mathcal{G}}^{\wedge}(x) \, dS(x) \tag{13.100}$$

$$= \frac{1}{\|\mathcal{F}\|} \sum_{\substack{|h| > 0 \\ h \in \Lambda^{-1}}} \frac{J_1(2\pi|h|\tau)}{\pi\tau|h|} \frac{1}{-4\pi^2 h^2} \int_{\substack{|x| = N \\ x \in \mathbb{R}^2}} \left(\frac{\partial}{\partial \nu_x} e^{-2\pi i h \cdot x} \right) F_{\mathcal{G}}^{\wedge}(x) \, dS(x).$$

From the metaharmonic theory (i.e., Theorem 6.30) it follows that the estimates

$$\int_{\substack{|x|=N \\ x \in \mathbb{R}^2}} e^{-2\pi i \ h \cdot x} \left(\frac{\partial}{\partial \nu} F_{\mathcal{G}}^{\wedge}(x) \right) \ dS(x) \tag{13.101}$$

$$= N \int_{\mathbb{S}^1} e^{-2\pi i |h| N \frac{h}{|h|} \cdot \xi} \left(\left(\frac{\partial}{\partial \nu} F_{\mathcal{G}}^{\wedge} \right) (N\xi) \right) \ dS(\xi)$$

$$= O \left(\frac{N}{(|h|N)^{\frac{1}{2}}} \left(\left| (\nabla F_{\mathcal{G}}^{\wedge}) \left(\frac{h}{|h|} \right) \right| + \left| (\nabla F_{\mathcal{G}}^{\wedge}) \left(-N \frac{h}{|h|} \right) \right| \right) \right)$$

$$= O \left(\frac{1}{(|h|N)^{\frac{1}{2}}} \right), \quad N \to \infty$$

and

$$\int_{|x|=N} \left(\frac{\partial}{\partial \nu} e^{-2\pi i h \cdot x} \right) F_{\mathcal{G}}^{\wedge}(x) \ dS(x) \tag{13.102}$$

$$= -2\pi i N \int_{\mathbb{S}^1} (h \cdot \xi) \ e^{-2\pi i |h| N \left(\frac{h}{|h|} \cdot \xi \right)} F_{\mathcal{G}}^{\wedge}(N\xi) \ dS(\xi)$$

$$= O \left(\frac{N}{(|h|N)^{\frac{1}{2}}} |h| \left(\left| F_{\mathcal{G}}^{\wedge} \left(N \frac{h}{|h|} \right) \right| + \left| F_{\mathcal{G}}^{\wedge} \left(-N \frac{h}{|h|} \right) \right| \right) \right)$$

$$= O \left(\left(\frac{|h|}{N} \right)^{\frac{1}{2}} \right), \quad N \to \infty$$

are valid for all $h \in \Lambda^{-1}$ with $|h| > 0$. Therefore we are able to conclude that

$$\int_{\substack{|x|=N \\ x \in \mathbb{R}^2}} G_\tau(\Delta; x) \left(\frac{\partial}{\partial \nu} F_{\mathcal{G}}^{\wedge}(x) \right) \ dS(x) = O \left(N^{-\frac{1}{2}} \right) \tag{13.103}$$

and

$$\int_{|x|=N} \left(\frac{\partial}{\partial \nu} G_\tau(\Delta; x) \right) F_{\mathcal{G}}^{\wedge}(x) \ dS(x) = O \left(N^{-\frac{1}{2}} \right) \tag{13.104}$$

hold true for *all* regular regions $\mathcal{G} \subset \mathbb{R}^2$ and $N \to \infty$.

Because of the estimate $|J_1(2\pi|h|\tau)| = O\left(|h|^{-\frac{1}{2}}\right)$ we are allowed to interchange integration and summation such that

$$\int_{\substack{|x| \leq N \\ x \in \mathbb{R}^2}} G_\tau(\Delta; x) \ \Delta F_{\mathcal{G}}^{\wedge}(x) \ dV(x) \tag{13.105}$$

$$= \frac{1}{\|\mathcal{F}\|} \sum_{\substack{|h| > 0 \\ h \in \Lambda^{-1}}} \frac{J_1(2\pi|h|\tau)}{\pi\tau|h|} \frac{1}{-4\pi^2 h^2} \int_{\substack{|x| \leq N \\ x \in \mathbb{R}^2}} e^{-2\pi i h \cdot x} \left(\Delta_x F_{\mathcal{G}}^{\wedge}(x) \right) \ dV(x).$$

We know that

$$\int_{\substack{|x|\leq N \\ x\in\mathbb{R}^2}} e^{-2\pi ih\cdot x} \Delta_x F_{\mathcal{G}}^{\wedge}(x) \ dV(x) \qquad (13.106)$$

$$= \int_{\substack{|x|\leq N \\ x\in\mathbb{R}^2}} e^{-2\pi ih\cdot x} \int_{\substack{y\in\mathcal{G} \\ y\in\mathbb{R}^2}} F(y)(-4\pi^2 y^2) e^{-2\pi ix\cdot y} \ dV(y) \ dV(x).$$

Moreover, the Fourier inversion formula (i.e., Theorem 13.1) guarantees that, for every $\varepsilon > 0$, there exists an integer $M(= M(\varepsilon))$ such that, for all $N \geq M$

$$\left| \int_{\substack{|x|\leq N \\ x\in\mathbb{R}^2}} e^{-2\pi ih\cdot x} \left(\int_{\mathcal{G}} F(y)(-4\pi^2 y^2) \ e^{-2\pi ix\cdot y} \ dV(y) \right) dV(x) \right| \leq \varepsilon \qquad (13.107)$$

holds uniformly for all $h \in \Lambda^{-1}$ with $|h| \geq R_0$, $R_0 > \sup_{x\in\overline{\mathcal{G}}} |x|$. Hence, for all $N \geq M$, we find

$$\left| \sum_{\substack{|h|>R_0 \\ h\in\Lambda^{-1}}} \frac{J_1(2\pi|h|\tau)}{|h|} \frac{1}{-4\pi^2 h^2} \int_{|x|\leq N} e^{-2\pi ih\cdot x} \Delta_x F_{\mathcal{G}}^{\wedge}(x) \ dV(x) \right| \qquad (13.108)$$

$$\leq \varepsilon\, C \sum_{\substack{|h|>R_0 \\ h\in\Lambda^{-1}}} \frac{1}{|h|^{2+\frac{3}{2}}}$$

$$\leq \varepsilon\, C'.$$

Therefore we obtain from (13.96) and (13.108) for $N \to \infty$

$$\frac{1}{\pi\tau^2} \int_{\substack{|x|\leq N \\ x\in\mathbb{R}^2}} \mathcal{X}_{\overline{\mathbb{B}_\tau^2}+\Lambda}(x) F_{\mathcal{G}}^{\wedge}(x) \ dV(x) \qquad (13.109)$$

$$= \frac{1}{\|\mathcal{F}\|} \int_{\substack{|x|\leq N \\ x\in\mathbb{R}^2}} \int_{y\in\mathcal{G}} F(y) \ e^{-2\pi ix\cdot y} \ dV(y) \ dV(x)$$

$$+ \frac{1}{\|\mathcal{F}\|} \sum_{\substack{0<|h|\leq R_0 \\ h\in\Lambda^{-1}}} \frac{J_1(2\pi|h|\tau)}{\pi\tau|h|} \frac{1}{-4\pi^2 h^2}$$

$$\times \int_{\substack{|x|\leq N \\ x\in\mathbb{R}^2}} e^{-2\pi ih\cdot x} \int_{\substack{y\in\mathcal{G} \\ y\in\mathbb{R}^2}} F(y)(-4\pi^2 y^2) \ e^{-2\pi ix\cdot y} \ dV(y) \ dV(x)$$

$$+ o(1).$$

The application of the Fourier inversion formula (i.e., Theorem 13.1) then yields the following lattice ball identity.

Theorem 13.5. *Let Λ be an arbitrary lattice in $\mathbb{R}^2$. Let $\mathcal{G} \subset \mathbb{R}^2$ be a regular region. Suppose that F is of class $C^{(2)}(\overline{\mathcal{G}})$, $\overline{\mathcal{G}} = \mathcal{G}\cup\partial\mathcal{G}$. Then, for all sufficiently small (fixed) values $\tau > 0$, the limit*

$$\lim_{N\to\infty} \frac{1}{\pi\tau^2} \int_{\substack{|x|\leq N \\ x\in\mathbb{R}^2}} \mathcal{X}_{\overline{\mathbb{B}_\tau^2}+\Lambda}(x) \int_{\mathcal{G}} F(y) \ e^{-2\pi ix\cdot y} \ dV(y) \ dV(x). \qquad (13.110)$$

exists, and we have

$$\frac{1}{\|\mathcal{F}\|} \underset{\substack{h \in \overline{\mathcal{G}} \\ h \in \Lambda^{-1}}}{\sum}{}' \frac{J_1(2\pi|h|\tau)}{\pi\tau|h|} F(h) \tag{13.111}$$

$$= \lim_{N \to \infty} \frac{1}{\pi\tau^2} \int_{\substack{|x| \le N \\ x \in \mathbb{R}^2}} \mathcal{X}_{\overline{\mathbb{B}^2_\tau} + \Lambda}(x) \int_{\mathcal{G}} F(y) \, e^{-2\pi i x \cdot y} \, dV(y) \, dV(x).$$

Replacing $\Lambda \subset \mathbb{R}^2$ by its inverse lattice $\Lambda^{-1} \subset \mathbb{R}^2$ we finally obtain

Corollary 13.8. *Under the assumptions of Theorem 13.5*

$$\underset{\substack{g \in \overline{\mathcal{G}} \\ g \in \Lambda}}{\sum}{}' \frac{J_1(2\pi|g|\tau)}{\pi\tau|g|} F(g) \tag{13.112}$$

$$= \lim_{N \to \infty} \frac{1}{\|\mathcal{F}\|} \frac{1}{\pi\tau^2} \int_{\substack{|x| \le N \\ x \in \mathbb{R}^2}} \mathcal{X}_{\overline{\mathbb{B}^2_\tau} + \Lambda^{-1}}(x) \int_{\mathcal{G}} F(y) \, e^{-2\pi i x \cdot y} \, dV(y) \, dV(x).$$

Corollary 13.8 represents the weighted lattice ball variant of the Hardy–Landau identity in $\mathbb{R}^2$. Clearly, with $J_1(r) = \frac{r}{2} + \dots$ we find

Corollary 13.9. *Let Λ be an arbitrary lattice in $\mathbb{R}^2$. Let $\mathcal{G} \subset \mathbb{R}^2$ be a regular region. Suppose that F is of class $\mathrm{C}^{(2)}(\overline{\mathcal{G}})$, $\overline{\mathcal{G}} = \mathcal{G} \cup \partial\mathcal{G}$. Then*

$$\underset{\substack{g \in \overline{\mathcal{G}} \\ g \in \Lambda}}{\sum}{}' F(g) \tag{13.113}$$

$$= \lim_{\substack{\tau \to 0 \\ \tau > 0}} \lim_{N \to \infty} \frac{1}{\|\mathcal{F}\|} \frac{1}{\pi\tau^2} \int_{\substack{|x| \le N \\ x \in \mathbb{R}^2}} \mathcal{X}_{\overline{\mathbb{B}^2_\tau} + \Lambda^{-1}}(x) \int_{\mathcal{G}} F(y) \, e^{-2\pi i x \cdot y} \, dV(y) \, dV(x).$$

Once again, for a convex region $\mathcal{G}$ containing the origin with curve ∂G such that its normal field ν is continuously differentiable and its curvature is non-vanishing, the series and the limit in (13.113) can be interchanged; hence, Theorem 13.4 again follows as the immediate result.

14

Planar Distribution of Lattice Points

CONTENTS

14.1 Qualitative Hardy–Landau Induced Geometric Interpretation 386
    Non-Uniform Radial Lattice Point Distribution 386
    Non-Uniform Radial and Angular Lattice Point Distribution 387
    Non-Uniform Radial and Angular Lattice Ball Distribution 390
14.2 Constant Weight Discrepancy 391
    One-Dimensional τ-Integral Mean 392
    Asymptotic Expansions of Power Moments 393
14.3 Almost Periodicity of the Constant Weight Discrepancy 396
    Parseval Identity 398
    Fourier Coefficients 403
14.4 Angular Weight Discrepancy 406
    Discrepancy for Products of Radial and Angular Functions 406
    Angular Number Theoretical Functions 407
14.5 Almost Periodicity of the Angular Weight Discrepancy 408
14.6 Radial and Angular Weights 409
    Periodical Radial Lattice Point Expansions 412
    Asymptotic Behavior of Weighted Lattice Point Sums 413
14.7 Non-Uniform Distribution of Lattice Points 415
    Distributions Generated by Cosine Functions 415
    Distributions Generated by Lattice Functions 417
    Distributions Generated by Step Functions 420
14.8 Quantitative Step Function Oriented Geometric Interpretation 421
    Circular Configurations 421
    Circular and Sectorial Configurations 425

In this chapter the planar non-uniform distribution of lattice points is explained in more detail. Our goal is to discuss this phenomenon for combined radial and angular patterns. Certain variants of Hardy–Landau identities act as the points of departure for our motivation. An essential tool for the description is the almost periodicity of the modified lattice point discrepancy.

The layout of this chapter is as follows: in Section 14.1, the qualitative study of non-uniform radial as well as angular distribution of lattice points in the plane is motivated within the framework of Hardy–Landau identities. Section 14.2 starts with a discussion of properties for the ordinary lattice point discrepancy in the jargon of the classical Hardy-Landau nomenclature. The resulting hints serve as the points of departure for a more general setup of ideas and concepts in the field of non-uniform distribution of planar lattice points. Section 14.3 deals with integral mean asymptotics and the almost periodicity of the modified lattice point discrepancy in the (B^2)-Besicovitch sense. Section 14.4 is concerned with lattice point expansions involving (sufficiently smooth) products of radial and angular functions. Section 14.5 is devoted to the almost (B^2)-Besicovitch periodicity for lattice point discrepancies involving angular weights. Section 14.6 discusses general aspects in asymptotic lattice point expansions for radial and/or angular weights. Section 14.7 illustrates the background of non-uniform distribution of lattice points for representative radial examples, namely weighted lattice point sums generated by cosine functions, lattice functions, and step functions, respectively. Finally, Section 14.8 provides some quantitative insight into the step function generated non-uniform (phase) distribution of lattice points in the plane by counting lattice point sums in certain circular and sectorial ring configurations.

For simplicity, throughout this chapter, we restrict ourselves to the unit lattice $\Lambda = \mathbb{Z}^2$. Moreover, we concentrate our discussion on geometrical configurations with circle rings centered at the origin, although any point of the plane independently of its position can be taken as center for our discussion.

14.1   Qualitative Hardy–Landau Induced Geometric Interpretation

We begin with the Hardy–Landau induced motivation of the non-uniform radial lattice point distribution.

Non-Uniform Radial Lattice Point Distribution

For $q = 2$, $\nu \in \mathbb{R}$ satisfying $\frac{1}{2} < \nu < 1$, our approach to the Hardy–Landau identities (as given in Section 10.5) enables us to develop the asymptotic expansion

$$\sum_{\substack{|g|\leq N \\ g\in\mathbb{Z}^2}} \frac{J_\nu(2\pi|g|R)}{|g|^\nu} \tag{14.1}$$

$$= 2\pi \sum_{\substack{|h|=R \\ h\in\mathbb{Z}^2}} \int_0^N J_\nu(2\pi r R)\, J_0(2\pi|h|r)\, r^{-\nu+1}\, dr$$

$$+ \frac{\pi^{\nu-1}}{2R^{-\nu+2}\,\Gamma(\nu)} \sum_{\substack{|h|<R \\ h\in\mathbb{Z}^2}} \left(1 - \left(\frac{|h|}{R}\right)^2\right)^{\nu-1}$$

$$+ o(1).$$

Obviously, the formula (14.1) is a variant of the two-dimensional Hardy–Landau identity, and it coincides with the Hardy–Landau identity for $\nu = 1$ as N tends toward ∞.

The theory of discontinuous integrals (cf. Section 7.5) furnishes further information about the ingredients occurring in the expansion (14.1). The following facts for $\nu \in \mathbb{R}$ with $\frac{1}{2} < \nu < 1$ come into play:

(i) if $4\pi^2 R^2 \notin \mathrm{Spect}_\Delta(\mathbb{Z}^2)$, then the sum on the left side of (14.1) is convergent as $N \to \infty$,

(ii) if $4\pi^2 R^2 \in \mathrm{Spect}_\Delta(\mathbb{Z}^2)$, then the integral

$$\int_0^N J_\nu(2\pi r R) J_0(2\pi r R) \; r^{-\nu+1} \; dr \tag{14.2}$$

tends towards ∞ as $N \to \infty$.

In consequence, the alternative conditions *(i)* and *(ii)* lead to the following geometrical interpretation of the asymptotic expansion (14.1): *for certain radii R, there must be more lattice points in circular rings, where $J_\nu(2\pi r R)$ is of positive sign, than in circular rings, where $J_\nu(2\pi r R)$ is negative.* It seems that the observation of this surprising feature - inherent in two-dimensional lattice point Hardy–Landau identities of the type (14.1) - is of particular interest in the planar lattice point theory. In fact, this chapter shows that it is more than an unexpected accident of radial non-uniform distribution of lattice points.

Non-Uniform Radial and Angular Lattice Point Distribution

Our work about planar Hardy–Landau lattice point summation demonstrates not only non-uniform radial distribution of lattice points for circular rings, but also the angular dependence on concentric sectors around the origin. Our interest is to demonstrate this observation for an already known example in a qualitative way.

To be more concrete, under the assumption

$$q \geq 2, \quad \nu - 4k > \frac{q-1}{2}, \quad k \in \mathbb{N}_0, \tag{14.3}$$

the Hardy–Landau identities for radial and angular product functions (see

Section 10.5) tell us that for all radii $R > 0$

$$
\lim_{N \to \infty} \left(\sum_{\substack{|g| \leq N \\ g \in \mathbb{Z}^q}} \frac{J_\nu(q; 2\pi |g| R)}{|g|^\nu} H_{4k}(q; g) \right.
$$

$$
\left. - \sum_{\substack{|h| = R \\ h \in \mathbb{Z}^q}} \int_{\substack{|x| \leq N \\ x \in \mathbb{R}^q}} \frac{J_\nu(q; 2\pi |x| R)}{|x|^\nu} H_{4k}(q; x) e^{-2\pi i h \cdot x} \, dV(x) \right)
$$

$$
= \frac{2 \, \pi^{\nu - 4k}}{\|\mathbb{S}^{q-1}\| R^{8k - \nu + q} \Gamma(\nu - 4k)} \sum_{\substack{|h| < R \\ h \in \mathbb{Z}^q}} \left(1 - \left(\frac{|h|}{R} \right)^2 \right)^{\nu - 4k - 1} H_{4k}(q; h).
$$

Herein, $H_{4k}(q; \cdot)$ is an arbitrary homogeneous, harmonic polynomial of degree $4k$ in q dimensions, and $J_\nu(q; \cdot)$, $\nu > 0$, is the Bessel function of order ν and dimension q.

The theory of discontinuous integrals provides us with further conditions imposed on the values ν, k, however, different for non-eigenvalues and eigenvalues $4\pi^2 R^2$. In more detail, the conditions

(i) $4\pi^2 R^2 \notin \mathrm{Spect}_\Delta(\mathbb{Z}^2)$:

$$
\nu - 4k > 0, \tag{14.5}
$$

(ii) $4\pi^2 R^2 \in \mathrm{Spect}_\Delta(\mathbb{Z}^2)$:

$$
\nu - 4k > 1, \tag{14.6}
$$

or

$$
\nu - 4k > 0, \quad \nu - 4k \ \ odd \ integer, \tag{14.7}
$$

secure the convergence of the finite sum of integrals

$$
\sum_{\substack{|h| = R \\ h \in \mathbb{Z}^q}} \int_{\substack{|x| \leq N \\ x \in \mathbb{R}^q}} \frac{J_\nu(q; 2\pi |x| R)}{|x|^\nu} H_{4k}(q; x) \, e^{-2\pi i h \cdot x} \, dV(x) \tag{14.8}
$$

for $N \to \infty$. Combining the conditions (14.5), (14.6), and (14.7) of the theory of discontinuous integrals with the restriction (14.3) establishing (14.4) we are led to the following astonishing statements:

(1) for all dimensions $q \geq 3$, we have $\nu - 4k > \frac{q-1}{2} \geq 1$, such that the expression $\nu - 4k - 1$ is always positive. All summands in (14.8) converge as $N \to \infty$. In consequence, *for $q \geq 3$, independent of the status of $4\pi^2 R^2$ a non-eigenvalue or an eigenvalue of* $\mathrm{Spect}_\Delta(\mathbb{Z}^2)$, *a Hardy–Landau induced non-uniform distribution of lattice points is not detectable.*

(2) however, for the dimension $q = 2$ and all values ν, k with $\frac{1}{2} < \nu - 4k < 1$, the convergence of the integrals

$$\int_0^N J_\nu(\pi r R) \, J_{4k}(\pi |h| r) \, r^{4k-\nu+1} \, dr, \quad h \in \mathbb{Z}^2, \tag{14.9}$$

leads to an alternative result, different for the classes of non-eigenvalues and eigenvalues $4\pi^2 R^2$, namely

(i) for $4\pi^2 R^2 \notin \mathrm{Spect}_\Delta(\mathbb{Z}^2)$ the integrals (14.9) are convergent as $N \to \infty$,

(ii) for $4\pi^2 R^2 \in \mathrm{Spect}_\Delta(\mathbb{Z}^2)$ the integrals (14.9) tend towards ∞ as $N \to \infty$.

Written out in formulas of the Hardy–Landau nomenclature (cf. Section 10.5) this leads to the following conclusions (for example, with $H_{4k}(2; x) = H_{4k}(2; x_1, x_2) = (x_1 + ix_2)^{4k}$):

Lemma 14.1. *For $q = 2$, $\frac{1}{2} < \nu - 4k < 1$, and $4\pi^2 R^2 \notin \mathrm{Spect}_\Delta(\mathbb{Z}^2)$, the limit*

$$\lim_{N \to \infty} \sum_{\substack{|g| \leq N \\ g \in \mathbb{Z}^2}} \frac{J_\nu(2\pi |g| R)}{|g|^\nu} H_{4k}(2; g) \tag{14.10}$$

exists and is equal to

$$\frac{\pi^{\nu-4k-1}}{R^{8k-\nu+2} \Gamma(\nu - 4k)} \sum_{\substack{|h| < R \\ h \in \mathbb{Z}^2}} \left(1 - \left(\frac{|h|}{R}\right)^2\right)^{\nu-4k-1} H_{4k}(2; h). \tag{14.11}$$

Lemma 14.2. *For $q = 2$, $\frac{1}{2} < \nu - 4k < 1$, and $4\pi^2 R^2 \in \mathrm{Spect}_\Delta(\mathbb{Z}^2)$, and $N \to \infty$ we have*

$$\sum_{\substack{|g| \leq N \\ g \in \mathbb{Z}^2}} \frac{J_\nu(2\pi |g| R)}{|g|^\nu} H_{4k}(2; g) \tag{14.12}$$

$$= 2\pi \frac{1}{R^{4k}} \sum_{\substack{|h| = R \\ h \in \mathbb{Z}^2}} H_{4k}(2; h) \int_0^N J_\nu(2\pi r R) \, J_{4k}(2\pi |h| r) \, r^{4k-\nu+1} \, dr$$

$$+ \frac{\pi^{\nu-4k-1}}{2 R^{8k-\nu+2} \Gamma(\nu - 4k)} \sum_{\substack{|h| < R \\ h \in \mathbb{Z}^2}} \left(1 - \left(\frac{|h|}{R}\right)^2\right)^{\nu-4k-1} H_{4k}(2; h)$$

$$+ \, o(1).$$

All in all, our two-dimensional Hardy–Landau type identities (i.e., Lemma 14.1 and Lemma 14.2) allow the following geometrical interpretation about the planar distribution of lattice points:

(i) for $k = 0$, the non-uniform radial lattice point distribution over circular rings is detectable dependent on the choice of the radii R, different for non-eigenvalues and eigenvalues $4\pi^2 R^2$,

(ii) for $k > 0$, the non-uniform radial as well as angular lattice point distribution is detectable dependent on the choice of the radii R, different for non-eigenvalues and eigenvalues $4\pi^2 R^2$ in the geometrical pattern determined by the sign behavior of the two-dimensional product function

$$(x_1, \, x_2)^T \mapsto \frac{J_\nu(2\pi \sqrt{x_1^2 + x_2^2}\; R)}{\sqrt{x_1^2 + x_2^2}^{\;\nu}} H_{4k}(2; x_1, x_2), \quad (x_1, \, x_2)^T \in \mathbb{R}^2. \quad (14.13)$$

In conclusion, *non-uniform radial as well as angular distribution turns out to be an inherent phenomenon of planar lattice point theory, whereas the higher dimensions $q \geq 3$ do not exhibit a similar phenomenon, at least not in accordance with the Hardy–Landau approach as presented in this work.*

Non-Uniform Radial and Angular Lattice Ball Distribution

Under the assumption

$$q \geq 2, \;\; 0 < \nu - 4k < 1, \; k \in \mathbb{N}_0, \quad (14.14)$$

the lattice ball theory of Hardy–Landau summation (cf. Section 11.5) provides identities analogous to Lemma 14.1 and Lemma 14.2.

Lemma 14.3. *For $4\pi^2 R^2 \notin \mathrm{Spect}_\Delta(\mathbb{Z}^q)$*

$$\frac{2\,\pi^{\nu-4k}}{R^{8k-\nu+q}\Gamma\,(\nu - 4k)\,\|\mathbb{B}_\tau^q\|}\frac{\tau^q}{} \quad (14.15)$$

$$\times \sum_{\substack{|g| < R \\ g \in \mathbb{Z}^q}} \frac{J_1(q; 2\pi|g|\tau)}{2\pi\tau|g|}\left(1 - \left(\frac{|g|}{R}\right)^2\right)^{\nu-4k-1} H_{4k}(q; g)$$

$$= \frac{1}{\|\mathbb{B}_\tau^q\|}\int_{x \in \mathbb{R}^q} \mathcal{X}_{\overline{\mathbb{B}_\tau^q}+\mathbb{Z}^q}(x)\,\frac{J_\nu(q; 2\pi|x|R)}{|x|^\nu}H_{4k}(q; x)\,dV(x).$$

Lemma 14.4. *For* $4\pi^2 R^2 \in \mathrm{Spect}_\Delta(\mathbb{Z}^q)$

$$\frac{1}{\|\mathbb{B}_\tau^q\|} \int_{x \leq N} \mathcal{X}_{\overline{\mathbb{B}_\tau^q} + \mathbb{Z}^q}(x) \frac{J_\nu(q; 2\pi|x|R)}{|x|^\nu} H_{4k}(q;x) \, dV(x) \tag{14.16}$$

$$= \frac{\|\mathbb{S}^{q-1}\|}{\|\mathbb{B}_\tau^q\|} \frac{\tau^q}{\|\mathcal{F}\|} \sum_{\substack{|h|=R \\ h \in \mathbb{Z}^q}} \frac{J_1(q; 2\pi|h|\tau)}{2\pi|h|\tau}$$

$$\times \int_{\substack{|x| \leq N \\ x \in \mathbb{R}^q}} \frac{J_\nu(q; 2\pi|x|R)}{|x|^\nu} H_{4k}(q;x) e^{-2\pi i h \cdot x} \, dV(x)$$

$$+ \frac{2 \pi^{\nu-n}}{R^{8k-\nu+q} \Gamma(\nu-4k)} \frac{\tau^q}{\|\mathbb{B}_\tau^q\|}$$

$$\times \sum_{\substack{|h|<R \\ h \in \mathbb{Z}^q}} \frac{J_1(q; 2\pi|h|\tau)}{2\pi\tau|h|} \left(1 - \left(\frac{|h|}{R}\right)^2\right)^{\nu-4k-1} H_{4k}(q;h)$$

$$+ o(1), \quad N \to \infty.$$

However, the condition (14.14) does not affect any dimension q in (14.15) and (14.16). In other words, the Hardy–Landau summation as presented here allows us to conclude that *non-uniform radial as well as angular lattice ball distribution is an inherent phenomenon of all dimensions* $q \geq 2$.

14.2   Constant Weight Discrepancy

In the sequel we discuss the role of the $\mathbb{Z}^2$-lattice point discrepancy under the particular auspices of the classical Hardy–Landau identity (10.98). In light of the asymptotic behavior of the Bessel function J_1 within the two-dimensional Hardy–Landau series (see Corollary 10.1), it is conventional to write the lattice point discrepancy in the form

$$P\left(\overline{\mathbb{B}_N^2}\right) = \sqrt{N} \, Q\left(\overline{\mathbb{B}_N^2}\right), \tag{14.17}$$

where $Q\left(\overline{\mathbb{B}_N^2}\right)$ is called the *modified* $\mathbb{Z}^2$-*lattice point discrepancy*.

In the context of the Hardy–Landau summation this means

$$Q\left(\overline{\mathbb{B}_N^2}\right) = \sqrt{N} \sum_{\substack{0<|h| \\ h \in \mathbb{Z}^2}} \frac{J_1(2\pi|h|N)}{|h|}. \tag{14.18}$$

Using $r(n) = \sum_{\substack{|h|^2=n \\ h \in \mathbb{Z}^2}} 1$, $n \in \mathbb{N}$, the identity (14.18) can be rewritten in the form

$$Q\left(\overline{\mathbb{B}_N^2}\right) = \sqrt{N}\sum_{n=1}^{\infty} r(n)\frac{J_1(2\pi\sqrt{n}N)}{\sqrt{n}}. \tag{14.19}$$

The partial sums of the series

$$\sum_{n=1}^{\infty} r(n)\frac{J_1(2\pi\sqrt{n}N)}{\sqrt{n}} \tag{14.20}$$

are uniformly bounded on every compact subinterval of the half-line $[1,\infty)$. In addition, from the asymptotic formula of the Bessel function J_1 (see (6.568)), we get for $N \to \infty$

$$Q\left(\overline{\mathbb{B}_N^2}\right) = \frac{1}{\pi}\sum_{n=1}^{\infty} \frac{r(n)}{n^{\frac{3}{4}}}\sin\left(2\pi\sqrt{n}N - \frac{\pi}{4}\right) + O\left(N^{-\frac{1}{2}}\right), \tag{14.21}$$

where the series on the right of (14.21) is uniformly bounded on every compact subinterval of the half-line $[1,\infty)$. This was already known to E. Landau [1927].

One-Dimensional τ-Integral Mean

As a preparatory result to establish an average description, we start from a one-dimensional τ-*integral mean* for the $\mathbb{Z}^2$-lattice point discrepancy (14.18) with sufficiently small $\tau > 0$

$$\frac{1}{2\tau}\int_{N-\tau}^{N+\tau} \frac{P\left(\overline{\mathbb{B}_r^2}\right)}{r}\,dr = \frac{1}{2\tau}\int_{N-\tau}^{N+\tau} \sum_{\substack{|h|\neq 0 \\ h\in\mathbb{Z}^2}} \frac{J_1(2\pi|h|r)}{|h|}\,dr. \tag{14.22}$$

An easy calculation shows that

$$\frac{P\left(\overline{\mathbb{B}_r^2}\right)}{r} = \frac{Q\left(\overline{\mathbb{B}_r^2}\right)}{\sqrt{r}} = \frac{1}{r}\int_{\substack{|x|=r \\ x\in\mathbb{R}^2}} \frac{\partial}{\partial\nu}G(\Delta;x)\,dS(x) \tag{14.23}$$

$$= \frac{1}{dr}\frac{1}{r}\int_{\substack{|x|=r \\ x\in\mathbb{R}^2}} G(\Delta;x)\,dS(x)$$

$$= -\frac{d}{dr}\sum_{\substack{|h|\neq 0 \\ h\in\mathbb{Z}^2}} \frac{J_0(2\pi|h|r)}{2\pi|h|^2}.$$

Thus it follows that

$$\frac{1}{2\tau} \int_{N-\tau}^{N+\tau} \frac{P\left(\overline{\mathbb{B}_r^2}\right)}{r} \, dr = \frac{1}{2\tau} \int_{N-\tau}^{N+\tau} \sum_{\substack{|h| \neq 0 \\ h \in \mathbb{Z}^2}} \frac{J_1(2\pi|h|r)}{|h|} \, dr \qquad (14.24)$$

$$= \frac{1}{2\tau} \int_{N-\tau}^{N+\tau} \frac{1}{r} \int_{\substack{|x|=r \\ x \in \mathbb{R}^2}} \frac{\partial}{\partial \nu} G(\Delta; x) dS(x) \, dr$$

$$= \frac{1}{4\pi\tau} \sum_{\substack{|h| \neq 0 \\ h \in \mathbb{Z}^2}} \frac{1}{|h|^2} \Big(J_0(2\pi|h|(N-\tau)) - J_0(2\pi|h|(N+\tau)) \Big).$$

We therefore obtain

Lemma 14.5. *Let the $\mathbb{Z}^2$-lattice point discrepancy $P\left(\overline{\mathbb{B}_N^2}\right)$ be defined in the classical way by*

$$P\left(\overline{\mathbb{B}_N^2}\right) = \sum_{\substack{|g| \leq N \\ g \in \mathbb{Z}^2}}{}' 1 - \pi N^2 = \sqrt{N} \; Q\left(\overline{\mathbb{B}_N^2}\right). \qquad (14.25)$$

Then

$$\frac{1}{2\tau} \int_{N-\tau}^{N+\tau} \frac{P\left(\overline{\mathbb{B}_r^2}\right)}{r} \, dr = \frac{1}{2\tau} \int_{N-\tau}^{N+\tau} \frac{Q\left(\overline{\mathbb{B}_r^2}\right)}{\sqrt{r}} \, dr = O(N^{-\frac{1}{2}}), \quad N \to \infty.$$
$$(14.26)$$

Proof. The asymptotic expansion (14.26) follows immediately from the estimate $|J_0(r)| \leq Cr^{-\frac{1}{2}}$ and the resulting absolute and uniform convergence of the lattice point sum (14.24). $\quad\square$

Of course, the asymptotic relation (14.26) does *not* mean that the discrepancy $P\left(\overline{\mathbb{B}_N^2}\right)$ behaves like $O(N^{\frac{1}{2}})$ for $N \to \infty$.

Asymptotic Expansions of Power Moments

The lattice point problem for the circle also allows integral mean square-asymptotic formulas of other shapes. From G.H. Hardy [1916] we know that

$$\frac{1}{N} \int_1^N \left| P\left(\overline{\mathbb{B}_{\sqrt{r}}^2}\right) \right| \, dr = O\left(N^{\frac{1}{4}+\varepsilon}\right), \quad N \to \infty, \qquad (14.27)$$

for *every* $\varepsilon > 0$. Furthermore, it can be shown that

$$\int_0^N \left(P\left(\overline{\mathbb{B}_{\sqrt{r}}^2}\right) \right)^2 \, dr = \left(\frac{1}{3\pi^2} \sum_{n=1}^{\infty} \frac{r^2(n)}{n^{\frac{3}{2}}} \right) N^{\frac{3}{2}} + S\left(\overline{\mathbb{B}_{\sqrt{N}}^2}\right), \qquad (14.28)$$

where the remainder term $S\left(\overline{\mathbb{B}^2_{\sqrt{N}}}\right)$ has been the subject of intensive research: for example, A. Cramér [1922] proved

$$S\left(\overline{\mathbb{B}^2_{\sqrt{N}}}\right) = O\left(N^{\frac{5}{4}+\epsilon}\right), \quad N \to \infty, \tag{14.29}$$

while E. Landau [1924] derived the estimate

$$S\left(\overline{\mathbb{B}^2_{\sqrt{N}}}\right) = O\left(N^{1+\epsilon}\right), \quad N \to \infty. \tag{14.30}$$

Refinements are given by many other authors, for example, A. Walfisz [1927], I. Katai [1965], E. Preissmann [1989], and W.G. Nowak [2004].

In addition, E. Landau [1927] proved the integral mean relation

$$\int_0^N \left(P\left(\overline{\mathbb{B}^2_{\sqrt{r}}}\right)\right)^2 dr = \frac{1}{3\pi^2} \frac{16\,\zeta^2\left(\frac{3}{2}\right)L^2\left(\frac{3}{2}\right)}{\zeta(3)\left(1+2^{-\frac{3}{2}}\right)} N^{\frac{3}{2}} + O\left(N^{1+\varepsilon}\right), \quad N \to \infty. \tag{14.31}$$

It should be noted that the first expression on the right side of (14.31) is of particular interest for our later work. This is the reason why we are interested in (the proof of)

Lemma 14.6. *For $s \in \mathbb{C}$ with $\Re(s) > 1$ we have*

$$\sum_{n=1}^{\infty} \frac{r^2(n)}{n^s} = \frac{16\,\zeta^2(s)\,L^2(s)}{\zeta(2s)\,(1+2^{-s})}, \tag{14.32}$$

where (see (4.137))

$$\zeta(s) = \prod_p (1-p^{-s})^{-1} \tag{14.33}$$

is the prime number representation of the ζ-series

$$\zeta(s) = \sum_{n=1}^{\infty} \frac{1}{n^s} \tag{14.34}$$

and

$$L(s) = \prod_{p\equiv 1(\mathrm{mod}4)} (1-p^{-s})^{-1} \prod_{p\equiv 3(\mathrm{mod}4)} (1+p^{-s})^{-1} \tag{14.35}$$

is the prime number decomposition of the (special) L-series given by

$$L(s) = \sum_{k=0}^{\infty} \frac{(-1)^k}{(2k+1)^s}. \tag{14.36}$$

Proof. On the one hand, we have

$$\zeta^2(s) = \prod_p \left(\frac{1}{1-p^{-s}}\right)^2, \quad \zeta(2s) = \prod_p (1-p^{-2s})^{-1} \tag{14.37}$$

such that

$$\frac{\zeta^2(s)}{\zeta(2s)} = \prod_p \frac{1-p^{-2s}}{(1-p^{-s})^2} = \frac{1+2^{-s}}{1-2^{-s}} \prod_{p>2} \frac{1+p^{-s}}{1-p^{-s}}. \quad (14.38)$$

Moreover,

$$L^2(s) = \prod_{p\equiv1(\mathrm{mod}4)} (1-p^{-s})^{-2} \prod_{p\equiv3(\mathrm{mod}4)} (1+p^{-s})^{-2}. \quad (14.39)$$

Thus we get

$$\frac{\zeta^2(s) L^2(s)}{\zeta(2s) (1+2^{-s})} = \frac{1}{1-2^{-s}} \prod_{p\equiv1(\mathrm{mod}4)} \frac{1+p^{-1}}{(1-p^{-s})^3} \prod_{p\equiv3(\mathrm{mod}4)} (1-p^{-2s}). \quad (14.40)$$

On the other hand, the function δ given by $\delta(n) = \frac{r(n)}{4} = d_1(n) - d_3(n)$ is multiplicative, i.e., $\delta(km) = \delta(k)\,\delta(m)$, $k, m \in \mathbb{N}$, $(k, m) = 1$. We know (cf. E. Landau [1927]) that

$$\delta\left(p^l\right) = \begin{cases} 1 & : & p = 2, \\ l+1 & : & p \equiv 1(\mathrm{mod}4), \\ 1 & : & p \equiv 3(\mathrm{mod}4), \; 2 \mid l, \\ 0 & : & p \equiv 3(\mathrm{mod}4), \; 2 \nmid l, \end{cases} \quad (14.41)$$

provided that l is a non-negative integer. Therefore, we obtain

$$\frac{1}{16} \sum_{n=1}^{\infty} \frac{r^2(n)}{n^s} = \sum_{n=1}^{\infty} \frac{\delta^2(n)}{n^s} = \prod_p \left(1 + \frac{\delta^2(p)}{p^s} + \frac{\delta^2(p^2)}{p^{2s}} + \cdots \right)$$

$$= \left(1 + \frac{1}{2^s} + \frac{1}{2^{2s}} + \cdots \right) \quad (14.42)$$

$$\times \prod_{p\equiv1(\mathrm{mod}4)} \left(1 + \frac{2^2}{p^s} + \frac{3^2}{p^{2s}} + \cdots \right)$$

$$\times \prod_{p\equiv3(\mathrm{mod}4)} \left(1 + \frac{1}{p^{2s}} + \frac{1}{p^{4s}} + \cdots \right).$$

Consequently we have

$$\frac{1}{16} \sum_{n=1}^{\infty} \frac{r^2(n)}{n^s} = \frac{1}{1-2^{-s}} \prod_{p\equiv1(\mathrm{mod}4)} \frac{1+p^{-s}}{(1-p^{-s})^3} \prod_{p\equiv3(\mathrm{mod}4)} (1-p^{-2s}), \quad (14.43)$$

where we have used the easily derivable power series representation

$$\sum_{n=0}^{\infty} (n+1)^2 x^n = \frac{1+x}{(1-x)^3}, \quad |x| < 1. \quad (14.44)$$

By comparison of (14.40) and (14.43) we obtain the desired result of Lemma 14.6. $\qquad\square$

Remark 14.1. *A numerical calculation (taking the value from A. Ivic et al. [2004]) shows that*

$$\frac{1}{3\pi^2} \sum_{n=1}^{\infty} \frac{r^2(n)}{n^{\frac{3}{2}}} = \frac{1}{3\pi^2} \frac{16 \, \zeta^2\left(\frac{3}{2}\right) L^2\left(\frac{3}{2}\right)}{\zeta(3)\left(1 + 2^{-\frac{3}{2}}\right)} = 1.69396\ldots. \tag{14.45}$$

Until now, the asymptotic expansions for power moments do not give any deeper insight into the phenomenon of planar distribution of lattice points. Of interest in the context of planar non-uniform lattice point distribution, how-ever, are the papers, e.g., by K.-M. Tsang [1992], D.R. Heath-Brown [1992], W. Zhai [2004] (see also the references therein). They provide results of the type

$$\int_0^N \left(P\left(\overline{\mathbb{B}_{\sqrt{r}}^2}\right)\right)^l dr \sim (-1)^l \, C_l \, N^{1+\frac{l}{4}}, \quad N \to \infty, \tag{14.46}$$

with l integers in the interval $[3,9]$, where C_l, $l \in [3,9]$, are explicitly known positive constants. In conclusion, the asymptotic assumptions for odd integers l admit an interesting interpretation: they can be geometrically understood in the sense that the radii r behave inconsistently such that there is some excess of the radii for which the discrepancy $P\left(\overline{\mathbb{B}_{\sqrt{r}}^2}\right)$ is negative over those for which it is positive.

14.3 Almost Periodicity of the Constant Weight Discrepancy

From (14.21) we get for the modified $\mathbb{Z}^2$-lattice point discrepancy

$$Q\left(\overline{\mathbb{B}_N^2}\right) = \frac{P\left(\overline{\mathbb{B}_N^2}\right)}{\sqrt{N}} = \frac{1}{\sqrt{N}}\left(\sum_{\substack{|g| \le N \\ g \in \mathbb{Z}^2}}' 1 - \pi N^2\right) \tag{14.47}$$

the asymptotic relation

$$Q\left(\overline{\mathbb{B}_N^2}\right) = \frac{1}{\pi} \sum_{n=1}^{\infty} \frac{r(n)}{n^{\frac{3}{4}}} \sin\left(2\pi\sqrt{n}N - \frac{\pi}{4}\right) + O\left(N^{-\frac{1}{2}}\right). \tag{14.48}$$

In view of the asymptotic expansion (14.48), A. Wintner [1941] posed the canonical question: *in what sense is the trigonometric series on the right side of (14.48) the Fourier series of the modified lattice point discrepancy* $Q\left(\overline{\mathbb{B}_N^2}\right)$? Our purpose is to answer this question by recapitulating that (14.18) is almost periodical in the (B^2)-Besicovitch sense and has precisely the Fourier series which one would expect to belong to (14.18). In fact, in virtue of (14.48), we have

$$Q\left(\overline{\mathbb{B}_t^2}\right) \simeq \frac{1}{\pi}\sum_{n=1}^{\infty}\frac{r(n)}{n^{\frac{3}{4}}}\sin\left(2\pi\sqrt{nt}-\frac{\pi}{4}\right), \tag{14.49}$$

where the relation "$\simeq$" in (14.49) - understood in the sense of the (B^2-)Besicovitch theory - is meant to be equivalent to the statement: the limit

$$\lim_{T\to\infty}\frac{1}{T}\int_1^T\left(Q\left(\overline{\mathbb{B}_t^2}\right)-\frac{1}{\pi}\sum_{n=1}^{N}\frac{r(n)}{n^{\frac{3}{4}}}\sin\left(2\pi\sqrt{nt}-\frac{\pi}{4}\right)\right)^2 dt \tag{14.50}$$

exists for every fixed N and tends to 0 as $N\to\infty$. In order to ensure the validity of (14.49) we have to verify the following double limit relation.

Theorem 14.1. *Let* $Q\left(\overline{\mathbb{B}_t^2}\right)$ *be given by (14.17). Then we have*

$$\lim_{\substack{T\to\infty\\N\to\infty}}\frac{1}{T}\int_1^T\left(Q\left(\overline{\mathbb{B}_t^2}\right)-\frac{1}{\pi}\sum_{n=1}^{N}\frac{r(n)}{n^{\frac{3}{4}}}\sin\left(2\pi\sqrt{nt}-\frac{\pi}{4}\right)\right)^2 dt = 0.$$

A. Besicovitch [1954] proved that Theorem 14.1 is valid if and only if one ascertains the following two statements (i.e., Lemma 14.7, and Lemma 14.8):

Lemma 14.7. *The Fourier coefficient*

$$\lim_{T\to\infty}\frac{1}{T}\int_1^T e^{2\pi i\mu t}Q\left(\overline{\mathbb{B}_t^2}\right)dt \tag{14.51}$$

exists for every real number μ, *and we have for* $n=1,2,\ldots$,

$$\lim_{T\to\infty}\frac{1}{T}\int_1^T e^{2\pi i\mu t}Q\left(\overline{\mathbb{B}_t^2}\right)dt$$
$$= \begin{cases} \frac{i}{2\pi}\frac{r(n)}{n^{\frac{3}{4}}}\,e^{i\frac{\pi}{4}} &, \quad \mu=\sqrt{n}, \\ -\frac{i}{2\pi}\frac{r(n)}{n^{\frac{3}{4}}}e^{-i\frac{\pi}{4}} &, \quad \mu=-\sqrt{n}, \\ 0 &, \quad \mu^2\neq n. \end{cases} \tag{14.52}$$

Lemma 14.8. *The Parseval identity holds true, i.e., the integral*

$$\lim_{T\to\infty}\frac{1}{T}\int_1^T\left(Q\left(\overline{\mathbb{B}_t^2}\right)\right)^2 dt \tag{14.53}$$

exists, and we have

$$\lim_{T\to\infty}\frac{1}{T}\int_1^T\left(Q\left(\overline{\mathbb{B}_t^2}\right)\right)^2 dt = \frac{1}{2\pi^2}\sum_{n=1}^{\infty}\frac{r^2(n)}{n^{\frac{3}{2}}}. \tag{14.54}$$

Remark 14.2. *The convergence of the series on the right of (14.54) is clear from the estimate (5.128).*

As already pointed out, in accordance with the (B^2)-Besicovitch theory the proof of Theorem 14.1 can be split into two steps which will be discussed subsequently: first, we realize the proof of Lemma 14.8; second, we pay special attention to the proof of Lemma 14.7.

Parseval Identity

For the verification of Lemma 14.8 we start from the identity (13.46)

$$\int_0^t \left(\sum_{\substack{|g| \le \sqrt{u} \\ g \in \mathbb{Z}^2}} 1 \right) du = \frac{\pi}{2}t^2 + \frac{t}{\pi} \sum_{n=1}^{\infty} \frac{r(n)}{n} J_2(2\pi\sqrt{nt}). \tag{14.55}$$

With the notational convention (cf. A. Cramér [1922])

$$\delta F(t) = F(t+1) - F(t) \tag{14.56}$$

we obtain

$$\int_t^{t+1} \left(\sum_{\substack{|g| \le \sqrt{u} \\ g \in \mathbb{Z}^2}} 1 \right) du = \pi t + \frac{1}{\pi} \sum_{n=1}^{\infty} \frac{r(n)}{n} \delta\left(t J_2(2\pi\sqrt{nt}) \right) + O(1). \tag{14.57}$$

Now we trivially have

$$\int_t^{t+1} \left(\sum_{\substack{|g| \le \sqrt{u} \\ g \in \mathbb{Z}^2}} 1 \right) du = \sum_{\substack{|g| \le \sqrt{t} \\ g \in \mathbb{Z}^2}} 1 \tag{14.58}$$

$$+ \int_t^{t+1} \left(\sum_{\substack{|g| \le \sqrt{u} \\ g \in \mathbb{Z}^2}} 1 - \sum_{\substack{|g| \le \sqrt{t} \\ g \in \mathbb{Z}^2}} 1 \right) du,$$

so that (because of $r(n) = O(n^\varepsilon)$ for every $\varepsilon > 0$)

$$\int_t^{t+1} \left(\sum_{\substack{|g| \le \sqrt{u} \\ g \in \mathbb{Z}^2}} 1 \right) du = \pi t + P\left(\overline{\mathbb{B}^2_{\sqrt{t}}} \right) + O(t^\varepsilon). \tag{14.59}$$

Hence, we find

$$P\left(\overline{\mathbb{B}^2_{\sqrt{t}}} \right) = \frac{1}{\pi} \sum_{n=1}^{\infty} \frac{r(n)}{n} \delta\left(t J_2(2\pi\sqrt{nt}) \right) + O(t^\varepsilon). \tag{14.60}$$

Observing the asymptotic expansion

$$J_2(r) = -\sqrt{\frac{2}{\pi r}} \cos\left(r - \frac{\pi}{4}\right) + \frac{15}{8}\sqrt{\frac{2}{\pi r^3}} \sin\left(r - \frac{\pi}{4}\right) + O\left(r^{-\frac{5}{2}}\right) \qquad (14.61)$$

we find

$$
\begin{aligned}
P\left(\overline{\mathbb{B}^2_{\sqrt{t}}}\right) &= -\frac{1}{\pi^2} \sum_{n=1}^{\infty} \frac{r(n)}{n^{\frac{5}{4}}} \delta\left(t^{\frac{3}{4}} \cos\left(2\pi\sqrt{nt} - \frac{\pi}{4}\right)\right) \qquad (14.62) \\
&+ \frac{15}{16\pi^3} \sum_{n=1}^{\infty} \frac{r(n)}{n^{\frac{7}{4}}} \delta\left(t^{\frac{1}{4}} \sin\left(2\pi\sqrt{nt} - \frac{\pi}{4}\right)\right) \\
&+ O(t^{\varepsilon}).
\end{aligned}
$$

In (14.62), the second series on the right side is of the form $O(t^{\varepsilon})$ since it can be easily seen that

$$
\begin{aligned}
\delta\left(t^{\frac{1}{4}} \sin\left(2\pi\sqrt{nt} - \frac{\pi}{4}\right)\right) &= \int_t^{t+1} \frac{d}{du}\left(u^{\frac{1}{4}} \sin\left(2\pi\sqrt{nu} - \frac{\pi}{4}\right)\right) du \\
&= O\left(\sqrt{n}\, t^{-\frac{1}{4}}\right). \qquad (14.63)
\end{aligned}
$$

In other words, we have

$$P\left(\overline{\mathbb{B}^2_{\sqrt{t}}}\right) = E\left(\overline{\mathbb{B}^2_{\sqrt{t}}}\right) + O(t^{\varepsilon}), \qquad (14.64)$$

where the series

$$E\left(\overline{\mathbb{B}^2_{\sqrt{t}}}\right) = -\frac{1}{\pi^2} \sum_{n=1}^{\infty} \frac{r(n)}{n^{\frac{5}{4}}} \delta\left(t^{\frac{3}{4}} \cos\left(2\pi\sqrt{nt} - \frac{\pi}{4}\right)\right) \qquad (14.65)$$

is absolutely and uniformly convergent on any compact subinterval of $[1, \infty)$. From (14.64) it follows that

$$
\begin{aligned}
\int_1^T \left(P\left(\overline{\mathbb{B}^2_{\sqrt{u}}}\right)\right)^2 du &= \int_1^T \left(E\left(\overline{\mathbb{B}^2_{\sqrt{u}}}\right)\right)^2 du \qquad (14.66) \\
&+ \int_1^T \left(E\left(\overline{\mathbb{B}^2_{\sqrt{u}}}\right)\right) O(u^{\varepsilon}) du \\
&+ O\left(T^{1+\varepsilon}\right),
\end{aligned}
$$

where

$$
\begin{aligned}
&\int_1^T \left(E\left(\overline{\mathbb{B}^2_{\sqrt{u}}}\right)\right)^2 du \qquad (14.67) \\
&= \frac{1}{\pi^4} \sum_{m=1}^{\infty} \sum_{n=1}^{\infty} \frac{r(m)r(n)}{(mn)^{\frac{5}{4}}} \\
&\times \int_1^T \delta\left(u^{\frac{3}{4}} \cos\left(2\pi\sqrt{mu} - \frac{\pi}{4}\right)\right) \delta\left(u^{\frac{3}{4}} \cos\left(2\pi\sqrt{nu} - \frac{\pi}{4}\right)\right) du.
\end{aligned}
$$

Moreover, term-by-term integration yields in connection with (14.65)

$$
\int_1^T \frac{\left(E\left(\overline{\mathbb{B}^2_{\sqrt{u}}} \right) \right)^2}{u} \, du \tag{14.68}
$$

$$
= \frac{1}{\pi^4} \sum_{m=1}^\infty \sum_{n=1}^\infty \frac{r(m)r(n)}{(mn)^{\frac{5}{4}}}
$$

$$
\times \int_1^T \frac{1}{u} \delta \left(u^{\frac{3}{4}} \cos \left(2\pi \sqrt{mu} - \frac{\pi}{4} \right) \right) \delta \left(u^{\frac{3}{4}} \cos \left(2\pi \sqrt{nu} - \frac{\pi}{4} \right) \right) \, du,
$$

where the double series on the right of (14.68) is absolutely convergent for every $T \geq 1$. By elementary manipulations we are able to deduce that the integral (for more details see A. Cramér [1922], A. Wintner [1941])

$$
I_{m,n}(T) \tag{14.69}
$$

$$
= \int_1^T \frac{1}{u} \delta \left(u^{\frac{3}{4}} \cos \left(2\pi \sqrt{mu} - \frac{\pi}{4} \right) \right) \delta \left(u^{\frac{3}{4}} \cos \left(2\pi \sqrt{nu} - \frac{\pi}{4} \right) \right) \, du
$$

can be estimated (by standard applications of the one-dimensional second mean-value theorem) as follows

$$
|I_{m,n}(T)| \leq \frac{C}{\sqrt{m} - \sqrt{n}} \min \left(\sqrt{mn} \ln(T), T + \sqrt{Tm} \right), \quad m > n, \tag{14.70}
$$

where C is independent of m, n and T is sufficiently large (for example, $T > 2$). Furthermore, we have

$$
I_{n,n}(T) = \pi^2 n \sqrt{T} + O\left(n^{\frac{3}{2}} \right) + O\left(\sqrt{n} \ln(T) \right) \tag{14.71}
$$

as $n \to \infty$, $T \to \infty$, where the first O-term is uniform in T. Moreover, it follows that

$$
I_{n,n}(T) = O\left(T^{\frac{3}{2}} \right), \quad T \to \infty, \tag{14.72}
$$

uniformly in n.

Following A. Wintner [1941], for every fixed T, we break the double sum on the right of (14.68) into a sum of four parts

$$
\sum_{m=1}^\infty \sum_{n=1}^\infty \frac{r(m)r(n)}{(mn)^{\frac{5}{4}}} I_{m,n}(T) = L_1(T) + L_2(T) + L_3(T) + L_4(T), \tag{14.73}
$$

where

$$L_1(T) = \sum_{\substack{m=n \\ m \leq T}} \frac{r(m)r(n)}{(mn)^{\frac{5}{4}}} I_{m,n}(T) , \qquad (14.74)$$

$$L_2(T) = \sum_{\substack{m=n \\ m > T}} \frac{r(m)r(n)}{(mn)^{\frac{5}{4}}} I_{m,n}(T) , \qquad (14.75)$$

$$L_3(T) = 2 \sum_{\substack{n < m \\ m \leq T}} \frac{r(m)r(n)}{(mn)^{\frac{5}{4}}} I_{m,n}(T) , \qquad (14.76)$$

$$L_4(T) = 2 \sum_{\substack{n < m \\ m > T}} \frac{r(m)r(n)}{(mn)^{\frac{5}{4}}} I_{m,n}(T) . \qquad (14.77)$$

These four functions of T can be dealt with separately (note that we follow almost literally A. Wintner [1941]). First, from (14.73) and (14.71),

$$L_1(T) = \sum_{n \leq T} \frac{r^2(n)}{n^{\frac{5}{2}}} \left(\pi^2 n T^{\frac{1}{2}} + O\left(n^{\frac{3}{4}}\right) + O\left(n^{\frac{1}{4}} \ln(T)\right) \right) . \qquad (14.78)$$

Hence, for every $\varepsilon > 0$,

$$L_1(T) = \pi^2 T^{\frac{1}{2}} \sum_{n=1}^{\infty} \frac{r^2(n)}{n^{\frac{3}{2}}} - \pi^2 T^{\frac{1}{2}} \sum_{n > T} \frac{(r(n))^2}{n^{\frac{3}{2}}} + O(T^{\varepsilon}). \qquad (14.79)$$

Consequently, it follows from $r(n) = O(n^{\varepsilon})$ that

$$L_1(T) = \pi^2 T^{\frac{1}{2}} \sum_{n=1}^{\infty} \frac{r^2(n)}{n^{\frac{3}{2}}} + O(T^{\varepsilon}). \qquad (14.80)$$

Similarly, from (14.73) and (14.72) in connection with $r(n) = O(n^{\varepsilon})$

$$L_2(T) = O\left(T^{\frac{3}{2}} \sum_{n > T} \frac{r^2(n)}{n^{\frac{5}{2}}} \right) = O\left(T^{\frac{3}{2}} \sum_{n > T} n^{\varepsilon - \frac{5}{2}} \right) , \qquad (14.81)$$

where $\varepsilon > 0$ is arbitrary, so that

$$L_2(T) = O(T^{\varepsilon}). \qquad (14.82)$$

From (14.73) and (14.70) we find

$$L_3(T) = O\left(\sum_{\substack{n < m \\ m \leq T}} \frac{r(m)r(n)}{m^{\frac{5}{4}} n^{\frac{5}{4}}} \frac{m^{\frac{1}{2}} n^{\frac{1}{2}}}{m^{\frac{1}{2}} - n^{\frac{1}{2}}} \ln(T) \right) . \qquad (14.83)$$

Hence, from $(m^{\frac{1}{2}} - n^{\frac{1}{2}})^{-1} = (m^{\frac{1}{2}} + n^{\frac{1}{2}})/(m-n)$ in connection with (14.70)

$$L_3(T) = O\left(\sum_{m=1}^{\lfloor T \rfloor} \frac{m^\varepsilon}{m^{\frac{3}{4}}} \sum_{n=1}^{m-1} \frac{n^\varepsilon}{n^{\frac{3}{4}}} \frac{m^{\frac{1}{2}} + n^{\frac{1}{2}}}{m-n} \ln(T)\right), \qquad (14.84)$$

where $\varepsilon > 0$ is arbitrary. Thus,

$$L_3(T) = O\left(\sum_{m=1}^{\lfloor T \rfloor} m^{\varepsilon - \frac{1}{4}} \sum_{n=1}^{m-1} \frac{n^{\varepsilon - \frac{3}{4}}}{m-n} \ln(T)\right). \qquad (14.85)$$

Since this implies that

$$L_3(T) = O\left(\sum_{m=1}^{\lfloor T \rfloor} \frac{\ln(m)}{m^{1-2\varepsilon}} \ln(T)\right), \qquad (14.86)$$

where $\varepsilon > 0$ is arbitrarily fixed, it follows that

$$L_3(T) = O(T^\varepsilon) \qquad (14.87)$$

for every $\varepsilon > 0$. Finally, from (14.73) and (14.70),

$$L_4(T) = O\left(\sum_{\substack{n<m \\ m>T}} \frac{r(m)r(n)}{m^{\frac{5}{4}} n^{\frac{5}{4}}} \frac{T + T^{\frac{1}{2}} m^{\frac{1}{2}}}{m^{\frac{1}{2}} - n^{\frac{1}{2}}}\right). \qquad (14.88)$$

Hence, from $r(n) = O(n^\varepsilon)$, we obtain

$$L_4(T) = O\left(\sum_{m=\lfloor T \rfloor}^\infty \sum_{n=1}^{m-1} \frac{m^\varepsilon n^\varepsilon}{m^{\frac{5}{4}} n^{\frac{5}{4}}} \frac{T^{\frac{1}{2}} + m^{\frac{1}{2}}}{m^{\frac{1}{2}} - n^{\frac{1}{2}}} T^{\frac{1}{2}}\right). \qquad (14.89)$$

Consequently, from $(m^{\frac{1}{2}} - n^{\frac{1}{2}})^{-1} = (m^{\frac{1}{2}} + n^{\frac{1}{2}})/(m-n)$, we obtain

$$L_4(T) = O\left(\sum_{m=\lfloor T \rfloor}^\infty m^{\varepsilon - \frac{1}{4}} \sum_{n=1}^{m-1} \frac{n^{\varepsilon - \frac{5}{4}}}{m-n} T^{\frac{1}{2}}\right), \qquad (14.90)$$

since $m^{\frac{1}{2}} + n^{\frac{1}{2}} = O\left(m^{\frac{1}{2}}\right)$ and $T^{\frac{1}{2}} = O\left(m^{\frac{1}{2}}\right)$ in view of $n \le m-1$ and of $m \ge \lfloor T \rfloor$. Since

$$\sum_{n=1}^{m-1} \frac{n^{\varepsilon - \frac{5}{4}}}{m-n} = O\left(m^{\delta - 1}\right), \qquad (14.91)$$

as $m \to \infty$, where $\varepsilon > 0$, by $\delta = \delta(\varepsilon) > 0$ (if arbitrarily small), it follows that

$$L_4(T) = O\left(\sum_{m=\lfloor T \rfloor}^\infty m^{\varepsilon - \frac{5}{4}} T^{\frac{1}{2}}\right) \qquad (14.92)$$

for every $\varepsilon > 0$. Therefore we have

$$L_4(T) = O\left(T^{\varepsilon - \frac{1}{4}} T^{\frac{1}{2}}\right) = O\left(T^{\varepsilon + \frac{1}{4}}\right). \tag{14.93}$$

Summarizing our results we obtain for (14.68)

$$\int_1^T \frac{\left(E\left(\overline{\mathbb{B}^2_{\sqrt{t}}}\right)\right)^2}{t}\, dt = \frac{1}{\pi^2} \sum_{n=1}^{\infty} \frac{r^2(n)}{n^{\frac{3}{2}}} T^{\frac{1}{2}} + O\left(T^{\varepsilon + \frac{1}{4}}\right) \tag{14.94}$$

for every $\varepsilon > 0$. Furthermore, we get from (14.64)

$$\sum_{\substack{|g| \le \sqrt{t} \\ g \in \mathbb{Z}^2}} 1 = \pi t + E\left(\overline{\mathbb{B}^2_{\sqrt{t}}}\right) + O(t^{\varepsilon}). \tag{14.95}$$

Observing that

$$E\left(\overline{\mathbb{B}^2_{\sqrt{t}}}\right) = t^{\frac{1}{4}} Q\left(\overline{\mathbb{B}^2_{\sqrt{t}}}\right) + O(t^{\varepsilon}), \tag{14.96}$$

we therefore obtain, from (14.94) together with (14.64) and (14.17),

$$\int_1^T \frac{\left(Q\left(\overline{\mathbb{B}^2_{\sqrt{t}}}\right)\right)^2}{t^{\frac{1}{2}}}\, dt = \frac{1}{\pi^2} \sum_{n=1}^{\infty} \frac{r^2(n)}{n^{\frac{3}{2}}} T^{\frac{1}{2}} + O\left(T^{\varepsilon + \frac{1}{4}}\right). \tag{14.97}$$

Hence, if $t^{\frac{1}{2}}$ is replaced by t,

$$\int_1^T \left(Q\left(\overline{\mathbb{B}^2_t}\right)\right)^2 dt = \frac{1}{2\pi^2} \sum_{n=1}^{\infty} \frac{r^2(n)}{n^{\frac{3}{2}}} T + O\left(T^{\varepsilon + \frac{1}{2}}\right). \tag{14.98}$$

This completes the proof of Lemma 14.8.

Remark 14.3. *Note that (14.98) means (see also (14.31)) that*

$$\int_1^T \frac{\left(P\left(\overline{\mathbb{B}^2_t}\right)\right)^2}{t}\, dt = \frac{1}{2\pi^2} \sum_{n=1}^{\infty} \frac{r^2(n)}{n^{\frac{3}{2}}}\, T + O\left(T^{\varepsilon + \frac{1}{2}}\right), \tag{14.99}$$

where

$$\frac{1}{2\pi^2} \sum_{n=1}^{\infty} \frac{r^2(n)}{n^{\frac{3}{2}}} = 2.54094\dots . \tag{14.100}$$

Fourier Coefficients

For the verification of Lemma 14.7 we follow the ideas of A. Cramér [1922] and A. Wintner [1941]. In order to start with the proof of the identity (14.52) the existence of the Fourier coefficients of Q must be established. To this end, we choose a real number μ and integrate term-by-term

$$\int_1^T e^{2\pi i \mu t} Q\left(\overline{\mathbb{B}_t^2}\right) dt \tag{14.101}$$

$$= \frac{1}{\pi} \sum_{n=1}^{\infty} \frac{r(n)}{n^{\frac{3}{4}}} \int_1^T e^{2\pi i \mu t} \sin\left(2\pi\sqrt{n}t - \frac{\pi}{4}\right) dt + O\left(T^{\frac{1}{2}}\right)$$

(note that summation and integration can be interchanged). The integral on the right of (14.101) can be rewritten in the form

$$\int_1^T e^{2\pi i \mu t} \sin\left(2\pi\sqrt{n}t - \frac{\pi}{4}\right) dt \tag{14.102}$$

$$= \frac{1}{2i} \int_1^T e^{2\pi i \mu t} \left(e^{2\pi i \sqrt{n}t - \frac{i\pi}{4}} - e^{-2\pi i \sqrt{n}t + \frac{i\pi}{4}}\right) dt.$$

This enables us to verify that there exists for every μ a constant C (dependent on μ) such that

$$\left| \int_1^T e^{2\pi i \mu t} \sin\left(2\pi\sqrt{n}t - \frac{\pi}{4}\right) dt \right| \tag{14.103}$$

$$\leq \frac{C}{|2\pi\mu + \sqrt{n}|} + \frac{C}{|2\pi\mu - \sqrt{n}|}, \quad \mu \neq \pm\sqrt{n}$$

and

$$\left| \int_1^T e^{2\pi i \mu t} \sin\left(2\pi\sqrt{n}t - \frac{\pi}{4}\right) dt \pm \frac{T}{2i} e^{\pm i\frac{\pi}{4}} \right| \tag{14.104}$$

$$\leq \frac{C}{|2\pi\mu \pm \sqrt{n}|}, \quad \mu \neq \pm\sqrt{n}.$$

For μ fixed we investigate the limit $T \to \infty$. Suppose first that μ is given in such a way that $\mu^2 \neq n$, $n \in \mathbb{N}$. Then, (14.103) is applicable for every $n \in \mathbb{N}$. Therefore, from (14.101), it follows that

$$\left| \int_1^T e^{2\pi i \mu t} Q\left(\overline{\mathbb{B}_t^2}\right) dt \right| \leq D \sum_{n=1}^{\infty} \frac{r(n)}{n^{\frac{3}{4}} n^{\frac{1}{2}}} + O\left(T^{\frac{1}{2}}\right), \tag{14.105}$$

$\mu \neq \pm\sqrt{n}$, $n \in \mathbb{N}$ (note that the positive constant D is dependent on μ alone). Suppose now that μ is chosen such that either $\mu = \sqrt{n_0}$ or $\mu = -\sqrt{n_0}$ for some $n_0 \in \mathbb{N}$. Then, on applying (14.104) to the n_0-th term of (14.101), and (14.103) to all remaining terms of (14.101), we see that

$$\left| \int_1^T e^{2\pi i \mu t} Q\left(\overline{\mathbb{B}_t^2}\right) dt \pm \frac{T}{2i} e^{\pm i\frac{\pi}{4}} \frac{1}{\pi} \frac{r(n_0)}{n_0^{\frac{3}{4}}} \right| < D \sum_{n=1}^{\infty} \frac{r(n)}{n^{\frac{3}{4}} n^{\frac{1}{2}}} + O\left(T^{\frac{1}{2}}\right) \tag{14.106}$$

if $\mu = \pm\sqrt{n_0}$. Collecting our results we get

$$\int_1^T e^{2\pi i\mu t} Q\left(\overline{\mathbb{B}_t^2}\right) dt = O\left(T^{\frac{1}{2}}\right), \quad \mu \neq \pm\sqrt{n_0} \tag{14.107}$$

and

$$\int_1^T e^{2\pi i\mu t} Q\left(\overline{\mathbb{B}_t^2}\right) dt = \pm\frac{T}{2\pi i}\frac{r(n_0)}{n_0^{\frac{3}{4}}}e^{\pm i\frac{\pi}{4}} + O\left(T^{\frac{1}{2}}\right), \quad \mu = \pm\sqrt{n_0}. \tag{14.108}$$

This shows Lemma 14.7.

Clearly, in connection with Lemma 14.6, the Parseval identity (14.54) can be rewritten as follows

$$\int_1^T \left(Q\left(\overline{\mathbb{B}_t^2}\right)\right)^2 dt = \frac{8}{\pi^2}\frac{\zeta^2(\frac{3}{2})L^2(\frac{3}{2})}{\zeta(3)\left(1 + 2^{-\frac{3}{2}}\right)} T + o(T) \tag{14.109}$$

for $T \to \infty$.

Remark 14.4. *Equation (14.109) is in evidence with the assertion that*

$$Q^2\left(\overline{\mathbb{B}_T^2}\right) = \frac{P^2\left(\overline{\mathbb{B}_T^2}\right)}{T} = o(1), \quad T \to \infty, \tag{14.110}$$

is **wrong**. *The incorrectness of (14.110) can be proved by assuming that the relation $Q^2\left(\overline{\mathbb{B}_T^2}\right) = o(1)$ is valid. We show that this statement amounts to a contradiction to (14.109). In fact, under the assumption (14.110) there exists, for every $\varepsilon > 0$, a value $A(= A(\varepsilon)) > 0$ such that*

$$Q^2\left(\overline{\mathbb{B}_y^2}\right) < \varepsilon \tag{14.111}$$

holds for all $y > A$; hence, for all $T > A$, we find

$$\begin{aligned}
\int_0^T Q^2\left(\overline{\mathbb{B}_y^2}\right) dy &= \int_0^A Q^2\left(\overline{\mathbb{B}_y^2}\right) dy + \int_A^T Q^2\left(\overline{\mathbb{B}_y^2}\right) dy \\
&< \int_0^A Q^2\left(\overline{\mathbb{B}_y^2}\right) dy + \varepsilon \int_0^T dy \\
&= \int_0^A Q^2\left(\overline{\mathbb{B}_y^2}\right) dy + \varepsilon T, \tag{14.112}
\end{aligned}$$

i.e., in contrast to the identity (14.109), we have

$$\frac{1}{T}\int_0^T Q^2\left(\overline{\mathbb{B}_y^2}\right) dy = o(1), \quad T \to \infty. \tag{14.113}$$

14.4 Angular Weight Discrepancy

The weighted Hardy–Landau identity (Corollary 13.5) is the initial key for a variety of properties, for example, integral mean asymptotics, the almost periodicity of the weighted (angular) counterpart of the lattice point discrepancy Q, and the non-uniform distribution of lattice points in the plane. In this section we start with the characterization of the $\mathbb{Z}^2$-lattice point discrepancy for radial and angular functions.

Discrepancy for Products of Radial and Angular Functions

From Corollary 13.5 we obtain

$$\sum_{\substack{\rho \leq |g| \leq N \\ g \in \mathbb{Z}^2}} {}' F(g) = \int_{\substack{\rho \leq |x| \leq N \\ x \in \mathbb{R}^2}} F(x) \, dV(x) + P\left(F; \overline{\mathbb{B}^2_{\rho,N}}\right) \tag{14.114}$$

for all ρ, N with $0 \leq \rho < 1 \leq N$, where the $\mathbb{Z}^2$-lattice point discrepancy $P\left(F; \overline{\mathbb{B}^2_{\rho,N}}\right)$ is given by

$$P\left(F; \overline{\mathbb{B}^2_{\rho,N}}\right) = \sum_{\substack{|h| \neq 0 \\ h \in \mathbb{Z}^2}} \int_{\substack{\rho \leq |x| \leq N \\ x \in \mathbb{R}^2}} F(x) \, e^{-2\pi i h \cdot x} \, dV(x) \,, \tag{14.115}$$

provided that F is twice continuously differentiable on $\overline{\mathbb{B}^2_{\rho,N}}$.

Accordingly, the *$\mathbb{Z}^2$-lattice point discrepancy of a (twice continuously differentiable) radial function G in $\overline{\mathbb{B}^2_{\rho,N}}$, $0 \leq \rho < 1 \leq N$, is given by*

$$P\left(G; \overline{\mathbb{B}^2_{\rho,N}}\right) = \sum_{\substack{\rho \leq |g| \leq N \\ g \in \mathbb{Z}^2}} {}' G(|g|) - 2\pi \int_\rho^N r G(r) \, dr, \tag{14.116}$$

while the *$\mathbb{Z}^2$-lattice point discrepancy of a (twice continuously differentiable) angular function H in $\overline{\mathbb{B}^2_{\rho,N}}$, $0 \leq \rho < 1 \leq N$, reads as follows*

$$P\left(H; \overline{\mathbb{B}^2_{\rho,N}}\right) = \sum_{\substack{\rho \leq |g| \leq N \\ g \in \mathbb{Z}^2}} {}' H\left(\frac{g}{|g|}\right) - \frac{N^2 - \rho^2}{2} \int_{\mathbb{S}^1} H(\xi) \, dS(\xi) \,. \tag{14.117}$$

Applying the Second Green Theorem we obtain

$$\sum_{\substack{|h|\neq 0 \\ h\in\mathbb{Z}^2}} \int_{\substack{\rho\leq|x|\leq N \\ x\in\mathbb{R}^2}} H(x)\, e^{-2\pi i h\cdot x}\, dV(x) \tag{14.118}$$

$$= \frac{N}{2\pi i} \sum_{\substack{|h|\neq 0 \\ h\in\mathbb{Z}^2}} \frac{1}{|h|} \int_{\mathbb{S}^1} \left(\xi\cdot\frac{h}{|h|}\right) H(\xi)\, e^{-2\pi i N h\cdot\xi}\, dS(\xi) + O(1)$$

for $N\to\infty$. From the asymptotic relations (i.e., Corollary 6.30) of the meta-harmonic theory we are able to deduce that

$$\int_{\mathbb{S}^1} \left(\xi\cdot\frac{h}{|h|}\right) H(\xi)\, e^{-2\pi i N h\cdot\xi}\, dS(\xi) \tag{14.119}$$

$$= (N|h|)^{-\frac{1}{2}} \left(e^{2\pi i N|h| - \frac{i\pi}{4}} H\left(-\frac{h}{|h|}\right) - e^{-2\pi i N|h| + \frac{i\pi}{4}} H\left(\frac{h}{|h|}\right)\right)$$

$$+ o\left((|h|N)^{-\frac{1}{2}}\right).$$

Thus it follows in parallel to (14.48) that

$$P\left(H;\overline{\mathbb{B}^2_{\rho,N}}\right) = \frac{\sqrt{N}}{\pi} \sum_{\substack{|h|\neq 0 \\ h\in\mathbb{Z}^2}} \frac{1}{|h|^{\frac{3}{2}}} H\left(\frac{h}{|h|}\right) \sin\left(2\pi|h|N - \frac{\pi}{4}\right) + O(1). \tag{14.120}$$

Angular Number Theoretical Functions

Introducing the *number theoretical function* $r(H;\cdot)$ given by

$$n \mapsto r(H;n) = \sum_{\substack{|h|^2 = n \\ h\in\mathbb{Z}^2}} H\left(\frac{h}{|h|}\right), \quad n\in\mathbb{N}, \tag{14.121}$$

we find for the modified $\mathbb{Z}^2$-lattice discrepancy

$$Q\left(H;\overline{\mathbb{B}^2_N}\right) = \frac{P\left(H;\overline{\mathbb{B}^2_{\rho,N}}\right)}{\sqrt{N}} \tag{14.122}$$

the asymptotic relation

$$Q\left(H;\overline{\mathbb{B}^2_N}\right) = \frac{1}{\pi}\sum_{n=1}^{\infty} \frac{1}{n^{\frac{3}{4}}}\, r(H;n)\,\sin\left(2\pi\sqrt{n}N - \frac{\pi}{4}\right) + O\left(N^{-\frac{1}{2}}\right) \tag{14.123}$$

(note that $Q\left(H;\overline{\mathbb{B}^2_N}\right)$ can be understood to be asymptotically independent of the radius ρ).

14.5   Almost Periodicity of the Angular Weight Discrepancy

As in the case of a constant function H, it is natural to ask within the framework of angular weight functions (cf. W. Freeden [1978b]): *in what sense is the series on the right side of (14.123) the Fourier series of the remainder term (14.122)?* In analogy to the constant weight function as discussed in Section 14.2, we are able to justify the almost periodicity in the $(\mathbb{B}^2)$-Besicovitch sense. In fact, as we already pointed out that the limit relation

$$\lim_{\substack{T \to \infty \\ N \to \infty}} \frac{1}{T} \int_1^T \left(Q(H; \overline{\mathbb{B}_t^2}) - \frac{1}{\pi} \sum_{n=1}^N \frac{r(H; n)}{n^{\frac{3}{4}}} \sin\left(2\pi\sqrt{nt} - \frac{\pi}{4}\right) \right) dt = 0$$

(14.124)

is equivalent to the following two statements:

Lemma 14.9. *For real μ, the Fourier coefficients read for $n = 1, 2, \ldots$*

$$\lim_{T \to \infty} \frac{1}{T} \int_1^T e^{2\pi i \mu t} Q(H; \overline{\mathbb{B}_t^2}) \, dt$$

$$= \begin{cases} \frac{i}{2\pi} \frac{r(H;n)}{n^{\frac{3}{4}}} e^{i\frac{\pi}{4}} & , \quad \mu = \sqrt{n}, \\ -\frac{i}{2\pi} \frac{r(H;n)}{n^{\frac{3}{4}}} e^{-i\frac{\pi}{4}} & , \quad \mu = -\sqrt{n}, \\ 0 & , \quad \mu \neq n^2. \end{cases}$$

(14.125)

Lemma 14.10. *The Parseval identity holds true, i.e.,*

$$\lim_{T \to \infty} \frac{1}{T} \int_1^T \left(Q(H; \overline{\mathbb{B}_t^2}) \right)^2 dt = \frac{1}{2\pi^2} \sum_{n=1}^\infty \frac{(r(H; n))^2}{n^{\frac{3}{2}}}.$$

(14.126)

Both properties (14.125) and (14.126) can be realized by exactly the same arguments as for $r(n)$, thereby observing the fact that $r(H; n)$ satisfies the estimate

$$r(H; n) = \sum_{\substack{|h|^2 = n \\ h \in \mathbb{Z}^2}} H\left(\frac{h}{|h|} \right) = O\left(\sum_{\substack{|h|^2 = n \\ h \in \mathbb{Z}^2}} 1 \right) = O\left(r(n) \right) = O(n^\varepsilon) \quad (14.127)$$

for *every* positive number ε (see also W. Freeden [1978b] for more details). In consequence, our approach canonically leads to the following statement.

Theorem 14.2. *Let H be twice continuously differentiable on the unit circle $\mathbb{S}^1$. Then*

$$Q\left(H; \overline{\mathbb{B}_N^2} \right) \simeq \frac{1}{\pi} \sum_{\substack{|h| \neq 0 \\ h \in \mathbb{Z}^2}} \frac{1}{|h|^{\frac{3}{2}}} H\left(\frac{h}{|h|} \right) \sin\left(2\pi |h| N - \frac{\pi}{4} \right).$$

(14.128)

where "$\simeq$" is meant in the (B^2)*-Besicovitch sense, i.e.,* (14.125) *and* (14.126) *hold true.*

14.6 Radial and Angular Weights

Let G be of class $\mathrm{C}^{(2)}([\rho, N])$, $0 \le \rho < 1 \le N$. Assume that H is of class $\mathrm{C}^{(2)}(\mathbb{S}^1)$. Then, from Corollary 13.5, it follows that

$$\sideset{}{'}\sum_{\substack{\rho \le |g| \le N \\ g \in \mathbb{Z}^2}} G(|g|) \, H\left(\frac{g}{|g|}\right) \tag{14.129}$$

$$= \int_\rho^N rG(r) \, dr \int_{\mathbb{S}^1} H(\xi) \, dS(\xi)$$

$$+ \sum_{\substack{|h| \ne 0 \\ h \in \mathbb{Z}^2}} \int_\rho^N rG(r) \int_{\mathbb{S}^1} H(\xi) e^{-2\pi i r |h| \left(\frac{h}{|h|} \cdot \xi\right)} \, dS(\xi) \, dr.$$

The last series on the right side of (14.129) is the discrepancy

$$P\left(GH; \overline{\mathbb{B}^2_{\rho, N}}\right) \tag{14.130}$$

$$= \int_\rho^N rG(r) \sum_{\substack{|h| \ne 0 \\ h \in \mathbb{Z}^2}} \int_{\mathbb{S}^1} H(\xi) e^{-2\pi i r |h| \left(\frac{h}{|h|} \cdot \xi\right)} \, dS(\xi) \, dr,$$

which can be rewritten in a formal way by

$$P\left(GH; \overline{\mathbb{B}^2_{\rho, N}}\right) = \int_\rho^N G(r) P'\left(H; \overline{\mathbb{B}^2_r}\right) dr, \tag{14.131}$$

where $P'\left(H; \overline{\mathbb{B}^2_r}\right)$ is given by

$$P'\left(H; \overline{\mathbb{B}^2_r}\right) = r \sum_{\substack{|h| \ne 0 \\ h \in \mathbb{Z}^2}} \int_{\mathbb{S}^1} H(\xi) e^{-2\pi i r |h| \left(\frac{h}{|h|} \cdot \xi\right)} \, dS(\xi). \tag{14.132}$$

Remark 14.5. $P'\left(H; \overline{\mathbb{B}^2_r}\right)$ *can be understood as the derivative of* (14.117) *with respect to the variable* r. *In fact, from* (14.117), *we get*

$$P\left(H; \overline{\mathbb{B}^2_{\rho, r}}\right) = \sum_{\substack{|h| \ne 0 \\ h \in \mathbb{Z}^2}} \int_{\rho \le |x| \le r} H\left(\frac{x}{|x|}\right) e^{-2\pi i h \cdot x} \, dV(x), \tag{14.133}$$

where $0 < \rho < 1 \leq r$. Applying the Second Green Theorem we obtain

$$\sum_{\substack{|h| \neq 0 \\ h \in \mathbb{Z}^2}} \int_{\substack{\rho \leq |x| \leq r \\ x \in \mathbb{R}^2}} H(x)\, e^{-2\pi i h \cdot x}\, dV(x) \tag{14.134}$$

$$= -\frac{r}{2\pi i} \sum_{\substack{|h| \neq 0 \\ h \in \mathbb{Z}^2}} \frac{1}{|h|} \int_{\mathbb{S}^1} \left(\xi \cdot \frac{h}{|h|} \right) H(\xi)\, e^{-2\pi i r h \cdot \xi}\, dS(\xi) + O(1)$$

for $N \to \infty$. Thus we are allowed to understand

$$P\left(H; \overline{\mathbb{B}_r^2}\right) = -\frac{r}{2\pi i} \sum_{\substack{|h| \neq 0 \\ h \in \mathbb{Z}^2}} \frac{1}{|h|} \int_{\mathbb{S}^1} \left(\xi \cdot \frac{h}{|h|} \right) H(\xi) e^{-2\pi i |h| r (\xi \cdot \frac{h}{|h|})}\, dS(\xi) \tag{14.135}$$

as an antiderivative of $P'\left(H; \overline{\mathbb{B}_r^2}\right)$ as given by (14.132).

By aid of (14.134) and (14.135) we are led to the following asymptotic relation.

Lemma 14.11. *For $N \to \infty$*

$$P\left(GH; \overline{\mathbb{B}_{\rho,N}^2}\right) = G(r)P\left(H; \overline{\mathbb{B}_N^r}\right)\Big|_\rho^N - \int_\rho^N G'(r)P(H; \overline{\mathbb{B}_r^2})\, dr \tag{14.136}$$

$$+ O(1),$$

where $P\left(H; \overline{\mathbb{B}_r^2}\right)$ is given by (14.135).

In what follows we are interested in asymptotic relations involving the modified discrepancy $Q\left(H; \overline{\mathbb{B}_r^2}\right)$ (see (14.122)).

Lemma 14.12. *Under the assumptions of (14.129) we have for $N \to \infty$*

$$\sideset{}{'}\sum_{\substack{\rho \leq |g| \leq N \\ g \in \mathbb{Z}^2}} G(|g|) H\left(\frac{g}{|g|}\right)$$

$$= \int_\rho^N rG(r)\, dr \int_{\mathbb{S}^1} H(\xi)\, dS(\xi) + G(r)P\left(H; \overline{\mathbb{B}_r^2}\right)\Big|_\rho^N$$

$$+ \left. G'(r) \left(\sum_{\substack{|h| \neq 0 \\ h \in \mathbb{Z}^2}} \frac{1}{4\pi^2 h^2} \int_{\mathbb{S}^1} H(\xi) e^{-2\pi r |h| (\xi \cdot \frac{h}{|h|})}\, dS(\xi) \right) \right|_\rho^N$$

$$- \int_\rho^N (G'(r)r)' \left(\sum_{\substack{|h| \neq 0 \\ h \in \mathbb{Z}^2}} \frac{1}{4\pi^2 h^2} \int_{\mathbb{S}^1} H(\xi) e^{-2\pi i r |h| (\xi \cdot \frac{h}{|h|})}\, dS(\xi) \right) dr.$$

Observing (14.122) we obtain from Lemma 14.11 for $N \to \infty$

$$P\left(GH; \overline{\mathbb{B}^2_{\rho,N}}\right) = G(N)\sqrt{N}\, Q\left(H; \overline{\mathbb{B}^2_N}\right) \tag{14.137}$$

$$- \int_1^N G'(r)\sqrt{r}\, Q\left(H; \overline{\mathbb{B}^2_r}\right)\, dr + O(1).$$

We introduce an *auxiliary function* $A : [1, \infty] \to \mathbb{R}$ by

$$A(r) = \int_1^r G'(u)\, Q\left(H; \overline{\mathbb{B}^2_u}\right)\, du, \tag{14.138}$$

which occupies a central role in our following investigations. Looking at the integral on the right side of (14.137) we find by partial integration

$$\int_1^N G'(r)\sqrt{r}\, Q\left(H; \overline{\mathbb{B}^2_r}\right)\, dr = \left(\sqrt{r}\, A(r)\right)\Big|_1^N - \frac{1}{2}\int_1^N r^{-\frac{1}{2}} A(r)\, dr. \tag{14.139}$$

Collecting our results we therefore obtain the following asymptotic relation.

Lemma 14.13. *Let G be of class* $\mathrm{C}^{(2)}([0, N])$. *Assume that H is of class* $\mathrm{C}^{(2)}(\mathbb{S}^1)$. *Then, for $N \to \infty$,*

$$\sum_{\substack{0 < |g| \leq N \\ g \in \mathbb{Z}^2}}{}' G(|g|) H\left(\frac{g}{|g|}\right) = \int_1^N rG(r)\, dr \int_{\mathbb{S}^1} H(\xi)\, dS(\xi)$$

$$+ G(N)\sqrt{N}\, Q\left(H; \overline{\mathbb{B}^2_N}\right) - \sqrt{N}\, A(N)$$

$$+ \frac{1}{2}\int_1^N r^{-\frac{1}{2}} A(r)\, dr + O(1),$$

where A is given by (14.138).

It is worth mentioning some special cases of Lemma 14.13:

(i) If G is constant, e.g., $G = 1$, then we have for $N \to \infty$

$$\sum_{\substack{0 < |g| \leq N \\ g \in \mathbb{Z}^2}}{}' H\left(\frac{g}{|g|}\right) = \frac{N^2}{2}\int_{\mathbb{S}^1} H(\xi)\, dS(\xi) \tag{14.140}$$

$$+ \sqrt{N}\, Q\left(H; \overline{\mathbb{B}^2_N}\right) + O(1).$$

(ii) If H is constant, e.g., $H = 1$, then we have for $N \to \infty$

$$\sum_{\substack{0<|g|\leq N \\ g\in\mathbb{Z}^2}} {}' G(|g|) \;=\; 2\pi \int_1^N rG(r)\,dr \tag{14.141}$$

$$+\, G(N)P\left(\overline{\mathbb{B}_N^2}\right) - \sqrt{N}\,A(N)$$

$$+\, \frac{1}{2}\int_1^N r^{-\frac{1}{2}} A(r)\,dr + O(1).$$

Observing the series expansion of $P'\left(\mathbb{B}_r^2\right)$ we obtain for $N\to\infty$

$$\sum_{\substack{0<|g|\leq N \\ g\in\mathbb{Z}^2}} {}' G(|g|) \;=\; 2\pi \int_1^N rG(r)\,dr + G(N)P\left(\overline{\mathbb{B}_N^2}\right) \tag{14.142}$$

$$+\, \frac{N}{2\pi}G'(N) \sum_{\substack{|h|\neq 0 \\ h\in\mathbb{Z}^2}} \frac{J_0(2\pi|h|N)}{|h|^2}$$

$$-\, \frac{1}{2\pi} \sum_{\substack{|h|\neq 0 \\ h\in\mathbb{Z}^2}} \frac{1}{|h|^2} \int_1^N \left(uG'(u)\right)' J_0(2\pi|h|u)\,du$$

$$+\, O(1).$$

Periodical Radial Lattice Point Expansions

Next we suppose, in addition, that G is a real-valued $C^{(2)}$-function with period $\beta > 0$ (note that the $C^{(2)}$-assumption can be weakened, but we omit these considerations) such that

$$G(r) = \sum_{\mu\in I} G^\wedge(\mu)\, e^{2\pi i\mu r}, \tag{14.143}$$

where the index set I is given by

$$I = \left\{\mu = \beta^{-1}h \mid h\in\mathbb{Z}\right\}. \tag{14.144}$$

Our objective is to determine the integral (14.138) under the representation (14.143) for the radial part G. In the sense of the (B^2)-Besicovitch theory we are allowed to conclude that

$$G'(r) \simeq \sum_{\mu\in I} (G')^\wedge(\mu)\, e^{2\pi i\mu r} \tag{14.145}$$

and

$$Q\left(H;\overline{\mathbb{B}_r^2}\right) \simeq \sum_{\mu\in J} Q^\wedge(n)\, e^{2\pi i\mu r}, \tag{14.146}$$

where
$$J = \{\mu \in \mathbb{R} \mid \mu^2 \in \mathbb{N}\} \tag{14.147}$$

and

$$(G')^\wedge(\mu) = \begin{cases} 2\pi i\mu \, G^\wedge(\mu) , & \mu = \beta^{-1}j, \quad j \in \mathbb{Z}, \\ 0 , & \text{otherwise} \end{cases} \tag{14.148}$$

$$Q^\wedge(\mu) = \begin{cases} -\frac{\sqrt{i}}{2\pi} \frac{r(H;n)}{n^{\frac{3}{4}}} , & \mu^2 = n \in \mathbb{N}, \ \mu > 0, \\[2mm] \frac{i\sqrt{i}}{2\pi} \frac{r(H;n)}{n^{\frac{3}{4}}} , & \mu^2 = n \in \mathbb{N}, \ \mu < 0, \\[2mm] 0 , & \text{otherwise.} \end{cases} \tag{14.149}$$

From the Parseval identity of the (B^2)-Besicovitch theory of almost periodical functions (see A. Besicovitch [1954]) we then get for $N \to \infty$

$$A(N) = \frac{N}{2} \sum_{\mu \in I \cap J} \left((G')^\wedge(\mu) \, \overline{Q^\wedge(\mu)} + \overline{(G')^\wedge(\mu)} \, Q^\wedge(\mu) \right) + o(N). \tag{14.150}$$

Remark 14.6. *The coefficients*

$$\frac{1}{2} \left((G')^\wedge(\mu) \, \overline{Q^\wedge(\mu)} + \overline{(G')^\wedge(\mu)} \, Q^\wedge(\mu) \right) \tag{14.151}$$

are different from 0 only if μ is a member of $I \cap J$, i.e., $\mu \in I \cap J$.

Asymptotic Behavior of Weighted Lattice Point Sums

In order to evaluate (14.150) in a more explicit way we need some results of elementary number theory (cf. G.H. Hardy, E.M. Wright [1958], A. Dressler [1967, 1972]); i.e., for the interpretation of $I \cap J$ in (14.150) we have to distinguish:

(i) there are integers n_1, n_2, and k with $k \neq 0$ satisfying

$$n_1^2 + n_2^2 = \frac{k^2}{\beta^2}. \tag{14.152}$$

(ii) for all sets of integers n_1, n_2, and $k \neq 0$

$$n_1^2 + n_2^2 \neq \frac{k^2}{\beta^2}. \tag{14.153}$$

Condition *(ii)* is not of deeper interest here. Following G.H. Hardy, E.M.

Wright [1958] Condition *(i)* can be characterized as follows: The equation (14.152) is solvable in the integers n_1, n_2, and $k \neq 0$ if and only if β^2 is of the form

$$\beta^2 = \frac{n^2}{l_1^2 + l_2^2}, \tag{14.154}$$

where $l_1, l_2, n \neq 0$ are integers (such that the numerator and the denominator of (14.154) contain only common prime factors of first order). In the case of (14.154), all sums of squares $m_1^2 + m_2^2$ with $m_1, m_2 \in \mathbb{Z}$ satisfying (14.152) are given by the equation

$$m_1^2 + m_2^2 = \frac{k^2}{\beta^2}, \tag{14.155}$$

where

$$k = l \, n, \quad l \in \mathbb{N}. \tag{14.156}$$

Going back to (14.151) we are therefore confronted with the following situation: let l denote the smallest of all integers $j \in \mathbb{N}$, for which $\beta^{-2} j^2$ admits a representation as sum of squares $n_1^2 + n_2^2$ of two integers n_1, n_2, i.e., $\beta^{-2} l^2 = n_1^2 + n_2^2$. Then, the greatest common divisor of $n_1^2 + n_2^2$ and l^2 does contain at most prime factors of order 1. The integers j, for which the coefficients (14.151) might be different from 0, are the numbers ml with $m \in \mathbb{Z}$. All in all, we have

$$A(N) \tag{14.157}$$

$$= -\frac{\sqrt{\beta}}{2\sqrt{l}} N \sum_{\substack{m \neq 0 \\ m \in \mathbb{Z}}} \frac{1}{|m|^{\frac{1}{2}}} r\left(H; l^2 \beta^{-2} m^2\right) \left(G^\wedge (\beta^{-1} lm) e^{i\frac{\pi}{4}} + \overline{G^\wedge(\beta^{-1} lm)} e^{-i\frac{\pi}{4}}\right)$$

$$+ \; o(N).$$

In connection with Lemma 14.13 the asymptotic expansion (14.157) leads to the following identity

$$G(N) \sqrt{N} \, Q\left(H; \overline{\mathbb{B}_N^2}\right) - \sqrt{N} \, A(N) + \frac{1}{2} \int_1^N r^{-\frac{1}{2}} A(r) \, dr \tag{14.158}$$

$$= -\frac{2}{3} N^{\frac{1}{2}} \sum_{\mu \in I \cap J} \left((G')^\wedge(\mu) \overline{Q^\wedge(\mu)} + \overline{(G')^\wedge(\mu)} Q^\wedge(\mu)\right) + o(N^{\frac{3}{2}}).$$

Summarizing our results we therefore obtain

Theorem 14.3. *Assume that the function H is of class* $\mathrm{C}^{(2)}(\mathbb{S}^1)$. *Suppose that $G \in \mathrm{C}^{(2)}(\mathbb{R})$ is a real-valued function with period β of the form*

$$G(r) = \sum_{j \in \mathbb{Z}} a_j \, e^{2\pi i \beta^{-1} jr}, \tag{14.159}$$

i.e., in the notation of (14.143) and (14.144)

$$a_j = G^\wedge(\beta^{-1} j), \quad j \in \mathbb{Z}. \tag{14.160}$$

Let β be representable in the form

$$\beta = \frac{j}{\sqrt{n_1^2 + n_2^2}} \tag{14.161}$$

with $j \in \mathbb{N}, n_1, n_2 \in \mathbb{Z}$. Furthermore, let l denote the smallest of the positive integers j, for which (14.161) holds true. Then

$$\sum_{\substack{n \leq N^2 \\ n \in \mathbb{N}}} {}' G(\sqrt{n})\, r(H; n) = \frac{a_0 N^2}{2} \int_{\mathbb{S}^1} H(\xi)\, dS(\xi) + C_H(\beta)\, N^{\frac{3}{2}} + o\left(N^{\frac{3}{2}}\right),$$

$$\tag{14.162}$$

where

$$C_H(\beta) = \frac{\sqrt{\beta}}{3\sqrt{l}} \sum_{\substack{m \neq 0 \\ m \in \mathbb{Z}}} \frac{1}{|m|^{\frac{1}{2}}}\, r\left(H; l^2 \beta^{-2} m^2\right) \left(a_{lm} e^{i\frac{\pi}{4}} + \overline{a_{lm}} e^{-i\frac{\pi}{4}}\right). \tag{14.163}$$

Of course, it is possible that $C_H(\beta) = 0$ holds true. In fact, if β is not of the form (14.161), then we always have $C_H(\beta) = 0$. For example, with $H = 1$,

$$\sum_{\substack{n \leq N^2 \\ n \in \mathbb{N}}} {}' G(\sqrt{n})\, r(n) = a_0 \pi N^2 + o\left(N^{\frac{3}{2}}\right) \tag{14.164}$$

if β is different from the numbers $\frac{1}{\sqrt{p}}$ with $r(p) > 0$, $p \in \mathbb{N}$.

14.7 Non-Uniform Distribution of Lattice Points

Next the non-uniform distribution of lattice points in a "circular and sectorial ring configuration" of the plane is illustrated for simple geometrical cases.

Distributions Generated by Cosine Functions

Let us start with the radial function G with period $\beta = p^{-\frac{1}{2}}$, $p \in \mathbb{N}$, given by

$$G(r) = \cos\left(2\pi \sqrt{p}\, r\right). \tag{14.165}$$

From Theorem 14.3 we obtain

$$\sum_{\substack{n \leq N^2 \\ n \in \mathbb{N}}} {}' \cos\left(2\pi \sqrt{pn}\,\right) r(H; n) = C_H\left(p^{-\frac{1}{2}}\right) N^{\frac{3}{2}} + o\left(N^{\frac{3}{2}}\right) \tag{14.166}$$

with

$$C_H\left(p^{-\frac{1}{2}}\right) = \frac{p^{-\frac{1}{4}}}{3\sqrt{2}}\, r\left(H; p\right) = \frac{p^{-\frac{1}{4}}}{3\sqrt{2}} \sum_{\substack{|h|^2 = p \\ h \in \mathbb{Z}^2}} H\left(\frac{h}{|h|}\right). \tag{14.167}$$

(Radial dependence) If we choose $H = \text{const}$, e.g., $H = 1$, then we find with the period $\beta = p^{-\frac{1}{2}}$, $p \in \mathbb{N}$, the following results:

(i) we get $C_H \left(p^{-\frac{1}{2}} \right) \neq 0$, *provided that* $p \in \mathbb{N}$ *(fixed) is chosen such that* $r(p) > 0$, *and it follows that*

$$\sum_{\substack{n \leq N^2 \\ n \in \mathbb{N}}}{}' \cos\left(2\pi\sqrt{pn}\right) r(n) = \frac{p^{-\frac{1}{4}}}{3\sqrt{2}}\, r(p)\, N^{\frac{3}{2}} + o\left(N^{\frac{3}{2}}\right) \qquad (14.168)$$

(ii) we certainly have $C_H \left(p^{-\frac{1}{2}} \right) = 0$, *provided that* $p \in \mathbb{N}$ *is chosen such that* $r(p) = 0$, *and it follows that*

$$\sum_{\substack{n \leq N^2 \\ n \in \mathbb{N}}}{}' \cos\left(2\pi\sqrt{pn}\right) r(n) = o\left(N^{\frac{3}{2}}\right). \qquad (14.169)$$

Already this simple argument allows the interpretation that the lattice points are not uniformly distributed over concentric circles around the origin.

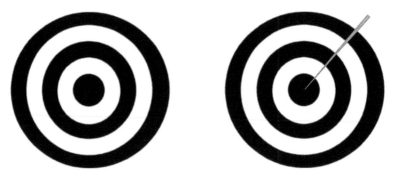

FIGURE 14.1
Circular rings around the origin of fixed width (left), a special sector within circular circles of fixed width (right).

(Angular dependence) If we especially choose the angular function H given in the form

$$H : \xi \mapsto H(\xi) = \begin{cases} e^{\frac{1}{1-\cos(\delta)} - \frac{1}{1-\cos(\delta)-(1-\xi\cdot\eta)}} & , \quad 1 - \xi\cdot\eta < 1 - \cos(\delta) \\ 0 & , \quad 1 - \xi\cdot\eta \geq 1 - \cos(\delta), \end{cases} \qquad (14.170)$$

(such that for all directions $\eta \in \mathbb{S}^1$ and all sufficiently small $\delta > 0$ the function H in (14.170) satisfies the assumption of Theorem 14.3) as well as the period β of (14.165) to be given in the form $\beta = \frac{1}{\sqrt{p}}$, $p \in \mathbb{N}$ (fixed) and $r(p) > 0$, then we find:

(iii) for every $\eta \in \mathbb{S}^1$ with $\sqrt{p}\,\eta \in \mathbb{Z}^2$ and each (sufficiently) small δ we have $C_H(p^{-\frac{1}{2}}) \neq 0$, i.e.,

$$\sideset{}{'}\sum_{\substack{n \leq N^2 \\ n \in \mathbb{N}}} \cos\left(2\pi\sqrt{pn}\,\right)\, r(H;n) = \frac{p^{-\frac{1}{4}}}{3\sqrt{2}}\, N^{\frac{3}{2}} + o\left(N^{\frac{3}{2}}\right). \tag{14.171}$$

(iv) for every direction $\eta \in \mathbb{S}^1$ with $\sqrt{p}\,\eta \notin \mathbb{Z}^2$ we are able to determine a (sufficiently small) number δ such that $C_H(\beta) = 0$, i.e.,

$$\sideset{}{'}\sum_{\substack{n \leq N^2 \\ n \in \mathbb{N}}} \cos\left(2\pi\sqrt{pn}\,\right)\, r(H;n) = o\left(N^{\frac{3}{2}}\right). \tag{14.172}$$

This simple argument indicates that in the concentric circular rings of width $\beta = p^{-\frac{1}{2}}$ ($p \in \mathbb{N}$ with $r(p) > 0$) a larger number of lattice points can be detected for a direction $\eta \in \mathbb{S}^2$ with $\sqrt{p}\,\eta \in \mathbb{Z}^2$ than for the other direction (see Figure 14.1).

Distributions Generated by Lattice Functions

Let β be of the form (14.161). The one-dimensional lattice functions $G\left(\Delta^k; \beta^{-1}\cdot\right)$, $k \in \mathbb{N}$, possess the Fourier expansions

$$G\left(\Delta^k; \beta^{-1}r\right) \sum_{\substack{h \neq 0 \\ h \in \mathbb{Z}}} \frac{e^{2\pi i h \beta^{-1}r}}{(2\pi i h)^{2k}}. \tag{14.173}$$

Consequently, we obtain from Theorem 14.3 for every partial sum

$$G^{(T)}\left(\Delta^k; \beta^{-1}r\right) = \sum_{\substack{0 < |h| \leq T \\ h \in \mathbb{Z}}} \frac{e^{2\pi i h \beta^{-1}r}}{(-4\pi^2 h^2)^k} \tag{14.174}$$

the expression

$$C_H^{(T)}(\beta) = \frac{\sqrt{2}}{3}\sqrt{\beta}\left(\frac{1}{-4\pi^2}\right)^k \sum_{\substack{1 \leq |m| \leq T \\ m \in \mathbb{Z}}} \frac{r\left(H; l^2\beta^{-2}m^2\right)}{|lm|^{2k+\frac{1}{2}}}, \tag{14.175}$$

such that the limit $T \to \infty$ implies the following corollary.

Corollary 14.1. *Assume that H is of class $\mathrm{C}^{(2)}(\mathbb{S}^1)$. Suppose that G is the real-valued radial function with period β of the form (14.161)*

$$G\left(\Delta^k; \beta^{-1}r\right) = \sum_{\substack{|h| \neq 0 \\ h \in \mathbb{Z}}} \frac{e^{2\pi i h \beta^{-1}r}}{(-4\pi^2 h^2)^k}. \tag{14.176}$$

Let β be representable in the form

$$\beta = \frac{j}{\sqrt{n_1^2 + n_2^2}} \tag{14.177}$$

with $j \in \mathbb{N}, n_1, n_2 \in \mathbb{Z}$. Furthermore, let l denote the smallest of the positive integers j, for which (14.161) is satisfied. Then

$$\sum_{\substack{0 \leq \sqrt{n} \leq N \\ n \in \mathbb{N}}} {}'G(\sqrt{n}) r(H; n) = C_H(\beta) \, N^{\frac{3}{2}} + o\left(N^{\frac{3}{2}}\right), \tag{14.178}$$

where

$$C_H(\beta) = \frac{\sqrt{2}}{3} \sqrt{\beta} \left(\frac{1}{-4\pi^2}\right)^k \sum_{\substack{|m| \neq 0 \\ m \in \mathbb{Z}}} \frac{r_H\left(l^2 \beta^{-2} m^2\right)}{|lm|^{2k+\frac{1}{2}}}. \tag{14.179}$$

Note that, for $k \geq 2$, Corollary 14.1 immediately follows from Theorem 14.3 without considering the (infinitely often differentiable) partial sums (cf. C. Müller, A. Dressler [1972]).

Example 14.1. *Choosing $H = \mathrm{const}$, e.g., $H = 1$, we are able to express the coefficient (14.179) by known functions in analytic theory of numbers. The explicit calculation can be based on the same apparatus known from the mean value formula (14.31). The main problem is to calculate the Dirichlet series (cf. Lemma 14.6)*

$$\sum_{m=1}^{\infty} \frac{r\left(dm^2\right)}{m^s}, \quad s \in \mathbb{C}, \quad \Re(s) > 1, \tag{14.180}$$

with $d = \beta^{-2} l^2 = n_1^2 + n_2^2, \quad n_1, n_2 \in \mathbb{Z}$.

Our point of departure is the prime decomposition of the positive integer d of the form

$$d = \prod_{p \mid d} p^{v_p(d)}, \tag{14.181}$$

We note that $v_p(d)$ is necessarily even for $p \equiv 3(\mathrm{mod}\ 4)$, otherwise we know from the Fermat–Euler identity (see Theorem 5.4)

$$r(n) = 4 \prod_{\substack{p \mid n \\ p \equiv 1(\mathrm{mod}\ 4)}} (1 + v_p(d)) \prod_{\substack{p \mid n \\ p \equiv 3(\mathrm{mod}\ 4)}} \frac{1 + (-1)^{v_p(d)}}{2}, \tag{14.182}$$

that d is not representable as a sum of squares. Furthermore, we know that

$\delta(n) = \frac{1}{4}r(n)$ *is a multiplicative number theoretical function. Thus, for $s \in \mathbb{C}$ with $\Re(s) > 1$, we get*

$$\sum_{m=1}^{\infty} \frac{\delta(dm^2)}{m^s} = \prod_{p \nmid d} \sum_{j=0}^{\infty} \frac{\delta(p^{2j})}{p^{js}} \prod_{r|d} \sum_{g=0}^{\infty} \frac{\delta(r^{v_g(d)+2g})}{r^{gs}} . \tag{14.183}$$

It follows from (14.182) for $v_g(d) \geq 0$ that

$$\delta(r^{v_g(d)+2g}) = \begin{cases} v_g(d) + 2g + 1, & if \quad r \equiv 1 \pmod 4 \\ 1, & if \quad r \equiv 3 \pmod 4 \ or \ r = 2. \end{cases} \tag{14.184}$$

Therefore we obtain

$$\sum_{g=0}^{\infty} \frac{\delta(r^{v_g(d)+2g})}{r^{gs}} = \begin{cases} \frac{1+v_r(d)+(1-v_r(d))r^{-s}}{(1-r^{-s})^2}, & if \quad r \equiv 1 \pmod 4 \\ \frac{1}{1-r^{-s}}, & if \quad r \equiv 3 \pmod 4 \ or \ r = 2, \end{cases} \tag{14.185}$$

where we have used the power series

$$\sum_{g=0}^{\infty} x^g = \frac{1}{1-x}, \qquad \sum_{g=0}^{\infty} (2g+1)x^g = \frac{1+x}{(1-x)^2}, \tag{14.186}$$

which are valid for all $x \in \mathbb{R}$ with $|x| < 1$.

Next we decompose (cf. C. Müller, A. Dressler [1972]) the right side of (14.183) in the form

$$\prod_{\substack{p|d \\ p \equiv 1 (\mathrm{mod} 4)}} \cdots \prod_{\substack{r|d \\ r \equiv 3 (\mathrm{mod} 4)}} \cdots \prod_{\substack{p_1 \nmid d \\ p_1 \equiv 1 (\mathrm{mod} 4)}} \cdots \prod_{\substack{r_1 \nmid d \\ r_1 \equiv 3 (\mathrm{mod} 4)}} \cdots . \tag{14.187}$$

From (14.185) we get for the right side of (14.183)

$$\prod_{p \nmid d} \sum_{j=0}^{\infty} \frac{\delta(p^{2j})}{p^{js}} \prod_{r|d} \sum_{g=0}^{\infty} \frac{\delta(r^{v_g(d)+2g})}{r^{gs}} \tag{14.188}$$

$$= \prod_{\substack{p|d \\ p \equiv 1 (\mathrm{mod} 4)}} \frac{1 + v_p(d) + (1 - v_p(d))p^{-s}}{1 + p^{-s}} \frac{1}{1 + 2^{-s}} \frac{\zeta^2(s)L(s)}{\zeta(2s)}.$$

Especially for $s = \frac{5}{2}$ (i.e., $k = 2$) we find

$$C(\beta) = -\frac{2\sqrt{2}}{3\pi^2} \frac{1}{1 + 2^{-\frac{5}{2}}} \frac{\zeta^2\left(\frac{5}{2}\right) L\left(\frac{5}{2}\right)}{\zeta(5)} D(\beta). \tag{14.189}$$

Hence, with $\beta^2 = l^2 d^{-1}$, we obtain

$$D(\beta) = l^{-2} d^{-\frac{1}{4}} \prod_{\substack{r|d \\ r \equiv 1 (\mathrm{mod} 4)}} \frac{1 + v_r(d) + (1 - v_r(d))r^{-\frac{5}{2}}}{1 - r^{-\frac{5}{2}}} . \tag{14.190}$$

In particular, we get $D(1) = 1$. Moreover, we have

$$C(1) = -\frac{2\sqrt{2}}{3\pi^2} \frac{1}{1 + 2^{-\frac{5}{2}}} \frac{\zeta^2\left(\frac{5}{2}\right) L\left(\frac{5}{2}\right)}{\zeta(5)}. \tag{14.191}$$

It should be remarked that the periods β of the form (14.177) are dense in the interval $[1, \infty)$. Because of (14.190), however, the corresponding values $C(\beta)$ do not admit a "completion" to a continuous function.

Distributions Generated by Step Functions

Theorem 14.3 offers results involving one-dimensional step functions responsible for the radial contribution.

Theorem 14.4. *Suppose that the half "width" $\tau \in \mathbb{R}$ and the "phase" $w \in \mathbb{R}$ satisfy $0 \le w < \beta$, $0 < \tau \le \frac{\beta}{4}$. Let G be the β-periodical function (with β^2 of the form (14.161)) given by*

$$G(r) = \begin{cases} 1 & , & 0 < r < 2\tau \\ 0 & , & 2\tau \le r \le \beta \end{cases} \tag{14.192}$$

and

$$G(r \pm \beta) = G(r). \tag{14.193}$$

Assume that H is of class $C^{(2)}(\mathbb{S}^1)$. Then, for $N \to \infty$, we have

$$\sum_{\substack{0 < |g| \le N \\ g \in \mathbb{Z}^2}} {}' G(|g| - w) H\left(\frac{g}{|g|}\right) \tag{14.194}$$

$$= \left(\int_{\mathbb{S}^1} H(\xi)\, dS(\xi)\right) \frac{\tau}{\beta} N^2 + C_H(w; \tau; \beta) N^{\frac{3}{2}} + o(N^{\frac{3}{2}}),$$

where

$$C_H(w; \tau; \beta^2) \tag{14.195}$$

$$= \frac{4\sqrt{\beta}}{3\pi} \sum_{n=1}^{\infty} \frac{r(H; l^2\beta^{-2}n^2)}{(l\, n)^{\frac{3}{2}}} \sin\left(\frac{2\pi\, l\, n}{\beta}\tau\right) \cos\left(\frac{2\pi\, l\, n}{\beta}(\tau + w) - \frac{\pi}{4}\right).$$

Proof. The β-periodical function G admits the Fourier series

$$G(r) = \frac{c_0}{2} + \sum_{m=1}^{\infty} \left(c_m \cos\left(\frac{2\pi m}{\beta}r\right) + d_m \sin\left(\frac{2\pi m}{\beta}r\right)\right), \tag{14.196}$$

where the expansion coefficients c_m, d_m, respectively, are given by

$$c_m = \frac{1}{\pi m} \sin\left(\frac{4\pi m}{\beta}\tau\right), \quad d_m = \frac{2}{\pi m} \sin\left(\frac{2\pi m}{\beta}\tau\right). \tag{14.197}$$

Its partial sums $G^{(T)}$, $T \in \mathbb{N}$, given by

$$G^{(T)}(r-w) = \frac{2\tau}{\beta} + \frac{2}{\pi} \sum_{\substack{1 \le k \le T \\ k \in \mathbb{Z}}} \frac{\sin\left(\frac{2\pi k}{\beta}\tau\right)}{k} \cos\left(\frac{2\pi k}{\beta}(r - w - \tau)\right) \quad (14.198)$$

form infinitely often differentiable functions (with respect to the variable r). Hence, from Theorem 14.3, we obtain the asymptotic expansion

$$\sum_{\substack{\rho \le |g| \le N \\ g \in \mathbb{Z}^2}} {}' G^{(T)}(|g| - w) \, H\left(\frac{g}{|g|}\right) \quad (14.199)$$

$$= \left(\int_{\mathbb{S}^1} H(\xi) \, dS(\xi)\right) \frac{\tau}{\beta} N^2 + C_H^{(T)}\left(w; \tau; \beta^2\right) N^{\frac{3}{2}} + o\left(N^{\frac{3}{2}}\right),$$

where

$$C_H^{(T)}(w; \tau; \beta^2) \quad (14.200)$$

$$= \frac{4\sqrt{\beta}}{3\pi} \sum_{\substack{1 \le n \le T \\ n \in \mathbb{N}}} \frac{r(H; l^2 \beta^{-2} n^2)}{(l\,n)^{\frac{3}{2}}} \sin\left(2\pi \frac{n\,l}{\beta}\tau\right) \cos\left(\frac{2\pi n\,l}{\beta}(\tau + w) - \frac{\pi}{4}\right).$$

Consequently, Theorem 14.4 follows by taking the limit $T \to \infty$. $\qquad\qquad \square$

14.8 Quantitative Step Function Oriented Geometric Interpretation

Obviously, the planar sets

$$\mathbb{B}^2_{m+w-\tau, m+w+\tau} = \left\{ x \in \mathbb{R}^2 \mid m + w - \tau < |x| < m + w + \tau \right\} \quad (14.201)$$

form circular rings, which are dependent on the parameters m (ring number), τ (half width), and w (phase), where $m \in \mathbb{N}_0$; $w, \tau \in \mathbb{R}$ with $0 \le w < 1$, $0 < \tau \le \frac{1}{4}$. The union of all circular rings is denoted by

$$\mathbb{B}(w; \tau) = \bigcup_{m \in \mathbb{N}_0} \overline{\mathbb{B}^2_{m+w-\tau, m+w+\tau}}. \quad (14.202)$$

Circular Configurations

It follows from Theorem 14.4 (with period $\beta = 1$) that

$$\sum_{\substack{0 < |g| \le N \\ g \in \mathbb{B}(w;\tau) \\ g \in \mathbb{Z}^2}} {}' 1 = 2\pi\tau N^2 + O\left(N^{\frac{3}{2}}\right). \quad (14.203)$$

Thus, the identity (14.203) tells us that the *mean density*

$$\lim_{N \to \infty} \left(\frac{1}{\pi N^2} \sum_{\substack{0 < |g| \le N \\ g \in \mathbb{B}(w;\tau) \\ g \in \mathbb{Z}^2}} {}^{\prime} 1 \right) = 2\tau \qquad (14.204)$$

depends only on the *width* 2τ of the circular rings, but it is independent of the *phase* w.

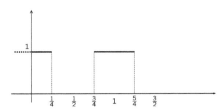

FIGURE 14.2
Configuration generating circular rings with $\tau = \frac{1}{4}$, $w = 0$.

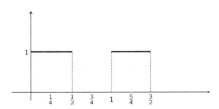

FIGURE 14.3
Configuration generating circular rings with $\tau = \frac{1}{4}$, $w = \frac{1}{4}$.

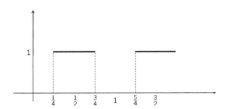

FIGURE 14.4
Configuration generating circular rings with $\tau = \frac{1}{4}$, $w = \frac{1}{2}$.

From Section 14.7 we are able to deduce that the non-uniform distribution of lattice points in the plane; i.e., the dependence of the total number of lattice points on the phase w is strongly existent in the remainder term on the right hand side of (14.203). More concretely, Theorem 14.4 shows (with $\beta = 1$, $l = 1$, $H = 1$, and $\mathbb{B}(w, \tau)$ given by (14.202)) that

$$\underset{\substack{0<|g|\le N \\ g\in \mathbb{B}(w;\tau) \\ g\in \mathbb{Z}^2}}{\sum}{}' 1 = 2\pi\tau N^2 + C(w;\tau)N^{\frac{3}{2}} + o(N^{\frac{3}{2}}), \quad N \to \infty. \tag{14.205}$$

such that

$$\lim_{N\to\infty} \left(\frac{1}{N^{\frac{3}{2}}} \underset{\substack{0<|g|\le N \\ g\in \mathbb{B}(w,\tau) \\ g\in \mathbb{Z}^2}}{\sum}{}' 1 \; - \; 2\pi\tau\sqrt{N} \right) \tag{14.206}$$

$$= \frac{4}{3\pi} \underset{\substack{|g|>0 \\ g\in\mathbb{Z}^2}}{\sum} \frac{1}{|g|^3} \, \sin\left(2\pi|g|^2\tau\right) \, \cos\left(2\pi|g|^2(\tau+w) - \frac{\pi}{4}\right).$$

In fact, the function $w \mapsto C(w;\tau) = C(w;\tau;1)$, $w \in [-\frac{1}{2}, \frac{1}{2}]$, given by

$$C(w;\tau) = \frac{4}{3\pi} \sum_{n=1}^{\infty} \frac{r(n^2)}{n^{\frac{3}{2}}} \, \sin\left(2\pi\, n\tau\right) \, \cos\left(2\pi\, n(\tau+w) - \frac{\pi}{4}\right) \tag{14.207}$$

is non-constant and continuous; hence, (14.206) is a phase-dependent limit relation.

Example 14.2. *Our particular interest is to discuss circular rings with the "width" $2\tau = \frac{1}{2}$, i.e., $\tau = \frac{1}{4}$, in dependence on the phase w (see Figure 14.2). Elementary manipulations give*

$$C\left(w; \frac{1}{4}\right) \tag{14.208}$$

$$= \frac{4}{3\pi} \sum_{k=0}^{\infty} \frac{r((2k+1)^2)}{(2k+1)^{\frac{3}{2}}} (-1)^k \, \cos\left(2\pi(2k+1)\left(w+\frac{1}{4}\right) - \frac{\pi}{4}\right).$$

Of course, $C\left(\cdot, \frac{1}{4}\right)$ is $\mathbb{Z}$-periodic, i.e.,

$$C\left(w+1; \frac{1}{4}\right) = C\left(w; \frac{1}{4}\right). \tag{14.209}$$

Even more, we see that

$$C\left(w + \frac{1}{2}; \frac{1}{4}\right) = -C\left(w; \frac{1}{4}\right). \tag{14.210}$$

Consequently it suffices to consider $C\left(w; \frac{1}{4}\right)$ *on the interval* $\left[-\frac{1}{4}, \frac{1}{4}\right]$.

A simple calculation yields

$$C\left(0; \frac{1}{4}\right) = \frac{2\sqrt{2}}{3\pi} \sum_{k=0}^{\infty} r\left((2k+1)^2\right) \frac{1}{(2k+1)^{\frac{3}{2}}} > 0 . \tag{14.211}$$

Furthermore, it is not difficult to see that

$$C\left(\frac{1}{2}; \frac{1}{4}\right) = -C\left(0; \frac{1}{4}\right) < 0. \tag{14.212}$$

Because of the continuity of $C\left(\cdot \; ; \frac{1}{4}\right)$ *there must be a point* $w_0 \in [0, \frac{1}{2}]$ *such that* $C\left(w_0; \frac{1}{4}\right) = 0.$

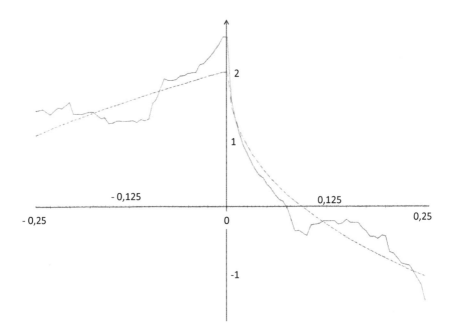

FIGURE 14.5
$C_\delta(w; \frac{1}{4})$ (continuous curve) and its approximation $C_\delta^{appr}(w; \frac{1}{4})$ (dashed curve).

Figure (14.5) provides a numerical illustration of (14.208) on the interval $\left[-\frac{1}{4}, \frac{1}{4}\right]$. *The evaluations are based on the integral mean values.*

$$C_\delta\left(w; \frac{1}{4}\right) = \frac{1}{2\delta} \int_{w-\delta}^{w+\delta} C\left(u; \frac{1}{4}\right) du, \tag{14.213}$$

which represent approximations for $C\left(w; \frac{1}{4}\right)$ *for small positive values* δ. *As a matter of fact,* $C_\delta\left(w; \frac{1}{4}\right)$, $w \in \left[-\frac{1}{2}, \frac{1}{2}\right)$, *can easily be calculated as series expansion*

$$C_\delta\left(w; \frac{1}{4}\right) = \frac{2}{3\pi^2\delta} \sum_{n=0}^{\infty} \frac{r((2k+1)^2)}{(2k+1)^{\frac{5}{2}}}(-1)^k \tag{14.214}$$

$$\times \sin\left(2\pi(2k+1)\delta\right) \; \cos\left(2\pi(2k+1)\left(w + \frac{1}{4}\right) - \frac{\pi}{4}\right).$$

In Figure 14.5 the continuous line shows the graph of (the first 4000 elements of the series expansion of) $C_\delta\left(w; \frac{1}{4}\right)$ *on* $\left[-\frac{1}{4}, \frac{1}{4}\right]$ *(with* $\delta = 10^{-3}$*). Figure 14.5 shows that the maximum of* $C\left(\cdot; \frac{1}{4}\right)$ *is attained at a value* $w_{\max}$ *close to* $w = 0$.
The dashed line (illustrating the graph) of $C_\delta^{appr}\left(w; \frac{1}{4}\right)$ *on* $\left[-\frac{1}{4}, \frac{1}{4}\right]$ *is obtained from* $C_\delta(w; \frac{1}{4})$ *by replacing* $r((2k+1)^2)$ *in (14.214) just by 4. Obviously, the trend of the "phase dependence" of* $C\left(\cdot; \frac{1}{4}\right)$ *is reflected by the graph of this function, too.*

Altogether, our geometrically motivated results lead to the observation that there is a radially phase-dependent non-uniform distribution of lattice points in the plane. Expressed in terms of rings of width $\frac{1}{2}$ we are confronted with the situation that the radial distribution of lattice points may be expected to be particularly strong provided that

$$\left\{x \in \mathbb{R}^2 \big| w_{\max} < |x| - \lfloor|x|\rfloor < w_{\max} + \varepsilon\right\} \tag{14.215}$$

with $\varepsilon > 0$ appropriately small.

Circular and Sectorial Configurations

Even more generally, it has been exposed in Subsection 14.7 that the non-uniform distribution of lattice points has two geometrical components - not only the radius, but also the angle. If H is a twice continuously differentiable (angular) function, then we obtain from Theorem 14.4

$$\lim_{N\to\infty}\left(\frac{1}{N^{\frac{3}{2}}} \sum_{\substack{0<|g|\leq N \\ g\in B(w,\tau) \\ g\in\mathbb{Z}^2}} H\left(\frac{g}{|g|}\right) - \tau\sqrt{N}\left(\int_{\mathbb{S}^1} H(\xi)\, dS(\xi)\right)\right) \tag{14.216}$$

$$= \frac{4}{3\pi} \sum_{\substack{|g|>0 \\ g\in\mathbb{Z}^2}} \frac{1}{|g|^3} H\left(\frac{g}{|g|}\right) \sin\left(2\pi|g|^2\tau\right) \cos\left(2\pi|g|^2(\tau+w) - \frac{\pi}{4}\right).$$

In consequence, the radial and simultaneously angular dependence can be investigated for the distribution of lattice points of $\mathbb{Z}^2$.

Example 14.3. *Choosing H especially in the form*

$$H\left(\frac{x}{|x|}\right) = \frac{1}{\sqrt{x_1^2 + x_2^2}^{4k}} (x_1 + i\, x_2)^{4k} \,, \quad x = (x_1, x_2)^T \in \mathbb{R}^2 \backslash \{0\}, \quad (14.217)$$

we are led to the identity

$$\lim_{N \to \infty} \left(\frac{1}{N^{\frac{3}{2}}} \sum_{\substack{0 < \sqrt{n_1^2 + n_2^2} \leq N \\ (n_1, n_2)^T \in \mathbb{B}(w, \tau) \\ (n_1, n_2)^T \in \mathbb{Z}^2}} \frac{(n_1 + i\, n_2)^{4k}}{\sqrt{n_1^2 + n_2^2}^{4k}} \right) \qquad (14.218)$$

$$= \frac{4}{3\pi} \sum_{\substack{n_1^2 + n_2^2 > 0 \\ (n_1, n_2)^T \in \mathbb{Z}^2}} \frac{(n_1 + in_2)^{4k}}{\sqrt{n_1^2 + n_2^2}^{4k+3}}$$

$$\times \sin\left(2\pi(n_1^2 + n_2^2)\tau\right) \cos\left(2\pi(n_1^2 + n_2^2)(\tau + w) - \frac{\pi}{4}\right),$$

provided that $k \in \mathbb{N}$ and $\mathbb{B}(w, \tau)$ is the configuration of circular rings with fixed width described by (14.201) and (14.202) (note that $k = 0$ leads back to (14.206)).

Example 14.4. *For $\delta \in (0, \pi]$, $\eta \in \mathbb{S}^1$, we consider the function H of the form*

$$\xi \mapsto H(\xi \cdot \eta) = \begin{cases} 1 & , \quad 1 - \xi \cdot \eta \leq 1 - \cos(\delta), \\ 0 & , \quad 1 - \xi \cdot \eta > 1 - \cos(\delta). \end{cases} \qquad (14.219)$$

Its expansion in terms of "circular harmonics" reads as follows

$$H(\xi \cdot \eta) = \frac{\delta}{\pi} + \frac{2}{\pi} \sum_{n=1}^{\infty} \underbrace{\frac{\sin(n\delta)}{n} \cos(n \arccos(\xi \cdot \eta))}_{= P_n(2; \xi \cdot \eta)}, \quad \xi \in \mathbb{S}^1. \qquad (14.220)$$

Hence, (14.215) is valid for all truncated expansions $H^{(T)}$ given by

$$H^{(T)}(\xi \cdot \eta) = \frac{\delta}{\pi} + \frac{2}{\pi} \sum_{n=1}^{T} \frac{\sin(n\delta)}{n} \cos(n \arccos(\xi \cdot \eta)), \quad \xi \in \mathbb{S}^1. \qquad (14.221)$$

The limits $\lim_{T\to\infty}\sum_{|g|>0}\sum_{n=1}^{T}$ *can be interchanged. Thus we finally arrive at the limit relations*

$$\lim_{N\to\infty}\frac{1}{N^2}\sum_{\substack{0<|g|\le N \\ g\in\mathbb{B}(w;\tau)\cap\mathbb{S}(\delta;\eta) \\ g\in\mathbb{Z}^2}}1 \;=\; 2\tau\delta \qquad (14.222)$$

and

$$\lim_{N\to\infty}\left(\frac{1}{N^{\frac{3}{2}}}\sum_{\substack{0<|g|\le N \\ g\in\mathbb{B}(w;\tau)\cap\mathbb{S}(\delta;\eta) \\ g\in\mathbb{Z}^2}}1 \;-\; 2\tau\delta\sqrt{N}\right) \qquad (14.223)$$

$$=\frac{4}{3\pi}\sum_{\substack{0<|g| \\ g\in\mathbb{S}(\delta;\eta) \\ g\in\mathbb{Z}^2}}\frac{1}{|g|^3}\sin(2\pi|g|^2\tau)\cos\left(2\pi|g|^2(\tau+w)-\frac{\pi}{4}\right),$$

where, for $\delta\in(0,\pi]$*,* $\eta\in\mathbb{S}^1$*, the set (cf. Figure 14.1)*

$$\mathbb{S}(\delta;\eta)=\left\{x\in\mathbb{R}^2,\; x=|x|\xi,\xi\in\mathbb{S}^1\big|1-\xi\cdot\eta\le 1-\cos(\delta)\right\} \qquad (14.224)$$

denotes the sector in the direction of $\eta\in\mathbb{S}^1$ *of width* δ*.*

Remark 14.7. *It should be noted that, initiated by the work of C. Müller [1951], A. Dressler [1967, 1972], and C. Müller, A. Dressler [1972], the radial distribution of lattice points in the plane was the object of a series of papers (see also F. Fricker [1982] and the references therein for an overview).*

In our work, most of the activities in the theory of non-uniformly distributed lattice points in the plane was derived from a weighted variant of the Hardy–Landau identity due to W. Freeden [1978a], which was imbedded in Chapter 12 of this book. Even more, our approach also shows the "combined" non-uniform radial and angular distribution of the lattice points within geometrical configurations as illustrated by Figure 14.1 (see W. Freeden [1978b] for initial steps). The more detailed quantitative investigation of combined non-uniform radial and angular distribution of lattice points is a challenge for future work.

Finally it may be conjectured that the Hardy–Landau induced non-uniform distribution of lattice balls (see Section 11.5) seems to be realizable not only in the two-dimensional case, but also in all higher dimensions.

15

Conclusions

CONTENTS

15.1 Summary 429
15.2 Outlook 430

During the last decades, the subject of lattice point theory has not been overly popular in the mathematical community. At universities, its teaching has been usually oriented towards the work of the first half of the last century. In research, the diversification of mathematical disciplines in the second half of the last century resulting into a heterogenity of knowledge has brought increasingly specialized lattice point results in different fields, mostly in applications. All in all, the situation today shows somehow a loss of interest in the core of the lattice point theory, and only a few monographs (for example, F. Fricker [1982], E. Hlawka et al. [1991], E. Krätzel [1988], M.N. Huxley [1996]) exist for consolidation.

15.1 Summary

The present work represents the fruits of cross-fertilization of two subjects, namely analytic number theory and geomathematics. It constitutes an attempt to revitalize the sadly neglected subject of homogenization of recent knowledge in lattice point theory, at least in the field originated by harmonic and metaharmonic settings and structures. Our particular aim is to put together apparently disparate features as closely related building blocks of a common core. The unusual selection of the material and its presentation in a poly(meta)harmonic framework makes this book an addition to the textbook literature on analytic number theory. In fact, at least in the opinion of the author, the essence of lattice point theory is its interrelation to other branches in science. It is useful not only in number theory, but also in Fourier analysis, numerical integration, data handling and so on. The ideas, concepts, and structures in lattice point theory strongly influence a variety of areas in mathematical physics, and vice versa.

The content of the book is based on elliptic partial differential operators, especially on the iterated Laplace and on the iterated Helmholtz operators and arbitrary lattices; hence, it covers the field of elliptic partial differential operators with constant coefficients. Spherical harmonics and Bessel functions turn out to be essential instruments. Hardy–Landau identities and their extensions in the lattice point theory are the canonical manifestations. Harmonic and metaharmonic theory act as tools, but we also get the capacity to participate in other branches of mathematics. To be more concrete, there are important instances such as gravitation, geomagnetics, geothermal research, particle methods in fluid transport, etc., in which lattice point theory is needed as relevant and significant ingredient (in particular, for applications in geomathematical research the reader is referred to, e.g., W. Freeden et al. [2010], E.W. Grafarend et al. [2011], C. Gerhards [2010], I. Ostermann [2010], A. Palczewski et al. [1997], and the references therein).

15.2   Outlook

Evidently, a great number of questions still remains unanswered by our approach. A striking confinement is the restriction to (special types of) elliptic differential operators. It helps us to apply the Fredholm theory for (singular) integral equations. Moreover, it avoids distributional characterizations of the lattice functions, but it also prevents us, for instance, from incorporating the Dirichlet divisor problem and its extensions. No doubt, the specific treatment of hyperbolic and/or parabolic differential equations in lattice point theory is a great challenge for future work. In the opinion of the author, it demands its own nature which by no means can be studied in parallel with the elliptic theory of this work.

Bibliography

Abramowitz, M., Stegun, I.A., *Handbook of Mathematical Functions*, Dover Publications, Inc., New York, 1972.

Adhikari, S.D., Petermann, Y.-F.S., Lattice Points in Ellipsoids, *Acta Arith.*, (59): 329-338, 1991.

Ait-Haddou, R., Biard, L., Slawinski, M.A., Minkowski Isoperimetric-hodograph Curves, *Computer Aided Geometric Design*, (17): 835-861, 2000.

Apostol, T.M., An Elementary View of Euler's Summation Formula, *The American Math. Monthly* (106): 409-418, 1999.

Arenstorf, R. F., Johnson D., Uniform Distribution of Integral Points on 3-Dimensional Spheres via Modular Forms, *J. Number Theory*, (11): 218-238, 1979.

Aronszajn, N., Creese T.M., Lipkin L.J., *Polyharmonic Functions*, Clarendon Press, Oxford, 1983.

Besicovitch, A., *Almost Periodic Functions*, Cambridge University Press, 1954.

Butzer, P.L., Nessel, R., *Fourier Analysis and Approximation Theory*, Birkhäuser, Basel, 1971.

Butzer, P.L., Stens, R.L., The Euler–MacLaurin Summation Formula, the Sampling Theorem, and Approximate Integration over the Real Axis, *Linear Algebra and Its Applications*, (52/53): 141-155, 1983.

Cassels, J.W.S., *An Introduction to the Geometry of Numbers*, Springer, Berlin, Heidelberg, New York, 1968.

Chamizo, F., Iwaniec, H., On the Sphere Problem, *Rev. Mat. Iberoamericana*, (11): 417-429, 1995.

Chandrasekharan, K., Narasimhan, R., Hecke's Functional Equation and the Average Order of Arithmetical Functions, *Acta Arith.*, (6): 487-503, 1961.

Chandrasekharan, K., Narasimhan, R., Functional Equations with Multiple Gamma-factors and the Average Order of Arithmetical Functions, *Ann. of Math., (2)*, (76): 93-136, 1962.

Chen, J.-R., The Lattice Points in a Circle, *Sci. Sinica*, (12): 633-649, 1963.

Christ, T., Kalpokas, J., Steuding, J., Neue Resultate über die Wertverteilung der Riemannschen Zetafunktion auf der kritischen Geraden, *Math. Semesterber.*, (57): 201-229, 2010.

Cilleruelo, J., The Distribution of Lattice Points on Circles, *J. Number Theory*, (43): 198-202, 1993.

Courant, R., Hilbert, D., *Methoden der Mathematischen Physik I, II*, Springer, Berlin, 2nd ed., 1968.

Cramér, H., Über zwei Sätze des Herrn G.H. Hardy, *Math. Z.*, (15): 201-210, 1922.

Cui, J., Freeden, W., Equidistribution on the Sphere, *SIAM J. Sci. Stat. Comput.*, (18): 595-609, 1997.

Davis, P.J., *Interpolation and Approximation*, Blaisdell Publishing Company, Waltham, MA, 1963.

Davis, P.J., Rabinowitz, P., *Numerical Integration*, Blaisdell, Toronto, London, 1967.

Dressler, A., *Über die ungleichförmige Verteilung von Gitterpunkten in der Ebene*, Dissertation, RWTH Aachen, 1967.

Dressler, A., Über die ungleichförmige Verteilung von Gitterpunkten in ebenen Bereichen, *Math. Nachr.*, (52): 1-20, 1972.

Erdös, P., Gruber, M., Hammer, J., *Lattice Points*, Pitman Monographs and Surveys in Pure and Applied Mathematics, 39, Longman Scientific-Technical, John Wiley Inc., New York, 1989.

Erdös, P., Hall, R.R., On the Angular Distribution of Gaussian Integers with Fixed Norm, *Discrete Math.*, (200): 87-94, 1999.

Epstein, P.S., Zur Theorie allgemeiner Zetafunktionen I, *Math. Ann.*, (56): 615-644, 1903.

Epstein, P.S., Zur Theorie allgemeiner Zetafunktionen II, *Math. Ann.*, (63): 205-216, 1907.

Euler, L., Methodus universalis serierum convergentium summas quam proxime inveniendi, Commentarii Academiae Scientiarum Petropolitanae, (8): 3-9, *Opera Omnia* (XIV): 101-107, 1736.

Euler, L., Methodus universalis series summandi ulterius promota, Commentarii Academiae Scientarium Petropolitanae, (8): 147-158; *Opera Omnia* (XIV): 124-137, 1736.

Ewald, P.P., Die Berechnung optischer und elektrostatischer Gitterpotentiale, *Ann. Phys.*, (64): 253-287, 1921.

Erdelyi, A., *Higher Transcendental Functions* I, II, III, McGrawHill, New York, 1955.

Fomenko, O.M., On the Problem of Gauss, *Acta. Arith.*, (6): 277-284, 1961.

Freeden, W., *Eine Verallgemeinerung der Hardy–Landauschen Identität*, Dissertation, RWTH Achen, 1975.

Freeden, W., Über eine Verallgemeinerung der Hardy–Landauschen Identität, *Manuscr. Math.*, (24): 205-216, 1978a.

Freeden, W., Über gewichtete Gitterpunktsummen in kreisförmigen Bereichen, *Mitt. Math. Seminar Giessen*, (132): 1-22, 1978b.

Freeden, W., An Application of a Summation Formula to Numerical Computation of Integrals over the Sphere, *Bull. Géod.*, (52): 165-175, 1978c.

Freeden, W., Über eine Klasse von Integralformeln der Mathematischen Geodäsie, *Veröff. Geod. Inst. RWTH Aachen*, Habilitationsschrift, Report No. 27, 1979.

Freeden, W., Über die Gaußsche Methode zur angenäherten Berechnung von Integralen, *Math. Meth in the Appl. Sci.*, (2): 397-409, 1980a.

Freeden, W., On Integral Formulas of the (Unit) Sphere and Their Application to Numerical Computation of Integrals, *Computing*, (25): 131-146, 1980b.

Freeden, W., On Spherical Spline Interpolation and Approximation, *Math. Meth. in the Appl. Sci.*, (3): 551-575, 1981.

Freeden, W., Multidimensional Euler Summation Formulas and Numerical Cubature, *ISNM*, (57): 77-88, 1982.

Freeden, W., Interpolation by Multidimensional Periodic Splines, *J. Approx. Theory*, (55): 104-117, 1988.

Freeden, W., *Multiscale Modelling of Spaceborne Geodata*, B.G. Teubner, Stuttgart, Leipzig, 1999.

Freeden, W., Geomathematics: Its Role, Its Aim, and Its Potential, in *Handbook of Geomathematics*, (1): 3-43, Springer, Berlin, Heidelberg, 2010.

Freeden, W., Fleck,. J., Numerical Integration by Means of Adapted Euler Summation Formulas, *Numer. Math.*, (51): 37-64, 1987.

Freeden,W., Gutting, M., On the Completeness and Closure of Vector and Tensor Spherical Harmonics, *Integral Transforms and Special Functions*, (19): 713-734, 2008.

Freeden, W., Hermann, P., Some Reflections on Multidimensional Euler and Poisson Summation Formulas, *ISNM*, (75): 166-179, 1985.

Freeden, W., Hermann, P., Uniform Approximation by Spherical Spline Interpolation. *Math. Z.*, (193): 265-275, 1986.

Freeden, W., Michel, V., *Multiscale Potential Theory (with Applications to Geoscience)*, Birkhäuser, Boston, Basel, Berlin, 2004.

Freeden, W., Reuter, R., A Class of Multidimensional Periodic Splines, *Manuscr. Math.*, (35): 371-386, 1981.

Freeden, W., Reuter, R., Remainder Terms in Numerial Integration Formulas of the Sphere, *ISNM*, (61): 151-170, 1982.

Freeden, W., Schreiner, M., *Spherical Functions of Mathematical Geosciences, A Scalar, Vectorial, and Tensorial Setup.* Springer, Berlin-Heidelberg, 2009.

Freeden, W., Gervens, T., Schreiner, M., *Constructive Approximation on the Sphere (With Applications to Geomathematics)*, Oxford Science Publications, Clarendon Press, Oxford, 1998.

Freeden, W., Nashed, M.Z., Sonar, T. (eds.), *Handbook of Geomathematics*, Volume 1+2, Springer, Berlin, Heidelberg, 2010.

Fricker, F., Geschichte des Kreisproblems, *Mitt. Math. Sem. Giessen*, (111): 1-34, 1975.

Fricker, F., *Einführung in die Gitterpunktlehre*, Birkhäuser, Basel, 1982.

Funk, H., Beiträge zur Theorie der Kugelfunktionen, *Math. Ann.*, (77): 136-152, 1916.

Gauß, C.F., *Disquisitiones Arithmetica*, Leipzig, 1801.

Gauß, C.F., De nexu inter multitudinem classicum, in quas formae binariae secondi gradus distribuuntur, earumque determinantem, *Werke*, (2): 269-291, 1826.

Gerhards, C., *Spherical Multiscale Methods in Terms of Locally Supported Wavelts: Theory and Application to Geomagnetic Modeling*, PhD-Thesis, Geomathematics Group, TU Kaiserslautern, 2011.

Gradshteyn, I.S., Ryzhik, I.M., *Table of Integrals Series and Products*, Academic Press, New York, London, 1965.

Grafarend, E.W., Schmidt, M., Wild-Pfeiffer, F., *Gravitation-Geometric Geodesy Versus Physical Geodesy*, Springer, Berlin, Göttingen, Heidelberg, (in preparation).

Grosswald, E., *Representations of Integers as Sums of Squares*, Springer-Verlag, New York, 2006.

Günter, N.M., *Die Potentialtheorie und ihre Anwendung auf Grundaufgaben der Mathematischen Physik*, B.G. Teubner, Stuttgart, 1957.

Hardy, G.H., On the Expression of a Number as the Sum of Two Squares, *Quart. J. Math. (Oxford)*, (46): 263-283, 1915.

Hardy, G.H., The Average Order of the Arithmetical Functions $P(x)$ and $\Delta(x)$, *Proc. London Math. Soc. (2)*, (15): 192-213, 1916.

Hardy, G.H., Ramanujan, *Twelve Lectures on Subjects Suggested by Its Life and Work*, Chelsea Publishing Company, New York, 1940.

Hardy, G.H., *Divergent Series*, Clarendon Press, Oxford, 1949.

Hardy, G.H., Landau, E., The Lattice Points of a Circle, *Proceedings of the Royal Society, A.*, (105): 244-258, 1924.

Hardy, G.H., Wright, E.M., *Einführung in die Zahlentheorie*, R. Oldenbourg, München, 1958.

Hartmann, R., On Solutions of $\Delta V + V = 0$ in an Exterior Region, *Math. Z.*, (71): 251-257, 1959.

Hartmann, P., Wilcox, C., On Solutions of the Helmholtz Equation in Exterior Domains, *Math. Z.*, (75): 228-255, 1961.

Heath-Brown, D. R., The Distribution and Moments of the Error Term in the Dirichlet Divisor Problem, *Acta Arith.*, (60): 389-415, 1992.

Heath-Brown, D. R., Lattice Points in the Sphere, In: *Proc. Number Theory Conf. Zakopane*, K. Györy Eds., Vol. 2, 883-892, 1999.

Hecke, E., Über orthogonal-invariante Integralgleichungen, *Math. Ann.*, (78): 398-404, 1918.

Hecke, E., Analytische Arithmetik der positiv quadratischen Formen, *Kgl. Danske Vidensk. Selskab. Math.-Fys.-M.*, XVII, (12): 1-134, 1940.

Hilbert, D., *Grundzüge einer allgemeinen Theorie der linearen Integralgleichungen*, Teubner, Leipzig, 1912.

Hilbert, D., Cohn-Vossen, S. *Anschauliche Geometrie*, Springer, Berlin, 1932.

Hlawka, E., Zur Geometrie der Zahlen, *Math. Z.*, (49): 285-312, 1943.

Hlawka, E., Über Integrale auf konvexen Körpern I, *Monatshefte für Mathematik*, (54): 1-37, 1950.

Hlawka, E., Grundbegriffe der Geometrie der Zahlen, *Jber. Deutsch. Mathem. Vereinigung*, (57): 37-55, 1954.

Hlawka, E., *Theorie der Gleichverteilung*, B.I., Mannheim, 1979.

Hlawka, E., Über einige Sätze, Begriffe und Probleme in der Theorie der Gleichverteilung II, *Sitzungsberichte der Österreichischen Akademie der Wissenschaften*, (189): 437-490, 1980.

Hlawka, E., Gleichverteilung auf Produkten von Sphären, *J. Reine Angew. Math.*, (330): 1-45, 1981.

Hlawka, E., Näherungslösungen der Wellengleichung und verwandter Gleichungen durch zahlentheoretische Methoden, *Sitzungsberichte der Österreichischen Akademie der Wissenschaften*, (193) : 359-442, 1984.

Hlawka, E., Schoißengeier, J., Taschner, R., *Geometric and Analytic Number Theory*, Springer, Berlin, Heidelberg, New York, 1991.

Hua, L.-K., Die Abschätzung von Expotentialsummen und ihre Anwendung in der Zahlentheorie, *Enzyklopädie der Mathematischen Wissenschaften*, Bd. 12, Heft 13, Teil 1, 105-107, 1959.

Huxley, M.N., *Area, Lattice Points, and Exponential Sums*, LMS Monographs, 13, Oxford University Press, Oxford, 1996.

Huxley, M.N., Exponential Sums and Lattice Points III, *Proc. London Math. Soc.*, (87): 591-609, 2003.

Hobson, E.W., *The Theory of Spherical and Ellipsoidal Harmonics*, Reprint Chelsea Publishing Company, New York, 1955.

Ivanow, V.K.,A Generalization of the Voronoi–Hardy Identity, *Sibirsk. Math. Z.*, (3): 195-212, 1962.

Ivanow, V.K., Higher-dimensional Generalization of the Euler Summation Formula (Russian), *Izv. Vyss. Ucebn. Zaved. Mathematika, 6*, (37): 72-80, 1963.

Ivic, A., The Laplace Transform of the Square in the Circle and Divisor Problems, *Studia Scientiarum Mathematicarum Hungarica*, (32): 181-205, 1996.

Ivic, A., Krätzel, E., Kühleitner, Nowak, W.G., Lattice Points in Large Regions and Related Arithmetic Functions: Recent Developments in a Very Classical Topic, *arXiv*: 0410522v1, 1-39, 2004.

Iwaniec, H., Mozzochi, C.J., On the Divisor and Circle Problems, *J. Number Theory*, (29): 60-93, 1988.

Jacobi, C.G., Fundamenta nova theoriae functionum ellipticarum (1829), in: *Gesammelte Werke*, Erster Band, 1891.

Katai, I., The Number of Lattice Points in a Circle (in Russian), *Ann. Univ. Sci. Budapest Rolando Eötvös Sect. Math.*, (8): 39-60, 1965.

Katai, I., Környei, I., On the Distribution of Lattice Points on Circles, *E. Math. Ann. Univ. Sci. Budapest. Eötvös Sect. Math,*. (19): 87-91, 1977.

Kellogg, O.D., *Foundations of Potential Theory*, Frederick Ungar Publishing Company, New York, 1929.

Kolesnik, G., On the Method of Exponent Pairs, *Acta Arith.*, (45): 115-143, 1985.

Knopp, K., *Funktionentheorie II*, Bd. 703, Sammlung Göschen, W. de Gruyter, Berlin, 1971.

Krätzel, E., *Lattice Points*, Kluwer Dordrecht, Boston, London, 1988.

Krätzel, E., *Analytische Funktionen in der Zahlentheorie*, Teubner Stuttgart, Leipzig, Wiesbaden, 2000.

Kuipers, L., Niederreiter H., *Uniform Distribution of Sequences*. John Wiley & Sons, New York, 1974.

Landau, E., *Handbuch der Lehre von der Verteilung der Primzahlen*, Teubner, Leipzig, 1909.

Landau, E., Über die Gitterpunkte in einem Kreis (Erste Mitteilung), *Gött. Nachr.*, 148-160, 1915.

Landau, E., Über die Gitterpunkte in einem Kreis IV, *Gött. Nachr.*, 58-65, 1924.

Landau, E., Über die Gitterpunkte in mehrdimensionalen Ellipsoiden (Zweite Abhandlung), *Math. Z.*, (24): 299-310, 1925.

Landau, E., *Vorlesungen über Zahlentheorie*, Chelsea Publishing Compagny, New York, 1969 (reprint from the orignal version published by S. Hirzel, Leipzig, 1927).

Landau, E., *Ausgewählte Abhandlungen zur Gitterlehre*, VEB, Berlin 1962.

Lebedev, N.N., *Spezielle Funktionen und ihre Anwendungen*, Bibliographisches Institut, Mannheim, 1973.

Lekkerkerker, C.G., *Geometry of Numbers*, North Holland, Amsterdam, London, 1969.

Lense, J., Kugelfunktionen, *Mathematik und ihre Anwendungen in Physik und Technik*, Reihe A, Bd. 23, Akad. Verlagsgesellschaft, Leipzig, 1954.

Linnik, Y. L., *Ergodic Properties of Algebraic Fields*, Springer, Berlin, Heidelberg, New York, 1968.

Littlewood, J.E., Walfisz, A., The Lattice Points of a Circle, *Proc. Royal Soc. London Ser.* (A), (106): 478-487, 1924.

Maclaurin, C., *A Treatise of Fluxions*, Edinburgh, 1742.

Magnus, W., Fragen der Eindeutigkeit und des Verhaltens im Unendlichen für Lösungen von $\Delta U + k^2 U = 0$, *Abh. Math. Sem. Univ. Hamburg*, (16): 77-88, 1949.

Magnus, W., Oberhettinger, F., Formeln und Sätze für die Speziellen Funktionen der Mathematischen Physik, *Die Grundlagen der mathematischen Wissenschaften in Einzeldarstellungen*, Band 52, 2. Auflage, Springer, Berlin, Göttingen, Heidelbert, 1948.

Magnus, W., Oberhettinger, and F., Soni, R.P., Formulas and Theorems for the Special Functions of Mathematical Physics, in *Die Grundlehren der mathematischen Wissenschaften in Einzeldarstellungen*, Band 52, Springer, Berlin, 3. Auflage, 1966.

Malysev, A.V., The Distribution of Integer Points in a Four-Dimensional Sphere, *Doklady Akad. Nauk* SSSR, (114): 25-28, 1957.

Malysev, A.V., On Representations of Integers by Positive Quadratic Forms. Trudy V.A. *Steklov Math. Inst. AN SSSR* (65): 1-212, 1962.

Michlin, S.G., *Mathematical Physics, an Advanced Course*, North Holland, Amsterdam, London, 1970.

Michlin, S.G., *Lehrgang der Mathematischen Physik*, Akademie Verlag, Berlin, 2nd edition, 1975.

Minkowski, H., *Geometrie der Zahlen*, Teubner, Leipzig, 1896.

Mordell, L.J., Poisson's Summation Formula in Several Variables and Some Applications to the Theory of Numbers, *Cambr. Phil. Soc.*, (25): 412-420, 1928.

Mordell, L.J., Some Applications of Fourier Series in the Analytic Theory of Numbers, *Cambr. Phil. Soc.*, (24): 585-595, 1928.

Mordell, L.J., Poisson's Summation Formula and the Riemann Zeta Function, *J. London Math. Soc.*, (4): 285-296, 1929.

Mordell, L.J., On Some Arithmetical Results in the Geometry of Numbers, *Compositio Math.*, (1): 248-253, 1935.

Mordell, L.J., *Diophantine Equations*, Academic Press, 1969.

Müller, C., Über die Hardy–Landausche Identität und verwandte Fragen, *Jber. Deutsch. Mathem. Vereinigung*, (56): 23-24, 1951.

Müller, C., Über die ganzen Lösungen der Wellengleichung (nach einem Vortrag von G. Herglotz), *Math. Ann.*, (124): 235-264, 1952.

Müller, C., Eine Verallgemeinerung der Eulerschen Summenformel und ihre Anwendung auf Fragen der analytischen Zahlentheorie, *Abh. Math. Sem. Univ. Hamburg*, (19): 41-61, 1954a.

Müller, C., Eine Formel der analytischen Zahlentheorie, *Abh. Math. Sem. Univ. Hamburg*, (19): 62-65, 1954b.

Müller, C., Eine Erweiterung der Hardyschen Identität, *Abh. Math. Sem. Univ. Hamburg*, (19): 66-76, 1954c.

Müller, C., Die Grundprobleme der Geometrie der Zahlen, *Mathem.-Phys. Semesterberichte*, (5): 63-70, 1956.

Müller, C., Spherical Harmonics, *Lecture Notes in Mathematics*, 17, Springer, Berlin, 1966.

Müller, C., *Foundations of the Mathematical Theory of Electromagnetic Waves*, Springer, Berlin, 1969.

Müller, C., *Analysis of Spherical Symmetries in Euclidean Spaces*, Springer, New York, Berlin, Heidelberg, 1998.

Müller, C., Dressler, A., Über eine gewichtete Mittelung der Gitterpunkte in der Ebene, *J. Reine Angew. Mathematik*, (252): 82-87, 1972.

Müller, C., Freeden, W., Multidimensional Euler and Poisson Summation Formulas, *Result. Math.*, (3): 33-63, 1980.

Natanson, P.I., *Theorie der Funktionen einer reellen Veränderlichen*, Akademie-Verlag, Berlin, 1961.

Nielsen, N., *Handbuch der Theorie der Gammafunktion*, Teubner, Leipzig, 1906.

Niemeyer, H., Lokale und asympotische Eigenschaften der Lösung der Helmholtzschen Schwingungsgleichung, *Jber. Deutsch. Mathem. Vereinigung*, (65): 1-44, 1962.

Nowak, W. G., Lattice Points in a Circle: An Improved Mean-Squares Asymptotics, *Acta Arith.*, (113): 259-272, 2004.

Oberhettinger, F., *Fourier Expansions: A Collection of Formulas*, Academic Press, New York and London, 1973.

Oppenheim, A., Some Identities in the Theory of Numbers, *Proc. London Math. Soc.*, Series, 2, (26): 295-350, 1927.

Ostermann, I., *Modeling Heat Transport in Deep Geothermal Systems by Radial Basis Functions*. PhD-Thesis, Geomathematics Group, TU Kaiserslautern, 2011.

Ostrowski, A.M., Note on Poisson's Treatment of the Euler–Maclaurin Formula, Comment. *Math. Helv.*, (44): 202-206, 1969.

Palczewski, A., Schneider, J., Bobylev, A., A Consistency Result for a Discrete Velocity Model of the Boltzmann Equation, *SIAM J. Numer. Anal.*, (34): 1865-1883, 1997.

Pommerenke, C., Über die Gleichverteilung von Gitterpunkten auf m-dimensionalen Ellipsoiden, *Acta Arith.*, (5): 227-257, 1959.

Preissmann, E., Sur la moyenne quadratique du terme de reste du problème du cercle, *C.R. Acad. Sci. Paris Ser. I*, (306): 151-154, 1989.

Rademacher, H., Topics in Analytic Number Theory, *Die Grundlehren der mathematischen Wissenschaften in Einzeldarstellungen*, Band 169, Springer, Berlin, Heidelberg, New York, 1973.

Riemann, B., Über die Anzahl der Primzahlen unterhalb einer gegebenen Größe, *Monatsber. Preuss. Akad. Wiss. Berlin*, 671-680, 1859.

Reuter, R., Über Integralformeln der Einheitssphäre und harmonische Spline-funktionen, PhD-Thesis, *Veröff. Geod. Inst. RWTH Aachen*, Report No. 33, 1982.

Schulte, O., *Euler Summation Oriented Spline Interpolation*, PhD-Thesis, Geomathematics Group, TU Kaiserslautern, 2009.

Siegel, C.L., Über die Klassenzahl quadratischer Zahlkörper, *Acta Arith.*, (1): 83-86, 1935.

Sierpinski, W., O pewnem zagadnieniu z rachunku funckcyj asmptotycznych (Über ein Problem des Kalküls der asymptotischen Funktion (polnisch)), *Prace Math.-Fiz.*, (17): 77-118, 1906.

Sloan, I.H., Joe, S., *Lattice Methods for Multiple Integration*, Clarendon Press, Oxford, 1994.

Sommerfeld, A., *Partielle Differentialgleichnugen der Physik*, 6. Auflage, Akademische Verlagsgesellschaft, Leipzig, 1966.

Stein, E.M., Weiss, G., *Introduction to Fourier Analysis on Euclidean Spaces*, Princeton University Press, Princeton, NJ, 1971.

Stenger, F., Approximations via Whittaker's Cardinal Function, *J. Approximation Theory*, (17): 222-240, 1976.

Stenger, F., Numerical Methods Based on Whittaker Cardinal or Sinc Functions, *SIAM Rev.*, (23): 165-224, 1981.

Szegö, G., *Orthogonal Polynomials*, American Mathematical Society, Rhode Island, 1939.

Titchmarsh, E.C., The Lattice Points in a Circle, *Proc. London Math. Soc. Ser.* [2], (38): 96-115, *Corrigendum* 555, 1934.

Titchmarsh, E.C., *The Theory of Riemann Zeta-Function*, Oxford, 1951.

Tsang, K.-M., Higher Power Moments of $\Delta(x)$, $E(t)$, and $P(x)$, *Proc. London Math. Soc.*, (3)65: 65-84, 1992.

van der Corput, J.G., Zum Teilerproblem, *Math. Ann.*, (98): 697-716, 1928.

Vinogradov, I.M., Anzahl der Gitterpunkte in der Kugel, *Traveaus Inst. Phys.-Math. Stekloff (Leningrad)*, (9): 17-38, 1935.

Vinogradov, I.M., Verbesserungen der asymptotischen Formeln für die Anzahl der ganzzahligen Punkte in dreidimensionalen Bereichen (in Russisch), *Izvestija Akad. Nauk SSSR, Ser. Mat.*, (19): 3-9, 1955.

Vinogradov, I.M., *The Method of Trigonometrical Sums in the Theory of Numbers*, London, 1957.

Voronoi, G., Sur un probleme du calcul des fonctions asymptotiques, *JRAM* (126): 241-282, 1903.

Wahba, G., Spline Interpolation and Smoothing on the Sphere, *SIAM J. Sci. Stat. Comput.*, (2): 5-16, (also errata: *SIAM J. Sci. Stat. Comput.*, (3): 385-386), 1981

Walfisz, A., Über Gitterpunkte in mehrdimensionalen Ellipsoiden, *Math. Z.*, (19): 300-307, 1924.

Walfisz, A., Teilerprobleme, *Math. Z.*, (26): 66-88, 1927.

Walfisz, A., *Gitterpunkte in mehrdimensionalen Kugeln*, Panstwowe Wydawnictwo Naukowe, Warschau, 1957.

Walfisz, A., Gitterpunkte in mehrdimensionalen Kugeln, *Acta Arith.*, (6): 115-136, 193-215, 1960.

Walfisz, A., Über Gitterpunkte in vierdimensionalen Ellipsoiden, *Math. Z.*, (72): 259-278, 1960.

Wangerin, A., *Theorie des Potentials und der Kugelfunktionen* I, II, Walter de Gruyter & Co, Berlin, Leipzig, 1921.

Warner, F.G., *Foundations of Differentiable Manifolds and Lie Groups*, Scott Foresman and Comp., Clenville, London, 1971.

Watson, G.N., *A Treatise on the Theory of Bessel Functions*, 2nd ed., Cambridge University Press, Cambridge, 1944.

Weyl, H., Über die Gleichverteilung von Zahlen mod. Eins, *Math. Ann.*, (77): 313-352, 1916.

Wienholtz, E., Kalf, H., Kriecherbauer, T., *Elliptische Differentialgleichungen zweiter Ordnung*, Springer, Berlin Heidelberg, 2009.

Wienkamp, R., *Über eine Klasse verallgemeinerter Zetafunktionen*, Dissertation, RWTH Aachen, 1958.

Wintner, A., On the Lattice Problem of Gauss, *Amer. J. Math.*, (63): 619-627, 1941.

Whittaker, E.T., Watson, G.N., *A Course of Modern Analysis*, Cambridge: University Press, New York: The MacMillan Company, 1948.

Yin, W.-L., The Lattice Points in a Circle, *Sci. Sinica*, (11): 10-15, 1962.

Zhai, W., On Higher-Power Moments of $\Delta(x)$, *Acta Arithm.*, (114): 35-54, 2004.

Index

Abel function, 241
Abel mean, 241
Abel transform, 236
addition theorem
    2D spherical harmonics, 160
    qD Bessel functions, 194
    qD harmonic polynomials, 152
    qD homogeneous polynomials, 146
    qD spherical harmonics, 159, 173
angular, 15, 406
area
    2D sphere, 23
    3D sphere, 23
    qD sphere, 25
associated Legendre polynomial, 173

ball lattice, 92
Beltrami derivative, 15
Beltrami integral formula, 137, 184
Beltrami operator, 14
    Helmholtz, 180
    Laplace, 134, 180
Bernoulli numbers
    definition, 62
    recurrence relations, 62
    representations, 62
Bernoulli polynomials
    functional relations, 62
    recurrence relations, 62
    representation, 62
    trigonometric series, 62
Bernstein polynomial, 166
Bessel function
    2D half odd integer order, 200
    2D theory, 200
    qD definition, 193

qD integral representation, 193
qD power series, 195
qD recurrence relations, 196, 199
qD theory, 193
bilinear expansion
    1D lattice function, 52
    qD lattice function, 256
    qD sphere function, 185
Blichfeldt's theorem, 95

characteristic function, 92
characteristic lattice ball function, 325
closure
    1D periodical polynomials, 50
    qD periodical polynomials, 225
    qD spherical harmonics, 169
completeness
    1D periodical polynomials, 50
    qD periodical polynomials, 225
    qD spherical harmonics, 171
conjecture
    Hardy, 103
    Riemann, 68
convergence conditions
    1D Poisson summation formula, 82
    qD Poisson summation formula, 231
convergence theorems
    1D Poisson summation formula, 82
    qD Poisson summation formula, 312, 337
convex, 95, 359
convolution, 228
coordinates

cartesian, 9
polar, 14

differential equation
    of the Helmholtz operator, 193
    of the Legendre operator, 154
differential operators, 10
dilation, 228
Dirac functional, 78
Dirichlet problem, 176
discontinuous integral
    Weber–Schlafheitlin type, 246
distribution of lattice points
    2D angular dependence, 387
    2D radial dependence, 387
divergence, 11
    surface, 14
duplicator formula, 27, 28, 30, 65

eigenspectrum
    1D Laplace operator, 48
    2D Beltrami operator, 181
    qD Laplace operator, 225
equidistribution, 187
Euler summation formula
    1D Helmholtz operator, 81
    1D Laplace operator, 54, 59
    1D ordinary, 54
    1D shifted, 59, 81
    qD boundary conditions, 274
    qD Dirichlet's conditions, 275
    qD extended, 273, 297
    qD Helmholtz operator, 295
    qD Laplace operator, 270, 271,
        294
    qD Neumann's conditions, 276
    qD shifted, 273
Euler's Beta function, 20
Euler's constant, 29, 57, 58, 67, 288
Euler's product representation, 67
Euler–Fermat theorem, 110

factorial function, 19
    Gamma, 19
    Pochhammer, 28
Fermat's theorem, 110

field
    scalar valued, 10
    vector valued, 10
figure lattice, 92
first Green theorem, 123
Fourier inversion formula
    2D theory, 240
    qD theory, 229
Fourier transform, 227
function
    angular, 15, 406
    radial, 15, 406
    scalar valued, 10
    vector valued, 10
functional equation
    1D Beta function, 74
    1D Riemann Zeta function, 65
    1D Theta function, 75, 76
    2D Zeta function, 286
    qD Theta function, 346
    qD Zeta function, 291
fundamental cell
    1D lattice, 46
    qD inverse lattice, 90
    qD lattice, 89
fundamental solution
    iterated Laplacians, 129
    Laplace operator, 125
Funk–Hecke formula, 164

Gamma function, 19
    definition, 17
    duplicator formula, 28
    extended Stirling's formula, 57
    functional equation, 19
    Gauss' expression, 20
    Legendre relation, 27
    multiplicator formula, 28
    Stirling's formula, 26
Gauss mean, 238
Gauss theorem, 122
Gaussian function, 236
Gegenbauer polynomial, 158
gradient, 10
    surface, 14

Green's surface theorem
  first, 134
  second, 134
  third, 137
Green's theorem
  first, 123
  second, 123
  third, 128

Hankel functions
  2D theory, 212
  qD theory, 202
Hankel transform, 243
  inversion formula, 244
Hardy's conjecture, 103, 392
Hardy–Landau identity
  1D interval, 70, 76, 85
  2D circle, 314, 316
  2D extended, 366, 371, 377
  2D sphere, 342, 388
  2D weighted, 369, 375
  qD sphere, 320, 321, 341
Hardy–Landau summation
  qD lattice ball, 338
  qD lattice point, 317
harmonic, 124
Helmholtz–Beltrami operator, 180
Hlawka–Koksma formula, 188
homogeneous harmonic polynomials
  2D theory, 160
  qD theory, 144

integral formula
  2D Beltrami, 137
  qD Beltrami, 184
  qD iterated Beltrami, 185
integral transform
  Abel, 241
  Fourier, 228
  Gauß, 236
  Poisson, 241
  Weierstraß, 237
inverse lattice, 90
inversion formula
  2D Fourier, 362

qD Abel/Poisson, 241
qD Fourier, 229
qD Gauß/ Weierstraß, 236

Jacobi's formula, 119

Kelvin function
  2D theory, 212
  qD theory, 207
Kronecker's limit formula
  1D version, 67
  2D version, 287
  qD version, 294

L-function, 394, 395
Lagrange's formula, 119
Laplace derivative, 15
Laplace operator, 11
Laplace–Beltrami operator, 134, 180
Laplacian, 11
lattice
  1D integer, 46
  qD inverse, 90
  qD periodical, 88
lattice ball discrepancy, 328
  constant weight function, 329
  involving Laplace operator, 328
  non-constant weight function, 328
lattice ball function, 324
  characteristic, 325
  qD Helmholtz operator, 331
  qD iterated Helmholtz operator, 331
  qD iterated Laplace operator, 326
  qD Laplace operator, 325
lattice balls
  in circles, 338, 379
  in spheres, 338
lattice function
  1D $\mathbb{Z}$-periodical, 51
  1D Fourier expansion, 52
  1D Helmholtz operator, 79
  1D Laplace operator, 51
  1D bilinear expansion, 52
  1D explicit representation, 52

1D iterated Laplacian, 61
2D Laplace operator, 248
2D representation, 256
qD Helmholtz operator, 248
qD iterated Helmholtz operator, 255
qD iterated Laplace operator, 255
qD Laplace operator, 248
qD representation, 265
uniqueness, 51, 79
lattice point discrepancy, 278, 299
2D sphere, 100
qD sphere, 114
angular function, 406
constant weight function, 280
Helmholtz operator, 299
Laplace operator, 278
modified, 392
non-constant weight function, 278
radial function, 406
lattice points
in circles, 99
in spheres, 99
on circles, 105
on spheres, 118
Legendre polynomial
2D addition theorem, 160
2D theory, 160
qD associated, 173
qD asymptotic estimates, 165
qD definition, 153
qD explicit representation, 161
qD generating coefficients, 156
qD generating series, 157
qD integral relations, 155
qD Maxwell representation, 158
qD orthogonality, 153
qD recurrence relations, 154
qD Rodrigues formula, 154
qD theory, 153, 161
Legendre relation, 27
Lipschitz continuity, 133

Maxwell's representation, 158

mean
τ-integral, 392
Abel, 241
Gauss, 238
Poisson, 241
Weierstraß, 238
metaharmonic, 124
Minkowski's theorem, 96, 359
modified lattice point discrepancy, 392
asymptotic, 392
modulus of continuity, 133
multiplicator formula, 28

Parseval's identity
1D periodical polynomials, 50
qD regular region, 356, 358
qD spherical harmonics, 170
partial derivative, 10
periodical
$\mathbb{Z}$-lattice, 46
Λ-lattice, 224
Pochhammer's factorial, 28
point lattice, 92
point set
boundary, 10
closure, 10
pointwise expansion theorem
spherical harmonics, 173, 176, 179
Poisson differential equation, 129, 183, 253
Poisson integral formula, 175
Poisson mean, 241
Poisson summation formula
1D Helmholtz operator, 82
1D Laplace operator, 72
1D interval, 68
1D lattice, 72, 82
2D lattice ball variant, 322, 328, 384
2D lattice point variant, 377
qD Gauß–Weierstraß summability, 347
qD Helmholtz operator, 312

qD Laplace operator, 313, 338
polyharmonic, 124
polymetaharmonic, 124
polynomial
    Bernstein, 166
    Gegenbauer, 158
    harmonic, 147
    homogeneous, 147
    homogeneous harmonic, 147
    Legendre, 153
principal value, 239

radial, 15, 406
recurrence relations
    Bessel function, 199
    cylinder functions, 207
    Kelvin function, 211
    Legendre polynomial, 154
    modified Bessel function, 201
region, 10, 122
    convex, 359
    regular, 12, 122
    symmetric, 359
regular region, 12
remainder term, 304
restriction, 10
Riemann Zeta function, 63, 394
Riemann's conjecture, 68
Riemann–Lebesgue theorem, 31

scalar function, 10
scalar zonal function, 134
second Green theorem, 123
    extended, 123
sinc-function, 70, 77, 85, 200, 314
solid angle, 128, 349
sphere function, 134
    qD Helmholz–Beltrami, 180
    qD iterated Beltrami, 138
    qD Laplace–Beltrami, 134
spherical equidistribution, 187
spherical harmonics
    qD addition theorem, 159
    qD asymptotic relations, 178
    qD closure, 169

qD completeness, 169
qD eigenfunctions, 180
qD expansion theorem, 173, 176
qD Funk–Hecke formula, 163
qD Laplace representation, 164
qD orthogonal coefficients, 178
qD Parseval's identity, 171
qD Poisson formula, 175
qD theory, 158
spherically continuous, 239, 242
star-shaped, 95
stationary phase, 35
stationary point, 35, 36
Stirling's formula, 25, 58
    extended, 57
summation
    lattice ball, 323
    lattice point, 303
summation formula
    1D Euler, 55
    1D Poisson, 68, 70, 83
    qD Euler, 294
    qD lattice ball Euler, 328
    qD lattice point Euler, 270
    qD Poisson, 313, 346, 350
surface divergence, 14
surface gradient, 14
symmetrical, 95
symmetrically continuous, 244

Theta function, 76, 262, 344
    1D functional equation, 76
    1D theory, 75
    2D functional equation, 262
    2D theory, 262
    qD functional equation, 346
    qD theory, 344
third Green theorem, 128
translate, 10
translation, 228

unimodular, 91

volume
    (q–1)D sphere, 24
    qD ball, 25

Weierstraß $\mathcal{P}$-function, 263
Weierstraß function, 236, 241
Weierstraß mean, 238
Weierstraß transform, 237
weight function
    angular, 376
    radial, 376

Zeta function
    1D Riemann, 63, 395
    1D functional equation, 65
    2D Epstein, 282
    2D functional equation, 287
    qD Epstein, 288
    qD functional equation, 292

Milton Keynes UK
Ingram Content Group UK Ltd.
UKHW031138141024
449569UK00024B/1243